D0324007

DATE DUE

Thermodynamics of Systems in Nonequilibrium States

Ralph J. Tykodi

Emeritus Professor of Chemistry
University of Massachusetts Dartmouth

published by Thinkers' Press Inc.
2002

Thermodynamics of Systems in Nonequilibrium States
September 2002

ISBN: 1-888710-08-X

Requests for permissions and republication rights should be addressed in writing to:

Ralph J. Tykodi
632 Elm St.
South Dartmouth, MA 02748-2148
email: ralphloistykodi@attbi.com

ACKNOWLEDGMENTS

I am indebted in the following ways to various organizations for permission to reprint some of my previously published papers:

To the American Institute of Physics for permission to reprint from the *Journal of Chemical Physics* my three papers

R.J. Tykodi, *On the Thermocouple in a Magnetic Field,*
J. Chem. Phys. **47** (5), 1879 (1967),

R.J. Tykodi, *Nonequilibrium Thermodynamics: Current-Affinity Relations and Thermokinetic Potentials,*
J. Chem. Phys. **57** (1), 37-42 (1972),

R.J. Tykodi, *Thermodynamics of Steady States: Is the Entropy-Production Surface Convex in the Thermodynamic Space of Steady Currents?*
J. Chem. Phys. **80** (4), 1652-1655 (1984);

To the Chemical Society of Japan for permission to reprint from the *Bulletin of the Chemical Society of Japan* my paper

R.J. Tykodi, *Thermodynamics of Steady States: "Resistance Change" Transitions in Steady-State Systems,*
Bull. Chem. Soc. Jpn. **52** (2), 564-570 (1979);

To Elsevier Science for permission to reprint from *Physica* my paper

R.J. Tykodi, *Thermodynamics of Steady States: A Weak Entropy- Production Principle,*
Physica **72**, 341-354 (1972).

INVOCATION

Thermotica, goddess of Thermodynamics,
Daughter of Hephaestus, god of fire and metal working,
Hear my prayer:

May my work be free of logical and mathematical blunders,
And may it be of consequence,
I.e. may it be widely read and commented on.

CONTENTS

BOOK II. THE OLD IDEA

The Rate of Entropy Generation

BOOK III. IDEAS ABOUT THE OLD IDEA
Reprinted Papers

IN-PRESS ADDENDUM TO BOOK I, SECTION 14
Last Minute Additions

ABOUT ME

Ralph J. Tykodi

Early Years

I was born 18 April 1925 in Cleveland, Ohio—the first child of John and Mary (Vojtko) Tykodi. In due course, the family grew to include my two brothers and sister; in birth order, we were Ralph (1925), Tom (1929), Jack (1934), and Lory (1943). While I and my brothers were young, my mother worked as a nurse–receptionist for Dr. L.L. Chandler; from the time she started working for him until his death in his 80s in an automobile accident, Dr. Chandler was both family doctor and family friend—he delivered both of my brothers Tom and Jack. My father was an upholsterer; he worked on private cars during my early years, on commercial bus seats later, and on household furniture in his final period. In my pre-school years, since both my parents were working, I was often cared for by my maternal grandparents, Frank and Theresa (Kaczur) Vojtko. Grandmother and grandfather Vojtko spoke mainly Hungarian and never really achieved fluency in English. I am told that I learned to speak Hungarian before I learned to speak English. I was fairly fluent in spoken Hungarian until my late teens, from which time on my fluency declined through lack of use—I retain today but a few dozen words and some snatches of songs that were popular in my youth.

I grew up in an ethnic neighborhood on Cleveland's east side; the languages spoken were Hungarian, Polish, Russian, and some Slavic dialects. I remember the real shock I experienced upon hearing a gray-haired old lady speak in fluent colloquial English rather than in the broken, halting English characteristic of the elders in my neighborhood.

A person raised in a bilingual environment is constantly making comparisons of things as expressed in the two languages, i.e. the person is constantly translating expressions from one language into the other. This habit of constantly making comparisons stands the person in good stead later on if he or she becomes involved in intellectual pursuits in the arts or sciences. I was always thankful for being raised in a bilingual environment; it made further language acquisition seem natural and easy. I had a year and a half of French in junior high school, four years of Latin in grades 9–12, and two years of German at the university. Late in life, I mastered just enough Russian to be able to pick my way through Russian chess publications.

I don't remember when I learned to read—it was in first and second grade, I suppose—but from early on, I was an avid reader. My parents used to admonish me not to read so much—it would ruin my eyes, they said. A bout of scarlet fever at age 11 affected my eyes; and from that time on, I wore glasses of ever increasing thickness.

While living in Cleveland, I attended Lafayette Elementary School and Audubon Junior High School. In 1939 the family moved to Mansfield, Ohio, where I completed grades 9–12. I was a conscientious student and always got good grades. I found my elementary and secondary school teachers to be competent and supportive. I had a good memory, and that was my chief scholarly asset. I do remember, however, lying awake one night for several hours and working out in my head a *reductio ad absurdum* proof of a theorem in solid geometry.

I enjoyed the math, physics, and chemistry I was exposed to in high school (I didn't have any interest in biology), and I wondered if I could make a career in one of these subjects. Upon graduation from high school in 1943, I was (temporarily) exempted from military service because of my poor eyesight. In the spring of 1944, I was drafted into the U.S. Army on a special services basis; and I served two years and three months in the U.S. Army Medical Corps. As part of a Field Hospital unit, I landed in France on the day President Roosevelt died (12 April 1945). The war in Europe ended before our Hospital unit could be deployed, so we were scheduled to be transferred to the Pacific Theater, to be available for the expected invasion of the home islands of Japan. We were at sea in the Mediterranean, heading for the Panama Canal, when news of the Japanese surrender reached our ship. We were diverted to Boston. I served out the rest of my military service as a surgical technician in a Veterans Administration hospital in Martinsburg, W. Virginia. I consider myself very lucky to have gone through the war without being placed in any positions of danger.

Northwestern 1946–49

In my senior year at Mansfield High School, my Latin teacher, Ruth Dunham, encouraged me to apply for a scholarship at Northwestern University; I did so, and was awarded a partial tuition scholarship. In the fall of 1943, I enrolled in the Chemical Engineering program at Northwestern University—the Engineering program seemed to promise interesting work in a subject I liked (chemistry) with good employment prospects upon graduation.

After completing my military service, I went back to Northwestern University on the GI Bill of Rights and majored in chemistry, having rethought my priorities. I didn't think I was creative enough to make a go of it in mathematics, and I wondered if I could survive the rigors of undergraduate training in physics. I felt comfortable with the thought of doing things in chemistry, and have never regretted my decision—work in chemistry covers a very broad spectrum of activities, and one can work at it in a very theoretical or a very practical way.

Of the undergraduate laboratories in chemistry, I enjoyed the analytical laboratories the most. If one were analyzing an ore for its % copper, say, one did the analysis in triplicate; and if the results agreed with one another to a few parts per thousand, one was pretty sure that the mean % of the three trials was a good representation of the % copper in the ore. In the inorganic and physical chemistry laboratories, the emphasis was on verifying the literature value of some quantity. Student values of the quantity that differed much from the literature value were simply "wrong," and it was only very rarely that the source of the "wrongness" could be ascertained. I very early on decided that I was not meant to be an "organiker."

An amusing pattern developed while I was taking the physical chemistry course. The lecturer was Arthur A. Frost, a good-enough lecturer, and the text was one by Farrington Daniels. I believe there were three lectures a week, each lecture going from 12:30 P.M. to 1:20 P.M. At the time, I was working in the kitchen of a campus sorority house, washing up pots and pans in exchange for my meals. On a day when a physical chemistry lecture was offered, I would arrive at the kitchen at about 11:15 A.M. I would have my lunch and then scrub up pots and pans until about 12:15. Then I would race across campus to get to the lecture hall by 12:30. The midday meal, the splashing around in hot soapy water, and the often somewhat warm and stuffy air of the lecture hall combined to make me sleepy; and I frequently dozed off for a few minutes at a time during the lecture. I used to challenge myself to stay awake until 1:00 P.M. I would look at the clock on the wall of the lecture hall and say to myself, "12:55. Good, today I'm going to make it to 1:00." Suddenly there would be a loud CRASH—I would wake with a start, having dropped my pencil to the floor (the CRASH), and the time would be 1:05. I don't think I ever made it to 1:00 without dozing off for a few minutes.

During my undergraduate years at Northwestern (1946–1949), I was bothered by proofs in mathematics and derivations in physics and physical chemistry. My problem with proofs and derivations was much like the problem I had with chess puzzles: "White to move and win;" "Black to move and mate in three;" etc. If someone showed me the solution to the puzzle, I could see, after the fact, that the moves had the required structure; but I usually couldn't see, before the fact, that the required moves had to be so and so. Similarly, I could follow published or stated proofs and derivations, but I had difficulty constructing them on my own. My experience with proofs and derivations was a lot like my experience in learning to ride a bicycle: I kept trying and trying, but kept "tipping over." Finally one day, I kept my balance and rode around without tipping over; from that day on, I could ride a bicycle. Toward the end of my first year of graduate study, I suddenly found that I could "keep my balance"—from that point on, I could do proofs and derivations pretty well on my own.

While at Northwestern, I became a member of the Northwestern University fencing team, thereby starting a career in fencing that was to last for 50 years (I retired from competitive fencing in 1997).

On graduation day, I was inducted into Phi Beta Kappa.

Penn State 1949–54

After graduating from Northwestern University with a B.S. in Chemistry, I decided to continue on for the Ph.D. degree; and I accepted the offer of a graduate teaching assistantship from Penn State University where W. Conard Fernelius was chairman of the chemistry department at the time. I scored high enough in the battery of proficiency exams given to incoming graduate students to be excused from a survey course in physical chemistry taught by John G. Aston and required of all entering graduate students (except for those few who "placed out"). To prepare for the Ph.D. qualifying exams to be given at the end of the first year of graduate study, I enrolled in graduate courses in organic chemistry, chemical thermodynamics, and chemical kinetics. The organic course was a "necessary evil"—it being the only way I could force myself to survey enough organic chemistry to get by the qualifying exam. I had pretty much decided to become a physical chemist, so the thermodynamics and kinetics courses were part of my major.

The kinetics course was taught by Joseph H. Simons, he who invented the electrochemical method of synthesizing fluorocarbons (teflon, etc.). Simons was a small man who looked a lot like the actor Barry Fitzgerald. He had a pleasant classroom manner, and I was sufficiently impressed by him to choose him as my thesis advisor. Before we could outline a thesis project, however, Simons announced that he was moving to Florida to take up a university position there, so I turned to John Aston and arranged for him to be my thesis supervisor.

Aston taught the thermodynamics course using Lewis and Randall for the text. He projected a great enthusiasm for thermodynamics, but in his lectures he frequently lost his way. With loud exclamations of "Oh rats!" he would tell us to strike out part of his demonstration, to substitute something else for it, to change a sign here, to multiply by a factor there, and so on. Aston was notoriously absent-minded. Once, being a little late for class, I entered the appropriate building and saw Professor Aston walking down the hall, systematically opening every door he came to and peering inside—he had forgotten the room number of his class and was hunting for the students!

John Aston had single-handedly established a low-temperature laboratory at Penn State. In that lab, a technician routinely generated liquid air (bath temp. 75*K*–90*K*) and liquid hydrogen (bath temp. 13*K*–20*K*) for use in experiments; and inside some pieces of equipment, helium was liquefied *in situ* (via expansion across a needle valve—Joule–Thomson effect) (temp. range 1*K*–4*K*) for the experiment at hand. At the time in question, Aston was supervising the work of S.V.R. Mastrangelo on the properties of helium adsorbed on titanium dioxide at liquid helium temperatures. I joined Mastrangelo in the adsorption study, and my thesis work was the continuation of his.

The adsorption chamber was about the size of an ordinary drinking glass. It was packed with finely powdered titanium dioxide and connected to a huge array of tubes, wires, pumps, compressors, gas holders, gauges, and measuring instruments. We used as refrigerants liquid air, liquid hydrogen, and, generated inside the apparatus, liquid helium. To be able to control the temperature at various places inside the apparatus, we had a vacuum inside a vacuum inside a vacuum. This complex nest of vacua was a constant source of headaches—the stresses induced by cooling down and warming up the apparatus often led to cracks in glass tubing or sprung solder joints, and "hunting for the leak" was the name of the game for weeks on end.

Obtaining data for the adsorption system proved to be both exhausting and (at times) depressing. The length of a successful data-gathering session was controlled by the rate of loss of the liquid-hydrogen refrigerant. The technician made batches of liquid hydrogen 25 liters at a time, and we were more or less honor bound to continue the data-gathering session until all the liquid hydrogen had evaporated—sessions of 36–48 hours were the norm, with a few minutes sleep snatched now and then while waiting for the system to settle into a steady state after the adjustment of some parameter or other. When the system was "down" for a long time due to failure to locate a leak in one of the vacuum systems, I would get depressed and sometimes wondered if I would ever collect enough data to pad out an acceptable thesis.

The low-temperature work had a dramatic quality to it. All the pipes, tubes, and wires leading to the adsorption chamber were the means whereby I could compel the temperature of that small region of space to drop far below room temperature. I used a comparison cathetometer that had a telescope mounted on a long vertical shaft; the shaft had a horizontal rod affixed to it with which I could rotate the shaft and swing the telescope to bear on the mercury meniscus in each arm of a precision manometer used to measure the helium gas pressure inside the adsorption chamber. The setup looked for all the world like the periscope of a submarine. As part of the data-gathering process, I would sit on a low stool and sight through the telescope to get a reading on one arm of the manometer; then I would rotate the shaft, via the handle, so the scope would bear on the other arm of the manometer. While carrying out these operations, I felt like some kind of Captain Nemo, diving to the bottom of the ocean—except that I was staying topside and sending a probe to the bottom of the thermal ocean. The lowest temperature that we ever achieved in the apparatus was about 2.8*K*.

Although work in the low-temperature lab was alternately exhausting and depressing, I did find time for some social activities: my friends and I climbed Nittany mountain, swam at Whipple's Dam, made forays into New York City for entertainment or culture, played a lot of bridge, and exchanged invitations to parties. Several of the male members of our group met their wives-to-be at Penn State: myself, Harry Pinch, Herb Siegal, Alan Shriesheim, and young faculty member Bill Steele. Upon graduation, Alan Shriesheim carved out an exemplary career at Exxon, first in research, then in management; he later left Exxon to become the Director of the Argonne National Laboratory. Bill Steele took over as Director of the Low-Temperature Lab when John Aston retired and left Penn State.

My wife-to-be, Lois Benham, and I were teaching assistants in the same undergraduate analytical chemistry laboratory. At the time, Lois was seeking a Master's degree in chemistry; she later decided that that was not what she wanted and spent the last part of her stay at Penn State working as a numerical analyst on an Office of Naval Research Project at the university.

I was granted the Ph.D. degree by Penn State in the summer of 1954. I stayed on a few months at the low-temperature lab as a Post-Doctoral Fellow. Then I treated myself to a "Sabbatical" from the first of January to the end of August 1955. I had saved enough money so that by living frugally I could support myself for that interval of time. I used the time to study some math and physics: I read in Arnold Sommerfeld's *Lectures on Theoretical Physics*, and I sat in on a course in modern algebra and a course in quantum mechanics.

From the time I completed Aston's course in thermodynamics based on the Lewis and Randall text, I considered myself a professional thermodynamicist; and I read extensively in the subject. I was impressed by the early editions of Guggenheim's book [1]; I enjoyed the books by Zemansky [2] and Epstein [3]; I found the engineering approach used by Keenan in his 1941 book [4] very invigorating, and I prefer that book over his later book with Hatsopoulos [5]. I read Percy Bridgman's discussions of operationalism [6,7] and was sold on the idea. In dealing with the writings of Gibbs [8], I felt uncomfortable with his use of infinitesimals of various orders and with virtual variations, so I tended to draw inspiration from other sources.

Max Born, in his book *My Life & My Views*, bemoaned the fact that in conventional treatments of thermodynamics the physics and the mathematics are all tangled together. He wanted to shed all the talk about cycles, engines, piston-and-cylinder devices, and so on, and deal with the abstract mathematical description of the logical relations inherent in the subject. Born urged his friend Caratheodory to carry out such a program of abstract analysis, and Caratheodory did so. Today there is some disagreement over the validity and utility of the Caratheodory analysis—Truesdell, in his *Rational Thermodynamics* [9], is highly critical of Caratheodory's work.

For my own part, I find the tangling of the physics and mathematics of the subject one of its most endearing traits. To be able to establish general relations by carrying out manipulations on real or model systems is to me a great delight. In dealing with thermodynamic matters, I take a macroscopic, in-the-large, external approach: the nature of the surroundings for the system of interest is studied so that from changes in the surroundings the properties of the system may be inferred. The system itself is defined in terms of its composition, its internal constraints, and its boundary conditions (or external constraints). In place of Gibbs's virtual changes, I deal with real changes brought about by changes in the internal or boundary constraints (in this regard, following the treatments of Callen [10] and Reiss [11]). In my view, thermodynamics is the study of those aspects of the behavior of macroscopic systems that are fully determined by the composition of the system together with its internal and external constraints. The behavior that is fully determined by the indicated items is, of course, the time-average behavior of the system for given composition and (internal and external) constraints. In this view, the local fluctuations of intensive quantities at internal points of the system are not matters for thermodynamic discussion but belong to the field of statistical mechanics.

Writers on chess often adopt the conceit that there is a goddess, *Caissa*, who looks after things chessic and who inspired the likes of Morphy, Pillsbury, Rubinstein, etc. Chess is thought to have been invented about 570 A.D., so Caissa has had a busy time of it for about 1500 years; and presumably in recent times she has smiled on the likes of Fischer and Kasparov (I wonder what she thinks of Judit Polgar and of the silicon-based players Deep Blue and Fritz?).

I like to think that thermodynamics, too, has its own goddess to look after things thermodynamic and to inspire devotees of the discipline. Let us call her

Thermotica—she is a daughter of the Greek god Hephaestus, god of fire and metal working [the Romans identified the Greek god Hephaestus with their own god Vulcan (Volcanus)]. Thermotica didn't have much to do until about the time of the American Revolution, when she showed James Watt (in his dreams) how to perfect the workings of the steam engine. After that, she was busy enough inspiring the likes of Carnot, Clausius, Thomson, Planck, G.N. Lewis, etc.; and she had long discussions (in his dreams) with J. Willard Gibbs. Since World War II Thermotica has been busy inspiring her devotees to extend the ideas and techniques of thermodynamics to cover nonequilibrium situations. In my own case, I sometimes dream about thermodynamic matters—perhaps Thermotica is trying to get through to me?

IIT 1955–65

Returning to my *curriculum vitae* (road of life), in the spring of 1955, I had an interview with Martin Kilpatrick, chairman of the chemistry department at the Illinois Institute of Technology (IIT) in Chicago. As a result of the interview, I was hired by Kilpatrick and spent the next ten years in the chemistry department at IIT. In August 1955, Lois Benham and I were married. After the honeymoon, we packed our things and set off for Chicago to take up my teaching post at IIT.

Martin Kilpatrick had put together a faculty that was quite respectable in both teaching and research. Among the senior staff in the department was Scott Wood, a thermodynamicist of some note. A few years later, Rubin Battino joined the department and became a research collaborator of Wood's, Scott being something of a mentor for Rubin. The two men eventually coauthored two books on thermodynamics. During my stay at IIT, I had many interesting discussions about thermodynamic matters with both Scott and Rubin. Hired along with me in 1955 was Peter Lykos, a quantum chemist. Peter became my best friend at IIT. He had an intense interest in all parts of physical chemistry, and I had a number of very useful discussions with him about thermodynamic matters. Peter got into computer usage at an early date and eventually wound up with a joint appointment as Director of the IIT Computation Center and Professor of Chemistry.

In my last year at Penn State, I had made a survey of the thermodynamic writings of J.N. Bronsted. Bronsted had reformulated the principles of thermodynamics into what he considered a simpler system, a system that he called *energetics* [12]. I liked the feel of Bronsted's formulation and thought I saw a way of making it more general. During that same year, I also read some writings on irreversible thermodynamics by members of the "Belgian school": Prigogine, de Groot, etc. I decided to try to formulate the basic ideas of irreversible thermodynamics in terms of my macroscopic, in-the-large, operational way of looking at things thermodynamic.

My research program at IIT, therefore, was: to improve on Bronsted's energetics and to set up a "my way" description of irreversible processes. I was unable to improve on Bronsted's formulation of energetics, the idea I had expected to pursue did not lead to anything useful. Eventually, I abandoned all consideration of energetics, inasmuch as the classical, Gibbs-style formulation of thermodynamics can cover all that energetics promised to cover—and can cover even more.

On the other hand, I was successful in formulating an operational, in-the-large description of the thermodynamics of systems in steady states. The rate of entropy generation in a steady state is well defined in the classical sense, and I demonstrated a number of features of the rate of entropy generation in a series of papers in J. Chem. Phys. 1959–60. I originally thought that the rate of entropy production for steady states had some simple and general extremum properties, but in due course I found counterexamples to each of my proposed extremum requirements.

Irreversible thermodynamics (nonequilibrium thermodynamics) was a popular subject at the time—many journal articles on the subject were appearing, and books on the subject were being published. After some shopping around among publishers, I finally contracted with Macmillan to do a book on the thermodynamics of steady states (the book appeared in 1967, after I had left IIT).

At the time in question (late 1950s, early 1960s), the Armour Research Foundation was a subdivision of IIT. A junior staff member at Armour, Ted Erikson, wanted to work on an M.S. degree at IIT; and he asked me to be his thesis advisor. As thesis topic, we chose the forced vaporization of water and planned some experiments to follow up on the pioneering work of Alty [13]. Ted designed some nifty bits of apparatus and carried through the investigation in exemplary fashion. In addition to his professional work, Ted was also a long-distance swimmer; among other swims, he did a swim of 50+ miles across a part of Lake Michigan, and he did a round-trip swim of the English Channel.

I tried to get an experimental program going to measure the Soret effect of a soluble salt in a saturated solution of another, sparingly soluble salt; but the design of our Soret cell (graduate student Morton Lieberman was working with me) was somewhat imperfect and we never got reproducible results. Lieberman grew discouraged, abandoned the project, and went to work in the laboratory of Philip Wahlbeck, where he ultimately produced a satisfactory Ph.D. thesis.

About this time, things were changing at IIT. Martin Kilpatrick had retired, and Arthur Martel had replaced him as Head of the Chemistry Department. Martel sought to increase the size of the department, to bring in more staff and more graduate students. To carry out his plan in the face of the sort of budgetary constraints he was operating under at the Institute, he sought to boost the amount of grant money that was coming into the department. We were all urged to write grant proposals which, if funded, would support us over the summer and for part of the academic year and which would support an appropriate number of graduate students. I was carrying on my research program alone and did not desire to try to establish a program that would employ several graduate students. Although I had tenure, I was in various ways made to feel *persona non grata* by the IIT administration. So in the summer of 1965, I left IIT.

Thermotica's Devotees

Before detailing the next phase of my academic career, let me comment on some of Thermotica's devotees whom I met along the way.

I heard Clifford Truesdell give an invited lecture at IIT. Truesdell was dressed in a tuxedo and gave a bravura performance. I admire Truesdell's erudition: he is fluent in Latin and in several modern languages, and his historical studies of the works of Euler, Gauss, Cauchy, and others are true tours de force. In studying the writings of Truesdell, I was amazed to learn of the existence of a group of modern "natural philosophers" (Truesdell, Coleman, Noll, Gurtin, Serrin, –) who by applying the techniques of modern mathematics have made of the subject *continuum mechanics* a highly abstract discipline. The "*philosophes*" include the thermodynamics of continua under their purview—it seems that Thermotica can be served in many, widely disparate, ways. I mentioned earlier that Truesdell is highly critical of the thermodynamic work of Caratheodory [9].

Sometime in the early 1960s, Brown University hosted an event honoring Lars Onsager (he of the Onsager reciprocal relations). I attended the event; and at the reception after the formal part of the event, I found myself standing next to Boris Leaf and Lars Onsager, who were engaged in conversation. I had read a paper by Leaf in which he compared Bronsted's energetics with classical thermodynamics. I introduced myself and talked to Leaf about that paper. I then mentioned to Onsager that I had written some things about irreversible thermodynamics, but that my work had not been very popular (my papers were rarely quoted by other workers in the field). Onsager said, "You say it's not popular, but is it right?" I replied that I certainly thought so. "If it's right, it will be popular," he said. The years have rolled on since that conversation, but my work has remained, in my view, both right and not popular.

While at IIT, I met Myron Tribus and Ed Jaynes. Tribus was then Dean of the Thayer School of Engineering at Dartmouth College. Jaynes had written two papers relating information theory to the foundations of statistical mechanics. Tribus saw in the work of Jaynes an elegant way to bring thermal physics into the engineering curriculum: information theory → statistical mechanics → thermal physics, and he wrote a textbook to show how it could be done [14]. Tribus was on a lecture tour, pushing his new approach; and he was doing a tandem presentation with Jaynes (Jaynes was the Great One and Tribus was His prophet). I was very interested in the ideas of Jaynes and Tribus, and at a later time I attended a five-day workshop on information theory and thermal physics at Dartmouth College, at which Jaynes was the principal presenter. Jaynes made a convincing case for the use of Bayesian methods in probability and said he was working on a small book on probability and its uses in science. At the time of his death in 1998, the "little book" had grown to two volumes but was still unpublished. I understand that some of Jaynes's students are working on the material and preparing it for posthumous publication. I met Jaynes on a few other occasions and had a chance to discuss my work on steady states with him; he had some laudatory things to say about my 1967 book in a review article that he did on the principle of minimum entropy production (the article is reprinted in the volume of his collected papers edited by Rosenkrantz [15]—the two famous papers on information theory and statistical mechanics are reprinted there, too).

I heard Joseph Keenan give a presentation at a conference on the teaching of thermodynamics. He was challenged by a colleague to give a definition of *heat* and he came up with "*heat* is that which passes from one body to another by virtue of a temperature difference." I admired the argumentation in his 1941 book; and as I mentioned earlier, I prefer the 1941 book over the later Keenan–Hatsopoulos book [5]. I corresponded with Keenan about a point that came up in one of my papers; and in an example he used to bolster the discussion, I was surprised at the amount of detail he included for the temperature and pressure controlling features of the model system he proposed.

At a workshop on thermodynamics at MIT one summer, George Hatsopoulos was one of the principal presenters. He demonstrated a fine grasp of thermodynamic matters and was all for extending thermodynamics to cover as much of science as possible. He considered a system in a gravitational field of such a nature that the local temperature was a function of the local gravitational energy, and he proposed as a test of disparity in temperature of two locations the hooking up of the locations to a heat engine to see if the pair of temperatures could drive the heat engine—if they couldn't, then they were equal. I am afraid that his exuberance ran away with him in this case, since for the analogous situation of a gas in an isothermal gravitational field (the barometric formula) two points of different pressure can be hooked up to a turbine to see if they will drive the turbine—they won't, but the pressures are different nevertheless.

I heard Mark Zemansky lecture on a couple of occasions—a small man but a large presence—and I was part of a luncheon group once for which he was the principal raconteur. He favored the classical, engineering-style of presentation for thermodynamic matters and was rather cool toward abstract treatments of the subject. He said of a book by Landsberg [16] (very quantum mechanical and full of discussions of the properties of interior points versus those of boundary points for regions in various abstract thermodynamic spaces), "Landsberg states in the preface to his book that thermodynamics is a very difficult subject, and the book reads as though he wrote it to prove that assertion."

I never met Peter Landsberg; but I corresponded with him about a few thermodynamic topics, and we coauthored a paper.

I met Joseph Kestin a few times and corresponded with him on topics both literary and thermodynamic. At the time of his death, we were discussing via correspondence the thermodynamic description of a strained solid.

I heard Herbert Callen lecture several times, and I have spent many hours studying his fine book (both the 1st and 2nd editions) on thermodynamics [10,17]. Callen's description of a thermodynamic process as taking a system, upon the relaxation of a constraint, from an initial equilibrium state to a final equilibrium state was the model I drew upon for my treatment of steady states. In an obituary notice for Callen, it was stated that in his later years he suffered from Alzheimer's Disease. It is sad to think of that bright intellect gradually dimming out from the ravages of Alzheimer's; but we have no control over our manner of departing this life, and pretty much we have to play the cards dealt us by Fate.

I heard E.A. Guggenheim lecture once, but I just can't remember anything either about him or about the lecture.

I heard several lectures by Ilya Prigogine, and on an occasion when he visited

IIT I had a chance to discuss my work on steady states with him—I was unable to sell him on my way of doing thermodynamics.

Among the many persons whom I have never met but whose writings on thermodynamics I found stimulating are Laszlo Tisza [18] and Howard Reiss [11].

UMASS Dartmouth 1965–95

Now back to the story of my academic career. I left IIT in the summer of 1965. At the time, we had two children, Paul (born 1959) and Laura (born 1963); and Lois was pregnant with our third child, Scott (born 1966). Both my parents were dead, so Lois's parents, George and Myra (Smith) Benham, were the only grandparents our children had; Lois was an only child., so our children were also the only grandchildren that her parents had. The Benhams were fond of the grandchildren and wished we were located closer to them (they lived in the outskirts of Boston, in Somerville, MA—within sight of the Tufts University campus) so that they could visit with the children more frequently. A new technological institute, Southeastern Massachusetts Technological Institute (SMTI), was then being established in Dartmouth, Mass.—at a later date, under a reorganization plan for higher education in the state, SMTI became part of a state-wide university system, and its name was changed to University of Massachusetts Dartmouth (UMD). The institution was rapidly adding faculty, and in a few years it reached a steady state of about 300 faculty and 5000 students. I joined the faculty of SMTI in the fall of 1965 as an Associate Professor of Chemistry. Two years later I became a Full Professor, and I served as Associate Dean of the College of Arts and Sciences 1969–72. I did not enjoy administrative work, and in 1972 I returned to the chemistry department as Professor of Chemistry, a position I held until I retired in 1995, whereupon I was awarded the title of Emeritus Professor of Chemistry.

My book, *Thermodynamics of Steady States* [19], was published in 1967 and drew very little notice. I published a series of papers extending the material in the book, the papers also drawing little notice. I kept trying to generalize the principle of minimum entropy generation, which holds for linear current–affinity relations; but every "obvious" generalization that I tried fell victim to one or more counterexamples. I had pretty much exhausted my store of ideas about the rate of entropy generation, and, simply as something to do to keep busy, I started looking at the entropy itself for systems in nonequilibrium states. I was surprised to find that for a large domain of nonequilibrium states a simple extension of the ideas of equilibrium thermodynamics makes it possible to treat such states in the classical manner. I wrote a set of papers outlining my ideas and applying those ideas to a number of standard nonequilibrium situations. I submitted the papers to several appropriate journals, but each submission resulted in rejection (the gist of those papers appears in Book I of this work).

I do not believe in Onsager's dictum, "If it's right, it will be popular." At any particular time for a particular subject, there are fashions in research; and work that is not done along the prevailing fashionable lines can be both right and not popular. I consider this work to be my thermodynamic legacy to thermodynamicists of the present and the future; and in writing it, I do homage to Thermotica, goddess of Thermodynamics—may she appear in the dreams of all good thermodynamicists and suggest to them that they read this work.

Retirement 1995–

In the years after my retirement in 1995, in addition to my work on the thermodynamics of systems in nonequlibrium states, I rather unexpectedly became involved in the editing of a series of chess books dealing with the chess writings of the Australian chess master and chess journalist C.J.S. Purdy (1906–1979). I did a great deal of the compiling and editing that was necessary in producing a multi-volume **Purdy Library of Chess** published by *Thinkers' Press*, Davenport, Iowa.

REFERENCES

1. E.A. Guggenheim, *Thermodynamics*, 2nd ed. (North-Holland, Amsterdam, 1950).
2. M.W. Zemansky, *Heat and Thermodynamics*, 4th ed. (McGraw–Hill, New York, 1957).
3. P.S. Epstein, *Textbook of Thermodynamics* (Wiley, New York, 1937).
4. J.H. Keenan, *Thermodynamics* (Wiley, New York, 1941).
5. G.N. Hatsopoulos and J.H. Keenan, *Principles of General Thermodynamics* (Wiley, New York, 1965).
6. P.W. Bridgman, *Nature of Thermodynamics* (Harper, New York, 1961).
7. P.W. Bridgman, *Reflections of a Physicist* (Philosophical Library, New York, 1955).
8. J.W. Gibbs, *The Scientific Papers of J. Willard Gibbs. Vol. 1. Thermodynamics* (Dover, New York, 1961).
9. C. Truesdell, *Rational Thermodynamics* (Springer, New York, 1984).
10. H.B. Callen, *Thermodynamics* (Wiley, New York, 1960).
11. H. Reiss, *Methods of Thermodynamics* (Blaisdell, New York, 1965).
12. J.N. Bronsted, *Principles and Problems in Energetics* (Interscience, New York, 1955).
13. T. Alty, Proc. Roy. Soc. (London), **A131**, 554 (1931).
14. M. Tribus, *Thermostatics and Thermodynamics* (Van Nostrand, New York, 1961).
15. R.D. Rosenkrantz, ed., *E.T. Jaynes: Papers on Probability, Statistics and Statistical Physics* (Reidel, Dordrecht, 1983).
16. P.T. Landsberg, *Thermodynamics with Quantum Statistical Illustrations* (Interscience, New York, 1961).
17. H.B. Callen, *Thermodynamics and an Introducton to Thermostatistics,* 2nd ed. (Wiley, New York, 1985).
18. L. Tisza, *Generalized Thermodynamics* (MIT, Cambridge, 1966).
19. R.J. Tykodi, *Thermodynamics of Steady States* (Macmillan, New York, 1967).

FOREWORD

One of the many ways of structuring classical thermodynamics is to take the concepts of heat and work as primitives and to found on those primitives the set of laws that defines the subject, each law asserting the existence of a suitable "function of state":

Zeroth Law: The function of state "temperature" exists and has certain properties.

Thermal balance or thermal equilibrium between two systems in contact across a diathermal surface requires equality in temperature for the two systems, and the thermal balance relation is transitive:

$$T_1 = T^* \ \& \ T_2 = T^* \Rightarrow T_1 = T_2.$$

First Law: The function of state "internal energy" exists and has certain properties:

$$dU = dQ + dW.$$

Second Law: The function of state "entropy" exists and has certain properties:

$$dS \geq dQ / T.$$

The structure reared on the given primitives and the indicated set of laws is adequate for dealing with equilibrium states and transitions between equilibrium states (processes). With the publication of J. Willard Gibbs's paper ON THE EQUILIBRIUM OF HETEROGENEOUS SUBSTANCES (1875–1878) the structure of classical thermodynamics was essentially complete—the big ideas were all in place, but there still remained some rounding out to do: the third law, the introduction of activities and fugacities to deal with nonideal systems, etc.

In attempting to expand this structure to cover systems in nonequilibrium states, we need to be able to define heat, work, temperature, internal energy, and entropy for the nonequilibrium states we desire to investigate. From the days of William Thomson (1824–1907) to the present, there have been many efforts to carry out such a program, but none has been deemed wholly successful. In a change of strategy, Onsager [1,2], Prigogine [3], and others [4] focused attention on the rate of entropy generation (they called it the "entropy production") for an irreversible process and established that a principle of minimum entropy production holds in a near-equilibrium domain of steady states for such processes. As I stated in the preceding autobiographical sketch, I developed a rigorous treatment of the thermodynamics of steady states; as part of that treatment, I studied the properties of the rate of entropy generation in steady states. My book, *Thermodynamics of Steady States* (1967) [5], was written at a time when I thought the principle of minimum entropy production had no exceptions in the domain of steady states; the book is reprinted here as Book II of this work.

As I continued my study of the properties of the rate of entropy generation, I soon found that there were exceptions to the validity of the principle of minimum entropy production in the domain of steady states—the principle failed to hold for some heat conduction processes and for some chemical reactions. I sought to generalize in various ways the principle of minimum entropy production for steady states, and Book III of this work reprints the more important papers that came out of that study—I failed to find a universally valid extremum principle for the domain of steady states, but I did learn a good deal more about the properties of the rate of entropy generation for steady states.

There came a time when I felt I had exhausted my stock of ideas about the rate of entropy generation for steady-state processes, and I turned back to a study of the entropy itself and the generated entropy for processes taking a system from one nonequilibrium state to another. I thought again about the problem of defining heat, work, temperature, internal energy, and entropy for systems in nonequilibrium states.

For steady states and slowly varying steady states, it is possible to give an operational definition of temperature by making use of an equilibrium system as a thermal probe and invoking the Zeroth Law: two points in a nonequilibrium system that are each in thermal balance with a particular setting of the thermal probe are assigned the same temperature, the temperature of that setting of the probe. For systems in nonequilibrium states such that the local properties of the system are varying rapidly, the operational approach to a definition of temperature is not available, and the temperature concept, if used at all, becomes a pencil-and-paper concept based on equations of evolution, such as Fourier's law ($\nabla^2 T - (1/\kappa)\partial T/\partial t = 0$). I wonder if the thermodynamic approach is practicable and/or useful for systems for which the operational definition of temperature is not available.

By linking a system in a nonequilibrium state to a reference equilibrium-based configuration, the entropy associated with the nonequilibrium state can be given a classical representation; and an appropriate version of the Second Law can be invoked for processes undergone by the system. To carry out such a linkage requires that it be possible to define local values of the conventional thermodynamic intensities: temperature, pressure, chemical potential, etc.—which brings us back to the problem of defining temperature in a nonequilibrium state.

In Book I of this work, I carry out a program of the type just mentioned and get a satisfactory macroscopic, in-the-large, operational treatment of a certain domain of nonequilibrium states; and I discuss the possibility of modifying the treatment so as to encompass a larger domain of nonequilibrium states. Again I wonder if such modification is practicable and/or useful.

For systems with well-defined local values of the conventional thermodynamic intensities, I offer a complete solution to the problem of migrational equilibrium in a spatial field, whether the field be of equilibrium or nonequilibrium type—the key to the solution is a principle, the Thomson-Gibbs Corollary, that is more general than the principle of minimum entropy production.

Approximately 100 years after the death of Gibbs (1839-1903) we have at hand a formulation of thermodynamics based on severely classical ideas and techniques that is adequate for the treatment of equilibrium states and for steady states and slowly varying steady states in a particular domain. To extend the thermodynamic style of analysis outside the given domain requires skills that I do not possess, so I must leave would-be explorers of those far regions at the frontier of the given domain and from there can only wave them on and wish them well.

Format

Figures and equations are numbered consecutively in each section of Book I, with the following form of cross-referencing: Fig. 3.1 refers to Fig. 1 of Section 3, Eq. (2.5) refers to Eq. (5) of Section 2, etc.

REFERENCES

1. L. Onsager, Phys. Rev. **37**, 405 (1931).
2. L. Onsager, Phys. Rev. **38**, 2265 (1931).
3. I. Prigogine, *Introduction to the Thermodynamics of Irreversible Processes*, 2nd ed. (Wiley, New York, 1961).
4. S. de Groot and P. Mazur, *Non-Equilibrium Thermodynamics* (North-Holland, Amsterdam, 1962).
5. R.J. Tykodi, *Thermodynamics of Steady States* (Macmillan, New York, 1967).

BOOK I

THE NEW IDEA

Entropy and Generated Entropy

ONE

INTRODUCTION

From the 1760s on, when James Watt converted the steam engine into a practical work-delivering device, the problem of heat-to-work conversion exercised many minds. In the hands of Carnot, Joule, Thomson, Clausius, Gibbs, Planck, G.N. Lewis, and many others, what started as an investigation into the efficiency of the heat-to-work conversion process branched out into today's far-reaching discipline of thermodynamics. The mature form of modern classical macroscopic thermodynamics is elegantly expounded in any number of texts [1-5]. The classical macroscopic thermodynamics displayed in the texts just cited is a thermodynamics of equilibrium states and of processes leading from one equilibrium state to another.

The world about us is notoriously full of systems in nonequilibrium configurations undergoing irreversible processes. It has been a long-standing challenge to classical thermodynamics to expand its domain of discourse to include systems in nonequilibrium states and the processes undergone by such systems. The work of Onsager [6,7], Prigogine [8], and others [9,10] has shown that there is a domain of near-equilibrium steady states in which considerations of entropy production lead to useful thermodynamic information about the irreversible processes at play in a system of interest. Outside the domain of near-equilibrium steady states, relations that can be relied on in the near-equilibrium domain cease to be generally valid; and adequate treatments of far-from-equilibrium situations are matters of ongoing research [11-15].

In Book I of this work, I build upon considerations espoused by William Thomson and J. Willard Gibbs a thermodynamic formalism for dealing with systems in both equilibrium and nonequilibrium states. Key features of the formalism are the assignment of an entropy and an internal energy to the *equilibrium-based image* of a system of interest, whether that system be in an equilibrium or a nonequilibrium state, and the follow-up of the consequences of that assignment. Although the evaluation of the generated entropy (Γ) attendant upon a given process is an important part of the procedures flowing from the formalism, the *rate* at which entropy is generated $(\dot{\Gamma})$ in a process is not a matter of concern, except when dealing with the time evolution of a system passing through a continuum of nonequilibrium transient states.

Book I deals with an unfamiliar way of extending classical macroscopic thermodynamics to include nonequilibrium states; in dealing with unfamiliar matters, it is often useful to introduce simple model systems to illustrate the ideas involved and to test the validity of those ideas, As manageable test cases, I use thermal and concentration fields of constant gradient—the formalism is not limited to fields of constant gradient, I merely use such fields because they readily supply explicit answers to important general questions.

For systems that have specified fields of the relevant intensities, I introduce the idea of the *equilibrium-based image* of the system. The domain of states for which the internal energy of the *equilibrium-based image* is an adequate representation of the internal energy of the system constitutes the Dartmouth domain. The domain of equilibrium states is a subdomain of the Dartmouth domain; and the Dartmouth domain covers part of, but not all of, the domain of nonequilibrium states. Systems having *equilibrium-based images* satisfy, in the Dartmouth domain, a restricted form (the Dartmouth version) of the second law.

Gibbs's Suggestion

A typical thermodynamic process starts with the system in an equilibrium state; upon the relaxation or unlocking of a constraint, the system undergoes a change to another state of equilibrium [16]. The entropy changes associated with the process satisfy the relation (second law)

$$\Gamma \equiv \Delta S(system) + \Delta S(surroundings) \geq 0 \ , \qquad (1)$$

where Γ is the generated entropy for the process.

Suppose we are interested in a change for which the initial state or the final state or both are nonequilibrium states. Can we evaluate entropy changes for such a process? In his 1873 paper "A Method of Geometrical Representation of the Thermodynamic Properties of Substances by Means of Surfaces," J. Willard Gibbs said, "When a body is not in a state of thermodynamic equilibrium, its state is not one of those which is represented by our surface [the (equilibrium) volume-entropy-energy surface]. The body, however, as a whole has a certain volume, entropy, and energy, which are equal to the sums of the volumes, etc., of its parts." [17].

In a footnote to the passage just cited, Gibbs said that he was using the word "energy" in such a way as to include the macroscopic kinetic energy if the parts of the system were macroscopically in motion [showed sensible *vis viva*]. Gibbs thus used the word "energy", for the case at hand, to stand for the *total energy* of the system, i.e. internal energy plus macroscopic kinetic energy plus ....

I shall distinguish between the internal energy U and the total energy "U" of a system, where

$$\text{“}U\text{”} \equiv U + \ macroscopic\ kinetic\ energy\ + \ldots. \qquad (2)$$

With this notational convention, the first law takes the form

$$\Delta\text{“}U\text{”} = Q + W; \qquad (3)$$

and we have the convenient notational compactions:

$$\text{“}H\text{”} \equiv H + \textit{macroscopic kinetic energy} + \ldots\,, \tag{4}$$

$$\text{“}A\text{”} \equiv A + \textit{macroscopic kinetic energy} + \ldots\,, \tag{5}$$

$$\text{“}G\text{”} \equiv G + \textit{macroscopic kinetic energy} + \ldots. \tag{6}$$

Consider the case of a glass of water stirred vigorously with a spoon and then left to relax to a quiescent state. Let the air in the room function as a constant temperature, constant pressure heat reservoir (thermostat) for the relaxation process. Let the system of interest be the water plus the glass, and treat both water and glass as incompressible substances. Then for the relaxation process,

$$W = 0, \tag{7}$$

$$\Delta\text{“}U\text{”} = Q; \tag{8}$$

and the heat exchanged between the room and the water-plus-glass system is equal to the sum of the changes in internal energy, in macroscopic kinetic energy, and in gravitational energy for the system.

For systems that are not in thermodynamic equilibrium, I follow Gibbs's suggestion in a particular way, a way that may or may not be what Gibbs had in mind.

The Equilibrium-based Image for a System

Consideration of the thermodynamics of a system in an equilibrium or nonequilibrium state starts with *given* fields of intensities (thermal field, pressure field, gravitational field, etc.) The values of the intensities at the various points in the spatial field under consideration are posited as a feature of the model system being investigated or are imposed by the constraints acting on the system or are inferred from readings of measuring devices posted at various places in the field or… —the point is that the values of the field quantities are taken as *given* at the start of the analysis.

As an example, consider a gas at a point x in a spatial field. The temperature T_x and the pressure p_x are to be given as part of the specification of the system. The nature of the thermal field may simply be posited ("Let us consider a system with a temperature distribution over the spatial field such that ...") or it may be imposed on the system by having the system interact with thermostats in a specified way or it may be inferred from an array of thermal sensors (thermocouples, resistance thermometers, etc.) posted at various places in the thermal field or .... The analysis starts by taking as temperatures the values of the field intensities at the points (x) of the field, *however those values were specified*, and building the theoretical structure on that foundation.

For the system under consideration, given the relevant fields of intensities, construct an *equilibrium-based image* of the system: assign local values of thermodynamic quantities to the points of the system by making use of equilibrium formulas based upon the given intensities; then integrate those assigned local values over the volume of the system. Thus, for a gas with given temperature T_x and given pressure p_x at point x in the spatial field, define molar values of entropy $\overline{S}_x$ and internal energy $\overline{U}_x$ in terms of the formulas $\overline{S}_x(T_x, p_x)$, $\overline{U}_x(T_x, p_x)$ appropriate to equilibrium situations. Then integrate those assigned local values over the volume of the system to get *equilibrium-based-image* values of the total entropy and the total internal energy. (I think this is what Gibbs had in mind in the quotation cited above.)

When we use the field intensities to define local values of thermodynamic quantities via equilibrium formulas and given intensities, I shall say that we are defining and dealing with the *equilibrium-based image* of the system under investigation—we map the *equilibrium-based image* onto the field of intensities. At the local level, the *equilibrium-based image* will, by construction, satisfy all the usual thermodynamic relations; thus for a gas with given temperature T_x and given pressure p_x at point x in the spatial field, the relation

$$d\overline{U}_x = T_x d\overline{S}_x - p_x d\overline{V}_x \qquad (9)$$

will hold, with $T_x = \partial\overline{U}_x / \partial\overline{S}_x\big)_{\overline{V}_x}$ and $-p_x = \partial\overline{U}_x / \partial\overline{V}_x\big)_{\overline{S}_x}$, where

$$\overline{V}_x = \overline{V}_x(T_x, p_x)$$

refers to the equilibrium equation of state.

I shall investigate in Book I of this work the thermodynamic properties of the *equilibrium-based images* of systems in nonequilibrium states—those thermodynamic properties will be found to satisfy some very general relations in a restricted domain, the *Dartmouth domain*, of nonequilibrium states.

Some Terminology and Notation, Dartmouth Domain

I shall say of a system that has the relevant specified fields of intensities that it *has an* ***equilibrium-based image***; and if a system does not have those specified fields of intensities, I shall say that it *does not have an* ***equilibrium-based image.*** I am uncertain as to whether a liquid in a state of turbulent flow has an *equilibrium-based image*, i.e. I am uncertain as to whether it is possible to specify the local temperature, pressure, and macroscopic kinetic energy at each point in a turbulent liquid (we shall see in Section 3 an example of a system that does not have an *equilibrium-based image*). I use regular symbols to represent properties of the *equilibrium-based image*. Where it proves necessary to distinguish between the properties of the system itself and the properties of the *equilibrium-based image* of the system, I indicate properties of the system itself by boldface symbols. Thus, for example, if the internal energy of the *equilibrium-based image* differed from the internal energy of the system itself, I would write $U(image) \neq \mathbf{U}(system)$. Note for future reference that the *equilibrium-based image* of an equilibrium state is identical to the equilibrium state itself.

Consider the case of a monatomic ideal gas in a thermal field, and let the gas have an *equilibrium-based image*. Now suppose that the thermal gradient is so extreme that the principle of equipartition of kinetic energy does not hold locally [18] and $\overline{\mathbf{U}}_x \neq \overline{U}_x = (3/2)RT_x$. Then

$$U = \int \overline{U}_x dn_x = \int (3/2) RT_x dn_x \neq \mathbf{U}\ , \tag{10}$$

where dn_x is the amount of the gas in a volume element surrounding the point x.

For thermodynamic systems that have *equilibrium-based images*, let the domain of states for which $U \approx \mathbf{U}$ (to a given level of accuracy) be called the *Dartmouth domain* (I live in the town of Dartmouth, Massachusetts). In the Dartmouth domain, a system in a given state, whether that state be an equilibrium or a nonequilibrium one, and its *equilibrium-based image* have the same field intensities at each point and (approximately) the same overall internal energies. The domain of equilibrium states is a subdomain of the Dartmouth domain. In addition to containing the domain of equilibrium states, the Dartmouth domain also contains part of, but not all of, the domain of nonequilibrium states.

In Book I, I study the thermodynamics of *equilibrium-based images* in the Dartmouth domain.

Strategy

In dealing with systems in nonequilibrium states, the strategy is to pair the system of interest with another system, the *equilibrium-based image* of the original system, that other system having the same fields of relevant intensities and (approximately) the same internal energy as the system of interest and also an entropy calculable via equilibrium formulas. The features of the paired system (the *equilibrium-based image*) are then investigated over the Dartmouth domain to seek out general results and interesting results.

All changes of state, whether infinitesimal or finite, are to be considered real changes, not virtual ones.

Laws for Equilibrium-based Images

Let a system that has an *equilibrium-based image* undergo a process. Flag features that are directly attributable to the process with a subscript upright arrow ($\uparrow$). Let $\Delta_\uparrow$ indicate changes in the surroundings that are directly attributable to the process. [It is not necessary to flag changes in the system since it is precisely the changes attributable to the process that we focus upon in studying the system.]

In the Dartmouth domain, the first law applies in the regular way to the *equilibrium-based image*:

$$\Delta \mathrm{U} = \mathrm{Q} + \mathrm{W}, \tag{11}$$

where ΔU is the difference in internal energy of the *images* of the final and initial states of the system of interest.

A limited version of the second law, appropriate for the Dartmouth domain, is as follows:

If the system and the surroundings in both the initial and final states have ***equilibrium-based images,*** *then, in the Dartmouth domain, the relation*

$$\Gamma_{\uparrow} \equiv \Delta S(system) + \Delta_{\uparrow} S(surroundings) \geq 0 \tag{12}$$

must be true. $\Gamma_{\uparrow}$ *is that part of* Γ *directly attributable to the process, and the changes* $\Delta_{\uparrow}$ *are for effects directly attributable to the process.*

I shall refer to the statement containing relation (12), together with the sentence following that statement, as the *Dartmouth version* (of the second law) or the *Dartmouth condition.*

Relation (12) is a reflection of the *distributive* character of the generated entropy Γ. If it is possible to resolve an overall process into a set of part-processes, then the generated entropy for each part-process has the same non-negative character that the generated entropy for the overall process has:

$$\Gamma(part - process) \geq 0 \; . \tag{13}$$

In other words, the second law holds locally as well as globally. It is not possible to devise two coupled part-processes such that entropy is destroyed in one part-process while a greater amount of entropy is generated in the other so that the global requirement is met, even though there is no local requirement [19-26].

Differentially Tailored Process

In a differentially tailored process, one brings about a sought-after change in state for a system by making differential changes in the boundary conditions together with adjustments to the constraints operating on the system, repeating the procedure a sufficient number of times if necessary.

For example, to transfer an amount dn of substance from a mass reservoir at α through an intermediate region to a mass reservoir at β in differentially tailored fashion, proceed as follows. In the initial state, let a piston exert a pressure p_α on the material in mass reservoir α; and let another piston exert a pressure p_β on the material in mass reservoir β.

i) Lock the piston at α in place.

ii) Increase the pressure exerted by the piston at α to $p_\alpha + dp_\alpha$ (the excess pressure dp_α is accommodated by the stops holding the piston in place).

iii) Unlock the piston at α and let it exert a steady pressure $p_\alpha + dp_\alpha$ on the material in the reservoir.

iv) When the piston at α has swept out a volume dV_α (originally containing the amount *dn* of substance), again lock the piston in place.

v) Material will leave the mass reservoir at α, enter the intermediate region, and eventually find its way into the mass reservoir at β. As material leaves the reservoir at α, the pressure in the reservoir will decrease until it eventually falls to p_α.

vi) The process will cease of itself when the pressure at α has fallen to p_α, the reservoir at β has gained an amount *dn* of substance, and the intermediate region has regained its original structure.

vii) When the process ceases, reduce the pressure exerted by the locked piston at α to p_α and unlock the piston.

Another example. An ideal monatomic gas resides in a cylinder that has adiabatically insulated side walls, end temperatures T_0, T_λ, and a movable piston at the T_λ end which exerts a pressure *p* on the gas (*vide infra*, Section 2). To expand the gas in differentially tailored fashion:

i) Lock the piston in place.

ii) Decrease the pressure exerted by the piston to $p - dp$

iii) Unlock the piston and let it exert a steady pressure $p - dp$ on the gas.

iv) The gas will expand until its pressure drops to $p - dp$; the piston will thereupon come to rest.

v) Repeat steps i)-iv) until the gas reaches the sought-after final volume.

The differential changes referred to here are real changes, not virtual ones.

A Useful Notational Convention

Let Z be a quantity characterizing a given thermodynamic situation, and let a thermodynamic process result in a change of another quantity or parameter ζ relating to the situation; then let [27]

$$\lim_{\zeta\to 0}\frac{Z(\zeta,\ldots)-Z(0,\ldots)}{\zeta}\equiv\left.\frac{\partial Z}{\partial\zeta}\right)_{\ldots}\Bigg|_{\zeta=0}\equiv\left.\frac{\delta Z}{\tilde{\delta}\zeta}\right)_{\ldots}, \tag{14}$$

the state $\zeta = 0$ to be attained via a sequence of steady states such that $\zeta \to 0$.

REFERENCES

1. A. B. Pippard, *The Elements of Classical Thermodynamics* (Cambridge University Press, Cambridge, 1957).
2. H. B. Callen, *Thermodynamics*, 2nd ed. (Wiley, New York, 1985).
3. E. A. Guggenheim, *Thermodynamics*, 8th ed. (Elsevier, New York, 1986).
4. M. Bailyn, *A Survey of Thermodynamics* (AIP, New York, 1994).
5. K. S. Pitzer, *Thermodynamics*, 3rd ed. (McGraw-Hill, New York, 1995).

6. L. Onsager, Phys. Rev. **37,** 405 (1931).
7. L. Onsager, Phys. Rev. **38,** 2265 (1931).
8. I. Prigogine, *Introduction to the Thermodynamics of Irreversible Processes*, 2nd ed. (Wiley, New York, 1961).
9. S. de Groot and P. Mazur, *Non-Equilibrium Thermodynamics* (North-Holland, Amsterdam, 1962).
10. R. J. Tykodi, *Thermodynamics of Steady States* (Macmillan New York, 1967).
11. R. J. Donnelly, R. Herman, and I. Prigogine, editors, *Non-Equilibrium Thermodynamics, Variational Techniques, and Stability* (University of Chicago Press, Chicago, 1966).
12. C. Truesdell, *Rational Thermodynamics*, 2nd, ed. (Springer, New York, 1984).
13. S. Sieniutycz and P. Salamon, editors, *Nonequilibrium Theory and Extremum Principles* (Taylor & Francis, New York, 1990).
14. D. Jou, J. Casas-Vazquez, and G. Lebon, *Extended Irreversible Thermodynamics* (Springer, Berlin, 1993).
15. See the series of papers following-up on Ref. 10; references to all the papers in the series appear in the last member of the series: R. J. Tykodi, J. Non-Equilib. Thermodyn. **16,** 267 (1991). In the paper just cited, there are a few errors in subscripting practice: in Eqs. (20), interchange the subscripts i,k on the L and K coefficients; in Eqs. (21), interchange the subscripts i,k throughout (the paper is reprinted in Book III).
16. Ref. 2, Ch. 1.
17. J. W. Gibbs, *The Scientific Papers of J. Willard Gibbs: Vol. I. Thermodynamics* (Dover, New York, 1961) p. 39.
18. D. J. Evans and G. P. Morriss, *Statistical Mechanics of Nonequilibrium Liquids* (Academic Press, San Diego, 1990) Ch. 9,10.
19. J. I. Belandria, J. Chem. Educ. **72,** 116 (1995).
20. R. Battino and S.E. Wood, J. Chem. Educ. **74,** 256 (1997).
21. R. D. Freeman, J. Chem. Educ. **74,** 281 (1997).
22. L. K. Nash, J. Chem. Educ. **74,** 282 (1997).
23. W. Olivares and P. J. Colmenares, J. Chem. Educ. **74**, 282 (1997).
24. R. J. Tykodi, J. Chem. Educ. **74,** 286 (1997).
25. J. I. Belandria, J. Chem. Educ. **74,** 286 (1997).
26. R. D. Freeman, J. Chem. Educ. **74,** 290 (1997).
27. See Book II, Ch.3.

TWO

MONATOMIC IDEAL GAS IN A THERMAL FIELD OF CONSTANT GRADIENT, DARTMOUTH DOMAIN

I use the model of an ideal gas in a thermal field of constant gradient to illustrate the applicability of the concepts mentioned above to nonequilibrium situations. I investigate several processes and show that the Dartmouth version of the second law holds for each process. I display a Carnot-like cycle for the gas in the thermal field, and I investigate the restrictions on heat-to-work conversion for a cycle using a working fluid in nonequilibrium states.

Let a right circular cylinder serve as a container with an interior volume V. Let the axis of the cylinder be of length λ, and let each of the circular end faces have inner area σ. Let the side walls of the cylinder be of rigid, adiabatically insulating material; and let the end faces be thin, rigid, and of diathermic material with negligible heat capacity.

Lay off a coordinate x along a line parallel to the axis of the cylinder; use a heat reservoir of temperature T_0 to maintain the face of the cylinder at $x = 0$ at the temperature T_0; likewise, maintain the face of the cylinder at $x = \lambda$ at the temperature T_λ. Fill the interior volume V with an amount n of an ideal monatomic gas, and allow the apparatus to reach a steady state. In the steady state, the pressure p is to be uniform throughout the cylinder (the mean free path between collisions for the gas particles is to be small compared to the diameter of the cylinder, i.e. there is to be no thermomolecular pressure effect [1]). Let the temperature on the cross sectional area at the coordinate value x be T_x, and assume that

$$T_x = T_0 + \left(\tfrac{x}{\lambda}\right)\left(T_\lambda - T_0\right) . \qquad (1)$$

See Fig. 1.

We shall limit our considerations to the Dartmouth domain. Remember that the properties displayed in ordinary type (as opposed to boldface type) are properties of the *equilibrium-based image* of the system.

Conventions

Use an overbar to represent a molar property: for example, the molar heat

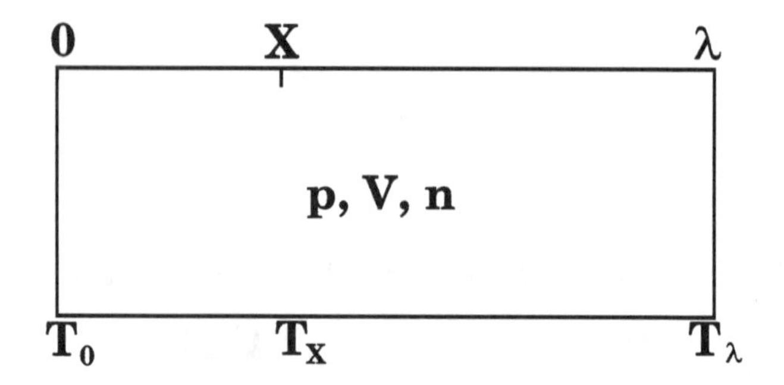

Fig. 1. Gas in a thermal field. An amount n of an ideal monatomic gas in a volume V at pressure p in a steady state; the steady-state thermal field is of constant gradient: $T_x = T_0 + \left(\frac{x}{\lambda}\right)\left(T_\lambda - T_0\right)$.

capacities of the gas are $\overline{C}_v = (3/2)R$, $\overline{C}_p = (5/2)R$, where R is the universal gas constant. For the gas, we have the relations $\overline{C}_p - \overline{C}_v = R$, $\gamma \equiv \overline{C}_p / \overline{C}_v = 5/3$, and $pV = nRT$. Where necessary, use the indexes i and f to identify initial and final states. Let unflagged quantities represent properties of the *equilibrium-based image* of the system; and where necessary, flag properties of the surroundings with the label *sur*.

The System

Let us now carry out an analysis of the thermodynamic properties of the gas in the steady-state thermal field.

In the steady state, the amount of gas dn_x contained between planes at x and $x + dx$ in the cylinder, i.e. the amount of gas contained in the volume σdx at the indicated place, is

$$dn_x = p\sigma dx / RT_x \; ; \tag{2}$$

and the total amount of gas n in the cylinder is

$$n = \int_{x=0}^{x=\lambda} dn_x = \frac{p\sigma}{R}\int_0^\lambda \frac{dx}{T_x} \; . \tag{3}$$

It follows from Eq. (1) that

$$dx = \left[\lambda / \left(T_\lambda - T_0\right)\right] dT_x \; ; \tag{4}$$

so

$$n = \frac{p\sigma\lambda}{R\left(T_\lambda - T_0\right)}\int_{T_0}^{T_\lambda} \frac{dT_x}{T_x} = \frac{pV}{R\left(T_\lambda - T_0\right)} \ln\frac{T_\lambda}{T_0} \; , \tag{5}$$

since $\sigma\lambda = V$.

In the steady state, then, the gas follows the equation of state

$$pV = nR\frac{T_\lambda - T_0}{\ln\left(T_\lambda / T_0\right)} = nR\theta \ , \tag{6}$$

where

$$\theta \equiv \theta\left(T_\lambda, T_0\right) \equiv \left(T_\lambda - T_0\right) / \ln\left(T_\lambda / T_0\right) \tag{7}$$

is the logarithmic mean of the temperatures T_λ, T_0. Logarithmic means are of some use in discussions of heat transfer [2].

For two non-negative real numbers a, b $(0 \le a, b < \infty)$ the logarithmic mean of the numbers lies between the ingredient numbers $\min(a,b) \le \theta(a,b) \le \max(a,b)$. Let the order of placement of *a,b* in $\theta(a,b)$ be significant; then, from the definition [Eq. (7)], it follows that

$$\theta(a,b) = \theta(b,a) \ ; \tag{8}$$

so we are free to show the variables in either order. For non-negative real numbers, the logarithmic mean lies between the arithmetic and geometric means of those numbers:

$$\frac{a+b}{2} \ge \frac{a-b}{\ln(a/b)} \ge (ab)^{1/2} \ . \tag{9}$$

The particles of the gas have kinetic energy only, so the molar internal energy $\overline{U}$ of such particles at temperature *T* is

$$\overline{U} = (3/2)RT = \overline{C}_v T \ . \tag{10}$$

The internal energy *U* of the amount *n* of gas in the steady state with end temperatures T_0, T_λ is

$$U = \int_{x=0}^{x=\lambda} \overline{U}_x dn_x = \int_{x=0}^{x=\lambda} \overline{C}_v T_x dn_x = \frac{\overline{C}_v p\sigma}{R}\int_0^\lambda dx$$

$$= \frac{\overline{C}_v p\sigma\lambda}{R} = \frac{\overline{C}_v pV}{R} = n\overline{C}_v\theta \ . \tag{11}$$

The molar entropy of the gas particles at temperature T_x, pressure *p* is

$$\overline{S}_x = \overline{S}_0 + \overline{C}_p \ln\left(T_x / T_0\right) , \tag{12}$$

where $\overline{S}_0 = \overline{S}_0\left(T_0, p\right)$—I show the variable-dependence of $\overline{S}_0$ explicitly only when needed. The total entropy *S* of the amount *n* of gas in the steady-state thermal field is, then,

$$S = \int_{x=0}^{x=\lambda} \overline{S}_x dn_x = \frac{p\sigma}{R} \int \left[\overline{S}_0 + \overline{C}_p \ln(T_x / T_0)\right](dx / T_x)$$

$$= \frac{pV}{R(T_\lambda - T_0)} \int_{T_0}^{T_\lambda} \left[\overline{S}_0 + \overline{C}_p \ln(T_x / T_0)\right](dT_x / T_x)$$

$$= n\left[\overline{S}_0 + (\overline{C}_p / 2)\ln(T_\lambda / T_0)\right]. \tag{13}$$

Thermally Depolarizing and Repolarizing the System

Let us now take the system in the steady state with end temperatures T_0, T_λ and instantaneously slip a thin, rigid, adiabatically insulating membrane over each end of the cylinder so as to isolate the system from interactions with the surroundings. The system will relax, at constant U, V, n, to a state of equilibrium. Since the internal energy remains constant during the process, the internal energy at equilibrium will be [Eq. (11)]

$$U = n\overline{C}_v \theta ; \tag{14}$$

and the uniform temperature at equilibrium will be $\theta(T_0, T_\lambda)$. Since n, V, θ are all the same as in the steady state, the equilibrium pressure will be p—the same as in the steady state.

For the equilibrium state, then,

$$S = n\left[\overline{S}_0 + \overline{C}_p \ln(\theta / T_0)\right]. \tag{15}$$

The generated entropy Γ for the equilibration step after the isolation operation is [Eqs. (13),(15)] (in this case $\Gamma_\uparrow = \Gamma$)

$$\Gamma = \Delta S$$

$$= n\left[\overline{S}_0 + \overline{C}_p \ln(\theta / T_0) - \overline{S}_0 - (\overline{C}_p / 2)\ln(T_\lambda / T_0)\right]$$

$$= n\overline{C}_p \ln\left[\theta / (T_0 T_\lambda)^{1/2}\right]. \tag{16}$$

In order that Γ be ≥ 0, it is required that

$$\theta(T_0, T_\lambda) \geq (T_0 T_\lambda)^{1/2} . \tag{17}$$

The logarithmic mean of the two temperatures T_0, T_λ is never less than their geometric mean. We have thus used a model system to demonstrate an instance of the mathematical inequality [3-5]

$$\frac{a+b}{2} \geq \frac{a-b}{\ln(a/b)} \geq (ab)^{1/2} \tag{18}$$

for any two non-negative real numbers a,b $(0 \leq a,b < \infty)$.

In isolating our system from interactions with the surroundings, we were—in a manner of speaking—thermally depolarizing it (destroying the thermal gradient and homogenizing the temperature). Let us now thermally repolarize the system and restore the original thermal gradient. But before doing that, let's discuss some features of the generated entropy Γ as they relate to our system and the processes it undergoes.

When the system is in a nonequilibrium steady state, there is a steady rate of entropy generation $\dot{\Gamma}$. Write the second law [relation (1.1)] in differential form:

$$d\Gamma = dS + dS_{sur} \geq 0 \ . \tag{19}$$

Then

$$\dot{\Gamma} = \dot{S} + \dot{S}_{sur} \geq 0 \ , \tag{20}$$

and in the steady state $(\dot{S} = 0)$

$$\dot{\Gamma} = \dot{S}_{sur} \geq 0 \ . \tag{21}$$

As long as we maintain the system in a nonequilibrium state by interaction with the surroundings, $\dot{\Gamma} \neq 0$, i.e. there is continuous dissipation going on which causes the generation of entropy. Now in acting on our system, we desire to keep track of those effects that influence the system in the given process as opposed to the overall effects that result in the total Γ. Our task is rather like looking at a badly adjusted TV set and trying to make out the pattern of the program through the obscuring "snow" on the screen. The ongoing dissipation for the situation under analysis forms an obscuring background against which we try to make out the traces left by the process of interest.

With respect to the internal energy of the system, we wish to keep track of those inputs of energy that actually affect the state of the system as opposed to those inputs that merely flush through the system.

Let the rate of influx of heat to a heat reservoir be $\dot{Q}^{(r)}$. Then, for the system diagrammed in Fig. 1, over a time interval from t_1 to t_2, the first law takes the form

$$\Delta U = Q + W \ , \tag{22}$$

where

$$Q = -\int_{t_1}^{t_2} \left[\dot{Q}_0^{(r)}(t) + \dot{Q}_\lambda^{(r)}(t) \right] dt \ . \tag{23}$$

Note that in a steady state, at a time t_q, $\dot{Q}_0^{(r)}(t_q) + \dot{Q}_\lambda^{(r)}(t_q) = 0$.

Let us now return to our analysis of the thermal repolarization process. Upon isolation, our system came to rest in an equilibrium state characterized by *V*, *n*, uniform temperature θ, uniform pressure *p*. Now instantaneously remove the adiabatic membranes from the end faces of the cylinder and reconnect those faces to the original heat reservoirs of temperatures T_0, T_λ. The system will thermally repolarize into the original gradient-bearing steady state characterized by the quantities $\theta(T_0, T_\lambda), p, V, n$.

For the polarization process:

$$p_f = p_i \equiv p \ , \tag{24}$$

$$\Delta U = n\overline{C}_v\theta - n\overline{C}_v\theta = 0 \ , \tag{25}$$

the repolarization process is thus a constant internal energy process;

$$S_f = n\left[\overline{S}_0 + (\overline{C}_p / 2)\ln(T_\lambda / T_0)\right], \tag{26}$$

$$S_i = n\left[\overline{S}_0 + \overline{C}_p \ln(\theta / T_0)\right]; \tag{27}$$

$$\begin{aligned}\Delta S &= S_f - S_i \\ &= n\left[(\overline{C}_p / 2)\ln(T_\lambda / T_0) - \overline{C}_p \ln(\theta / T_0)\right] \\ &= n\overline{C}_p \ln\left[(T_0 T_\lambda)^{1/2} / \theta\right],\end{aligned} \tag{28}$$

just the inverse of the entropy change that occurred when we isolated the system.

In the final gradient-bearing steady state, those elements of mass with temperatures greater than θ imported energy and those with temperatures less than θ exported energy. Let us assume that all the imported energy came from the heat reservoir with temperature $\max(T_0, T_\lambda)$ and that all the exported energy passed to the heat reservoir with temperature $\min(T_0, T_\lambda)$. Let the heat supplied by each reservoir be Q_0, Q_λ, where

$$Q_0 + Q_\lambda = 0 \tag{29}$$

since $\Delta U = 0$. Let x^* be the value of the *x*-coordinate for which $T_x = \theta$. Then (assume that $T_\lambda > T_0$)

$$Q_\lambda = \int_{x=x^*}^{x=\lambda} \overline{C}_v (T_x - \theta) dn_x = \frac{\overline{C}_v p\sigma\lambda}{R(T_\lambda - T_0)} \int_\theta^{T_\lambda} \frac{T_x - \theta}{T_x} dT_x$$

$$= \frac{\overline{C}_v pV}{R(T_\lambda - T_0)} \int_\theta^{T_\lambda} (dT_x - \theta d\ln T_x)$$

$$= \frac{n\overline{C}_v \theta}{(T_\lambda - T_0)} \left[T_\lambda - \theta - \theta \ln(T_\lambda / \theta)\right]. \tag{30}$$

So

$$\Delta_\uparrow S_{sur} = -T_0^{-1} Q_0 - T_\lambda^{-1} Q_\lambda = Q_\lambda \left(T_0^{-1} - T_\lambda^{-1}\right)$$

$$= \frac{n\overline{C}_v \theta}{(T_\lambda - T_0)} \left[T_\lambda - \theta - \theta \ln(T_\lambda / \theta)\right]\left(T_0^{-1} - T_\lambda^{-1}\right)$$

$$= \frac{n\overline{C}_v \theta}{T_0} \left[1 - \frac{\theta}{T_\lambda} + \frac{\theta}{T_\lambda} \ln(\theta / T_\lambda)\right]. \tag{31}$$

Let $f(y) \equiv 1 - y + y \ln y$. Then $f(1) = 0$, $f'(1) = 0$, $f''(1) = 1 > 0$; so $f(y) \geq 0$. The expression in square brackets in Eq. (31) is thus ≥ 0.

Represent the part of Γ that is directly attributable to the thermal polarization process by $\Gamma_\uparrow$; then

$$\Gamma_\uparrow = \Delta S + \Delta_\uparrow S_{sur}$$

$$= n\overline{C}_p \ln\left[(T_0 T_\lambda)^{1/2} / \theta\right] + \frac{n\overline{C}_v \theta}{T_0} \left[1 - \frac{\theta}{T_\lambda} + \frac{\theta}{T_\lambda} \ln(\theta / T_\lambda)\right]$$

$$= n\overline{C}_v \left\{\gamma \ln\left[(T_0 T_\lambda)^{1/2} / \theta\right] + \frac{\theta}{T_0} \left[1 - \frac{\theta}{T_\lambda} + \frac{\theta}{T_\lambda} \ln(\theta / T_\lambda)\right]\right\}. \tag{32}$$

The first term inside the curly braces is negative and the second term is positive. To see if the second term overbears the first term, let us look at a special case: $T_0 = 100K$, $T_\lambda = 400K$, $\theta(T_0, T_\lambda) = 216.4K$. Here

$$\Gamma_\uparrow = n\overline{C}_v(-0.1316 + 0.2741) = 0.1425 n\overline{C}_v. \tag{33}$$

Upon comparing, for this special case, $\Gamma_\uparrow$ for the thermal depolarization process [Eq. (16)] with $\Gamma_\uparrow$ for the repolarization process [Eq. (32)], we find that

$$\Gamma_{\uparrow}(\text{depolarization}) = 0.1316n\overline{C}_v \ , \tag{34}$$

$$\Gamma_{\uparrow}(\text{repolarization}) = 0.1425n\overline{C}_v \ . \tag{35}$$

In a sense, then, the adiabatic depolarization process was less dissipative than the repolarization process at constant U.

Pressure-Volume Work

Replace the end wall of the apparatus at $x = \lambda$ by a thin, snugly fitting, gas-tight, rigid piston of diathermic material with negligible heat capacity. Let the piston exert a pressure p_{ex} on the gas, and let the piston move in a frictionless fashion. Extend the side walls of adiabatically insulating material sufficiently so that the piston has adequate room for its excursions. Connect a heat reservoir of temperature T_λ to the piston in such a way that the heat reservoir can exchange energy with the gas inside the cylinder—the reservoir maintains the piston material at the temperature T_λ.

Now let the system be in a steady state with end temperatures T_0, T_λ, and let the piston undergo a displacement from $x = \lambda$ to $x = \lambda + d\lambda$. Then the work imparted to the system by the surroundings is

$$dW = -p_{ex}dV \ . \tag{36}$$

Let the work interaction take place over a time interval dt. Then the first law statement for the gain of internal energy by the system during that interval is

$$dU = dQ + dW = dQ - p_{ex}dV \ , \tag{37}$$

where

$$dQ = -\left(\dot{Q}_0^{(r)} + \dot{Q}_\lambda^{(r)}\right)dt \ . \tag{38}$$

If the system interacts with the surroundings during a time interval from t_1 to t_2, then

$$\Delta U = Q + W$$

$$= -\int_{t_1}^{t_2}\left[\dot{Q}_0^{(r)}(t) + \dot{Q}_\lambda^{(r)}(t)\right]dt - \int p_{ex}dV \ . \tag{39}$$

Let us now calculate the work involved in a differentially tailored expansion of the gas while the gas is in a steady state with end temperatures T_0, T_λ. In each infinitesimal step of the expansion process, $p_{ex} \approx p = nR\theta / V$. With the end temperatures T_0, T_λ fixed, $\theta\left(T_0, T_\lambda\right)$ stays constant during the process; so

$$W = -\int p_{ex} dV = -\int_{V_i}^{V_f} nR\theta(dV / V) = -nR\theta \ln(V_f / V_i) \,. \quad (40)$$

Since θ remains fixed for the process and $U = n\overline{C}_v\theta$, it follows that for the differentially tailored expansion

$$\Delta U = 0 \,, \quad (41)$$

$$Q = -W = nR\theta \ln(V_f / V_i) \,. \quad (42)$$

Now

$$S_f = n\left[\overline{S}_0(p_f) + (\overline{C}_p / 2)\ln(T_\lambda / T_0)\right], \quad (43)$$

$$S_i = n\left[\overline{S}_0(p_i) + (\overline{C}_p / 2)\ln(T_\lambda / T_0)\right]; \quad (44)$$

so

$$\Delta S = n\left[\overline{S}_0(p_f) - \overline{S}_0(p_i)\right] = n\left[-R\ln(p_f / p_i)\right]$$
$$= nR\ln(V_f / V_i) = Q / \theta \,. \quad (45)$$

In order to evaluate $\Gamma_\uparrow$, we need to know the source of the heat input Q—how much comes from the reservoir at T_0 and how much from the reservoir at T_λ. Let's evaluate the maximum value for $\Gamma_\uparrow$.

max$\Gamma_\uparrow$ rule

If we assume that for $Q > 0$ the energy ultimately comes from the heat reservoir whose temperature is $\max(T_0, T_\lambda)$ and that for $Q < 0$ the energy all ultimately goes into the heat reservoir whose temperature is $\min(T_0, T_\lambda)$, then we get the maximum value for $\Gamma_\uparrow$.

Consider a differentially tailored expansion at constant θ such that $V_f > V_i$ and consequently $Q > 0$. Then, by the max $\Gamma_\uparrow$ rule, for the case $T_\lambda > T_0$,

$$\Delta S = Q / \theta \,, \quad (46)$$

$$\Delta_\uparrow S_{sur} = -Q / T_\lambda \,, \quad (47)$$

$$\max \Gamma_\uparrow = \Delta S + \Delta_\uparrow S_{sur}$$

$$= Q(\theta^{-1} - T_\lambda^{-1}) = \theta(\theta^{-1} - T_\lambda^{-1})\Delta S = (1 - \theta T_\lambda^{-1})\Delta S > 0 \,. \quad (48)$$

Even though the expansion is carried out in differentially tailored fashion, it pays a price in generated entropy.

All we can say at this point about the actual value of $\Gamma_{\uparrow}$ for a differentially tailored expansion at constant θ is that

$$0 \le \Gamma_{\uparrow} \le \max\Gamma_{\uparrow} \,. \tag{49}$$

To be more specific would require additional information: there is no *a priori* way of assessing the individual contributions Q_0, Q_λ to Q for a differentially tailored expansion at constant θ. To determine those individual contributions requires making experimental measurements or solving the equations governing heat transfer [2] in the transient states of the expansion process.

A Carnot-like Cycle.

Let us take the apparatus we have just been dealing with (gas, cylinder, movable piston at $x = \lambda$) around a Carnot-like cycle in the following fashion (all expansions and compressions are to be carried out in differentially tailored fashion).

A) Start with the gas in a steady state with end temperatures T_0, T_λ, and volume V_1.

B) Expand the gas at constant $\theta(T_0, T_\lambda)$ to a volume V_2.

C) Instantaneously slip a thin, rigid, adiabatically insulating membrane over the end-face at $x = 0$ and over the inner piston-face. Allow the gas to relax to an equilibrium state with $T_C = \theta$.

D) Adiabatically and reversibly expand the gas from volume V_2 to volume V_3 and temperature T_D.

E) Instantaneously remove the adiabatically insulating membranes and apply heat reservoirs to set the end temperatures at T_0, T'_λ, where T'_λ is chosen so that

$$\theta' \equiv \theta'(T_0, T'_\lambda) \equiv (T'_\lambda - T_0) / \ln(T'_\lambda / T_0) = T_D.$$

F) Compress the gas at constant θ' to a volume V_4 that suffices to close the cycle (see step H).

G) Again instantaneously place adiabatically insulating membranes over the end-face at $x = 0$ and over the inner piston-face. Allow the gas to relax to an equilibrium state with $T_G = \theta'$.

H) Compress the gas reversibly and adiabatically back to the initial volume V_1. The volume V_4 was chosen so that at the end of this adiabatic compression the temperature of the gas T_H is equal to θ.

I) Instantaneously remove the insulating membranes and apply heat reservoirs to set the end temperatures at T_0, T_λ.

For the cycle as given:

a) Heat interactions occur at steps B,F.

b) Work interactions occur at steps B,D,F,H.

c) Steps E and I are thermal polarization steps at constant U and do not involve any net transfer of heat to or from the surroundings.

d) Contributions to $\Gamma_{\uparrow\downarrow}$ come from steps B,C,E,F,G,I ($\Gamma_{\uparrow\downarrow}$ is the sum of $\Gamma_{\uparrow}$ contributions around the cycle, i.e. it stands for $\oint d\Gamma_{\uparrow}$).
From the first law applied to the cycle, we have

$$0 = \Delta U = Q + W \tag{50}$$

and

$$-W = Q = Q_B + Q_F \ . \tag{51}$$

The work delivered to the surroundings, $-W$, is thus equal to the sum of all the heat interactions around the cycle. We see, then, that

$$-W = nR\theta \ln\left(V_2 / V_1\right) + nR\theta' \ln\left(V_4 / V_3\right)$$

$$= nR\theta \ln\left(V_2 / V_1\right) - nR\theta' \ln\left(V_3 / V_4\right) \ . \tag{52}$$

During reversible adiabatic expansion or compression of an ideal monatomic gas, the product $TV^{\gamma-1}$ remains constant; so in steps D and H we have

$$\theta V_2^{\gamma-1} = \theta' V_3^{\gamma-1} \ , \tag{53}$$

$$\theta V_1^{\gamma-1} = \theta' V_4^{\gamma-1} \ , \tag{54}$$

and consequently [divide Eq. (53) by Eq. (54)]

$$V_2 / V_1 = V_3 / V_4 \ . \tag{55}$$

Let the efficiency η of the heat-to-work conversion be defined as

$$\eta \equiv -W / Q_B \ ; \tag{56}$$

then

$$\eta = \frac{nR\theta \ln\left(V_2 / V_1\right) - nR\theta' \ln\left(V_3 / V_4\right)}{nR\theta \ln\left(V_2 / V_1\right)} = \frac{\theta - \theta'}{\theta} \ . \tag{57}$$

Let us examine a specific case: $T_0 = 300K$, $T_\lambda = 800K$, $T'_\lambda = 400K$; then

$$\theta(300K, 800K) = 509.8K, \ \theta'(300K, 400K) = 347.6K, \tag{58}$$

and

$$\eta = \frac{509.8K - 347.6K}{509.8K} = 0.318 \ . \tag{59}$$

On the other hand, a reversible Carnot engine operating between the temperature levels $800K$ and $400K$ would have an efficiency of

$$\eta_{\text{Carnot}} = \frac{800K - 400K}{800K} = 0.500 \ . \tag{60}$$

For the kind of cycle we described (T_0 = constant, $T_\lambda \to T'_\lambda$), we expect that

$$\eta\left(T_0, T_\lambda, T'_\lambda\right) \leq \eta_{\text{Carnot}}\left(T_\lambda, T'_\lambda\right) , \tag{61}$$

regardless of the value of T_0. To make relation (61) seem plausible, let's calculate η for $T_0 = 1000K$, $T_\lambda = 800K$, $T'_\lambda = 400K$, for $T_0 = 100K$, $T_\lambda = 800K$, $T'_\lambda = 400K$, and for $T_0 = 1K$, $T_\lambda = 800K$, $T'_\lambda = 400K$:

$$\eta\left(T_0 = 1000K\right) = 0.269, \ \eta\left(T_0 = 100K\right) = 0.357,$$

$$\eta\left(T_0 = 1K\right) = 0.443. \tag{62}$$

The numerical exercise suggests that, for fixed T_λ, T'_λ,

$$\lim_{T_0 \to 0} \eta\left(T_0, T_\lambda, T'_\lambda\right) = \eta_{\text{Carnot}}\left(T_\lambda, T'_\lambda\right) . \tag{63}$$

That such is indeed the case is fairly easy to demonstrate:

$$\eta = \frac{\theta - \theta'}{\theta} = 1 - \frac{\theta'}{\theta} = 1 - \frac{T'_\lambda - T_0}{T_\lambda - T_0} \bullet \frac{\ln T_\lambda - \ln T_0}{\ln T'_\lambda - \ln T_0} . \tag{64}$$

As $T_0 \to 0$, the first factor in the ratio of thetas goes to T'_λ / T_λ; and the second factor (upon application of l'Hospital's rule) goes to 1. So

$$\lim_{T_0 \to 0} \eta\left(T_0, T_\lambda, T'_\lambda\right) = 1 - \frac{T'_\lambda}{T_\lambda} = \frac{T_\lambda - T'_\lambda}{T_\lambda} = \eta_{\text{Carnot}}\left(T_\lambda, T'_\lambda\right) . \tag{65}$$

We thus see that the fact that the working fluid is in a nonequilibrium state during some of the steps of the cycle reduces the efficiency of the heat-to-work conversion relative to that of a reversible Carnot cycle.

Next, let's calculate the entropy change for the gas in each of steps B and F [Eq. (45)]:

$$\Delta_B S = nR \ln\left(V_2 / V_1\right),$$

$$\Delta_F S = nR\ln(V_4 / V_3) = -nR\ln(V_3 / V_4) \tag{66}$$

and [Eq. (55)]

$$\Delta_B S + \Delta_F S = nR\ln(V_2 / V_1) - nR\ln(V_3 / V_4) = 0 \ . \tag{67}$$

Relations (66) and (67) are the same as those that hold for a reversible Carnot cycle.

And now let's list the contributions to $\Gamma_{\uparrow\downarrow}$:

B) $0 \le \Gamma_{\uparrow} \le \max\Gamma_{\uparrow}$.

C) $n\overline{C}_p \ln\left[\theta / (T_0 T_\lambda)^{1/2}\right]$.

E) $n\overline{C}_v \left\{ \gamma \ln\left[(T_0 T'_\lambda)^{1/2} / \theta'\right] + \frac{\theta'}{T_0}\left[1 - \frac{\theta'}{T_\lambda} + \frac{\theta'}{T_\lambda}\ln(\theta' / T_\lambda)\right]\right\}$.

F) $0 \le \Gamma_{\uparrow} \le \max\Gamma_{\uparrow}$.

G) $n\overline{C}_p \ln\left[\theta' / (T_0 T'_\lambda)^{1/2}\right]$

I) $n\overline{C}_v \left\{ \gamma \ln\left[(T_0 T_\lambda)^{1/2} / \theta\right] + \frac{\theta}{T_0}\left[1 - \frac{\theta}{T_\lambda} + \frac{\theta}{T_\lambda}\ln(\theta / T_\lambda)\right]\right\}$.

The total generated entropy attributable to the cyclic process, $\Gamma_{\uparrow\downarrow}$, is the sum of the contributions just listed.

Discussion

In a reversible Carnot cycle, let the entropy input to the engine be $\Delta S(hot)$, $\Delta S(cold)$ at the hot, cold temperature, with $\Delta S(hot) + \Delta S(cold) = 0$. Then the functioning of a reversible Carnot engine is such that

$$-W = \Delta S(hot)[T(hot) - T(cold)] \ . \tag{68}$$

When a "charge" of entropy, $\Delta S(hot)$, drops from a hotter temperature environment to a colder one, work is delivered to the surroundings. Correspondingly, to lift a "charge" of entropy, $\Delta S(cold)$, from a colder temperature to a hotter one requires input of work from the surroundings:

$$W = \Delta S(cold)[T(hot) - T(cold)] \ . \tag{69}$$

Our Carnot-like cycle behaves in an analogous fashion [Eqs. (56),(57)]:

$$-W = \Delta_B S(\theta - \theta') \, . \tag{70}$$

The mode of functioning of our Carnot-like cycle is precisely the same as that of a reversible Carnot cycle, but the efficiency of the Carnot-like cycle is less than that of a reversible Carnot cycle operating between appropriate temperature levels because the working fluid for our cycle is in nonequilbrium states during parts of the cycle.

REFERENCES

1. See Book II, Ch. 5.
2. F. Kreith and M. S. Bohm, *Principles of Heat Transfer*, 4th ed. (Harper & Row, New York, 1986).
3. R. J. Tykodi, Am. J. Phys. **64,** 644 (1996).
4. A. R. Plastino, A. Plastino, and H. G. Miller, Am. J. Phys. **65**, 1102 (1997).
5. L. R. Berrone and C. D. Galles, Am. J. Phys. **66,** 87 (1998).

THREE

MORE GENERAL CONSIDERATIONS

Now that we have seen how the ideas of Section 1 apply to the ideal gas in a thermal field, let's turn again to some general considerations.

The Hypothesis of Local Equilibrium

In the discipline commonly called *thermodynamics of irreversible processes*, use is made of a *hypothesis of local equilibrium* [1-4] which asserts that in the domain of nonequilibrium states there is a subdomain consisting of the states of systems that may be considered to be locally in equilibrium even though they are globally not in equilibrium—the local properties of such systems satisfy standard equilibrium formulas. Thus, for example, in the case of a 1-component gas considered at some interior point characterized by temperature T, chemical potential μ, entropy density s, internal energy density u, and amount of substance concentration c, the equilibrium formula

$$du = Tds + \mu dc \tag{1}$$

is used in the form

$$\dot{s} = \frac{1}{T}\dot{u} - \frac{\mu}{T}\dot{c} \tag{2}$$

to evaluate the entropy production at the point in question [5].

The states (equilibrium or nonequilibrium) of a system satisfying the hypothesis of local equilibrium evidently fall into the Dartmouth domain—such states are associated with well-defined fields of intensities, and the internal energy of the *equilibrium-based image* of the system closely matches the internal energy of the system. Texts such as Reference 2 make frequent mention of local equilibrium temperature in nonequilibrium situations but give very little discussion about how to determine it, in the operational sense.

The Dartmouth domain and the domain of states for systems satisfying the hypothesis of local equilibrium are pretty much coextensive—it would be a highly unusual state, involving features such as those discussed in the appendix to Section 13, that would fall inside the Dartmouth domain and, simultaneously, outside the domain of states for systems satisfying the hypothesis of local equilibrium. The "Dartmouth" terminology is, I think, more flexible and less committing than the "local equilibrium" terminology: the definition of the Dartmouth domain requires only the existence of well-defined fields of intensities and near-equality of the internal energy of the system and its *equilibrium-based image*—nothing is implied about the nature of the local states. Outside the Dartmouth domain, the concept of *equilibrium-based image* is still a useful concept—see Section 13.

We shall find that nonequilibrium states that fall outside the Dartmouth domain are of two types, either the system fails to have well-defined fields of intensities (i.e. does not have an *equilibrium-based image*) or the internal energy of the *equilibrium-based image* fails to adequately represent the internal energy of the system because the internal energy of the system depends explicitly on gradients of the intensities as well as on the intensities themselves—see Section 13 for further discussion.

The Temperature Concept for Systems in Nonequilibrium States

The temperature concept in classical equilibrium thermodynamics stems from the ideas of thermal interaction, thermal balance or thermal equilibrium, and the transitivity of the thermal balance relation as embodied in the zeroth law: *two bodies in thermal equilibrium with a third body are in thermal equilibrium with each other*. The classical temperature concept applies in a natural way to systems in steady states, provided that the systems have *equilibrium-based images*.

To assign local temperatures to the points of such a system in a steady state, use as a probe a system in "equilibrium," and establish a condition of thermal balance between the point in question and the "equilibrium-state" probe. Then assign the temperature of the probe to the point in question.

As an example, consider the case of a cylinder containing an amount n of a monatomic ideal gas in a volume V. The end temperatures of the cylinder are set at T_0, T_λ by interactions with heat reservoirs (thermostats). The end and side walls of the cylinder are made of metal, and the side wall is adiabatically insulated. In the steady state, the gas has a uniform pressure p. See Fig. 2.1 and the discussion in Section 2— in the present case, the side wall of the cylinder is to be of metal.

Prepare a heat reservoir (thermostat) whose temperature can be varied and which is in thermal contact with a fine, metallic, adiabatically insulated wire. Let a short length of the wire be free of adiabatic insulation at the other end. On the exterior wall of the cylinder, along a line parallel to the axis of the cylinder, establish a number of test ports by drilling shafts through the insulation and drilling seats for the probe wire in the metal wall. When not in use, the shaft through the insulation can be plugged with insulation and the metal seat can be filled with a snippet of wire.

Pick one of the ports and plug the test wire into the seat. Let the rate of influx of heat to the test heat reservoir be $\dot{Q}^{(r)}$. Adjust the temperature of the heat reservoir so that $\dot{Q}^{(r)} = 0$, and assign that temperature to the inner cross sectional area of the cylinder defined by the port being tested.

Take two cylinders of the type just discussed: cylinder 1 having an amount of gas n, volume V, pressure p, end temperatures T_0, T_λ; cylinder 2 having an amount of gas n, volume V, pressure p', end temperatures $T_0, T_\lambda{}'$. Establish thermal test ports in the wall of each cylinder. Suppose that we find, in the steady state for the two cylinders, that the temperatures we assign to the inner cross sectional areas of the cylinders match for one pair of test ports: $T_j = T_{k'}$. Now connect those two ports j, k' by a wire of the probe type (insulated against lateral heat losses). We should find that the steady state of each cylinder is undisturbed by the linking up of the two cylinders (no loss or gain of energy for either system via the linkage) because of the zeroth law: *two systems each in thermal balance with a third system (the test heat reservoir) are in thermal balance with each other.*

In practical work, we replace the heat reservoir–and–wire probe with more manageable thermal sensors (resistance thermometers, thermocouples, etc.).

The temperature concept is well defined for systems in equilibrium states and for systems in steady states, provided that the systems have *equilibrium-based images*. The concept can be extended to cover systems for which the local thermal environment is changing slowly.

For systems in which the local thermal environment is changing rapidly enough so that the response–time of a sensor becomes a critical factor, the direct evaluation of local temperature via a thermal sensor is no longer satisfactory; and it is necessary—if we wish to use the temperature concept at all—to rely on theoretical equations describing the time–evolution of temperature [the Fourier equation, $\nabla^2 T - (1/\kappa)\partial T/\partial t = 0$, for example] from some properly defined initial state.

[See Reference 2 for a discussion of the temperature concept in *extended thermodynamics of irreversible processes*.]

Observation

The use of an "equilibrium-probe" to define local temperatures for systems in nonequilibrium states is akin to the procedure used by Muschik [6,7] to define "contact temperatures." Muschik's concept of contact temperature is more ambitious, however, in that Muschik assigns a single contact temperature to a finite system in a nonequilibrium state (I speak only of local temperatures in a system in a nonequilibrium state).

A Sequence of Cylinder Clusters (Constant-Gradient Systems)

Take a cylinder containing an amount n_1 of a monatomic ideal gas in a volume V_1. The cross sectional area of the inner space of the cylinder is σ. The end temperatures of the cylinder are set at T_0, T_λ by interactions with appropriate heat reservoirs (thermostats). The end and side walls are made of metal, and the side wall is adiabatically insulated. In the steady state the gas has a uniform pressure p. The inner length of the cylinder is λ_1, the temperature distribution in the gas in the axial direction is linear, so the thermal gradient in the axial direction is everywhere equal to $\left(T_\lambda - T_0\right)/\lambda_1$—see Fig. 2.1 and the discussion in Section 2. Let the cylinder form cluster 1 of a sequence of cylinder clusters.

Replace the original cylinder by two cylinders each of the same inner cross sectional area σ, with end temperatures T_0, T_λ and with the same sort of insulated side wall. The inner length of each cylinder is $\lambda_2 = \lambda_1 / 2$. Each cylinder has an inner volume $V_2 = V_1 / 2$ and is filled with an amount $n_2 = n_1 / 2$ of the monatomic ideal gas. In the steady state, the gas in each cylinder has a uniform pressure p. The thermal gradient in the axial direction everywhere inside each cylinder is $gradient_2 = 2gradient_1 = 2(T_\lambda - T_0) / \lambda_1$. Let the two cylinders together form cluster 2 of a sequence of cylinder clusters.

Replace each cylinder of cluster 2 by two new cylinders of half the inner length, same inner cross sectional area, same pattern of insulation, half the amount of the monatomic ideal gas, same end temperatures T_0, T_λ, same pressure p in the steady state. There are four cylinders in the cluster, and each cylinder has $n_3 = n_1 / 4$, $V_3 = V_1 / 4$, $\lambda_3 = \lambda_1 / 4$, $gradient_3 = 4gradient_1 = 4(T_\lambda - T_0) / \lambda_1$. Let the four cylinders together form cluster 3 of a sequence of cylinder clusters.

Continue generating clusters for the sequence of cylinder clusters in the manner just given. The *i*th cluster of the sequence has 2^{i-1} cylinders. Each cylinder of the *i*th cluster has $n_i = n_1 / 2^{i-1}$, $gradient_i = 2^{i-1} gradient_1 = 2^{i-1}(T_\lambda - T_0) / \lambda_1$, $p_i = p$.

For each cylinder in the early clusters of the sequence, $pV_i = n_i R\theta$, where θ is the logarithmic mean of the temperatures T_0, T_λ, i.e.

$$\theta \equiv \theta(T_\lambda, T_0) \equiv (T_\lambda - T_0) / \ln(T_\lambda / T_0) \quad \text{—see Section 2.}$$

The steady-state internal energy of the *equilibrium-based image* of the gas in a cylinder of the *i*th cluster is $U_i = n_i \overline{C}_v \theta$, and the total internal energy of the *equilibrium-based image* for the entire cluster is

$$2^{i-1} U_i = 2^{i-1} n_i \overline{C}_v \theta = n_1 \overline{C}_v \theta = U_1$$

—each cluster of the sequence has the same overall internal energy for its *equilibrium-based image* (but see below).

As we continue to generate clusters in the sequence, does $2^{i-1} U_i$ always equal $n_1 \overline{C}_v \theta = U_1$? There comes a place in the sequence where the distance between a cylinder face of temperature T_λ and the corresponding face of temperature T_0 is of the same order of magnitude as the mean free path of the particles of the gas. Under such conditions, a particle of the gas leaving one face of the cylinder has a significant probability of reaching the other face without colliding with another gas particle along the way. For such circumstances, for the cluster in question, $\mathbf{U}(gas) \neq n_1 \overline{C}_v \theta$. The failure of $n_1 \overline{C}_v \theta$ to adequately represent $\mathbf{U}(gas)$ for the later clusters of the sequence is not due to an explicit dependence of the local internal energy of the gas on spatial derivatives of the intensities, but rather to the fact that the system no longer has an *equilibrium-based image*: there is no longer a thermal field with well-defined local values of temperature for the gas, and we may have to distinguish between pressure in a radial direction and pressure in the axial direction—there is no longer a well-defined U_i. The later clusters of the sequence fall outside the Dartmouth domain because they do not have *equilibrium-based images*.

[In the discussion just completed, suppose that for a cylinder in a later cluster (say the kth) in the sequence $\mathbf{U}_k \neq n_k\overline{C}_v\theta$ and that the system does not have an *equilibrium-based image*, i.e. the system lacks a well-defined thermal field. Although the system lacks an *equilibrium-based image*, it does have an internal energy $\mathbf{U}_k$, the sum of the kinetic energies of all the gas particles, and a volume $\mathbf{V}_k$. Suppose we are able to express $\mathbf{U}_k$ and $\mathbf{V}_k$ in terms of local molar values $\overline{\mathbf{U}}_x$ and $\overline{\mathbf{V}}_x$ so that

$$\mathbf{U} = \int \overline{\mathbf{U}}_x dn_x \text{ and } \mathbf{V} = \int \overline{\mathbf{V}}_x dn_x .$$

Use the equilibrium formula $\overline{\mathbf{S}}_x = \overline{\mathbf{S}}_x\left(\overline{\mathbf{U}}_x, \overline{\mathbf{V}}_x\right)$ to establish a local molar entropy $\overline{\mathbf{S}}_x$. What, then, is the physical significance in the present instance of the quantities $1/\mathbf{T}_x$ and $\mathbf{p}_x/\mathbf{T}_x$ defined by $1/\mathbf{T}_x \equiv \partial\overline{\mathbf{S}}_x / \partial\overline{\mathbf{U}}_x)_{\overline{\mathbf{V}}_x}$ and

$$\mathbf{p}_x / \mathbf{T}_x \equiv \partial\overline{\mathbf{S}}_x / \partial\overline{\mathbf{V}}_x\Big)_{\overline{\mathbf{U}}_x} ?$$

As far as I can see, the quantities $1/\mathbf{T}_x$ and $\mathbf{p}_x/\mathbf{T}_x$ for the given situation have no physical significance whatever.]

There thus seem to be two ways for systems in nonequilibrium states to fall outside the Dartmouth domain: they either fail to have *equilibrium-based images* or they have internal energies that depend explicitly on spatial derivatives of the intensities that characterize the system. On further reflection, we are faced with the question, Can a system have an internal energy that depends explicitly on spatial derivatives of the intensities and also have an *equilibrium-based image*? In other words, Do all systems outside the Dartmouth domain fail to have *equilibrium-based images*?

Consider a system in a thermal field—the system of Fig.2.1, say. Suppose we try to make the thermal gradient so severe that $\mathbf{U} \neq n\overline{C}_v\theta$ by making the temperature difference $T_\lambda - T_0$ as large as practicable, say

$$T_\lambda = 3000K,\ T_0 = 300K,\ \theta(3000K, 300K) = 1173K,$$

$\lambda = 10$ cm, $p = 10$ bar. If we were dealing with a real substance—helium or neon, for example—we would have to worry about finding a container stable to high temperatures and impervious to the gas in question, about the possibility of exciting higher electronic states of the atoms or of fully ionizing the atoms, and about the importance of the radiant energy (electromagnetic "heat radiation") traversing the system. Let us, however, continue to deal with an ideal gas of structureless particles, a gas with kinetic energy only (we still have to deal with the radiant energy in the system).

Let us suppose that $\mathbf{U} \neq n\overline{C}_v\theta$. If the system were to have an *equilibrium-based image*, then the observation just stated would lead us to presume that an adequate treatment of $\mathbf{U}$ would require allowance for the explicit dependence of $\mathbf{U}$ on spatial derivatives of the local temperature. Another possibility, however, is that a well-defined local temperature does not exist for the system. The local thermal environment could be so inhomogeneous that it would defy description by a simple scalar "temperature" and would require more of a "tensor-type" description, with different components in different directions. If such were the case, it would be difficult, if not impossible, to use the zeroth law to compare "temperatures" within the system or between the system in question and another system.

If the local thermal environment is such that "temperature" comparisons cannot be made via the zeroth law, the system in question is better thought of as failing to have a well-defined thermal field—and therefore as failing to have an *equilibrium-based image*.

To explore the points just alluded to requires that the explorer have competence in dealing with the kinetic theory and statistical mechanics of non-uniform gases—a competence that I lack. Perhaps some readers who possess the necessary competence will be sufficiently interested in the matter to investigate the points at issue.

REFERENCES

1. S. de Groot and P. Mazur, *Non-Equilibrium Thermodynamics* (North-Holland, Amsterdam, 1962) Chs. III, IX.
2. D. Jou, J. Casas-Vazquez, and G. Lebon, *Extended Irreversible Thermo dynamics* (Springer, Berlin, 1993) Chs. 1,6. In the example illustrated in Fig. 6.2 of the Reference, the left-hand system is essentially an "equilibrium" thermal probe. When there is thermal balance between the "probe" and the right-hand system, the temperature of the "probe" will, according to the analysis in the Reference, be θ_r. My "temperature" is thus equivalent to the "θ" of the Reference.
3. S. Wisniewski, B. Staniszewski, and R. Szymanik, *Thermodynamics of Nonequilibrium Processes* (Reidel, Dordrecht, 1976) pp. 22–24.
4. G. D. C. Kuiken, *Thermodynamics of Irreversible Processes* (Wiley, New York, 1994) Ch. 3.
5. H.B. Callen, *Thermodynamics*, 2nd ed. (Wiley, New York, 1985) Ch. 14.
6. W. Muschik and G. Brunk, Int. J. Engng. Sci. **15**, 377 (1977).
7. W. Muschik, *Aspects of Non-Equilibrium Thermodynamics* (World Scientific, Singapore, 1990) Ch. 4.

FOUR

ADDITIONAL OPERATIONS WITH IDEAL GASES

I continue to illustrate the ideas of Section 1 via operations with ideal gases.

The Porous-Plug Process

In the porous-plug process, a gas flows in an adiabatically insulated tube across a porous plug from an initial state (T_i, p_i) to a final state (T_f, p_f) under constant enthalpy conditions. Let the flow take place in a steady-state manner, and let one mole of the gas pass from the initial state to the final state (neglect the macroscopic kinetic energy of the gas). Then, for the porous-plug flow process,

$$\overline{H}_f = \overline{H}_i \tag{1}$$

and

$$\left.\frac{\partial T}{\partial p}\right)_{\overline{H}} = \frac{T\partial\overline{V} / \partial T\big)_p - \overline{V}}{\overline{C}_p} \,. \tag{2}$$

Equations (1),(2) are displayed in nearly all textbooks of thermodynamics, but the entropy change for the flow process is rarely displayed—Kestin [1] is one of the few authors to take up the problem. What we want is $\partial\overline{S} / \partial p\big)_{\overline{H}}$. It follows from the relation

$$d\overline{H} = Td\overline{S} + \overline{V}dp \tag{3}$$

that

$$\left.\frac{\partial\overline{S}}{\partial p}\right)_{\overline{H}} = -\frac{\overline{V}}{T}. \tag{4}$$

The change in entropy for the porous-plug flow process is, then,

$$\Delta\overline{S} = \overline{S}_f - \overline{S}_i = -\int_{p_i}^{p_f} \left(\overline{V} / T\right) dp > 0 \tag{5}$$

(remember that $p_f < p_i$).

The flow process is an adiabatic irreversible process, so the generated entropy Γ is just $\Delta\overline{S}$:

$$\Gamma = \Delta\overline{S} = -\int_{p_i}^{p_f} \left(\overline{V} / T\right) dp > 0 \ . \tag{6}$$

For an ideal gas, $\overline{V} / T = R / p$, and

$$\Gamma = -R\ln\left(p_f / p_i\right) = R\ln\left(p_i / p_f\right) > 0. \tag{7}$$

Monatomic Ideal Gas in a Thermal Field

The monatomic ideal gas in a thermal field of constant gradient was dealt with extensively in Section 2; let us now look at some additional processes involving such a gas. First let's investigate the expansion of such a gas into a vacuum—see Fig. 1. Two identical cylinders of volume V and inner cross sectional area σ, with rigid side walls of adiabatically insulating material and thin, rigid end walls of diathermic material with negligible heat capacity are connected by an adiabatically insulated narrow tube containing a stopcock. Use heat reservoirs to maintain the end temperatures of the cylinders at temperatures T_0, T_λ as indicated in Fig. 1. Fill one cylinder

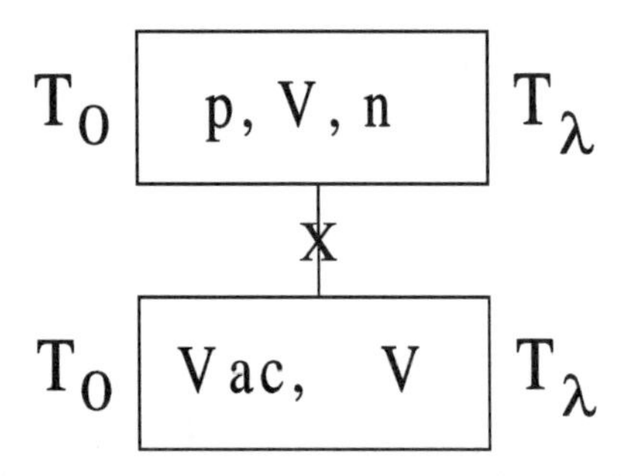

Fig. 1. Monatomic ideal gas in a thermal field: end temperatures T_0, T_λ; pressure p, volume V, amount of substance n. A gas-containing chamber is connected to a "twin" evacuated tube of volume V, end temperatures T_0, T_λ, by an adiabatically insulated narrow tube containing a stopcock.

with an amount n of a monatomic ideal gas, and evacuate the other cylinder. Let the composite system settle into a steady state with the ideal gas having a uniform pressure p. Then open the stopcock and allow the gas to expand into the evacuated space. Eventually, a new steady state ensues with uniform pressure $p/2$ in each cylinder.

Where necessary, use primes to designate quantities associated with the cylinder that was originally evacuated. The logarithmic mean, θ, of the temperatures T_0, T_λ is defined by $\theta \equiv \left(T_\lambda - T_0\right) / \ln\left(T_\lambda / T_0\right)$—see Section 2. Then (again see Section 2),

$$U_f = 2\left[(n/2)\overline{C}_v\theta\right], \tag{8}$$

$$U_i = n\overline{C}_v\theta , \tag{9}$$

$$\Delta U = 0, \;\; W = 0, \;\; Q = 0 , \tag{10}$$

$$S_f = 2\left\{(n/2)\left[\overline{S}_0(p/2) + \left(\overline{C}_p/2\right)\ln\left(T_\lambda/T_0\right)\right]\right\} , \tag{11}$$

$$S_i = n\left[\overline{S}_0(p) + \left(\overline{C}_p/2\right)\ln\left(T_\lambda/T_0\right)\right], \tag{12}$$

$$\Delta S = -nR\ln\left[(p/2)/p\right] = nR\ln 2 , \tag{13}$$

$$\Delta_{\uparrow} S_{sur} = -\left(Q_\lambda + Q'_\lambda\right)T_\lambda^{-1} - \left(Q_0 + Q'_0\right)T_0^{-1} , \tag{14}$$

$$\begin{aligned}\Gamma_{\uparrow} &= \Delta S + \Delta_{\uparrow} S_{sur} \\ &= nR\ln 2 - \left(Q_\lambda + Q'_\lambda\right)T_\lambda^{-1} - \left(Q_0 + Q'_0\right)T_0^{-1} .\end{aligned} \tag{15}$$

We see from Eq. (10) that the separate heat exchanges $Q_0, Q_\lambda, Q'_0, Q'_\lambda$ between the cylinders and the heat reservoirs satisfy the condition

$$0 = Q = Q_0 + Q_\lambda + Q'_0 + Q'_\lambda , \tag{16}$$

and Eq. (16) can be used to effect some simplification in Eq. (15).

To evaluate $\Gamma_{\uparrow}$ we would need to determine each of the four terms on the right-hand side of Eq. (16) by experiment or by solving the pertinent equations of heat transfer for the transient states between the steady initial and final states. Inasmuch as I am not prepared to act on either of these options, let's follow another pathway between the initial and final steady states and evaluate $\Gamma_{\uparrow}$ for the alternative pathway $\left[\Gamma_{\uparrow}(\text{alternative path}) \text{ probably } \neq \Gamma_{\uparrow}(\text{original path})\right]$.

Start in the original initial steady state and place rigid, adiabatically insulating caps over the end faces of both cylinders to isolate the system from the surroundings. The gas will relax to a state of equilibrium with uniform temperature θ (see Section 2). For that step

$$\Gamma_{\uparrow} = n\overline{C}_p \ln\left[\theta/\left(T_0 T_\lambda\right)^{1/2}\right]. \tag{17}$$

Neglect any effect that the isolation procedure might have on the radiation field in the evacuated cylinder. Now open the stopcock and allow the gas to expand adiabatically into the evacuated space. We now have two cylinders, each with amount *n/2* of gas at temperature θ and pressure *p/2*. For this step

$$\Gamma_{\uparrow} = \Delta S = nR\ln 2 \; . \tag{18}$$

Finally, remove the adiabatically insulating end caps and reconnect the original heat reservoirs to the end faces to restore the thermal gradients. For this step, assume that $T_\lambda > T_0$, then (see Section 2)

$$\Gamma_{\uparrow} = n\overline{C}_v\left\{\gamma\ln\left[\left(T_0T_\lambda\right)^{1/2}/\,\theta\right] + \frac{\theta}{T_0}\left[1 - \frac{\theta}{T_\lambda} + \frac{\theta}{T_\lambda}\ln\left(\theta / T_\lambda\right)\right]\right\} . \tag{19}$$

Summing Eqs. (17),(18),(19) gives

$$\Gamma_{\uparrow}(\text{overall}) = nR\ln 2 + \frac{n\overline{C}_v\theta}{T_0}\left[1 - \frac{\theta}{T_\lambda} + \frac{\theta}{T_\lambda}\ln\left(\theta / T_\lambda\right)\right] . \tag{20}$$

Now let's bring about the "same" change in state via a differentially tailored expansion of the gas in the cylinder to a total volume twice that of the original volume—see Fig. 2: $p \to p/2,\ V \to 2V,\ \theta \to \theta$.

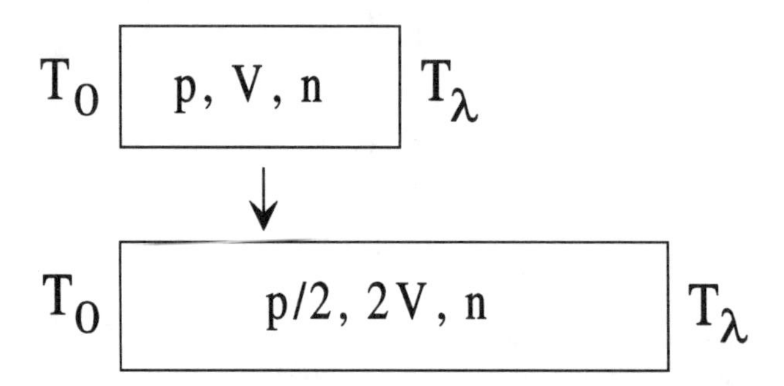

Fig. 2. Maintaining the end temperatures at T_0, T_λ and keeping the side walls adiabatically insulating, expand the gas in differentially tailored fashion to $2V$.

For the differentially tailored expansion (see Section 2),

$$U_f = n\overline{C}_v\theta \; , \tag{21}$$

$$U_i = n\overline{C}_v\theta \; , \tag{22}$$

$$\Delta U = 0 \; , \tag{23}$$

$$Q = -W = nR\theta \ln 2 \ , \tag{24}$$

$$S_f = n\left[\overline{S}_0(p/2) + \left(\overline{C}_p/2\right)\ln\left(T_\lambda/T_0\right)\right] , \tag{25}$$

$$S_i = n\left[\overline{S}_0(p) + \left(\overline{C}_p/2\right)\ln\left(T_\lambda/T_0\right)\right] , \tag{26}$$

$$\Delta S = nR\ln 2 \ . \tag{27}$$

Use double primes to designate the individual heat inputs Q_0'', Q_λ''; then

$$Q = Q_0'' + Q_\lambda'' = nR\theta\ln 2 \ , \tag{28}$$

$$\Delta_{\uparrow} S = -T_0^{-1}Q_0'' - T_\lambda^{-1}Q_\lambda'' \ , \tag{29}$$

$$\begin{aligned}\Gamma_{\uparrow} &= \Delta S + \Delta_{\uparrow} S \\ &= nR\ln 2 - T_0^{-1}Q_0'' - T_\lambda^{-1}Q_\lambda'' \ .\end{aligned} \tag{30}$$

To get an accurate expression for $\Gamma_{\uparrow}$ in this case, we would have to evaluate Q_0'' and Q_λ'' separately. Let's assume instead, as a special case, that $T_\lambda > T_0$ and $Q_0'' = 0$; then

$$Q_\lambda'' = nR\theta\ln 2 \ , \tag{31}$$

$$\Delta_{\uparrow} S = -T_\lambda^{-1}Q_\lambda'' = -T_\lambda^{-1}nR\theta\ln 2 \ , \tag{32}$$

$$\begin{aligned}\Gamma_{\uparrow} &= \Delta S + \Delta_{\uparrow} S \\ &= \left(1 - \theta T_\lambda^{-1}\right)nR\ln 2 \ .\end{aligned} \tag{33}$$

We note that

$$\Gamma_{\uparrow}(\text{Eq.}33) < \Gamma_{\uparrow}(\text{Eq.}20) \ , \tag{34}$$

and it seems plausible that

$$\Gamma_{\uparrow}(\text{Eq.}30) < \Gamma_{\uparrow}(\text{Eq.}15) \tag{35}$$

since we expect the expansion into the vacuum to be more dissipative than the differentially tailored expansion.

The analyses we have just gone through illustrate some standard features of the thermodynamic characteristics of nonequilibrium situations. The initial steady state of the gas was the same in both the expansion-into-the-vacuum process and the differentially tailored expansion process; and the final steady state had the same U_f and S_f for both processes. With respect to their *equilibrium-based images*, the final steady states for both processes are the same—even though the thermal gradients are not the same. Although U_f and S_f are the same for the two processes, $\dot{\Gamma}(\text{final steady state})$ differs for the two cases.

Assume that $T_\lambda > T_0$. Then, for the expansion-into-the-vacuum case,

$$\dot{\Gamma}(\text{final steady state}) = 2\left[\dot{Q}_0^{(r)}\left(T_0^{-1} - T_\lambda^{-1}\right)\right] ; \tag{36}$$

and for the differentially tailored expansion case,

$$\dot{\Gamma}(\text{final steady state}) = \dot{Q}_0^{(r)}\left(T_0^{-1} - T_\lambda^{-1}\right) . \tag{37}$$

In Figs. 1,2, the short cylinders are of length λ ; and the long cylinder (Fig. 2) is of length 2λ. The inner cross sectional area of each cylinder is σ. In terms of Fourier's law for each steady state, with κ being the thermal conductivity of the ideal gas for the case in question, we have

$$\dot{Q}_0^{(r)}(\text{Eq.}38) = \frac{\kappa(\text{Eq.}38)\sigma\left(T_\lambda - T_0\right)}{\lambda} , \tag{38}$$

$$\dot{Q}_0^{(r)}(\text{Eq.}39) = \frac{\kappa(\text{Eq.}39)\sigma\left(T_\lambda - T_0\right)}{2\lambda} . \tag{39}$$

The concentration of the ideal gas (mol L^{-1}) and the logarithmic mean temperature θ for the gas are the same in all three cylinders of the final steady states, so

$$\kappa(\text{Eq.}38) = \kappa(\text{Eq.}39) ; \tag{40}$$

therefore

$$\frac{\dot{\Gamma}(\text{Eq.}38)}{\dot{\Gamma}(\text{Eq.}39)} = \frac{2\dot{Q}_0^{(r)}(\text{Eq.}38)}{\dot{Q}_0^{(r)}(\text{Eq.}39)} = 4 . \tag{41}$$

In essence, in Eq. (38) we have heat flow through two thermal resistances in parallel, whereas in Eq. (39) we have heat flow through the same two thermal resistances in series.

It is relatively easy to determine $\dot{\Gamma}$ for a particular steady state, whereas it can be quite difficult to determine $\Gamma_{\uparrow}$ accurately for a process involving a nonequilibrium initial or final state (or both). The hard part of the determination of $\underset{\uparrow}{\Gamma}$ is the evaluation of the individual heat inputs by the acting heat reservoirs—the total heat input Q can readily be evaluated via the first law $(Q = \Delta U - W)$, but evaluation of the individual heat inputs (Q_0, Q_λ, for example) requires more information.

Mixing

In Fig. 1 let the evacuated cylinder be filled with an amount n of a different monatomic ideal gas at pressure p. After everything has settled into a steady state, open the stopcock and allow the two gases to mix. With respect to the initial and final steady states,

$$\Delta U = 0, \quad Q = 0, \quad W = 0 , \tag{42}$$

$$\underset{\uparrow}{\Gamma} = \Delta S = 2(nR \ln 2) > 0 . \tag{43}$$

Next make an array of interconnected (with stopcocks) cylinders, each with end temperatures T_0, T_λ and adiabatically insulating side walls and containing a different ideal gas. Let each ideal gas have a molar heat capacity that is independent of temperature (so long as this condition is satisfied, the ideal gas need not be monatomic). Let the ith cylinder be of volume V_i and contain an amount n_i of gas at pressure p; and let the gas satisfy the equation of state $pV_i = n_i R\theta$. Let $n \equiv \sum_i n_i$ be the total amount of gas in the array of cylinders, $V \equiv \sum_i V_i$ be the total volume, and $x_i \equiv n_i / n$ be the mole fraction of substance i in the final steady state.

When the gases in the array of cylinders have settled into a steady state, open all the connecting stopcocks and allow the gases to mix into the common volume V. With respect to the initial and final steady states, the conventional mixing formula at constant pressure holds in this case too:

$$\underset{\uparrow}{\Gamma} = \Delta S = -nR \sum_i x_i \ln x_i \geq 0. \tag{44}$$

Concentration-Field Analogues for Thermal-Field Processes

Many of the thermal-field processes that we have considered thus far have concentration-field analogues. For example, take two cylinders as in Fig. 1. Make all parts of the apparatus of diathermic material, and immerse the entire apparatus in a constant-temperature heat bath. Let the end faces of the cylinders be capped by diathermic membranes permeable to component 1 and impermeable to component 2. Attach constant-pressure (p_0, p_λ) mass reservoirs containing component 1 to the end faces at $x = 0, x = \lambda$. Fill one cylinder with an amount n_2 of component 2, and keep the other cylinder free of component 2. Let the entry ports to the connecting link between the cylinders be permeable to component 2 and impermeable to component 1. Let components 1 and 2 be ideal gases.

Let the system settle into a steady state with steady flows of component 1 through the two cylinders; then open the stopcock, and allow component 2 to expand into the second cylinder. This process is the analogue of the thermal-field expansion-into-the-vacuum process that we dealt with above relative to Fig. 1.

As another example, take the (initial state) situation just described and add an amount n_3 of a third ideal gas to the second cylinder (the one devoid of component 2). Let the end membranes of both cylinders be permeable to component 1 and impermeable to components 2 and 3. Let the entry ports of the connecting link be permeable to both components 2 and 3 and impermeable to component 1. After the system settles into a steady state with steady flows of component 1 through the cylinders, open the stopcock and let components 2 and 3 mix into the now available volume $2V$. This process is the analogue of the thermal-field mixing process that we dealt with above relative to Fig. 1.

Many other thermal-field processes have concentration-field analogues.

REFERENCE

1. J. Kestin, *A Course in Thermodynamics*, Vol. I (Blaisdell, Waltham, 1966) p. 556.

FIVE

MONOTHERMAL CONCENTRATION FIELD OF CONSTANT GRADIENT

I carry through operations in a concentration field of constant gradient, operations analogous to those I carried through for the thermal field in Section 2; and I verify that the Dartmouth condition holds in each case.

Take a long cylindrical tube and position at its center a snugly fitting rubber block (B). Mark off an x-coordinate along a line parallel to the axis of the cylinder, with the exposed faces of the rubber block at positions $x = 0$ and $x = \lambda$. Let the inner cross sectional area of the cylinder be σ. Expose the block of rubber to a gas soluble in the rubber, with gas pressures p_α, p_β at the exposed faces—see Fig. 1. Immerse the entire apparatus in a

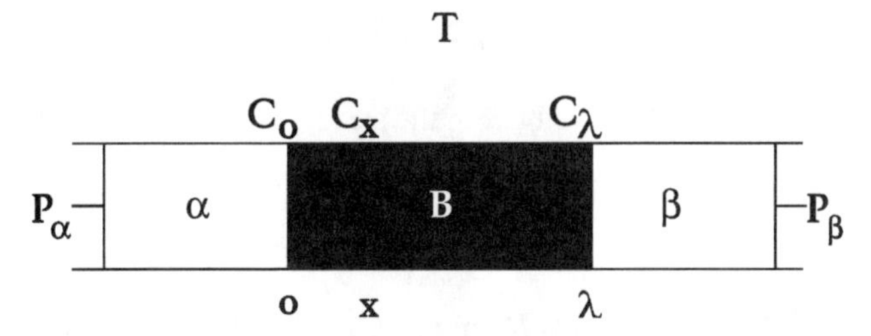

Fig. 1. Monothermal concentration field of constant gradient. A gas in steady flow through a block of rubber (B). The concentration of the dissolved gas at point x in the block of rubber is $C_x = C_0 + \left(\frac{x}{\lambda}\right)\left(C_\lambda - C_0\right)$.

thermostat (heat reservoir) of temperature T. Let the walls and movable pistons of the apparatus (Fig. 1) be of diathermic material. Establish a steady state in the system in which gas at p_α dissolves in the rubber at $x = 0$, diffuses through the rubber to $x = \lambda$, and then volatilizes at $x = \lambda$ into the gas at p_β. Neglect the macroscopic kinetic energy associated with the flow of the gas.

Let the (molar) concentration of the diffusing gas inside the rubber block at point x be C_x, and let the concentration field of the gas inside the rubber block be such that

$$C_x = C_0 + \left(\tfrac{x}{\lambda}\right)\left(C_\lambda - C_0\right) . \tag{1}$$

Then,

$$dC_x = \left[\left(C_\lambda - C_0\right)/\lambda\right]dx \text{ and } dx = \left[\lambda/\left(C_\lambda - C_0\right)\right]dC_x . \quad (2)$$

The amount of gas dn_x between planes at x and $x + dx$ is

$$dn_x = \sigma C_x dx = \left[V/\left(C_\lambda - C_0\right)\right]C_x dC_x , \quad (3)$$

where $V = \sigma\lambda$ is the volume occupied by the rubber block. The total amount n of gas dissolved in the rubber block is

$$n = \int_{x=0}^{x=\lambda} dn_x = \frac{V}{C_\lambda - C_0}\int_{C_0}^{C_\lambda} C_x dC_x = V\langle C\rangle , \quad (4)$$

where $\langle C\rangle \equiv \frac{1}{2}\left(C_\lambda + C_0\right)$.

Let us take the partial molar entropy $\bar{S}_x$ of the dissolved gas at position x in the concentration field to be

$$\bar{S}_x = \bar{S}_0\left(T, C_0\right) - R\ln\left(C_x/C_0\right) . \quad (5)$$

The entropy of the dissolved gas in the steady-state concentration field is

$$S = \int_{x=0}^{x=\lambda} \bar{S}_x dn_x = \frac{V}{C_\lambda - C_0}\int_{C_0}^{C_\lambda} \bar{S}_x C_x dC_x$$

$$= n\left\{\left(\bar{S}_0 + R\ln C_0\right) - R\left[\ln\left(C_\lambda^{r_\lambda} C_0^{r_0}\right) - \tfrac{1}{2}\right]\right\} , \quad (6)$$

where

$$r_\lambda \equiv \frac{C_\lambda^2/2}{C_\lambda^2/2 - C_0^2/2}, \quad r_0 \equiv \frac{-C_0^2/2}{C_\lambda^2/2 - C_0^2/2}, \quad r_\lambda + r_0 = 1. \quad (7)$$

Homogenizing the Concentration Field.

Now instantaneously slip flexible membranes, impermeable to the gas in question, over the exposed faces of the rubber block at $x = 0$ and $x = \lambda$. The dissolved gas in the block will thereupon relax to an equilibrium state of uniform concentration $\langle C\rangle$. Assume that the volume of the rubber block does not change during the process, i.e. assume that $\Delta V = 0$. Assume further that the thermodynamic properties of the rubber block are unaffected by the relaxation process.

The entropy of the dissolved gas in the final equilibrium state is

$$S_f = n\left[\overline{S}_0 - R\ln\left(\langle C\rangle / C_0\right)\right] \cdot \tag{8}$$

For the relaxation process, then [Eqs. (6),(8)],

$$\Delta S = nR\left[\ln\left(C_\lambda^{r_\lambda} C_0^{r_0} / \langle C\rangle\right) - \tfrac{1}{2}\right]. \tag{9}$$

For our rubber-block-with-dissolved-gas system, the first law tells us that

$$\Delta U = Q \tag{10}$$

since $W = 0$. Then

$$\Delta_{\uparrow} S = -Q / T = -\Delta U / T \tag{11}$$

and

$$\begin{aligned} \Gamma_{\uparrow} &= \Delta S + \Delta_{\uparrow} S \\ &= \Delta S - T^{-1}\Delta U = -T^{-1}\Delta A \ . \end{aligned} \tag{12}$$

Since we are assuming that the thermodynamic properties of the rubber block are unaffected by the relaxation process, the change in the Helmholtz energy, ΔA, is all due to the dissolved gas. If in the initial state we take the partial molar Helmholtz energy of the dissolved gas at position x to be

$$\overline{A}_x = \overline{A}_0\left(T, C_0\right) + RT\ln\left(C_x / C_0\right) , \tag{13}$$

we imply [Eqs. (5),(13)] that the partial molar internal energy $\overline{U}_x$ of the gas is independent of position in the concentration field; so

$$\Delta U = 0 \tag{14}$$

and

$$\begin{aligned} \Gamma_{\uparrow} &= \Delta S \\ &= nR\left[\ln\left(C_\lambda^{r_\lambda} C_0^{r_0} / \langle C\rangle\right) - \tfrac{1}{2}\right] \geq 0 \ . \end{aligned} \tag{15}$$

The Dartmouth condition requires in this instance that

$$\ln\left(C_\lambda^{r_\lambda} C_0^{r_0} / \langle C\rangle\right) - \tfrac{1}{2} \geq 0 \ . \tag{16}$$

Set $C_\lambda \equiv \xi C_0$ $\left(0 \leq \xi < \infty\right)$, then relation (16) takes the form

$$\ln\left[2 / \left(1 + \xi\right)\right] + \left[\xi^2 / \left(\xi^2 - 1\right)\right]\ln \xi - \tfrac{1}{2} \geq 0 \ . \tag{17}$$

Let

$$f_1(\xi) \equiv \ln\left[2/(1+\xi)\right] + \left[\xi^2/(\xi^2-1)\right]\ln\xi \ . \tag{18}$$

By using the calculus (making use of l'Hospital's rule where necessary), we can show that $f_1(\xi)$ has a minimum at $\xi = 1$ $\left(\text{i.e. } C_\lambda = C_0\right)$ and that

$$\lim_{\xi \to 1} f_1(\xi) = \tfrac{1}{2} \ ; \tag{19}$$

so relations (15)-(17) are valid relations.

For future comparison, note that

$$f_1\left(\tfrac{1}{2}\right) - \tfrac{1}{2} = 0.0187, \ f_1(1) - \tfrac{1}{2} = 0, \ f_1(2) - \tfrac{1}{2} = 0.0187 \ . \tag{20}$$

Repolarizing the System.

Let us now restore the original concentration field of Fig. 1. Take the system with uniform concentration $\langle C \rangle$ of the dissolved gas in the rubber block, and remove the flexible membranes from the faces at $x = 0$ and $x = \lambda$ that we installed in the previous part—thereby exposing those faces to the gas at pressures p_α, p_β. Eventually we recover the initial steady state [Eq. (1)] from which we started.

In the final steady state, for each position x for which C_x is greater than $\langle C \rangle$, there has been a gain of amount of substance; and for each position x for which C_x is less than $\langle C \rangle$, there has been a loss of amount of substance. Assume that the gains all came from mass reservoir α and that the losses all passed into mass reservoir β. The amount of substance $\left[n_\alpha\right]$ that came from mass reservoir α is:

$$\begin{aligned} \left[n_\alpha\right] &= \frac{V}{C_\lambda - C_0} \int_{C_0}^{\langle C \rangle} \left(C_x - \langle C \rangle\right) dC_x \\ &= \frac{n}{4} \frac{C_0 - C_\lambda}{C_0 + C_\lambda} \ . \end{aligned} \tag{21}$$

An equal amount of substance passed into the mass reservoir β.

The system that we are dealing with in the repolarization process consists of the rubber block with dissolved gas and the mass reservoirs at α and β. Again assume that $\Delta V = 0$ and that the thermodynamic properties of the rubber block are unaffected by the process in question. Let Δ_B refer to changes in the properties of the gas dissolved in the rubber block. Then for the changes directly attributable to the repolarization process, we have

$$\Delta U = \Delta_B U + \left(\overline{U}_\beta - \overline{U}_\alpha\right)\left[n_\alpha\right] = Q + \left(p_\alpha \overline{V}_\alpha - p_\beta \overline{V}_\beta\right)\left[n_\alpha\right], \tag{22}$$

$$\Delta_B U + \left(\overline{H}_\beta - \overline{H}_\alpha\right)\left[n_\alpha\right] = Q\,, \tag{23}$$

$$\Delta S = \Delta_B S + \left(\overline{S}_\beta - \overline{S}_\alpha\right)\left[n_\alpha\right], \tag{24}$$

$$\Delta_\uparrow S = -T^{-1}Q = -T^{-1}\left\{\Delta_B U + \left(\overline{H}_\beta - \overline{H}_\alpha\right)\left[n_\alpha\right]\right\}, \tag{25}$$

$$\begin{aligned}\Gamma_\uparrow &= \Delta S + \Delta_\uparrow S \\ &= \Delta_B S + \left(\overline{S}_\beta - \overline{S}_\alpha\right)\left[n_\alpha\right] - T^{-1}\left\{\Delta_B U + \left(\overline{H}_\beta - \overline{H}_\alpha\left[n_\alpha\right]\right)\right\} \\ &= -T^{-1}\left\{\Delta_B A + \left(\mu_\beta - \mu_\alpha\right)\left[n_\alpha\right]\right\} \\ &= -nR\left[\ln\left(C_\lambda^{r_\lambda} C_0^{r_0} / \langle C\rangle\right) - \tfrac{1}{2}\right] - \frac{n}{4}\frac{C_0 - C_\lambda}{C_0 + C_\lambda}\left(\mu_\beta - \mu_\alpha\right)T^{-1}\,. \end{aligned} \tag{26}$$

Assume that the diffusion process through the rubber block is a much slower process than the dissolving and volatilization processes at $x = 0$ and $x = \lambda$, so that

$$\mu_\alpha \approx \mu_0,\ \ \mu_\beta \approx \mu_\lambda,\ \text{ and}$$

$$\mu_\beta - \mu_\alpha \approx \mu_\lambda - \mu_0 = RT\ln\left(C_\lambda / C_0\right), \tag{27}$$

where we assume that $\mu_x = \mu_0\left(T, C_0\right) + RT\ln\left(C_x / C_0\right)$. Then

$$\Gamma_\uparrow \approx nR\left\{-\left[\ln\left(C_\lambda^{r_\lambda} C_0^{r_0} / \langle C\rangle\right) - \tfrac{1}{2}\right] - \frac{1}{4}\frac{C_0 - C_\lambda}{C_0 + C_\lambda}\ln\left(C_\lambda / C_0\right)\right\}. \tag{28}$$

Upon introducing the notation $C_\lambda \equiv \xi C_0\ \left(0 \le \xi < \infty\right)$, we have

$$\Gamma_\uparrow \approx nR\left\{-\ln\left[2/(\xi+1)\right] - \left[\xi^2/\left(\xi^2-1\right)\right]\ln\xi + \tfrac{1}{2} - \tfrac{1}{4}\frac{1-\xi}{1+\xi}\ln\xi\right\}. \tag{29}$$

Let

$$f_2(\xi) \equiv -\ln\left[2/(\xi+1)\right] - \left[\xi^2/\left(\xi^2-1\right)\right]\ln\xi +$$

$$\tfrac{1}{2} - \tfrac{1}{4}\frac{1-\xi}{1+\xi}\ln\xi\,; \tag{30}$$

then by direct calculation [and by making use of relation (19)], we find that

$$f_2\left(\tfrac{1}{2}\right) = 0.390,\ f_2(1) = 0,\ f_2(2) = 0.390\ . \tag{31}$$

The function $f_2(\xi)$ thus appears to have a minimum at $\xi = 1$ $\left(\text{i.e. } C_\lambda = C_0\right)$, and we can use the calculus to verify that such is the case. We find, therefore, that

$$\Gamma_{\uparrow} \geq 0 \tag{32}$$

for the repolarization process—as is required by the Dartmouth condition.

Upon comparing relations (20) and (31), we find that the homogenization process is less dissipative (generates less entropy) than the corresponding repolarization process. We noted a similar result in the case of a gas in a thermal field—see Section 2.

A more accurate treatment of the homogenization and repolarization processes would allow for the dissipation of the macroscopic kinetic energy of the material in the mass reservoirs at α and β in the homogenization step and the need for extra work in the repolarization step to accelerate the material in the mass reservoirs to steady-state velocity.

SIX

A FIRST-ORDER CHEMICAL RELAXATION PROCESS

I consider a simple isothermal first-order chemical relaxation process, and I show that at all times for the process $\dot{\Gamma} \geq 0$.

Consider the reversible chemical reaction

$$A = B\ . \tag{1}$$

Let the reaction take place in a solution of volume V acted upon by a constant atmospheric pressure p in a thermostat of temperature T. Prepare an initial mixture of A and B such that in approaching equilibrium the reaction will proceed from left to right. Assume that the volume V stays constant in time.

At times $t = 0$, $t = t$, $t = \infty$, let the concentration of A in mol L^{-1} be C_0, C, C_∞; and, for the same set of times, let the concentration of B be C_0', C', C_∞' (i.e. $C = [A]$, $C' = [B]$).

The initial nonequilibrium state of the system, C_0, C_0', is to relax via first-order kinetics to the final equilibrium state, C_∞, C_∞'. Assume that concentrations are adequate approximations to thermodynamic activities, so that the equilibrium constant for the reaction, for example, is

$$K = C_\infty' / C_\infty\ . \tag{2}$$

From the stoichiometry of the reaction, it follows that

$$C_0 + C_0' = C + C' = C_\infty + C_\infty'\ . \tag{3}$$

The rate of change of the concentration of A is

$$-dC / dt = k\left(C - C_\infty\right)\ . \tag{4}$$

Upon integration, Eq. (4) leads to

$$\left(C - C_\infty\right) / \left(C_0 - C_\infty\right) = e^{-kt}\ . \tag{5}$$

The concentrations of A and B at time t are, therefore,

$$C = \left(C_0 - C_\infty\right)e^{-kt} + C_\infty \ , \tag{6}$$

$$C' = C'_\infty - \left(C_0 - C_\infty\right)e^{-kt} \ . \tag{7}$$

At time t, the change in the Gibbs energy brought about by the reaction [Eq. (1)] is

$$\Delta_r G = RT \ln\left(\frac{C'}{C} \bullet \frac{C_\infty}{C'_\infty}\right) \ . \tag{8}$$

The rate of change of the Gibbs energy of the reacting system, $\dot{G}$, is just the rate at which moles of B are forming (or moles of A are disappearing) times $\Delta_r G$:

$$\dot{G} = \left(-VdC / dt\right)\Delta_r G$$

$$= Vk\left(C - C_\infty\right)RT \ln\left(\frac{C'}{C} \bullet \frac{C_\infty}{C'_\infty}\right)$$

$$= Vk\left(C_0 - C_\infty\right)e^{-kt} RT \ln\left[\frac{C'_\infty - \left(C_0 - C_\infty\right)e^{-kt}}{\left(C_0 - C_\infty\right)e^{-kt} + C_\infty} \bullet \frac{C_\infty}{C'_\infty}\right] . \tag{9}$$

Introduce the notation

$$y \equiv \left(C_0 - C_\infty\right)e^{-kt}, \ \ dy = -kydt \ . \tag{10}$$

In terms of y, Eq. (9) takes the form

$$\dot{G} = VRTky \ln\left[\frac{1 - \left(y / C'_\infty\right)}{1 + \left(y / C_\infty\right)}\right] . \tag{11}$$

Now, the rate of entropy generation, $\dot{\Gamma}_\uparrow$, stemming from the reaction is [1,2]

$$\dot{\Gamma}_\uparrow = -\dot{G} / T$$

$$= VRky \ln\left[\frac{1 + \left(y / C_\infty\right)}{1 - \left(y / C'_\infty\right)}\right] \geq 0 \ . \tag{12}$$

An initial assumption (left-to-right reaction) was that $C_0 - C_\infty > 0$. From the defining Eq. (10), it follows that $0 \leq y \leq C_0 - C_\infty$; so

$$0 \le \frac{y}{C'_\infty} \le \frac{C_0 - C_\infty}{C'_\infty} = \frac{C'_\infty - C'_0}{C'_\infty} \le 1 \ . \tag{13}$$

The argument of the logarithm in Eq. (12) is thus clearly ≥ 1 and $\dot{\Gamma}_\uparrow \ge 0$.

To find the total generated entropy $\Gamma_\uparrow$ for the relaxation from the initial nonequilibrium state to the final equilibrium state, integrate $\dot{\Gamma}_\uparrow dt$ from $t = 0$ to $t = \infty$:

$$\Gamma_\uparrow = \int_0^\infty \dot{\Gamma}_\uparrow dt$$

$$= -VR\int_{C_0 - C_\infty}^{0} \ln\left[\frac{1 + (y / C_\infty)}{1 - (y / C'_\infty)}\right] dy$$

$$= VR\left[C_0 \ln(C_0 / C_\infty) + C'_0 \ln(C'_0 / C'_\infty)\right] \ge 0 \ . \tag{14}$$

The final form of Eq. (14) can also be obtained by observing that $\Gamma_\uparrow$ is simply $-\Delta G / T$, where ΔG is the change in Gibbs energy in going from the initial nonequilibrium state to the final equilibrium state.

Let the chemical potentials of A and B be

$$\mu = \mu^o + RT \ln C, \ \ \mu' = \mu'^o + RT \ln C' \ ; \tag{15}$$

then

$$\Delta G = VC_\infty\left(\mu^o + RT \ln C_\infty\right) + VC'_\infty\left(\mu'^o + RT \ln C'_\infty\right) -$$
$$VC_0\left(\mu^o + RT \ln C_0\right) - VC'_0\left(\mu'^o + RT \ln C'_0\right) . \tag{16}$$

Upon substituting Eq. (16) into the relation $\Gamma_\uparrow = -\Delta G / T$ and carrying through some algebraic manipulations (note that $\mu^o + RT \ln C_\infty = \mu'^o + RT \ln C'_\infty$), we can arrive at the final form of Eq. (14).

This example of a first-order chemical relaxation process shows a system passing through a continuum of nonequilibrium transient states. As the system evolves forward in time from t to $t + dt$, the change in state must satisfy the Dartmouth condition, $d\Gamma_\uparrow \ge 0$; so, since $dt > 0$, it necessarily follows that $\dot{\Gamma}_\uparrow \equiv d\Gamma_\uparrow / dt \ge 0$. For situations involving the passage of a system through a continuum of nonequilibrium transient states, it is useful to evaluate $\dot{\Gamma}_\uparrow$ and to verify that $\dot{\Gamma}_\uparrow \ge 0$.

REFERENCES

1. R. J. Tykodi, *Thermodynamics of Steady States* (Macmillan, New York, 1967) pp. 11,12,39 (i.e. Book II, with internal pagination).
2. R. J. Tykodi, J. Chem. Educ. **72**, 103 (1995).

SEVEN

BÉNARD-RAYLEIGH INSTABILITY

If a thin layer of liquid sandwiched between two rigid surfaces is heated from below (i.e. the lower surface is kept hotter than the upper surface) and the resulting vertical temperature gradient is steadily increased, the transfer of heat through the layer by conduction suddenly gives way to a transfer by convection at a critical value of the thermal gradient; the sudden change from conduction to convection is called the Bénard-Rayleigh instability [1-4]. Let the lower surface have temperature T_0, the upper surface have temperature T_λ, and let the depth of the liquid layer be λ. Then in terms of a dimensionless number, the Rayleigh number Ra,

$$\mathrm{Ra} = \frac{\lambda^3 g\alpha}{\kappa\nu}\left(T_0 - T_\lambda\right) \tag{1}$$

(where g is the acceleration due to gravity, and α, κ, ν are the coefficients of volume expansion, thermometric conductivity, and kinematic viscosity, respectively), the onset of convection occurs at a critical value, Ra_c, of the Rayleigh number:

$$\mathrm{Ra}_c = 1708 \ . \tag{2}$$

The convecting liquid forms a pattern of cells where in each cell there is an upwelling of liquid in one part of the cell and a downwelling of liquid in another part. The sudden transition from conduction to convection is a *hysteretic transition*: once the convection pattern is established, if the temperature gradient is then reduced below the critical value, the convective pattern will persist below that critical value [5]—the system thus demonstrates a form of hysteresis.

Thermodynamic Considerations

Let us carry through a naive thermodynamic analysis of the *conduction* $\rightarrow$ *convection* transition—refer to Fig. 1. Consider the temperature T and the density ρ to be linearly distributed through the liquid system:

$$T_z = T_0 + \left(\tfrac{z}{\lambda}\right)\left(T_\lambda - T_0\right) , \tag{3}$$

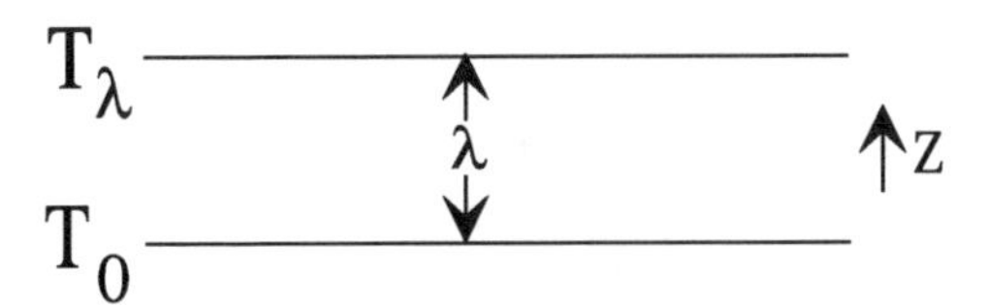

Fig. 1. Thin layer of liquid of depth λ subject to a thermal gradient $(T_0 - T_\lambda)/\lambda$.

$$\rho_z = \rho_0 + \alpha(T_z - T_0) . \tag{4}$$

Let lower case letters represent specific properties of the system: u, specific internal energy; s, specific entropy; c, specific heat capacity; etc. Then

$$u_z = u_0 + c(T_z - T_0) + gz , \tag{5}$$

$$s_z = s_0 + c\ln(T_z / T_0) , \tag{6}$$

where we assume that the specific heat capacity c is constant. The average value of the specific internal energy in the thermal field, $< u >$, is

$$< u >= u_0 + \tfrac{1}{2}\left[c(T_\lambda - T_0) + g\lambda\right] . \tag{7}$$

For the conducting state, let's assume that

$$\text{“}U_i\text{”} = U_i \approx m < u > ; \tag{8}$$

and for the convecting state, that

$$\text{“}U_f\text{”} \approx U_i + E_k , \tag{9}$$

where "U" $\equiv U +$ macroscopic kinetic energy (see Section 1), m is the mass of the liquid, and E_k is the macroscopic kinetic energy of the convecting liquid, i.e. we assume that the thermal part of the total energy is essentially the same both before and after the transition and that the difference in energy is essentially all due to the energy of motion of the convecting liquid:

$$\Delta\text{“}U\text{”} = Q_0 + Q_\lambda \approx E_k . \tag{10}$$

In dealing with the entropy, let's assume that the entropy of the conducting state is approximately the same as that of a liquid layer of constant density (equal to the average density of our liquid layer) in the thermal field described by Eq. (3) (see Section 2):

$$S_i \approx m\left[s_0 - c\ln T_0 - c + c\ln\left(T_\lambda^{r_\lambda} T_0^{r_0}\right)\right] , \tag{11}$$

where

$$r_\lambda \equiv \frac{T_\lambda}{T_\lambda - T_0}, \quad r_0 \equiv \frac{-T_0}{T_\lambda - T_0}, \quad r_\lambda + r_0 = 1 \ . \tag{12}$$

In the convecting state, let's assume that at any time half the liquid is at temperature T_0 and half is at temperature T_λ:

$$S_f \approx \tfrac{m}{2} s_0 + \tfrac{m}{2}\left[s_0 + c\ln\left(T_\lambda / T_0\right)\right] + S^* \ , \tag{13}$$

where S^* is any combinatorial contribution to S_f. Then

$$\Delta S \approx mc\left[1 + \ln\left(T_\lambda^{1/2 - r_\lambda} T_0^{1/2 - r_0}\right)\right] + S^* \ . \tag{14}$$

Example: $T_0 = 310K, \ T_\lambda = 300K, \ r_\lambda = -30, \ r_0 = 31;$

$$\Delta S \approx -9 \times 10^{-5} mc + S^* \approx S^* \ . \tag{15}$$

With our assumptions, there is scarcely any change in the thermal part of the entropy; yet we do expect a sizable increase in entropy since the transition is a hysteretic one and thus should be quite dissipative.

Let's assume that the energy needed to develop the macroscopic motion in the convecting state all comes from the heat reservoir of temperature T_0; then

$$\Delta_{\uparrow} S = -T_0^{-1} Q_0 \ , \tag{16}$$

$$\Gamma_{\uparrow} = \Delta S + \Delta_{\uparrow} S \approx S^* - E_k T_0^{-1} \geq 0 \ . \tag{17}$$

There thus seems to be a need for a combinatorial contribution to the entropy of the liquid in the convecting state. The convecting state should thus be a degenerate state—there should be several possible convecting patterns, all of equal energy, from among which the system is forced to choose when making the *conduction* $\rightarrow$ *convection* transition. For example, if in each convecting cell the upwelling plume of liquid could just as well have been a downwelling plume, and vice versa, then there would be a contribution of $(mR / M)\ln 2$ to S^*, where M is the molecular weight of the liquid and R is the gas constant.

An accurate treatment of the entropy in the initial and final states, however, may obviate the need for a combinatorial contribution S^*. For those cases where the vertical velocity of the convecting liquid was observed experimentally [2], the velocity was only a few tenths of a centimeter per second. A liquid of molecular weight 2.0×10^{-2} kg mol^{-1} moving with a velocity of 1 centimeter per second has a molar kinetic energy of 1×10^{-6} J mol^{-1}; so the entropy of the convecting state need be but slightly larger than that of the conducting state to meet the $\Gamma_{\uparrow} > 0$ requirement.

Clearly, accurate calculations of the thermodynamic properties of the liquid in both the conducting and convecting states are necessary before we can say with certainty that there is or is not a need for a combinatorial contribution S^* to the entropy of the convecting state.

REFERENCES

1. S. Chandrasekhar, *Hydrodynamic and Hydromagnetic Stability* (Oxford University Press, Oxford, 1961) Ch. 2.
2. C. Normand and Y. Pomeau, Rev. Mod. Phys.**49**, 581 (1977).
3. M. Dubois in J. Casas-Vasquez and G. Lebon, editors, *Stability of Thermodynamic Systems* (Springer, Berlin, 1982) pp. 177-191.
4. E.L. Koschmieder, *Bénard Cells and Taylor Vortices* (Cambridge University Press, Cambridge, 1993).
5. Ref. 4, p. 156.

EIGHT

CONDITIONS OF MIGRATIONAL EQUILIBRIUM

In this Section, I discuss the general conditions of migrational equilibrium for a substance in a nonequilibrium spatial field; and I obtain explicitly the conditions of migrational equilibrium for the thermomolecular pressure effect and for a gas in a monothermoal concentraiton field.

One of the major achievements of equilibrium thermodynamics has been the stating of the conditions of migrational equilibrium for a substance in a spatial field, i.e. the substance is free to move from point to point in the spatial field but it remains (macroscopically) at rest. Examples of migrational equilibrium in a spatial field are: i) a chemical species in an isothermal phase field, where the species is free to distribute itself over several contiguous phases—the balance condition in this case is that the molar chemical potential (μ) of the species have the same value in each of the contiguous phases; ii) a gas of molecular weight M in an isothermal gravitational field of strength g at rest at a height h—the balance condition in this case is that $\mu + Mgh$ have the same value at each height in a vertical column of the gas; iii) etc.

The finding of the conditions of migrational equilibrium for a substance in an equilibrium spatial field was accomplished by J. Willard Gibbs in his paper ON THE EQUILIBRIUM OF HETEROGENEOUS SUBSTANCES (1875–1878) [1]. It was noted early on that a number of nonequilibrium situations show "balance conditions" very much like equilibrium conditions—the thermocouple, the thermomolecular pressure effect, thermal diffusion, thermoösmosis, etc.—and W. Thomson [2] (for the thermocouple) and E. Eastman [3,4] (for a family of non-isothermal "effects") tried to apply classical thermodynamic procedures to the task of finding the conditions of migrational equilibrium for substances in nonequilibrium spatial fields. Both Thomson and Eastman implicitly assumed that entropy was well defined in nonequilibrium states and explicitly assumed that the second law of thermodynamics could be applied selectively to parts of processes for nonequilibrium situations, i.e. they assumed that reversible effects could be separated from irreversible effects for the process in question and that the second law could be applied selectively to the reversible parts of the process. The logic of the procedure [2-5] followed by Thomson and Eastman is somewhat shaky, but the two men did succeed in obtaining differential expressions pertaining to states of migrational equilibrium for substances in nonequilibrium spatial fields—expressions that later research showed to be very likely correct or that were of use to experimentalists in the reporting of experimental results.

In this section, I take the basic idea of Thomson and render it rigorous by distinguishing between the properties of a system in a nonequilibrium state and the *equilibrium-based image* of that system. In an appropriate subdomain of the domain of nonequilibrium states, the entropy of the *equilibrium-based image* of a system is well defined and satisfies a restricted form of the second law; that restricted form of the second law allows for a rigorous deduction of the conditions of migrational equilibrium for a substance in a nonequilibrium spatial field—the balance conditions thus found are in integral form, analogous to the equilibrium balance conditions $\mu = const$ and $\mu + Mgh = const$ mentioned above, and therefore are a step up from the differential forms established by others.

For a change between two equililbrium states, the second law says

$$\Gamma \equiv \Delta S(system) + \Delta S(surroundings) \geq 0 \, , \tag{1}$$

where Γ is the generated entropy for the process. Also, let us rewrite here the Dartmouth condition from Section 1:

If the system and the surroundings in both the initial and final states have *equilibrium-based images, then, in the Dartmouth domain, the relation*

$$\Gamma_{\uparrow} \equiv \Delta S(system) + \Delta_{\uparrow} S(surroundings) \geq 0 \tag{2}$$

must be true. $\Gamma_{\uparrow}$ *is that part of* Γ *directly attributable to the process, and the changes* $\Delta_{\uparrow}$ *are for effects directly attributable to the process.*

We shall need these results for the following discussion.

Suppose that a system at equilibrium $\left[\dot{\Gamma} = 0\right]$ is in a state of migrational equilibrium in a spatial field with respect to some quantity, i.e. the quantity is free to move from point to point in the spatial field but it remains (macroscopically) at rest.

J. Willard Gibbs, in his paper ON THE EQUILIBRIUM OF HETEROGENEOUS SUBSTANCES [1], established that for states of equilibrium $\left[\dot{\Gamma} = 0\right]$ the balance condition for the migrational equilibrium of a substance in a spatial field is to be found by requiring that

$$d\Gamma / dn \equiv \lim_{\Delta n \to 0} d\Gamma / \Delta n = 0 \tag{3}$$

[minimizing the generated entropy is equivalent to maximizing the bound entropy, i.e. the entropy of the system plus the entropy of the surroundings]. Relation (3) is a corollary of relation (1). Write relation (1) in differential form:

$$d\Gamma \equiv dS(system) + dS(surroundings) \geq 0 \tag{4}$$

[Note that $\int d\Gamma = \Gamma$, not $\Delta\Gamma$—just as $\int dQ = Q$, not ΔQ.]

Select a state of migrational equilibrium with respect to some substance, and select two points, A & B, in the field of migrational equilibrium. In differentially tailored fashion transfer an amount $\Delta_B n$ of the substance from point A to point B, where $\Delta_B n$ is the *accumulation* of the substance at point B. When the transfer is from A to B, $\Delta_B n$ is positive; when the transfer is from B to A, $\Delta_B n$ is negative.

Regardless of the direction of the transfer, $d\Gamma \geq 0$. From a state of migrational equilibrium in each case, transfer various amounts of the substance in differentially tailored fashion from A to B, generating a series of positive values for the accumulation $\Delta_B n$. Form the ratios $d\Gamma / \Delta_B n$, all positive. Next, from a state of migrational equilibrium in each case, transfer various amounts of the substance from B to A, generating a series of negative values for the accumulation $\Delta_B n$. Form the ratios $d\Gamma / \Delta_B n$, all negative. In each sequence, let $\Delta_B n$ approach zero. Then, with positive values of $d\Gamma / \Delta_B n$ and negative values of $d\Gamma / \Delta_B n$ converging on the same state, by continuity it must be the case that

$$\lim_{\Delta_B n \to 0-} d\Gamma / \Delta_B n = \lim_{\Delta_B n \to 0+} d\Gamma / \Delta_B n = 0 \tag{5}$$

i.e. $d\Gamma / dn = 0$.

Talk of moving an amount of substance dn about in a field of migrational equilibrium is to be treated as an idiomatic way of describing the limiting operations of Eq. (5).

Thomson-Gibbs Corollary

Suppose, now, that a system in a nonequilibrium steady state $\left[\dot{\Gamma} \neq 0\right]$ is in a state of migrational equilibrium in a spatial field with respect to some quantity. What is the appropriate balance condition for the substance in the given nonequilibrium spatial field? By precisely the same argument as that used to establish Eq. (3) [Eq. (5)], we find that

$$d\Gamma_{\uparrow} / dn \equiv \lim_{\Delta n \to 0} d\Gamma_{\uparrow} / \Delta n = 0 \tag{6}$$

for the differentially tailored transfer of an amount dn of the substance in question from one part of the spatial field to another—relation (6) is a corollary of relation (2). I shall call relation (6) the *Thomson-Gibbs Corollary*. The balance condition for a substance in migrational equilibrium in a nonequilibrium spatial field is thus found by applying the Thomson-Gibbs Corollary. [Talk of moving about an amount dn of substance is again to be treated as an idiomatic way of describing the limiting operation of Eq. (6).]

Why *Thomson-Gibbs* Corollary? In his discussion of the thermocouple, William Thomson [2] was clearly reaching toward some equivalent of relations (2) and (6). The procedure for applying the corollary is precisely that of J. Willard Gibbs [1], merely shifted to the Dartmouth domain and made relative to the Dartmouth condition.

Conventions

Use an overbar to indicate a molar or partial molar property—the molar or partial molar volume (according to context) of substance *j*, for example, would be $\overline{V}^{(j)}$. Where necessary, use the indexes *i* and *f* to identify initial and final states. Let unflagged quantities represent properties of the system; and where necessary, flag properties of the surroundings with the label *sur*. Let the rate of influx of heat into a heat reservoir be $\dot{Q}^{(r)}$.

Unless the text states otherwise, we are limiting our considerations to the Dartmouth domain. Remember that the properties displayed in ordinary type (as opposed to boldface type) are properties of the *equilibrium-based image* of the system.

The Thermomolecular Pressure Effect

Take two piston-and-cylinder containers and connect them with a capillary tube. Let the container material be diathermic, and cover the outer wall of the capillary tube with adiabatically insulating material. Surround each of the piston-and-cylinder containers with a thermostat (heat reservoir). Put a gas in the device and let everything settle into a steady state at thermostat temperatures T_α, T_β. Select the pressures p_α, p_β so that there is no net (macroscopic) tendency for the gas to flow in either direction, i.e. let the gas be in a state of migrational equilibrium for the set of values $p_\alpha, T_\alpha, p_\beta, T_\beta$. See Fig. 1.

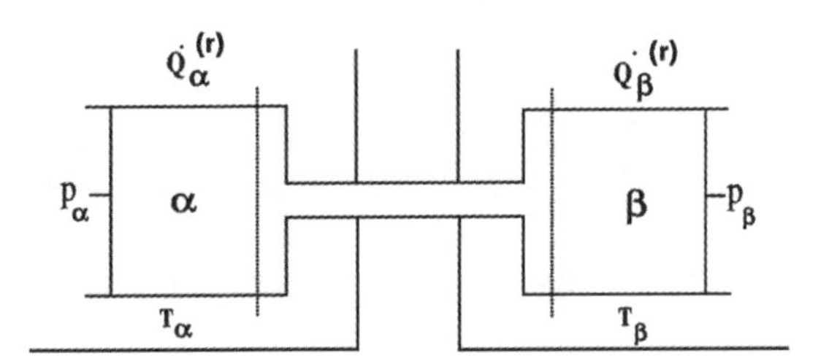

Fig. 1. Thermomolecular pressure effect. Two piston-and-cylinder containers are connected by a capillary tube. The containers are immersed in thermostats. The container material is diathermic; the walls of the capillary tube are adiabatically insulating. A gas in the device shows, in the steady state at migrational equilibrium, pressures p_α, p_β at temperatures T_α, T_β.

For steady states of migrational equilibrium, it is found experimentally that for pressures sufficiently low so that the mean free path between collisions for the gas particles is large compared to the diameter of the capillary tube the gas follows the Knudsen limiting relation $p_\alpha / p_\beta = (T_\alpha / T_\beta)^{1/2}$. When the pressures p_α, p_β are high, the pressure is uniform throughout the gas: $p_\alpha = p_\beta$.

Start with the system in a steady state of migrational equilibrium. Now over an interval of time from t_1 to t_2 force an amount *dn* of the gas from the cylinder at β to the cylinder at α in a differentially tailored fashion. Let the times t_1, t_2 be such that the process starts after time t_1 and is complete by time t_2, i.e.

$$\dot{Q}_\alpha^{(r)}(t_1) + \dot{Q}_\beta^{(r)}(t_1) = 0, \quad \dot{Q}_\alpha^{(r)}(t_2) + \dot{Q}_\beta^{(r)}(t_2) = 0 \ . \tag{7}$$

The total heat dQ exchanged between the system and the surroundings is given by

$$dQ = -\int_{t_1}^{t_2} \left[\dot{Q}_\alpha^{(r)}(t) + \dot{Q}_\beta^{(r)}(t) \right] dt, \tag{8}$$

and we are faced with the problem of apportioning dQ between the two heat reservoirs to get the individual contributions dQ_α, dQ_β. Consider dQ_α and dQ_β to be "undetermined quantities" for the moment, to be evaluated at a later stage in the proceedings.

In Fig. 1, the gas to the left of the dotted line at α behaves as though it were in an equilibrium state at T_α, p_α; and the gas to the right of the dotted line at β behaves as though it were in an equilibrium state at T_β, p_β.

Let us apply the first law to our process:

$$(\overline{U}_\alpha - \overline{U}_\beta)dn = dQ_\alpha + dQ_\beta + (p_\beta \overline{V}_\beta - p_\alpha \overline{V}_\alpha)dn \ . \tag{9}$$

Then

$$\overline{H}_\alpha - \overline{H}_\beta = dQ_\alpha / dn + dQ_\beta / dn \ . \tag{10}$$

Now let

$$dQ_\alpha / dn \equiv \overline{H}_\alpha - \left[H_\alpha \right], \quad dQ_\beta / dn \equiv \left[H_\beta \right] - \overline{H}_\beta \ , \tag{11}$$

where $\left[H_\omega \right] (\omega = \alpha, \beta)$ is an appropriate nonequilibrium molar "migrational enthalpy."

[The flow at β is from the "equilibrium" part into the capillary, whereas the flow at α is from the capillary into the "equilibrium" part.] We get, thereupon,

$$\left[H_\alpha \right] = \left[H_\beta \right] \equiv \left[H \right]. \tag{12}$$

Consider next the entropy changes that are attributable to the process.

$$dS = (\overline{S}_\alpha - \overline{S}_\beta)dn, \tag{13}$$

$$d_\uparrow S_{sur} = -T_\alpha^{-1} dQ_\alpha - T_\beta^{-1} dQ_\beta \ , \tag{14}$$

and

$$d\Gamma_{\uparrow} = \left(\bar{S}_{\alpha} - \bar{S}_{\beta}\right)dn - T_{\alpha}^{-1}dQ_{\alpha} - T_{\beta}^{-1}dQ_{\beta}\ ; \tag{15}$$

so

$$\frac{d\Gamma_{\uparrow}}{dn} = \left(\bar{S}_{\alpha} - \bar{S}_{\beta}\right) + \frac{[H] - \bar{H}_{\alpha}}{T_{\alpha}} + \frac{\bar{H}_{\beta} - [H]}{T_{\beta}}$$

$$= -\frac{\bar{G}_{\alpha}}{T_{\alpha}} + \frac{\bar{G}_{\beta}}{T_{\beta}} + [H]\left(\frac{1}{T_{\alpha}} - \frac{1}{T_{\beta}}\right)\ . \tag{16}$$

In terms of the Planck function $\bar{Y} \equiv -\bar{G} / T$,

$$d\bar{Y} = \frac{\bar{H}}{T^2}dT - \frac{\bar{V}}{T}dp\ , \tag{17}$$

Eq. (16) takes the form

$$\frac{d\Gamma_{\uparrow}}{dn} = \bar{Y}_{\alpha} - \bar{Y}_{\beta} + [H]\left(\frac{1}{T_{\alpha}} - \frac{1}{T_{\beta}}\right)\ . \tag{18}$$

Upon invoking the Thomson-Gibbs Corollary [Eq. (6)], we get

$$\bar{Y}_{\alpha} + \left[H_{\alpha}\right]T_{\alpha}^{-1} = \bar{Y}_{\beta} + \left[H_{\beta}\right]T_{\beta}^{-1} \tag{19}$$

or

$$\bar{Y}_{\alpha} - \bar{Y}_{\beta} + [H]\left(T_{\alpha}^{-1} - T_{\beta}^{-1}\right) = 0 \tag{20}$$

as the condition of migrational equilibrium for the gas in the thermomolecular pressure apparatus.

Now let's evaluate the "undetermined quantities" dQ_{α}, dQ_{β} used in the above analysis. In a separate series of operations, induce a (slow) steady flow $\dot{n}$ of the gas from the cylinder at β to the cylinder at α, keeping $T_{\alpha}, p_{\alpha}, T_{\beta}$ constant (constant T, p', for short); and evaluate the derivatives

$$\left.\delta\dot{Q}_{\alpha}^{(r)} / \tilde{\delta}\dot{n}\right)_{T,p'} \text{ and } \left.\delta\dot{Q}_{\beta}^{(r)} / \tilde{\delta}\dot{n}\right)_{T,p'}$$

via a sequence of steady states such that $\dot{n} \to 0$ [see Eq. (1.14)]. Make the identifications

$$dQ_\alpha / dn = -\delta\dot{Q}_\alpha^{(r)} / \tilde{\delta}\dot{n}\Big)_{T,p'} \text{ and } dQ_\beta / dn = -\delta\dot{Q}_\beta^{(r)} / \tilde{\delta}\dot{n}\Big)_{T,p'} \quad (21)$$

or

$$dQ_\alpha = -\delta\dot{Q}_\alpha^{(r)} / \tilde{\delta}\dot{n}\Big)_{T,p'} dn \text{ and } dQ_\beta = -\delta\dot{Q}_\beta^{(r)} / \tilde{\delta}\dot{n}\Big)_{T,p'} dn \quad (22)$$

in Eqs. (9)–(20) where appropriate.

Let's look at the differential form of Eq. (20) under conditions of constant T_β, constant p_β (constant β, for short):

$$\frac{\overline{H}_\alpha}{T_\alpha^2} dT_\alpha - \frac{\overline{V}_\alpha}{T_\alpha} dp_\alpha = \frac{[H]}{T_\alpha^2} dT_\alpha - d[H]\left(\frac{1}{T_\alpha} - \frac{1}{T_\beta}\right) \text{ (constant } \beta), \quad (23)$$

$$\frac{\overline{V}_\alpha}{T_\alpha} dp_\alpha = \frac{\overline{H}_\alpha - [H]}{T_\alpha^2} dT_\alpha + d[H]\left(\frac{1}{T_\alpha} - \frac{1}{T_\beta}\right) \text{(constant } \beta) . \quad (24)$$

If we treat the gas as an ideal gas, then $\overline{V}_\alpha / T_\alpha = R / p_\alpha$; and

$$\left.\frac{\partial \ln p_\alpha}{\partial T_\alpha}\right)_\beta = \frac{\overline{H}_\alpha - [H]}{RT_\alpha^2} + \frac{1}{R}\left.\frac{\partial [H]}{\partial T_\alpha}\right)_\beta \left(\frac{1}{T_\alpha} - \frac{1}{T_\beta}\right) . \quad (25)$$

Consider now an equithermal condition with $T_\alpha = T_\beta$. In such a case, Eq. (25) becomes

$$\left.\left.\frac{\partial \ln p_\alpha}{\partial T_\alpha}\right)_\beta\right|_{T_\alpha = T_\beta} = \frac{\overline{H}_\alpha - [H]}{RT_\alpha^2} . \quad (26)$$

In Eq. (26) set $[H] - \overline{H}_\alpha$ equal to the molar heat of transport $Q*$. Then Eq. (26) becomes

$$\left.\left.\frac{\partial \ln p_\alpha}{\partial T_\alpha}\right)_\beta\right|_{T_\alpha = T_\beta} = -\frac{Q*}{RT_\alpha^2} , \quad (27)$$

a result that has often been cited in discussions of the thermomolecular pressure effect [6-8]. For a gas following the Knudsen relation

$$p_\alpha / p_\beta = \left(T_\alpha / T_\beta\right)^{1/2},$$

we find that $Q* = -\frac{1}{2} RT_\alpha$.

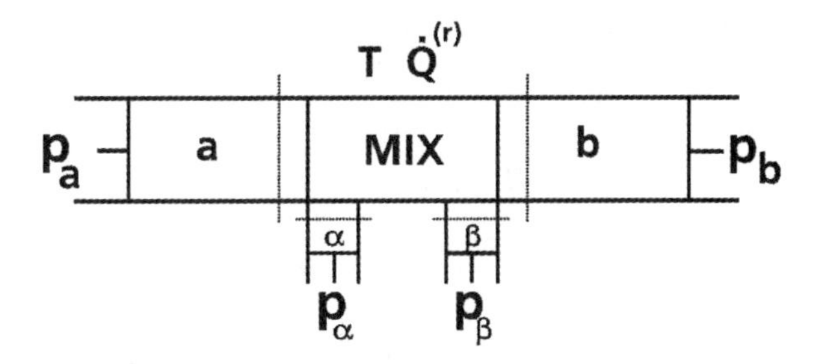

Fig. 2. Monothermal concentration field maintained by the steady flow of component 1.

A Monothermal Concentration Field

Consider next the problem of migrational equilibrium in a monothermal concentration field—the situation indicated schematically in Fig. 2. A spatial region marked MIX contains two fluid components, 1 and 2.

The MIX region communicates with terminal parts [9] ("equilibrium state" mass reservoirs) of the system at a and b via membranes permeable to component 1 alone and with terminal parts at α and β via membranes permeable to component 2 alone. The apparatus is made of diathermic material and rests in a thermostat (heat reservoir) of temperature T. Let us induce a steady flow of component 1 through the MIX region and seek the conditions for the migrational equilibrium of component 2 in the concentration field set up by the flow of component 1.

Start with the system in a steady state in which there is a steady flow of component 1 from the terminal part at a through the MIX region and into the terminal part at b and in which the pressures p_α, p_β are such that component 2 is in a state of migrational equilibrium (no macroscopic gain or loss of mass by either of the terminal parts at α, β). Let the rate of molar influx of component 1 to each of the terminal parts at a,b be $\dot{n}_a, \dot{n}_b$, with

$$\dot{n}_a + \dot{n}_b = 0 \; . \tag{28}$$

Let $\overline{K}_a, \overline{K}_b$ be the molar macroscopic kinetic energy associated with component 1 in terminal parts a,b.

Now over an interval of time from t_1 to t_2 force dn moles of component 2 out of the terminal part at β and into the terminal part at α in a differentially tailored fashion. Let the times t_1, t_2 be such that the process starts after time t_1 and is over by time t_2, i.e.

$$\dot{n}_a(t_1) + \dot{n}_b(t_1) = 0, \;\; \dot{n}_a(t_2) + \dot{n}_b(t_2) = 0 \; . \tag{29}$$

The generated entropy attributable to the monothermal process is [10,11] -d"G"/T, where d"G" is the change in the "total" Gibbs energy, "G" = G + macroscopic kinetic energy [see Eq. (1.6)], of the total system attributable to the process (let's include the kinetic energy for component 1 and neglect it for component 2). For our process, then,

$$d\Gamma_{\uparrow} = -d\text{"}G\text{"} / T$$
$$= -T^{-1}\left\{\left(\mu_{\alpha} - \mu_{\beta}\right)dn + \left(\mu_{a} + \overline{K}_{a}\right)dn_{a} + \left(\mu_{b} + \overline{K}_{b}\right)dn_{b}\right\}, \quad (30)$$

where μ is the relevant molar chemical potential and dn_a, dn_b are "undetermined quantities" (to be evaluated later) with

$$dn_{a} + dn_{b} = 0 \ . \quad (31)$$

We see, then, that

$$d\Gamma_{\uparrow} = -T^{-1}\left\{\left(\mu_{\alpha} - \mu_{\beta}\right)dn + \left[\left(\mu_{a} + \overline{K}_{a}\right) - \left(\mu_{b} + \overline{K}_{b}\right)\right]dn_{a}\right\} \quad (32)$$

and

$$d\Gamma_{\uparrow} / dn = -T^{-1}\left\{\mu_{\alpha} - \mu_{\beta} + \left[\left(\mu_{a} + \overline{K}_{a}\right) - \left(\mu_{b} + \overline{K}_{b}\right)\right]\left(dn_{a} / dn\right)\right\} . \quad (33)$$

Upon invoking the Thomson-Gibbs Corollary [Eq. (6)], we get

$$\mu_{\alpha} - \mu_{\beta} = -\left(dn_{a} / dn\right)\left[\left(\mu_{a} + \overline{K}_{a}\right) - \left(\mu_{b} + \overline{K}_{b}\right)\right] \quad (34)$$

as the condition of migrational equilibrium for component 2 in the concentration field set up by the flow of component 1.

If the flow of the two fluids is coupled $\left(dn_a / dn \neq 0\right)$, then the flow of component 1 induces a gradient in the chemical potential of component 2.

To evaluate the "undetermined quantities" dn_a, dn_b, induce, in a separate series of operations, a (slow) steady flow $\dot{n}$ of component 2 between its terminal parts at α and β by varying p_α, keeping T and the pressures for the other terminal parts constant (constant T, p', for short); and evaluate the derivatives

$$\left.\delta\dot{n}_{a} / \tilde{\delta}\dot{n}\right)_{T,p'} , \quad \left.\delta\dot{n}_{b} / \tilde{\delta}\dot{n}\right)_{T,p'}$$

via a sequence of steady states such that $\dot{n} \to 0$ [see Eq. (1.14)]. Make the identification

$$dn_{a} / dn = \left.\delta\dot{n}_{a} / \tilde{\delta}\dot{n}\right)_{T,p'} \quad (35)$$

or

$$dn_{a} = \left.\delta\dot{n}_{a} / \tilde{\delta}\dot{n}\right)_{T,p'} dn . \quad (36)$$

REFERENCES

1. J. W. Gibbs, *The Scientific Papers of J. Willard Gibbs: Vol. I. Thermodynamics* (Dover, New York, 1961), pp. 55-353.
2. W. Thomson, *Mathematical and Physical Papers, Vol. I* (Cambridge University Press, Cambridge, 1882), pp. 232-291.
3. E. Eastman, J. Am. Chem. Soc. **48**, 1482 (1926).
4. E. Eastman, J. Am. Chem. Soc. **49**, 794 (1927).
5. K. G. Denbigh, *The Thermodynamics of the Steady State* (Methuen, London, 1951), Chs. I,II,VI.
6. I. Prigogine, I., *Introduction to the Thermodynamics of Irreversible Processes*, 2nd ed. (Wiley, New York, 1961).
7. S. de Groot and P. Mazur, *Non-Equilibrium Thermodynamics* (North-Holland, Amsterdam, 1962).
8. R. J. Tykodi, *Thermodynamics of Steady States* (Macmillan, New York, 1967).
9. Book II, Ch. 1.
10. Book II, Ch. 2.
11. R. J. Tykodi, J. Chem. Educ. **72**, 103 (1995).

NINE

EVALUATING "UNDETERMINED QUANTITIES": A WATER-FLOW ANALOGUE TO A HEAT-FLOW PROBLEM

Let us next discuss the problem of evaluating "undetermined quantities."

Introduction

Consider a system in a steady state. Let the system undergo a process that terminates in another steady state. The system interacts with a series of reservoirs (of heat and/or mass) both in the steady states and during the process. The reservoirs are of two kinds, *active* and *quiescent* (i.e. *active* or *quiescent* in the initial and final nonequilibrium steady states). An active reservoir interacts with the system both in the initial and final steady states and during the process. The interactions with the active reservoirs in the initial and final states are what keeps the system from relaxing to equilibrium states. A quiescent reservoir interacts with the system only during the process and not while the system is in the initial or final steady state.

In the case of a quiescent reservoir, the exchange between the reservoir and the system during the process can be evaluated by a before–and–after examination of the reservoir.

In the case of an active reservoir, because of its ongoing interaction with the system in the initial and final steady states, before–and–after examination of the reservoir fails to reveal the net, lasting exchange between the system and the reservoir—net, lasting exchanges as opposed to exchanges that merely flush through the system. To evaluate the net, lasting exchanges between the system and an active reservoir, it is necessary to monitor the flow into or out of the reservoir during the process—a much more difficult task than the simple before–and–after examination of a quiescent reservoir.

In my treatment of the thermomolecular pressure effect in Section 8 (see Fig. 1 therein) for example, the mass reservoirs at α and β are quiescent reservoirs: it is easy to tell if they lost or gained mass during a process by a simple before–and–after inventory of the reservoirs. The heat reservoirs at α and β, on the other hand, are active reservoirs—they maintain a flow of heat through the system even in the initial and final steady states.

For a process occurring within the time interval from t_1 to t_2, the net, lasting heat Q communicated by the heat reservoirs at α and β to the system is given by

$$Q = -\int_{t_1}^{t_2} \left[\dot{Q}_\alpha^{(r)}(t) + \dot{Q}_\beta^{(r)}(t) \right] dt, \tag{1}$$

where $\dot{Q}_\omega^{(r)}$ $(\omega = \alpha, \beta)$ is the rate of influx of heat to the indicated reservoir; and the heat described in Eq. (1) enters into the statement of the first law for the system:

$$\Delta U = Q + W. \tag{2}$$

To be able to invoke the Thomson–Gibbs Corollary as a test of migrational equilibrium for the gas in the thermomolecular pressure apparatus, we need to be able to partition the total heat Q supplied by the reservoirs into separate contributions Q_α, Q_β from each reservoir.

Ideally, we would like to be able to deduce from our monitoring of each reservoir during a process the net, lasting contribution Q_α, Q_β of each reservoir as a function of the properties of the reservoir in question—say (for example)

$$Q_\alpha = F_\alpha\left(\ddot{Q}_\alpha^{(r)}\right) \text{ and } Q_\beta = F_\beta\left(\ddot{Q}_\beta^{(r)}\right). \tag{3}$$

Consider the system in question to be in a steady state at time t_1, and let the system undergo a process that results in another steady state. Let the system be in the final steady state at time t_q. Then, for the case in hand,

$$\dot{Q}_\alpha^{(r)}(t_1) + \dot{Q}_\beta^{(r)}(t_1) = 0, \quad \dot{Q}_\alpha^{(r)}(t_q) + \dot{Q}_\beta^{(r)}(t_q) = 0. \tag{4}$$

Now, we can write $(t_1 \le t \le t_q)$

$$\dot{Q}_\omega^{(r)}(t) = \dot{Q}_\omega^{(r)}(t_1) + \int_{t_1}^{t} \ddot{Q}_\omega^{(r)}(t)dt \quad (\omega = \alpha, \beta) \tag{5}$$

and

$$\int_{t_1}^{t_q} \dot{Q}_\omega^{(r)}(t)dt = \int_{t_1}^{t_q} \dot{Q}_\omega^{(r)}(t_1)dt + \int_{t_1}^{t_q} dt \int_{t_1}^{t} \ddot{Q}_\omega^{(r)}(t)dt. \tag{6}$$

Since

$$Q = -\int_{t_1}^{t_q} \left[\dot{Q}_\alpha^{(r)}(t) + \dot{Q}_\beta^{(r)}(t) \right] dt, \tag{7}$$

we have

$$Q = -\int_{t_1}^{t_q} \left[\dot{Q}_\alpha^{(r)}(t_1) + \dot{Q}_\beta^{(r)}(t_1) \right] dt - \int_{t_1}^{t_q} dt \int_{t_1}^{t} \ddot{Q}_\alpha^{(r)}(t)dt -$$

$$\int_{t_1}^{t_q} dt \int_{t_1}^{t} \ddot{Q}_{\beta}^{(r)}(t)dt = -\int_{t_1}^{t_q} dt \int_{t_1}^{t} \ddot{Q}_{\alpha}^{(r)}(t)dt - \int_{t_1}^{t_q} dt \int_{t_1}^{t} \ddot{Q}_{\beta}^{(r)}(t)dt, \quad (8)$$

where we have made use of Eq. (4). Let us divide the interval $t_1 \to t_q$ into $q-1$ sub-intervals by the points $t_1, t_2, \ldots, t_q$ in such a way that $\ddot{Q}_{\omega}^{(r)}(t)$ is continuous within each sub-interval, i.e. we are requiring that $\ddot{Q}_{\omega}^{(r)}(t)$
be at least piece-wise continuous. It follows then that

$$-\int_{t_1}^{t_q} dt \int_{t_1}^{t} \ddot{Q}_{\omega}^{(r)}(t)dt = -\int_{t_1}^{t_2} dt \int_{t_1}^{t} \ddot{Q}_{\omega}^{(r)}(t)dt -$$

$$\sum_{i=2}^{q-1} \int_{t_i}^{t_{i+1}} dt \left[\int_{t_i}^{t} \ddot{Q}_{\omega}^{(r)}(t)dt + \sum_{j=1}^{i-1} \int_{t_j}^{t_{j+1}} \ddot{Q}_{\omega}^{(r)}(t)dt \right]; \quad (9)$$

so

$$Q = -\int_{t_1}^{t_2} dt \int_{t_1}^{t} \ddot{Q}_{\alpha}^{(r)}(t)dt - \int_{t_1}^{t_2} dt \int_{t_1}^{t} \ddot{Q}_{\beta}^{(r)}(t)dt -$$

$$\sum_{i=2}^{q-1} \int_{t_i}^{t_{i+1}} dt \left[\int_{t_i}^{t} \ddot{Q}_{\alpha}^{(r)}(t)dt + \sum_{j=1}^{i-1} \int_{t_j}^{t_{j+1}} \ddot{Q}_{\alpha}^{(r)}(t)dt \right] -$$

$$\sum_{i=2}^{q-1} \int_{t_i}^{t_{i+1}} dt \left[\int_{t_i}^{t} \ddot{Q}_{\beta}^{(r)}(t)dt + \sum_{j=1}^{i-1} \int_{t_j}^{t_{j+1}} \ddot{Q}_{\beta}^{(r)}(t)dt \right]. \quad (10)$$

Equation (10) yields Q as a function of $\ddot{Q}_{\alpha}^{(r)}$ and $\ddot{Q}_{\beta}^{(r)}$, and we wonder if we can separate the expression into two satisfactory pieces

$$Q_{\alpha} = F_{\alpha}\left(\ddot{Q}_{\alpha}^{(r)}\right) \text{ and } Q_{\beta} = F_{\beta}\left(\ddot{Q}_{\beta}^{(r)}\right), \text{ with } Q = Q_{\alpha} + Q_{\beta}.$$

To get some orientation on the matter, let's look at a water-flow analogue of the heat-flow problem. The water-flow analogue shows all the features that we would have to deal with in the heat-flow case. We shall find that the results for the water-flow analogue indicate strongly that in the partitioning of the total heat dQ (say) into separate contributions dQ_{α}, dQ_{β} the separate contributions each depend on the characteristics of both the interacting reservoirs:

$$dQ_{\omega} = F_{\omega}\left(\ddot{Q}_{\alpha}^{(r)}, \ddot{Q}_{\beta}^{(r)}\right) \; (\omega = \alpha, \beta).$$

I shall obtain the conditions governing the stationary state for water in a standpipe or a reservoir that is fed by two other water reservoirs both by the principle of minimum work loss (which is a form of the Thomson-Gibbs Corollary) and by the principle of minimum power loss (which is a form of the principle of minimum entropy production).

Water-Flow Analogue

Let two water reservoirs, α, β, be connected to the bottom of a cylindrical standpipe. Let the heights of the reservoirs above a reference plane be different, and let $\dot{m}_\omega$ $(\omega = \alpha, \beta)$ be the rate of influx of mass to a water reservoir. Let the surrounding air function as a constant temperature, constant pressure heat reservoir (thermostat). At time t_1, let the flow of water under the influence of gravity be such that the level of the water in the standpipe is stationary, there being a steady flow of water through the standpipe from one water reservoir to the other:

$$\dot{m}_\alpha(t_1) + \dot{m}_\beta(t_1) = 0. \tag{11}$$

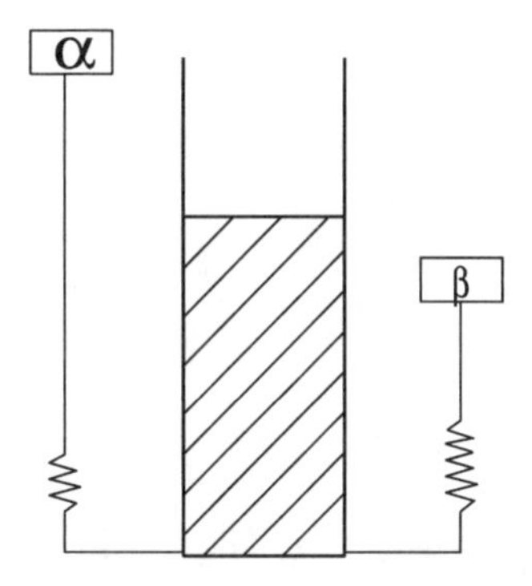

Fig. 1. Steady-state level of water in a standpipe fed by water reservoirs α and β.

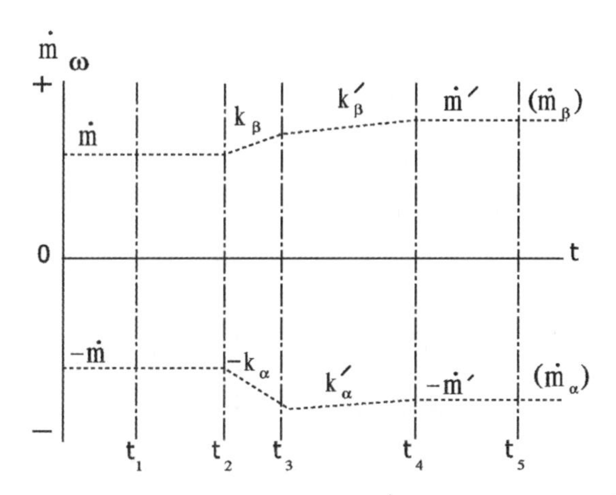

Fig. 2. Mass rate of inflow to reservoir α, $\dot{m}_\alpha$, and to reservoir β, $\dot{m}_\beta$, versus time.

Let each of the connectors between the standpipe and the reservoirs contain a valve, and let each of the valves be locked in a partially open position in the initial steady state. Now quickly open the valve in the α connector wider and lock it in its new position. The increased rate of inflow of water from the α reservoir will result in changes in the level of the water in the standpipe until a new steady state is reached; let the new steady state be well established by time t_5—see Fig. 1 and the (simplified) mass flow profile in Fig. 2.

During the change from the initial steady state to the final steady state, the mass of water in the standpipe will change by the amount Δm given by

$$\Delta m = -\int_{t_1}^{t_5}\left[\dot{m}_\alpha(t) + \dot{m}_\beta(t)\right]dt. \tag{12}$$

We would like to resolve Δm into separate contributions $\Delta_\alpha m, \Delta_\beta m$ from the water reservoirs at α, β:

$$\Delta m = \Delta_\alpha m + \Delta_\beta m. \tag{13}$$

The analogy to the heat-flow case should now be clear.

For the (simplified) mass flow profile of Fig. 2, let us catalogue the flow rates in the indicated time intervals in the form $\dot{m}_\alpha ; \dot{m}_\beta$:

$$t_1 \to t_2\colon\ -\dot{m};\dot{m}.$$

$$t_2 \to t_3\colon\ -\dot{m} - k_\alpha(t - t_2);\dot{m} + k_\beta(t - t_2).$$

$$t_3 \to t_4\colon\ -\dot{m} - k_\alpha(t_3 - t_2) + k'_\alpha(t - t_3);\dot{m} + k_\beta(t_3 - t_2) + k'_\beta(t - t_3).$$

$$\begin{aligned} t_4 \to t_5\colon\ & -\dot{m} - k_\alpha(t_3 - t_2) + k'_\alpha(t_4 - t_3) = -\dot{m}'; \\ & \dot{m} + k_\beta(t_3 - t_2) + k'_\beta(t_4 - t_3) = \dot{m}'. \end{aligned}$$

All the quantities $\dot{m}, \dot{m}', k_\alpha, k'_\alpha, k_\beta, k'_\beta$ are positive.

In the initial and final steady states, the rate of inflow of water into the standpipe must match the rate of outflow; this requires that

$$k_\alpha(t_3 - t_2) - k'_\alpha(t_4 - t_3) = k_\beta(t_3 - t_2) + k'_\beta(t_4 - t_3) \tag{14}$$

or, equivalently,

$$(k_\alpha - k_\beta)(t_3 - t_2) = (k'_\alpha + k'_\beta)(t_4 - t_3). \tag{15}$$

Now let's calculate $-\int_{t_1}^{t_5}\dot{m}_\alpha(t)dt$ and $-\int_{t_1}^{t_5}\dot{m}_\beta(t)dt$:

$$\begin{aligned} -\int_{t_1}^{t_5}\dot{m}_\alpha(t)dt = {} & \dot{m}(t_5 - t_1) + k_\alpha(t_3 - t_2)(t_5 - t_3) - \\ & k'_\alpha(t_4 - t_3)(t_5 - t_4) + \tfrac{1}{2}k_\alpha(t_3 - t_2)^2 - \tfrac{1}{2}k'_\alpha(t_4 - t_3)^2, \end{aligned} \tag{16}$$

$$-\int_{t_1}^{t_5}\dot{m}_\beta(t)dt = -\dot{m}(t_5 - t_1) - k_\beta(t_3 - t_2)(t_5 - t_3) -$$
$$k'_\beta(t_4 - t_3)(t_5 - t_4) - \tfrac{1}{2}k_\beta(t_3 - t_2)^2 - \tfrac{1}{2}k'_\beta(t_4 - t_3)^2. \tag{17}$$

Then,

$$\Delta m = -\int_{t_1}^{t_5}\left[\dot{m}_\alpha(t) + \dot{m}_\beta(t)\right]dt$$
$$= (k_\alpha - k_\beta)(t_3 - t_2)(t_5 - t_3) - (k'_\alpha + k'_\beta)(t_4 - t_3)(t_5 - t_4) +$$
$$\tfrac{1}{2}k_\alpha(t_3 - t_2)^2 - \tfrac{1}{2}k'_\alpha(t_4 - t_3)^2 - \tfrac{1}{2}k_\beta(t_3 - t_2)^2 - \tfrac{1}{2}k'_\beta(t_4 - t_3)^2$$
$$= (k_\alpha - k_\beta)(t_3 - t_2)(t_4 - t_3) + \tfrac{1}{2}k_\alpha(t_3 - t_2)^2 -$$
$$\tfrac{1}{2}k'_\alpha(t_4 - t_3)^2 - \tfrac{1}{2}k_\beta(t_3 - t_2)^2 - \tfrac{1}{2}k'_\beta(t_4 - t_3)^2$$
$$= \tfrac{1}{2}(k_\alpha - k_\beta)(t_3 - t_2)(t_4 - t_2). \tag{18}$$

Applying the formalism of Eqs. (4)-(10) to the present case

$$\left(Q \to \Delta m,\ Q_\omega \to \Delta_\omega m,\ \dot{Q}_\omega^{(r)} \to \dot{m}_\omega,\ \ddot{Q}_\omega^{(r)} \to \ddot{m}_\omega\right),$$

with $q = 5$, we may express relation (18) as

$$\Delta m = -\int_{t_2}^{t_3}dt\int_{t_2}^{t}\ddot{m}_\alpha(t)dt - \int_{t_3}^{t_4}dt\left[\int_{t_3}^{t}\ddot{m}_\alpha(t)dt + \int_{t_2}^{t_3}\ddot{m}_\alpha(t)dt\right] -$$
$$\int_{t_4}^{t_5}dt\left[\int_{t_2}^{t_3}\ddot{m}_\alpha(t)dt + \int_{t_3}^{t_4}\ddot{m}_\alpha(t)dt\right] -$$
$$\int_{t_2}^{t_3}dt\int_{t_2}^{t}\ddot{m}_\beta(t)dt - \int_{t_3}^{t_4}dt\left[\int_{t_3}^{t}\ddot{m}_\beta(t)dt + \int_{t_2}^{t_3}\ddot{m}_\beta(t)dt\right] -$$
$$\int_{t_4}^{t_5}dt\left[\int_{t_2}^{t_3}\ddot{m}_\beta(t)dt + \int_{t_3}^{t_4}\ddot{m}_\beta(t)dt\right], \tag{19}$$

where we have used the fact that $\ddot{m}_\omega = 0$ on intervals $t_1 \to t_2$ and $t_4 \to t_5$.

We observe from the mass flow profile in Fig. 2 that at each instant in the interval $t_1 \to t_5$ the rate of inflow of water into the standpipe equals or exceeds the rate of outflow:

$$-\dot{m}_\alpha \geq \dot{m}_\beta. \tag{20}$$

The rate excess, $|\dot{m}_\alpha| - |\dot{m}_\beta|$, results in the accumulation of the extra mass Δm in the standpipe; and it is evident that all of Δm comes from reservoir α:

$$\Delta_\alpha m = \Delta m,\ \ \Delta_\beta m = 0. \tag{21}$$

Juxtaposing Eqs. (19),(21), we can say that

$$\Delta_\alpha m = f_\alpha\left(\ddot{m}_\alpha, \ddot{m}_\beta\right); \tag{22}$$

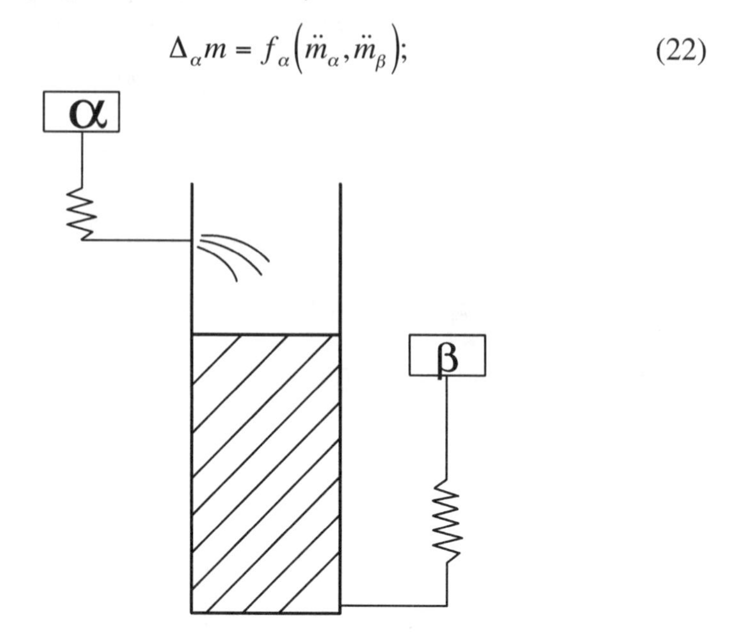

Fig. 3. Steady-state level of water in a standpipe fed by water reservoirs α and β.

so, in general, the net, long-lasting contribution from one of the mass reservoirs depends on the characteristics of both reservoirs. This observation will be reinforced by the example to follow.

Example II

Let us consider a different configuration for a standpipe and two water reservoirs—see Fig. 3. Let the water reservoir at α be connected to the top of a cylindrical standpipe—so that the water flowing from the α reservoir "cascades" into the standpipe; and let the water reservoir at β be connected to the bottom of the standpipe as before. At time t_1, let the flow of water under the influence of gravity be such that the level of the water in the standpipe is stationary:

$$\dot{m}_\alpha\left(t_1\right) + \dot{m}_\beta\left(t_1\right) = 0. \tag{23}$$

Now close down somewhat on the valve in the β connector, and lock the valve in its new setting. As a result, the rate of mass flow into reservoir β will first decrease and then increase until the outflow under the new conditions again matches the inflow. In the final steady state, the level of the water in the standpipe will be

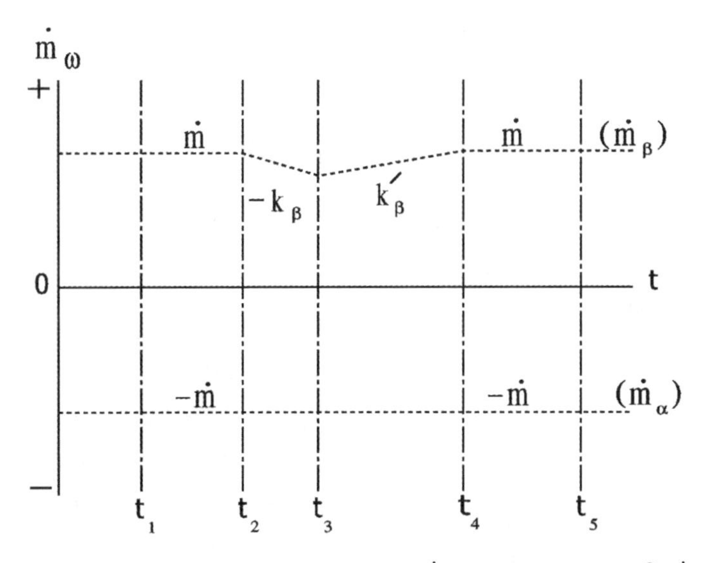

Fig. 4. Mass rate of inflow to reservoir α, $\dot{m}_\alpha$, and to reservoir β, $\dot{m}_\beta$, versus time.

higher than it was in the initial steady state; and the standpipe will have gained an amount of mass Δm in going from the initial steady state to the final steady state—see the (simplified) mass flow profile in Fig. 4:

$$\Delta m = -\int_{t_1}^{t_5} \left[\dot{m}_\alpha(t) + \dot{m}_\beta(t) \right] dt. \tag{24}$$

We again would like to resolve Δm into separate contributions $\Delta_\alpha m, \Delta_\beta m$ from the water reservoirs at α, β. Let us again catalogue the flow rates in the indicated time intervals in the form $\dot{m}_\alpha ; \dot{m}_\beta$:

$t_1 \rightarrow t_2$: $-\dot{m}; \dot{m}.$

$t_2 \rightarrow t_3$: $-\dot{m}; \dot{m} - k_\beta(t - t_2).$

$t_3 \rightarrow t_4$: $-\dot{m}; \dot{m} - k_\beta(t_3 - t_2) + k'_\beta(t - t_3).$

$t_4 \rightarrow t_5$: $-\dot{m}; \dot{m}.$

The quantities $\dot{m}, k_\beta, k'_\beta$ are all positive.

In the initial and final steady states, the rate of outflow of water from the standpipe must match the rate of inflow; this requires that

$$k_\beta(t_3 - t_2) = k'_\beta(t_4 - t_3). \tag{25}$$

Let's calculate $-\int_{t_1}^{t_5} \dot{m}_\alpha(t) dt$ and $-\int_{t_1}^{t_5} \dot{m}_\beta(t) dt$:

$$-\int_{t_1}^{t_5} \dot{m}_\alpha(t)dt = \dot{m}(t_5 - t_1), \tag{26}$$

$$-\int_{t_1}^{t_5} \dot{m}_\beta(t)dt = -\dot{m}(t_5 - t_1) + \tfrac{1}{2}k_\beta(t_3 - t_2)^2 + k_\beta(t_3 - t_2)(t_4 - t_3) - \tfrac{1}{2}k'_\beta(t_4 - t_3)^2. \tag{27}$$

Then

$$\begin{aligned}\Delta m &= -\int_{t_1}^{t_5}\left[\dot{m}_\alpha(t) + \dot{m}_\beta(t)\right]dt \\ &= \tfrac{1}{2}k_\beta(t_3 - t_2)^2 + k_\beta(t_3 - t_2)(t_4 - t_3) - \tfrac{1}{2}k'_\beta(t_4 - t_3)^2 \\ &= \tfrac{1}{2}k_\beta(t_3 - t_2)(t_4 - t_2).\end{aligned} \tag{28}$$

We may write relation (28) as

$$\Delta m = -\int_{t_2}^{t_3} dt \int_{t_2}^{t} \ddot{m}_\beta(t)dt - \int_{t_3}^{t_4} dt\left[\int_{t_3}^{t} \ddot{m}_\beta(t)dt + \int_{t_2}^{t_3} \ddot{m}_\beta(t)dt\right] - \int_{t_4}^{t_5} dt\left[\int_{t_2}^{t_3} \ddot{m}_\beta(t)dt + \int_{t_3}^{t_4} \ddot{m}_\beta(t)dt\right]. \tag{29}$$

Just as in Example I, we note from the mass flow profile in Fig. 4 that everywhere in the interval $t_1 \to t_5$

$$-\dot{m}_\alpha \geq \dot{m}_\beta. \tag{30}$$

The rate excess, $|\dot{m}_\alpha| - |\dot{m}_\beta|$, results in the accumulation of the extra mass Δm in the standpipe; and it is evident that all of Δm comes from reservoir α :

$$\Delta_\alpha m = \Delta m, \quad \Delta_\beta m = 0. \tag{31}$$

This time, we find that

$$\Delta_\alpha m = f_\alpha(\ddot{m}_\beta). \tag{32}$$

The contribution of the α reservoir is expressible entirely in terms of the properties of the β reservoir!

Finding

We have thus found for the water-flow model that we have been analyzing that the contribution of a water reservoir to the overall change of mass Δm in the standpipe in general depends on properties of both the reservoirs:

$$\Delta_\omega m = f_\omega\left(\ddot{m}_\alpha, \ddot{m}_\beta\right) \ (\omega = \alpha, \beta). \tag{33}$$

We expect something analogous to this to pertain to the two-heat-reservoir case—we expect in general that

$$Q_\omega = F_\omega\left(\ddot{Q}_\alpha^{(r)}, \ddot{Q}_\beta^{(r)}\right) \ (\omega = \alpha, \beta). \tag{34}$$

The monitoring of $\ddot{Q}_\alpha^{(r)}$ and $\ddot{Q}_\beta^{(r)}$ for a process is a difficult task, and the form of the function

$$F_\omega\left(\ddot{Q}_\alpha, \ddot{Q}_\beta\right)$$

is not known for the general case; so we are often reduced to using expedients such as the replacement of dQ_ω / dn by $-\delta\dot{Q}_\omega^{(r)} / \delta\tilde{\dot{n}}$—see Section 8.

A Stationary State in a Gravity-Driven Poiseuille-Field

In the cases to be analyzed in this section, two water reservoirs feed a standpipe or a third reservoir through connector tubes. Assume that the flow due to gravity of water through the connectors is Poiseuille flow [1], i.e. assume that the rate of mass flow through a connector is proportional to the pressure difference acting across the connector—that pressure difference in turn depending on the difference in height of the levels of the water that feeds into the connector. Let the surrounding air function as a constant temperature, constant pressure heat reservoir (thermostat).

We wish to explore the conditions that make the mass of water in a standpipe or in an intermediate reservoir stationary.

Principles of Minimum Work Loss and Minimum Power Loss

Let us treat water as an incompressible liquid; then the "isothermal" flow of water in our multi-reservoir systems is a constant entropy, constant volume, constant mass process—as far as the water is concerned. This observation allows us to state and use an energy form of the Dartmouth version of the second law:

If a system undergoing a process has an equilibrium-based image in both the initial and final states and if the entropy, volume, and mass of the system are the same in the initial and final states, then in the Dartmouth domain it must be true that

$$\Delta\text{“}U\text{”} \le 0 \ (const\ S, V, m) \tag{35}$$

for the process, where "U" ≡ U + macroscopic kinetic energy + gravitational energy + ..., i.e *"U" is the total energy of the system.* If we keep the diameters of the connector tubes small compared to the dimensions of the water reservoirs, then we can neglect the macroscopic kinetic energy of the water in the reservoirs; so for the cases we shall be discussing, "U" = U + *gravitational energy.*

For the situations we are describing,

$$\Delta\text{“}U\text{”} = -T\Gamma_{\uparrow} \leq 0, \tag{36}$$

where T is the temperature of the surrounding air thermostat. For our process, Δ"U" represents a "work loss": an energy change that under other circumstances could produce work occurs without the production of work.

Let the rate of flow of mass into a standpipe or into an intermediate water reservoir be $\dot{m}_x$. Then the stationary condition $(\dot{m}_x = 0)$ for the mass of water in the standpipe or intermediate reservoir is found via the principle of minimum work loss:

$$\left.\frac{d\text{“}U\text{”}}{dm_x}\right|_{S,V.m,\ldots} = 0 \tag{37}$$

which is a form of the Thomson-Gibbs Corollary—see Eq. (36).

If in studying the flow of water through a standpipe or through an intermediate reservoir, we choose to deal with sequences of steady states, then we find that

$$\text{“}\dot{U}\text{”} \leq 0 \quad (const\ S, V, m) \tag{38}$$

and

$$\text{“}\dot{U}\text{”} = -T\dot{\Gamma} \leq 0. \tag{39}$$

If we let $\dot{m}_x \rightarrow 0$ via a sequence of steady states, we get

$$\left.\frac{\delta\text{“}\dot{U}\text{”}}{\tilde{\delta}\dot{m}_x}\right|_{S,V,m,\ldots} = 0 \tag{40}$$

as an instance of the principle of minimum power loss—which is a form of the principle of minimum entropy production—see Eq. (39) [for a discussion of the principle of minimum entropy production, see Section 12].

Stationary Condition for a Standpipe

Let the apparatus be laid out as in Fig. 1, and let the top of the standpipe be closed by a movable frictionless, (essentially) massless platform that rests on the surface of the water in the standpipe. Let the system be in a steady state, the rate of inflow of water from reservoir α matching the rate of outflow to reservoir β. In the steady state, let the height of water in the standpipe be h_x. Let the α and β connector tubes be "twins," alike in length, cross sectional area, etc., with part of each tube coiled up like a garden hose. Let the flow of water through the connector tubes be Poiseuille flow. Choose the temperature of the surrounding air thermostat and the dimensions of the connector tubes so that in the steady state

$$\dot{m}_\alpha = h_x - h_\alpha, \tag{41}$$

$$\dot{m}_\beta = h_x - h_\beta, \tag{42}$$

$$\dot{m}_x = h_\alpha + h_\beta - 2h_x. \tag{43}$$

The mass-balance condition for the steady state is

$$\dot{m}_\alpha + \dot{m}_\beta + \dot{m}_x = 0. \tag{44}$$

Now drop a few grains of sand on the massless platform at height h_x in the standpipe. The extra pressure dp generated by the grains of sand at the inflow and outflow ports at the bottom of the standpipe will cause an initial decrease in the inflow of water from reservoir α and an initial increase in the outflow of water to reservoir β. The water in the standpipe will relax to a new steady state with a lower water level, and the standpipe will have undergone a change in mass of water dm_x. The reservoirs at α, β will have made net contributions $d_\alpha m, d_\beta m$ to dm_x:

$$dm_x = d_\alpha m + d_\beta m. \tag{45}$$

For the change from the initial steady state to the final steady state,

$$g^{-1} d\text{“}U\text{”} = -h_\alpha d_\alpha m - h_\beta d_\beta m + h_x dm_x \tag{46}$$

and

$$g^{-1} \left. \frac{d\text{“}U\text{”}}{dm_x} \right|_{T, h_\alpha, h_\beta} = -h_\alpha \frac{d_\alpha m}{dm_x} - h_\beta \frac{d_\beta m}{dm_x} + h_x, \tag{47}$$

where g is the acceleration due to gravity. The principle of minimum work loss leads to

$$h_x = h_\alpha \frac{d_\alpha m}{dm_x} + h_\beta \frac{d_\beta m}{dm_x}. \tag{48}$$

Equation (45) gives

$$1 = \frac{d_\alpha m}{dm_x} + \frac{d_\beta m}{dm_x}, \tag{49}$$

so

$$h_x = h_\alpha + \left(h_\beta - h_\alpha\right)\frac{d_\beta m}{dm_x}. \tag{50}$$

If we evaluate $d_\beta m / dm_x$ by the expedient

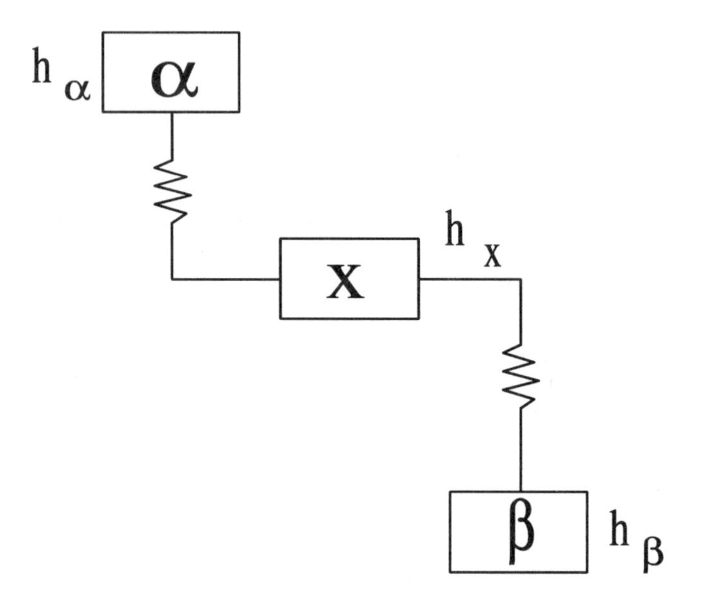

Fig. 5. Poiseuille-flow of water through connectors between water reservoirs α,β,x at heights h_α, h_β, h_x in a gravitational field.

$$\frac{d_\beta m}{dm_x} = -\left.\frac{\delta \dot{m}_\beta}{\tilde{\delta} \dot{m}_x}\right|_{T,h_\alpha,h_\beta} = -\left.\frac{\partial \dot{m}_\beta / \partial h_x}{\partial \dot{m}_x / \partial h_x}\right|_{T,h_\alpha,h_\beta,\dot{m}_x \to 0} = \frac{1}{2}, \tag{51}$$

then

$$h_x = h_\alpha + \tfrac{1}{2}\left(h_\beta - h_\alpha\right) = \tfrac{1}{2}\left(h_\alpha + h_\beta\right). \tag{52}$$

Stationary Condition for an Intermediate Reservoir

Let three water reservoirs α,β,x be positioned in a gravitational field as in Fig. 5, the heights of the reservoirs above a reference plane being h_α, h_β, h_x. Let the height of reservoir x be adjustable, and let the rate of influx of mass to a water reservoir be $\dot{m}_\omega$ $\left(\omega = \alpha,\beta,x\right)$. Let the α and β connector tubes be "twins," alike in length, cross sectional area, etc., with part of each tube coiled up like a garden hose. Let the flow of water through the connector tubes be Poiseuille flow. Choose the temperature of the surrounding air thermostat and the dimensions of the connector tubes so that in the steady state

$$\dot{m}_\alpha = h_x - h_\alpha, \tag{53}$$

$$\dot{m}_\beta = h_x - h_\beta, \tag{54}$$

$$\dot{m}_x = h_\alpha + h_\beta - 2h_x. \tag{55}$$

The mass-balance condition for the steady state is

$$\dot{m}_\alpha + \dot{m}_\beta + \dot{m}_x = 0. \tag{56}$$

The reservoirs α, β, x are each active reservoirs, so we shall deal with the time rate of change of reservoir properties.

Start with reservoir x at a height h_x such that $\dot{m}_x = 0$; then change the height to h_x' to induce a steady flow of mass into or out of the x reservoir. For the new steady state

$$\begin{aligned} g^{-1}\text{“}\dot{U}\text{”} &= h_\alpha \dot{m}_\alpha + h_\beta \dot{m}_\beta + h_x' \dot{m}_x \\ &= \left(h_\alpha - h_\beta\right)\dot{m}_\alpha + \left(h_x' - h_\beta\right)\dot{m}_x, \end{aligned} \tag{57}$$

where g is the acceleration due to gravity and we have drawn on Eq. (56). Repeat the procedure so as to generate a sequence of steady states converging on the state $\dot{m}_x = 0$ $\left(h_x' \to h_x\right)$. Then

$$g^{-1} \left.\frac{\delta\text{“}\dot{U}\text{”}}{\tilde{\delta}\dot{m}_x}\right|_{T,h_\alpha,h_\beta} = \left(h_\alpha - h_\beta\right) \left.\frac{\delta \dot{m}_\alpha}{\tilde{\delta}\dot{m}_x}\right|_{T,h_\alpha,h_\beta} + h_x - h_\beta, \tag{58}$$

and the state $\dot{m}_x = 0$ will be a state of minimum power loss relative to the neighboring steady-flow states with $\dot{m}_x \neq 0$; therefore

$$\left.\frac{\delta\text{“}\dot{U}\text{”}}{\tilde{\delta}\dot{m}_x}\right|_{T,h_\alpha,h_\beta} = 0 \tag{59}$$

and

$$h_x = h_\beta - \left(h_\alpha - h_\beta\right) \left.\frac{\delta \dot{m}_\alpha}{\tilde{\delta}\dot{m}_x}\right|_{T,h_\alpha,h_\beta} \tag{60}$$

for the state $\dot{m}_x = 0$. We may write [see Eqs. (53)-(55)]

$$\left.\frac{\delta \dot{m}_\alpha}{\tilde{\delta} \dot{m}_x}\right|_{T,h_\alpha,h_\beta} = \left.\frac{\partial \dot{m}_\alpha / \partial h_x}{\partial \dot{m}_x / \partial h_x}\right|_{T,h_\alpha,h_\beta,\dot{m}_x \to 0} = -\frac{1}{2}. \tag{61}$$

It follows, then, that (for the conditions as given) the state $\dot{m}_x = 0$ is characterized by

$$h_x = \tfrac{1}{2}\left(h_\alpha + h_\beta\right). \tag{62}$$

Summary

Let a system in a steady state interact with multiple heat reservoirs, and let the system undergo a process involving interactions with the heat reservoirs and terminating in another steady state. The net heat Q exchanged between the system and the reservoirs during the process is fixed by the first law:

$$\Delta U = Q + W. \tag{63}$$

Apportioning the total heat Q among the active reservoirs, q in number (say),

$$Q = \sum_{i=1}^{q} Q_i, \tag{64}$$

is a very difficult task; and we sought guidance in the matter by considering a water-flow analogy to heat flow.

We considered the case of a standpipe fed by two water reservoirs, flow through the system being driven by gravity. By making changes in the constraints operating on the system, we passed from one steady state for the water in the standpipe to another steady state—the change being accompanied by a loss or gain of mass, Δm, by the standpipe. We sought to resolve the net change in mass Δm into separate contributions $\Delta_\alpha m, \Delta_\beta m$ by the two water reservoirs. Our results appear to show that the contribution from one reservoir, $\Delta_\alpha m$ (say), depends on properties of both the reservoirs, i.e.

$$\Delta_\alpha m = f_\alpha\left(\ddot{m}_\alpha, \ddot{m}_\beta\right). \tag{65}$$

We interpreted the results of our analysis of the water-flow analogue to heat flow to mean that in the heat-flow case discussed above [Eq. (64)] the contribution of an active heat reservoir, Q_i (say), depends on properties of *all* the acting reservoirs:

$$Q_i = F_i\left(\ddot{Q}_1^{(r)}, \ddot{Q}_2^{(r)}, \ldots, \ddot{Q}_q^{(r)}\right). \tag{66}$$

The difficulty of dealing with relations of the form (66) led us in Section 8 to adopt expedients such as replacing

$$dQ_\omega \,/\, dn \text{ by } -\delta\dot{Q}_\omega^{(r)} \,/\, \tilde{\delta}\dot{n} \quad (\omega = \alpha, \beta).$$

In the final part of this section, we looked at the conditions determining the stationary state for the level of water in a standpipe or in an intermediate reservoir; and we related those conditions to the principles of minimum work loss and minimum power loss.

Final Observation

For a particular steady-state configuration of the standpipe, let one reservoir serve as source and the other as sink for the steady flow of water through the standpipe:

$$\dot{m}_{\text{source}} + \dot{m}_{\text{sink}} = 0. \tag{67}$$

The results of our analysis of the water-flow model would seem to indicate that $(\Delta m = \Delta_{\text{source}} m + \Delta_{\text{sink}} m)$:increasing inflow from source or decreasing outflow to sink $\Rightarrow \Delta_{\text{source}} m > 0$; decreasing inflow from source or increasing outflow to sink $\Rightarrow \Delta_{\text{sink}} m < 0$.

REFERENCE

1. G. K. Batchelor, *An Introduction to Fluid Dynamics* (Cambridge Univ. Press, Cambridge, 1970) Sect. 4.2.

TEN

MORE EXAMPLES OF MIGRATIONAL EQUILIBRIUM

I display explicitly the migrational equilibrium conditions for thermoösmosis, thermal diffusion, a concentration-field analogue of thermal diffusion, and the thermocouple.

Thermoösmosis

As in Fig. 5.1, position a snugly fitting block of rubber (B) in the center of a cylindrical tube. Let the part of the wall of the cylinder that is in contact with the rubber block be of adiabatically insulating material, and let the rest of the cylinder and the movable pistons at the ends of the cylinder be of diathermic material. Expose the block of rubber to a gas soluble in the rubber, with gas pressures p_α, p_β at the exposed faces. Put the α - end of the cylinder in a thermostat (heat reservoir) of temperature T_α, and put the β - end of the cylinder in a thermostat (heat reservoir) of temperature T_β—in such a way that the temperature gradient is localized inside the block of rubber. See Fig. 1.

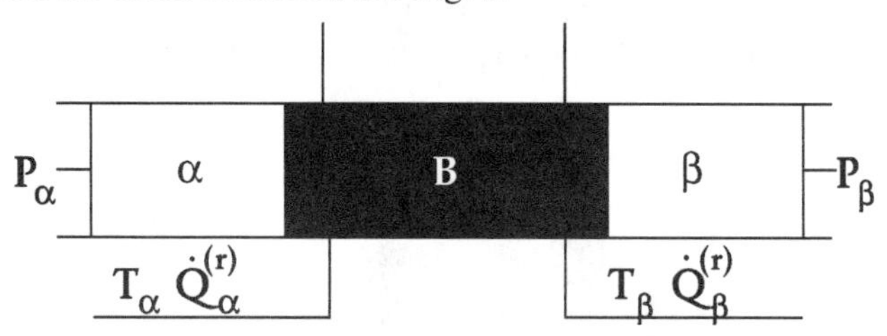

Fig. 1. Thermoösmosis. A gas in contact with a block of rubber (B), with the two gas-rubber interfaces in different thermostats. There is no lateral heat transfer to or from the part of the system outside the thermostats.

Let us determine the conditions of migrational equilibrium for the gas in the thermal field encompassing the rubber block. For given T_α, T_β, find a steady state with pressures p_α, p_β such that there is no macroscopic tendency for the gas to move from α to β or vice versa, i.e. the steady state with values $T_\alpha, T_\beta, p_\alpha, p_\beta$ is a state of migrational equilibrium for the gas. The fact that in a state of migrational equilibrium $p_\alpha \neq p_\beta$ when $T_\alpha \neq T_\beta$ is called the thermoösmotic effect.

Start in a steady state of migrational equilibrium and in differentially tailored fashion move *dn* moles of gas from region β through the rubber block into region α. For that process, the first law states that

$$\left(\overline{U}_{\alpha} - \overline{U}_{\beta}\right)dn = dQ_{\alpha} + dQ_{\beta} + \left(p_{\beta}\overline{V}_{\beta} - p_{\alpha}\overline{V}_{\alpha}\right)dn \ . \tag{1}$$

Let t_1 be a time before the process began and t_2 be a time after the process ended, then

$$\dot{Q}_{\alpha}^{(r)}(t_1) + \dot{Q}_{\beta}^{(r)}(t_1) = 0, \ \ \dot{Q}_{\alpha}^{(r)}(t_2) + \dot{Q}_{\beta}^{(r)}(t_2) = 0. \tag{2}$$

The total heat dQ exchanged between the system and the surroundings is given by

$$dQ = -\int_{t_1}^{t_2} \left[\dot{Q}_{\alpha}^{(r)}(t) + \dot{Q}_{\beta}^{(r)}(t)\right]dt, \tag{3}$$

and we are faced with the problem of apportioning dQ between the two heat reservoirs to get the individual contributions dQ_{α}, dQ_{β}. Consider dQ_{α} and dQ_{β} to be "undetermined quantities" for the moment, to be evaluated at a later stage in the proceedings.

Now divide Eq. (1) through by *dn* and rearrange the terms:

$$\overline{H}_{\alpha} - \overline{H}_{\beta} = dQ_{\alpha} / dn + dQ_{\beta} / dn \ , \tag{4}$$

where $\overline{H} \equiv \overline{U} + p\overline{V}$. Next set

$$dQ_{\alpha} / dn \equiv \overline{H}_{\alpha} - \left[H_{\alpha}\right] \text{ and } dQ_{\beta} / dn \equiv \left[H_{\beta}\right] - \overline{H}_{\beta}, \tag{5}$$

where $\left[H_{\omega}\right] \left(\omega = \alpha, \beta\right)$ is an appropriate nonequilibrium molar "migrational enthalpy."

[Note that the direction of flow at β is *into* the rubber block and at α is *out of* the rubber block.]

Upon combining Eqs. (4),(5), we get

$$\left[H_{\alpha}\right] = \left[H_{\beta}\right] \equiv \left[H\right]. \tag{6}$$

For the change in entropy of the system, we have

$$dS = \left(\overline{S}_{\alpha} - \overline{S}_{\beta}\right)dn \ ; \tag{7}$$

and for the change in the entropy of the surroundings attributable to the process,

$$d_{\uparrow}S_{sur} = -T_{\alpha}^{-1}dQ_{\alpha} - T_{\beta}^{-1}dQ_{\beta} \ . \tag{8}$$

So,

$$d\Gamma_{\uparrow} = \left(\overline{S}_{\alpha} - \overline{S}_{\beta}\right)dn - T_{\alpha}^{-1}dQ_{\alpha} - T_{\beta}^{-1}dQ_{\beta} \tag{9}$$

and

$$\begin{aligned}\frac{d\Gamma_{\uparrow}}{dn} &= \left(\overline{S}_{\alpha} - \overline{S}_{\beta}\right) + \frac{[H] - \overline{H}_{\alpha}}{T_{\alpha}} + \frac{\overline{H}_{\beta} - [H]}{T_{\beta}} \\ &= -\frac{\overline{G}_{\alpha}}{T_{\alpha}} + \frac{\overline{G}_{\beta}}{T_{\beta}} + [H]\left(\frac{1}{T_{\alpha}} - \frac{1}{T_{\beta}}\right) .\end{aligned} \tag{10}$$

In terms of the Planck function $\overline{Y} \equiv -\overline{G} / T$, Eq. (10) takes the form

$$\frac{d\Gamma_{\uparrow}}{dn} = \overline{Y}_{\alpha} - \overline{Y}_{\beta} + [H]\left(\frac{1}{T_{\alpha}} - \frac{1}{T_{\beta}}\right) . \tag{11}$$

Upon invoking the Thomson-Gibbs Corollary, we get

$$\overline{Y}_{\alpha} + \left[H_{\alpha}\right]T_{\alpha}^{-1} = \overline{Y}_{\beta} + \left[H_{\beta}\right]T_{\beta}^{-1} \tag{12}$$

or

$$\overline{Y}_{\alpha} - \overline{Y}_{\beta} + [H]\left(T_{\alpha}^{-1} - T_{\beta}^{-1}\right) = 0 \tag{13}$$

as the condition of migrational equilibrium for the thermoösmotic effect.

Consider the case of an ideal monatomic gas: the molar heat capacities $\overline{C}_{v}, \overline{C}_{p}$ are independent of temperature, and $\overline{H} = \overline{C}_{p}T$. For such a gas, Eq. (13) becomes

$$\begin{aligned}\frac{\ln\left(p_{\alpha} / p_{\beta}\right)}{\dfrac{1}{T_{\alpha}} - \dfrac{1}{T_{\beta}}} &= \frac{\left(\overline{C}_{p} / R\right)\ln\left(T_{\alpha} / T_{\beta}\right)}{\dfrac{1}{T_{\alpha}} - \dfrac{1}{T_{\beta}}} + \frac{\left\{[H] - \overline{H}_{\beta} + \overline{C}_{p}T_{\beta}\right\}}{R} \\ &= \frac{1}{R}\left\{[H] - \frac{\overline{C}_{p}T_{\alpha}T_{\beta}}{\theta\left(T_{\alpha}, T_{\beta}\right)}\right\} ,\end{aligned} \tag{14}$$

where $\theta\left(T_{\alpha}, T_{\beta}\right) \equiv \left(T_{\alpha} - T_{\beta}\right) / \ln\left(T_{\alpha} / T_{\beta}\right)$ is the logarithmic mean of the temperatures T_{α}, T_{β} [for a discussion of the properties of the logarithmic mean, see Section 2].

The expression $\overline{C}_{p}T_{\alpha}T_{\beta} / \theta\left(T_{\alpha}, T_{\beta}\right)$ in Eq. (14) can be considered to be a molar enthalpy intermediate between

$$\overline{H}_{\alpha} = \overline{C}_{p}T_{\alpha} \text{ and } \overline{H}_{\beta} = \overline{C}_{p}T_{\beta}.$$

Consider the example

$$T_\alpha = 400K, \;\; T_\beta = 200K, \quad \theta(400K, 200K) = 288.54K.$$

Then $T_\alpha T_\beta / \theta(T_\alpha, T_\beta) = 277.26K$; and $\overline{C}_p T_\alpha T_\beta / \theta(T_\alpha, T_\beta)$ behaves like a molar enthalpy of temperature $277.26K$.

The expression inside the curly braces in the second form of Eq. (14) is thus equivalent to some sort of "average" value of the difference $[H] - \overline{H}$. If we refer to the difference $[H] - \overline{H}$ as the molar heat of transport $Q*$ for the gas undergoing thermoösmosis, then the right-hand side of the second form of Eq. (14) is just R^{-1} times the "average" value of $Q*$.

Now let's evaluate the "undetermined quantities" dQ_α, dQ_β used in the above analysis. In a separate series of operations, induce a (slow) steady flow $\dot{n}$ of the gas from the terminal part [1] ("equilibrium-state" mass reservoir) at β to the terminal part at α, keeping $T_\alpha, p_\alpha, T_\beta$ constant (constant T, p', for short); and evaluate the derivatives

$$\delta\dot{Q}_\alpha^{(r)} / \tilde{\delta}\dot{n}\Big)_{T,p'} \quad \text{and} \quad \delta\dot{Q}_\beta^{(r)} / \tilde{\delta}\dot{n}\Big)_{T,p'}$$

via a sequence of steady states such that $\dot{n} \to 0$ [see Eq. (1.14)]. Make the identifications

$$dQ_\alpha / dn = -\delta\dot{Q}_\alpha^{(r)} / \tilde{\delta}\dot{n}\Big)_{T,p'} \quad \text{and}$$

$$dQ_\beta / dn = -\delta\dot{Q}_\beta^{(r)} / \tilde{\delta}\dot{n}\Big)_{T,p'} \tag{15}$$

or

$$dQ_\alpha = -\delta\dot{Q}_\alpha^{(r)} / \tilde{\delta}\dot{n}\Big)_{T,p'} dn \quad \text{and}$$

$$dQ_\beta = -\delta\dot{Q}_\beta^{(r)} / \tilde{\delta}\dot{n}\Big)_{T,p'} dn \tag{16}$$

in Eqs. (1)–(14) where appropriate.

Denbigh and Raumann [2], in a series of thermoösmotic measurements on several gases, found that the left-hand side of Eq. (14) was essentially constant for a particular gas for the range of temperatures and pressures that they investigated.

The analysis of the thermoösmotic effect is formally the same as that for the thermomolecular pressure effect (see Section 8). In the present case, however, if we choose to, we may resolve the overall heat of transport $Q* \equiv [H] - \overline{H}(gas)$ into a "heat of solution" effect $\Delta_{sol}\overline{H}$ and a "heat of transport along the temperature gradient" effect $Q*(solution)$ by adding and subtracting $\overline{H}(solution)$ to and from the expression for $Q*$:

$$Q* \equiv [H] - \overline{H}(gas) = [H] - \overline{H}(solution) + \overline{H}(solution) - \overline{H}(gas)$$

$$= \left[H \right] - \overline{H}\left(solution \right) + \Delta_{sol}\overline{H}$$

$$= Q*\left(solution \right) + \Delta_{sol}\overline{H} \ . \tag{17}$$

Bearman [3] noted that the sign of $Q*$ was the same as that of $\Delta_{sol}\overline{H}$ for the cases that he cited.

Thermal Diffusion in Gases

If a 2–component gaseous mixture is placed in a thermal field, the temperature gradient may induce some separation of the component gases—the separation of the two components by the thermal field is called the *thermal diffusion effect*.

Consider a 2–component gaseous mixture in an apparatus consisting of two chambers connected by a tubular linkage—refer to the components as component 1 and component 2. Each chamber (made of diathermic material) sits in a thermostat, and the tubular linkage is insulated against lateral heat losses. Each chamber communicates via semipermeable membranes with mass reservoirs containing the pure individual components of the mixture at suitable pressures. Let one thermostat be set at temperature T_α, and let the pressures in the mass reservoirs at α be $p_\alpha{}^{(1)}$ and $p_\alpha{}^{(2)}$. Let the other thermostat be set at temperature T_β, and let the pressures in the mass reservoirs at β be $p_\beta{}^{(1)}$ and $p_\beta{}^{(2)}$.

For given T_α, T_β, set the pressures $p_\omega{}^{(i)}$ $\left(\omega = \alpha,\beta;\ i = 1,2\right)$ to values such

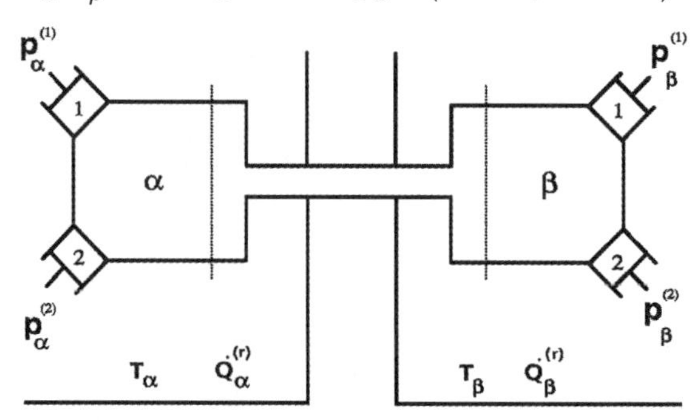

Fig. 2. Gaseous thermal diffusion. Pure components 1 and 2 are in mass reservoirs in contact with the gaseous mixture through semipermeable membranes at α and β.

that in the steady state the gas is in a condition of migrational equilibrium. See Fig. 2.

To establish the conditions of migrational equilibrium for the components of the gaseous mixture in the thermal field, carry out a differentially tailored series of steps involving one of the mass reservoirs, the net result of the series of steps being to move a small amount of the relevant component into or out of the mass reservoir, holding T_α, T_β, and the pressures $p_\omega{}^{(i)}$ for the other mass reservoirs constant. At the end of the differentially tailored process, the system is again in a steady state of migrational equilibrium described by the variables $T_\alpha, T_\beta, p_\omega{}^{(i)}$; and each mass reservoir may have experienced a change $dn_\omega{}^{(i)}$ of the amount of substance in that reservoir.

Upon applying the first law to the differentially tailored process, we have

$$dU = \sum_{\omega,i} \overline{U}_\omega{}^{(i)} dn_\omega{}^{(i)} = dQ_\alpha + dQ_\beta -$$

$$\sum_{\omega,i} p_\omega{}^{(i)} \overline{V}_\omega{}^{(i)} dn_\omega{}^{(i)} \tag{18}$$

or

$$\sum_{\omega,i} \overline{H}_\omega{}^{(i)} dn_\omega{}^{(i)} = dQ_\alpha + dQ_\beta, \tag{19}$$

where dQ_α, dQ_β are the amounts of heat supplied by the thermostats for the process,

$$\overline{H}_\omega{}^{(i)} \equiv \overline{U}_\omega{}^{(i)} + p_\omega{}^{(i)} \overline{V}_\omega{}^{(i)},$$

and ω is summed over α, β, whereas i is summed over 1,2.

The expressions dQ_α and dQ_β are once again "undetermined quantities" to be evaluated later.

Mass-conservation conditions show us that

$$dn_\alpha{}^{(1)} + dn_\beta{}^{(1)} = 0, \; dn_\alpha{}^{(2)} + dn_\beta{}^{(2)} = 0. \tag{20}$$

Now let

$$dQ_\alpha \equiv \left(\overline{H}_\alpha{}^{(1)} - \left[H_\alpha{}^{(1)}\right]\right) dn_\alpha{}^{(1)} + \left(\overline{H}_\alpha{}^{(2)} - \left[H_\alpha{}^{(2)}\right]\right) dn_\alpha{}^{(2)} \tag{21}$$

and

$$dQ_\beta \equiv \left(\overline{H}_\beta{}^{(1)} - \left[H_\beta{}^{(1)}\right]\right) dn_\beta{}^{(1)} + \left(\overline{H}_\beta{}^{(2)} - \left[H_\beta{}^{(2)}\right]\right) dn_\beta{}^{(2)}. \tag{22}$$

If we introduce relations (21),(22) into Eq. (19), we get

$$\sum_{\omega,i} \left[H_\omega{}^{(i)}\right] dn_\omega{}^{(i)} = 0. \tag{23}$$

Continuing the analysis, we have

$$dS = \sum_{\omega,i} \overline{S}_\omega{}^{(i)} dn_\omega{}^{(i)}, \tag{24}$$

$$d_\uparrow S_{sur} = -T_\alpha^{-1} dQ_\alpha - T_\beta^{-1} dQ_\beta$$

$$= -T_\alpha^{-1} \sum_i \left(\overline{H}_\alpha^{(i)} - \left[H_\alpha^{(i)}\right]\right) dn_\alpha^{(i)} -$$

$$T_\beta^{-1}\sum_i\left(\overline{H}_\beta^{(i)}-\left[H_\beta^{(i)}\right]\right)dn_\beta^{(i)}, \tag{25}$$

$$d\Gamma_{\uparrow} = dS + d_{\uparrow}S_{sur}$$

$$= \sum_{\omega,i}\overline{S}_\omega^{(i)}dn_\omega^{(i)} - T_\alpha^{-1}\sum_i\left(\overline{H}_\alpha^{(i)}-\left[H_\alpha^{(i)}\right]\right)dn_\alpha^{(i)} -$$

$$T_\beta^{-1}\sum_i\left(\overline{H}_\beta^{(i)}-\left[H_\beta^{(i)}\right]\right)dn_\beta^{(i)}. \tag{26}$$

In terms of the Planck function $\overline{Y} \equiv -\overline{G}/T$, we can write Eq. (26) as

$$d\Gamma_{\uparrow} = \sum_{\omega,i}\left(\overline{Y}_\omega^{(i)} + T_\omega^{-1}\left[H_\omega^{(i)}\right]\right)dn_\omega^{(i)}. \tag{27}$$

Let us suppose that we carried through the differentially tailored process in such a way that $dn_\alpha^{(1)} \neq 0$. Upon dividing Eq. (27) through by $dn_\alpha^{(1)}$, introducing the notation

$$r^{(21)} \equiv dn_\alpha^{(2)} / dn_\alpha^{(1)}, \tag{28}$$

making use of relation (20), and applying the Thomson–Gibbs Corollary, we have

$$\overline{Y}_\alpha^{(1)} + r^{(21)}\overline{Y}_\alpha^{(2)} - \left(\overline{Y}_\beta^{(1)} + r^{(21)}\overline{Y}_\beta^{(2)}\right) +$$

$$T_\alpha^{-1}\left(\left[H_\alpha^{(1)}\right] + r^{(21)}\left[H_\alpha^{(2)}\right]\right) - T_\beta^{-1}\left(\left[H_\beta^{(1)}\right] + r^{(21)}\left[H_\beta^{(2)}\right]\right) = 0. \tag{29}$$

We also note that Eq. (23) gives

$$\left[H_\alpha^{(1)}\right] + r^{(21)}\left[H_\alpha^{(2)}\right] - \left\{\left[H_\beta^{(1)}\right] + r^{(21)}\left[H_\beta^{(2)}\right]\right\} = 0. \tag{30}$$

If we introduce the notation

$$Z^{(1)} + r^{(21)}Z^{(2)} \equiv Z^{\dagger}, \tag{31}$$

then we may write Eqs. (29) and (30) as

$$\overline{Y}_\alpha^{\dagger} + T_\alpha^{-1}\left[H_\alpha^{\dagger}\right] = \overline{Y}_\beta^{\dagger} + T_\beta^{-1}\left[H_\beta^{\dagger}\right] \tag{32}$$

and

$$\left[H_\alpha^{\dagger}\right] = \left[H_\beta^{\dagger}\right] \equiv \left[H^{\dagger}\right]. \tag{33}$$

Compare relations (32),(33) to the corresponding relations in the treatment of the thermomolecular pressure effect, Eqs. (8.26),(8.19).

Now let's evaluate the "undetermined quantities" dQ_α, dQ_β used in the preceding analysis. In a separate series of operations, induce a (slow) steady flow $\dot{n}_\alpha^{(1)}$ of component 1 between its mass reservoirs at α and β by varying $p_\alpha^{(1)}$, keeping T_α, T_β, and the pressures $p_\omega^{(i)}$ for the other mass reservoirs constant (constant T, p', for short); and evaluate the derivatives

$$\delta\dot{Q}_\alpha^{(r)} / \tilde{\delta}\dot{n}_\alpha^{(1)}\Big)_{T,p'}, \; \delta\dot{Q}_\beta^{(r)} / \tilde{\delta}\dot{n}_\alpha^{(1)}\Big)_{T,p'}$$

via a sequence of steady states such that $\dot{n}_\alpha^{(1)} \to 0$ [see Eq. (1.14)]. Make the identifications

$$dQ_\alpha / dn_\alpha^{(1)} = -\delta\dot{Q}_\alpha^{(r)} / \tilde{\delta}\dot{n}_\alpha^{(1)}\Big)_{T,p'} \text{ and}$$

$$dQ_\beta / dn_\alpha^{(1)} = -\delta\dot{Q}_\beta^{(r)} / \tilde{\delta}\dot{n}_\alpha^{(1)}\Big)_{T,p'} \tag{34}$$

or

$$dQ_\alpha = -\delta\dot{Q}_\alpha^{(r)} / \tilde{\delta}\dot{n}_\alpha^{(1)}\Big)_{T,p'} dn_\alpha^{(1)} \text{ and}$$

$$dQ_\beta = -\delta\dot{Q}_\beta^{(r)} / \tilde{\delta}\dot{n}_\alpha^{(1)}\Big)_{T,p'} dn_\alpha^{(1)} \tag{35}$$

in Eqs. (18)–(33) where appropriate.

In the steady state of migrational equilibrium, the material in the gaseous mixture at α is in a state of migrational balance across each semipermeable membrane with respect to the component residing in that mass reservoir, i.e.

$$\overline{Y}_\alpha^{(i)}(mixture) = \overline{Y}_\alpha^{(i)}(reservoir); \tag{36}$$

and a similar state of affairs pertains to the gaseous mixture at β:

$$\overline{Y}_\beta^{(i)}(mixture) = \overline{Y}_\beta^{(i)}(reservoir). \tag{37}$$

It will be convenient for us to treat the functions $\overline{Y}_\omega^{(i)}$ from the point of view of the gaseous mixture at ω, and to write

$$\overline{Y}_\omega^{(i)} = \overline{Y}_\omega^{(i)}\left(T_\omega, p_\omega, X_\omega^{(1)}\right) \tag{38}$$

and

$$d\overline{Y}_\omega^{(i)} = \frac{\overline{H}_\omega^{(i)}}{T_\omega^2} dT_\omega - \frac{\overline{V}_\omega^{(i)}}{T_\omega} dp_\omega + \frac{\partial \overline{Y}_\omega^{(i)}}{\partial X_\omega^{(1)}}\Bigg)_{T_\omega, p_\omega} dX_\omega^{(1)}, \tag{39}$$

where p_ω is the total pressure in the gaseous mixture at

$\omega\left(p_\omega \approx p_\omega^{(1)} + p_\omega^{(2)}\right)$, $\overline{H}_\omega^{(i)}$ and $\overline{V}_\omega^{(i)}$ are partial molar properties, and $X_\omega^{(1)}$ is the mole fraction of component 1 in a sample of the gaseous mixture at ω.

Now let us differentiate Eq. (32) with respect to T_α, keeping $T_\beta, p_\beta^{(1)}, p_\beta^{(2)}$ constant—for short, call this differentiation under conditions of constant $\beta\left[\partial / \partial T_\alpha)_\beta\right]$:

$$\frac{\overline{H}_\alpha^\dagger - \left[H^\dagger\right]}{T_\alpha^2} + \left(\frac{1}{T_\alpha} - \frac{1}{T_\beta}\right)\frac{\partial\left[H^\dagger\right]}{\partial T_\alpha}\Bigg)_\beta +$$

$$\left(\overline{Y}_\alpha^{(2)} - \overline{Y}_\beta^{(2)}\right)\frac{\partial r^{(21)}}{\partial T_\alpha}\Bigg)_\beta - \frac{\overline{V}_\alpha^\dagger}{T_\alpha}\frac{\partial p_\alpha}{\partial T_\alpha}\Bigg)_\beta +$$

$$\left[\frac{\partial \overline{Y}_\alpha^{(1)}}{\partial X_\alpha^{(1)}}\right)_{T_\alpha, p_\alpha} + r^{(21)}\frac{\partial \overline{Y}_\alpha^{(2)}}{\partial X_\alpha^{(1)}}\Bigg)_{T_\alpha, p_\alpha}\Bigg]\frac{\partial X_\alpha^{(1)}}{\partial T_\alpha}\Bigg)_\beta = 0. \tag{40}$$

Consider the equithermal state for which $T_\alpha = T_\beta$. In that state, $\overline{Y}_\alpha^{(2)} = \overline{Y}_\beta^{(2)}$. The quantity $\partial p_\alpha / \partial T_\alpha)_\beta$ is essentially a thermomolecular pressure effect; we can make it vanish by making the pressure p_α large enough or by making the diameter of the tubular linkage connecting the two chambers large enough. Let us assume that in the present case $\partial p_\alpha / \partial T_\alpha)_\beta \approx 0$. Define a molar heat of transport Q^* by

$$Q^* \equiv \left[H^\dagger\right] - \overline{H}_\alpha^\dagger. \tag{41}$$

If we bring the several considerations just discussed to bear on Eq. (40), we get

$$\frac{\partial X_\alpha^{(1)}}{\partial T_\alpha}\Bigg)_\beta\Bigg|_{T_\alpha = T_\beta} = \frac{Q^*}{T_\alpha^2\left[\frac{\partial \overline{Y}_\alpha^{(1)}}{\partial X_\alpha^{(1)}}\right)_{T_\alpha, p_\alpha} + r^{(21)}\frac{\partial \overline{Y}_\alpha^{(2)}}{\partial X_\alpha^{(1)}}\Bigg)_{T_\alpha, p_\alpha}\Bigg]}. \tag{42}$$

If we use the ideal gas relation

$$\mu_\omega^{(i)}(mixture) = \mu_\omega^{(i)o}\left(T_\omega, p_\omega\right) + RT_\omega \ln X_\omega^{(i)}, \tag{43}$$

then Eq. (42) can be put into the form

$$-\frac{1}{X_\alpha^{(2)}}\frac{\partial \ln X_\alpha^{(1)}}{\partial \ln T_\alpha}\Bigg)_\beta\Bigg|_{T_\alpha = T_\beta} = \frac{Q^*}{RT_\alpha\left(X_\alpha^{(2)} - r^{(21)}X_\alpha^{(1)}\right)}. \tag{44}$$

The thermal diffusion factor **a** is defined to be

$$\mathbf{a} \equiv -\left(1 / X_\alpha^{(2)}\right)\left(\partial \ln X_\alpha^{(1)} / \partial \ln T_\alpha\right)_\beta . \tag{45}$$

Under equithermal conditions $\left(T_\alpha = T_\beta\right)$, the factor **a** is thus equal to the right-hand side of Eq. (44). See the discussion of thermal diffusion in Book II (the two treatments will match if the opposite sign convention is used for Q^* and if $r^{(21)}$ is set equal to zero).

Concentration-Field Analogue of the Thermal Diffusion Effect in Gases

The steady flow of heat through a 2–component gaseous mixture can cause some separation of the components—the thermal diffusion effect, which we discussed above. In the concentration-field analogue of the thermal diffusion effect, the steady flow of a third component through a 2–component gaseous mixture can cause some separation in the two stationary components. Let us carry out an analysis of the concentration-field analogue.

Consider the situation indicated schematically in Fig. 3. A spatial region marked MIX contains two gaseous components, 1 and 2; the MIX part of the apparatus is of length λ. The MIX region communicates with terminal parts at L and R via membranes permeable to gaseous component 3 alone, with terminal parts 1 at positions 0 and λ via membranes permeable to component 1 alone, and with terminal parts 2 at positions 0 and λ via membranes permeable to component 2 alone. The apparatus is made of diathermic material and rests in a thermostat (heat reservoir) of temperature T.

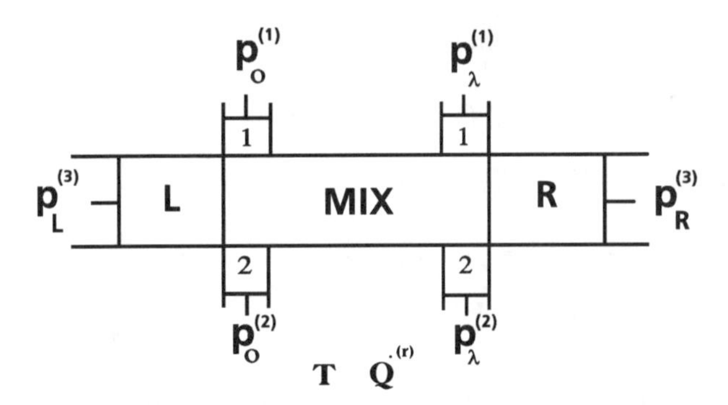

Fig. 3. Concentration field induced by the steady flow of component 3 through a 2–component MIX region.

Let us induce a steady flow of component 3 through the MIX region and seek the conditions for the migrational equilibrium of components 1 and 2 in the concentration field set up by the flow of component 3.

Start with the system in a steady state in which there is a steady flow of component 3 from the terminal part at L through the MIX region and into the terminal part at R $\left(\dot{n}_L^{(3)} + \dot{n}_R^{(3)} = 0\right)$ and in which the pressures

$$p_0^{(1)}, p_0^{(2)}, p_\lambda^{(1)}, p_\lambda^{(2)}$$

are such that components 1 and 2 are in states of migrational equilibrium.

Let $\overline{K}_L^{(3)}, \overline{K}_R^{(3)}$ be the molar macroscopic kinetic energy associated with component 3 in terminal parts L,R.

To establish the conditions of migrational equilibrium for components 1 and 2 in the concentration field, carry out a differentially tailored series of steps involving one of the mass reservoirs containing component 1 or component 2, the net result of the series of steps being to move a small amount of the relevant component into or out of the mass reservoir, holding the temperature and all the other mass-reservoir pressures constant. At the end of the differentially tailored process, the system is again in the original steady state, except for the fact that each mass reservoir may have experienced a differential change in its amount of substance. Mass-conservation considerations require that

$$dn_L^{(3)} + dn_R^{(3)} = 0,\ dn_0^{(1)} + dn_\lambda^{(1)} = 0,\ dn_0^{(2)} + dn_\lambda^{(2)} = 0. \tag{46}$$

In this case, $dn_L^{(3)}, dn_R^{(3)}$ are "undetermined quantities" to be evaluated later.

I shall distinguish between the internal energy U and the total energy "U" of a system, where

$$\text{"}U\text{"} \equiv U + \textit{macroscopic kinetic energy} + \ldots. \tag{47}$$

With this notational convention, the first law takes the form

$$\Delta\text{"}U\text{"} = Q + W; \tag{48}$$

and we have the convenient notational compactions:

$$\text{"}H\text{"} \equiv H + \textit{macroscopic kinetic energy} + \ldots, \tag{49}$$

$$\text{"}A\text{"} \equiv A + \textit{macroscopic kinetic energy} + \ldots, \tag{50}$$

$$\text{"}G\text{"} \equiv G + \textit{macroscopic kinetic energy} + \ldots. \tag{51}$$

The first law for the differentially tailored process gives

$$\sum\nolimits_{\omega,i} \text{"}\overline{U}_\omega^{(i)}\text{"} dn_\omega^{(i)} = dQ - \sum\nolimits_{\omega,i} p_\omega^{(i)} \overline{V}_\omega^{(i)} dn_\omega^{(i)}, \tag{52}$$

where "$\overline{U}_\omega^{(i)}$" $\equiv \overline{U}_\omega^{(i)} + \overline{K}_\omega^{(i)}$ and the indexes ω / i range over L,R/3 and $0, \lambda / 1,2$. Then

$$\sum_{\omega,i} \text{“}\overline{H}_{\omega}^{(i)}\text{”} dn_{\omega}^{(i)} = dQ. \tag{53}$$

For the entropy changes attributable to the process, we have

$$dS = \sum_{\omega,i} \overline{S}_{\omega}^{(i)} dn_{\omega}^{(i)}, \tag{54}$$

$$d_{\uparrow} S_{sur} = -T^{-1} dQ = -T^{-1} \sum \text{“}\overline{H}_{\omega}^{(i)}\text{”} dn_{\omega}^{(i)}, \tag{55}$$

$$d\Gamma_{\uparrow} = dS + d_{\uparrow} S_{sur}$$

$$= -T^{-1} \sum \text{“}\overline{G}_{\omega}^{(i)}\text{”} dn_{\omega}^{(i)}. \tag{56}$$

Let us suppose that we carried through the differentially tailored process in such a way that $dn_0^{(1)} \neq 0$. Upon dividing Eq. (56) through by $dn_0^{(1)}$, introducing the notation

$$r^{(21)} \equiv dn_0^{(2)} / dn_0^{(1)}, \; r^{(31)} \equiv dn_L^{(3)} / dn_0^{(1)}, \tag{57}$$

making use of Eq. (46), remembering that only component 3 has any macroscopic kinetic energy for the state in question, and applying the Thomson–Gibbs Corollary, we have

$$\left[\mu_L^{(3)} + \overline{K}_L^{(3)} - \left(\mu_R^{(3)} + \overline{K}_R^{(3)}\right)\right] r^{(31)} + \mu_0^{(1)} - \mu_\lambda^{(1)} +$$

$$\left(\mu_0^{(2)} - \mu_\lambda^{(2)}\right) r^{(21)} = 0. \tag{58}$$

Now let's evaluate the "undetermined quantities" $dn_L^{(3)}, dn_R^{(3)}$ used in the above analysis. In a separate series of operations, induce a (slow) steady flow $\dot{n}_0^{(1)}$ of component 1 between its terminal parts at 0 and λ by varying $p_0^{(1)}$, keeping T and the pressures for the other terminal parts constant (constant T, p', for short); and evaluate the derivatives

$$\left.\delta \dot{n}_L^{(3)} / \tilde{\delta} \dot{n}_0^{(1)}\right)_{T,p'}, \; \left.\delta \dot{n}_R^{(3)} / \tilde{\delta} \dot{n}_0^{(1)}\right)_{T,p'}$$

via a sequence of steady states such that $\dot{n}_0^{(1)} \to 0$ [see Eq. (1.14)]. Make the identifications

$$dn_L^{(3)} / dn_0^{(1)} = \left.\delta \dot{n}_L^{(3)} / \tilde{\delta} \dot{n}_0^{(1)}\right)_{T,p'} \text{ and}$$

$$dn_R^{(3)} / dn_0^{(1)} = \delta \dot{n}_R^{(3)} / \tilde{\delta} \dot{n}_0^{(1)} \Big)_{T,p'} \tag{59}$$

or

$$dn_L^{(3)} = \delta \dot{n}_L^{(3)} / \tilde{\delta} \dot{n}_0^{(1)} \Big)_{T,p'} dn_0^{(1)} \text{ and}$$

$$dn_R^{(3)} = \delta \dot{n}_R^{(3)} / \tilde{\delta} \dot{n}_0^{(1)} \Big)_{T,p'} dn_0^{(1)} \tag{60}$$

in Eqs. (52)–(58) where appropriate.

In the steady state, let the total pressure inside the MIX region at the 0-end of that region be $p_0(M)$. In states of migrational equilibrium for components 1 and 2, there is migrational balance across each membrane separating pure component 1 or 2 from the MIX region. We may, therefore, express the chemical potentials of components 1 and 2 in terms of the properties of the mixture of gases in the MIX region at the appropriate end (0-end or λ -end) of that region. At the 0-end of the MIX region, let

$$\mu_0^{(i)} = \mu_0^{(i)o}\left[T, p_0(M)\right] + RT \ln X_0^{(i)} \quad i = 1,2, \tag{61}$$

where $X_0^{(i)}$ is the mole fraction of component *i* in a sample from the MIX region at the 0-end.

Now let's differentiate Eq. (58) with respect to $p_L^{(3)}$, keeping $T, p_\lambda^{(1)}, p_\lambda^{(2)}, p_R^{(3)}$ constant—at constant T, λ, R, for short; and let's set

$$X_0^{(i)} = x_0^{(i)}\left(1 - X_0^{(3)}\right) \ (i = 1,2),$$

where $x_0^{(i)}$ is the *restricted* mole fraction of component *i* in the MIX region at the 0-end, i.e.

$$x_0^{(i)} \equiv n_0^{(i)} / \left(n_0^{(1)} + n_0^{(2)}\right) \ (i = 1,2)$$

for a sample of the gaseous mixture at the 0-end. Then,

$$r^{(31)}\left[\overline{V}_L^{(3)} + \frac{\partial\left(\overline{K}_L^{(3)} - \overline{K}_R^{(3)}\right)}{\partial p_L^{(3)}}\Bigg)_{T,\lambda,R}\right] +$$

$$\frac{\partial r^{(31)}}{\partial p_L^{(3)}}\Bigg)_{T,\lambda,R}\left[\mu_L^{(3)} + \overline{K}_L^{(3)} - \left(\mu_R^{(3)} + \overline{K}_R^{(3)}\right)\right] +$$

$$\frac{\partial r^{(21)}}{\partial p_L^{(3)}}\Bigg)_{T,\lambda,R}\left(\mu_0^{(2)} - \mu_\lambda^{(2)}\right) +$$

$$\left(\overline{V}_0^{(1)o} + r^{(21)}\overline{V}_0^{(2)o}\right)\frac{\partial p_0(M)}{\partial p_L^{(3)}}\Bigg)_{T,\lambda,R} +$$

$$RT\frac{\partial \ln x_0^{(1)}}{\partial p_L^{(3)}}\Bigg)_{T,\lambda,R} + r^{(21)}RT\frac{\partial \ln\left(1 - x_0^{(1)}\right)}{\partial p_L^{(3)}}\Bigg)_{T,\lambda,R} +$$

$$\left(1 + r^{(21)}\right)RT\frac{\partial \ln\left(1 - X_0^{(3)}\right)}{\partial p_L^{(3)}}\Bigg)_{T,\lambda,R} = 0. \tag{62}$$

For the equipiestic state $\left(p_L^{(3)} = p_R^{(3)}\right)$, we have

$$\mu_L^{(3)} = \mu_R^{(3)}, \ \ \overline{K}_L^{(3)} = \overline{K}_R^{(3)} = 0, \ \ \mu_0^{(2)} = \mu_\lambda^{(2)}. \tag{63}$$

In an equipiestic state, therefore, Eq. (62) reduces to

$$-RT\left[\left(1 - \frac{r^{(21)}x_0^{(1)}}{1 - x_0^{(1)}}\right)\frac{\partial \ln x_0^{(1)}}{\partial p_L^{(3)}}\Bigg)_{T,\lambda,R} +\right.$$

$$\left.\left(1 + r^{(21)}\right)\frac{\partial \ln\left(1 - X_0^{(3)}\right)}{\partial p_L^{(3)}}\Bigg)_{T,\lambda,R}\right]_{p_L^{(3)} = p_R^{(3)}}$$

$$= r^{(31)}\left[\overline{V}_L^{(3)} + \frac{\partial\left(\overline{K}_L^{(3)} - \overline{K}_R^{(3)}\right)}{\partial p_L^{(3)}}\Bigg)_{T,\lambda,R}\right]_{p_L^{(3)} = p_R^{(3)}} +$$

$$\left(\overline{V}_0^{(1)o} + r^{(21)}\overline{V}_0^{(2)o}\right)\frac{\partial p_0(M)}{\partial p_L^{(3)}}\Bigg)_{T,\lambda,R}\Bigg|_{p_L^{(3)} = p_R^{(3)}}. \tag{64}$$

Equation (64) shows how the separation of the stationary components in the MIX region depends on various kinds of behavior of the properties of the system.

The Thermocouple

Two different kinds of metallic wire, *a* and *b*, threading through a series of heat reservoirs (thermostats), are arranged as in Fig. 4. Let a steady state exist in the wire-thermostats layout. In that steady state, in the condition of migrational

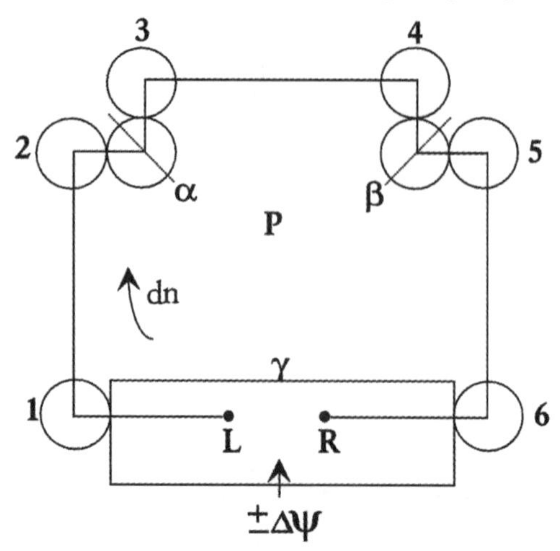

Fig. 4. The thermocouple. $1, \gamma, 6$—heat reservoirs of temperature T_γ; $2, \alpha, 3$—heat reservoirs of temperature T_α; $4, \beta, 5$—heat reservoirs of temperature T_β. Wire segments $L-1-2-\alpha$ and $\beta-5-6-R$ are of material b; wire segment $\alpha-3-4-\beta$ is of material *a*. Junctions between materials *a* and *b* occur in reservoirs α and β. There is an electric potential difference $\pm\Delta\Psi$ between the left (L) and right (R) electrodes of material *b* in reservoir γ. The ambient pressure is p. The parts of the wire segments outside of heat reservoirs are adiabatically insulated against lateral heat losses.

equilibrium for the electrons in the wire segments, there will be an electric potential difference $\pm\Delta\Psi$ (the Seebeck potential difference) between the electrodes L and R of material *b* in reservoir γ when $T_\alpha \neq T_\beta$. To obtain the conditions of migrational equilibrium for the electrons in the thermal field and the Thomson relations for the thermocouple [4], let us move an amount *dn* of electrons in a differentially tailored fashion from electrode L through the circuit to electrode R; and let us evaluate the thermodynamic consequences of that movement.

Use an overbar to represent a partial molar property of electrons in a given metal medium. For a particular property Z (internal energy, entropy, etc.), write the difference in $\overline{Z}$ for the electrons in the electrodes of material *b* in the reservoir γ as

$$\Delta_\gamma \overline{Z} \equiv \overline{Z}_R(b) - \overline{Z}_L(b) \; . \tag{65}$$

It is customary [5] to assume that a partial molar property of the electrons in a given metal medium is independent of the electric state of the medium; so, for the case in hand, we assume that $\Delta_\gamma \overline{Z} \neq f(\pm\Delta\Psi)$ and that, in fact,

$$\Delta_\gamma \overline{Z} \equiv \overline{Z}_R(b) - \overline{Z}_L(b) = 0 \; . \tag{66}$$

Now, for the process of moving *dn* moles of electrons around the circuit from electrode L to electrode R in a differentially tailored fashion, write the first law as

$$dn\Delta_\gamma \overline{U} = \sum_{i=1}^{6} dQ_i + dQ_\alpha + dQ_\beta + dQ_\gamma - pdn\Delta_\gamma \overline{V} + F(\pm\Delta\Psi)dn, \tag{67}$$

where F is the Faraday and the heat quantities $dQ_i, dQ_\alpha, dQ_\beta, dQ_\gamma$ are, for the moment, "undetermined quantities" to be evaluated later.

Choose the sign for $\pm\Delta\Psi$ so that, given the direction of the electronic current, the expression for the electric work, $F(\pm\Delta\Psi)dn$, will have the proper sign.

Upon dividing Eq. (67) through by dn and rearranging terms, we get

$$\Delta_\gamma \overline{H} = \sum_i dQ_i / dn + dQ_\alpha / dn + dQ_\beta / dn + F(\pm\Delta\Psi), \tag{68}$$

where $\overline{H} \equiv \overline{U} + p\overline{V}$ and $dQ_\gamma / dn = 0$ [the only heat effect influencing reservoir γ is the Joule heat produced by the electronic current, but that heat vanishes in the limit of zero current]. If we next observe that the molar Thomson coefficients $[C]$ for the electrons [4] are migrational molar heat capacities and that the molar Peltier heat absorbed at a junction, say between metals x and y, as the electronic current flows from metal x to metal y is Π^{xy} $(\Pi^{xy} = -\Pi^{yx})$, we have

$$dQ_1 / dn + dQ_2 / dn = \int_{T_\gamma}^{T_\alpha} [C] dT, \; dQ_\alpha / dn = \Pi_\alpha^{ba}, \text{ etc.} \tag{69}$$

The overall result is that Eq. (68) takes the form

$$0 = \int_{T_\alpha}^{T_\beta} \{[C_a] - [C_b]\} dT + \Pi_\alpha^{ba} + \Pi_\beta^{ab} + F(\pm\Delta\Psi), \tag{70}$$

where I have made use of assumption (66). Upon differentiating Eq. (70) with respect to T_α, at constant T_β, constant p [note that (70) is independent of T_γ], we get

$$-\{[C_a]_\alpha - [C_b]_\alpha\} + \Pi_\alpha^{ba} / T_\alpha + T_\alpha \partial(\Pi_\alpha^{ba} / T_\alpha) / \partial T_\alpha)_{p,T_\beta} + F\partial(\pm\Delta\Psi) / \partial T_\alpha)_{p.T_\beta} = 0, \tag{71}$$

where for convenience I have set $\Pi_\alpha^{ba} \equiv T_\alpha (\Pi_\alpha^{ba} / T_\alpha)$.

Now let's calculate the change in entropy for our wires:

$$dS = dn\Delta_\gamma \overline{S} \,. \tag{72}$$

The entropy changes in the surroundings directly attributable to our process are:

$$d_\uparrow S_{sur} = -\left[\sum_i T_i^{-1} dQ_i + T_\alpha^{-1} dQ_\alpha + T_\beta^{-1} dQ_\beta + T_\gamma^{-1} dQ_\gamma\right]. \tag{73}$$

So,

$$d\Gamma_{\uparrow} = dS + d_{\uparrow}S_{sur} \tag{74}$$

and

$$d\Gamma_{\uparrow} / dn = \Delta_{\gamma}\bar{S} - \left[\sum_i T_i^{-1}dQ_i / dn + T_{\alpha}^{-1}dQ_{\alpha} / dn + T_{\beta}^{-1}dQ_{\beta} / dn\right], \tag{75}$$

where again $dQ_{\gamma} / dn = 0$.

Upon invoking the Thomson-Gibbs Corollary and assumption (66), we get

$$\int_{T_\alpha}^{T_\beta}\left\{[C_a]-[C_b]\right\}d\ln T + \Pi_{\alpha}^{ba} / T_{\alpha} + \Pi_{\beta}^{ab} / T_{\beta} = 0 \ , \tag{76}$$

where I have again introduced the Thomson coefficients and the Peltier heats:

$$\frac{1}{T_\gamma}\frac{dQ_1}{dn} + \frac{1}{T_\alpha}\frac{dQ_2}{dn} = \int_{T_\gamma}^{T_\alpha}[C_b]d\ln T,$$
$$\frac{1}{T_\alpha}\frac{dQ_\alpha}{dn} = \frac{\Pi_{\alpha}^{ba}}{T_\alpha}, \quad etc. \tag{77}$$

If we apply the operation $\partial / \partial T_{\alpha}\big)_{p,T_\beta}$ to Eq. (76), we get

$$-\left\{[C_a]_\alpha - [C_b]_\alpha\right\}T_\alpha^{-1} + \partial\left(\Pi_{\alpha}^{ba} / T_{\alpha}\right) / \partial T_{\alpha}\Big)_{p,T_\beta} = 0 \ . \tag{78}$$

From Eqs. (71),(78), we get the Thomson relations for the thermocouple:

$$F\frac{\partial(\mp\Delta\Psi)}{\partial T_\alpha}\Bigg)_{p,T_\beta} = \frac{\Pi_{\alpha}^{ba}}{T_\alpha}, \tag{79}$$

$$FT_\alpha\frac{\partial^2(\mp\Delta\Psi)}{\partial T_\alpha^2}\Bigg)_{p,T_\beta} = [C_a]_\alpha - [C_b]_\alpha . \tag{80}$$

With Eqs. (79), (80) we complete our discussion of the conditions of migrational equilibrium for the thermocouple—see the discussions of Thomson [4], Eastman [6,7], Wagner [8], and Tykodi [9].

To evaluate the "undetermined quantities" $dQ_{..}$, use an electric engine to generate a sequence of steady-flow states; and make the identifications $dQ_{..} / dn = -\delta\dot{Q}_{..}^{(r)} / \delta\tilde{n}\Big)_{p,T_\alpha,\ldots}$ [see Ch. 9 of Book II].

And so on

The procedure for finding conditions of migrational equilibrium that I have just illustrated for the cases above can be used to get the conditions of migrational equilibrium for a wide variety of nonequilibrium effects—see the catalog of cases described in Reference 9.

Note that the procedure making use of the Thomson-Gibbs Corollary gives the conditions of migrational equilibrium in integral form and is thus a step up from the differential forms that were found historically one by one, in isolated fashion, for the members of the family of nonequilibrium "effects" (thermomolecular pressure effect, thermoelectric effect, Soret effect, etc.). It is an easy matter to derive the differential form from the integral form, whereas the reverse procedure would be difficult—to the best of my knowledge, such a reverse procedure has never been carried through.

As an illustration of the matters discussed above, consider the equilibrium vapor pressure for a 1-component substance. Let the substance have vapor pressure p at temperature T. The slope of the liquid-gas coexistence line—the vapor pressure curve—is given by the Clapeyron equation:

$$dp / dT = \left[\overline{H}(gas) - \overline{H}(liq)\right] / T\left[\overline{V}(gas) - \overline{V}(liq)\right]. \tag{81}$$

Equation (81) follows readily from the integral balance condition

$$\mu(gas) = \mu(liq). \tag{82}$$

Consider two points, an infinitesimal distance apart, on the vapor pressure curve. For the two points, we have

$$\mu(gas) = \mu(liq) \text{ and}$$

$$\mu(gas) + d\mu(gas) = \mu(liq) + d\mu(liq). \tag{83}$$

Equation (81) follows directly from Eqs. (83).

B. P. E. Clapeyron found the equivalent of Eq. (81) in 1832, long before Gibbs introduced the chemical potential μ into the thermodynamic literature (1876). Clapeyron used an argument based upon an infinitesimal Carnot cycle to deduce his result [10-12]. In the years between 1832 and 1876, no one dreamed of recovering from relation (81) the relations (83).

REFERENCES

1. Book II, Ch. 1.
2. K. G. Denbigh and G. Raumann, Proc. Roy. Soc. (London) **A210,** 518. (1952).
3. R. J. Bearman, J. Phys. Chem. **61,** 708 (1957).
4. W. Thomson, *Mathematical and Physical Papers,* Vol. I (Cambridge Univ. Press, Cambridge, 1882) pp. 232–291.
5. E. A. Guggenheim, *Thermodynamics*, 8th ed. (Elsevier, New York, 1986) Ch. 8.
6. E. Eastman, J. Am. Chem. Soc. **48,**1482 (1926).
7. E. Eastman, J. Am. Chem. Soc. **49,** 794 (1927) .
8. C. Wagner, Ann. Phys. **3,** 629 (1929).
9. Book II, Ch. 9.
10. P. S. Epstein, *Textbook of Thermodynamics* (Wiley, New York, 1937) Sect. 44.
11. M. N. Saha and B. N. Srivastava, *A Treatise on Heat*, 4th ed. (Indian Press, Allahabad, 1958) Sect. 8.14.
12. M. Bailyn, *A Survey of Thermodynamics* (AIP, New York, 1994) Sect. 3.18.

ELEVEN

RADIAL TEMPERATURE AND CONCENTRATION FIELDS

I consider the case of an ideal gas in a thermal or concentration field that has a gradient in the radial direction.

The System

Consider two right circular cylinders, one of radius r_1 and the other of radius r_3 $(r_1 < r_3)$. Let the cylinders be coaxial. Mark off a coordinate x along a line parallel to the common axis of the cylinders. Truncate the larger cylinder $(r = r_3)$ by planes perpendicular to the common axis at positions $x = x_a$ and $x = x_b$. Similarly, truncate the smaller cylinder ($r = r_1$) by planes at $x = x_1$ and $x = x_2$. The surfaces of the two cylinders together with the planes at x_a, x_b define an annular space with a volume V equal to (see Fig. 1)

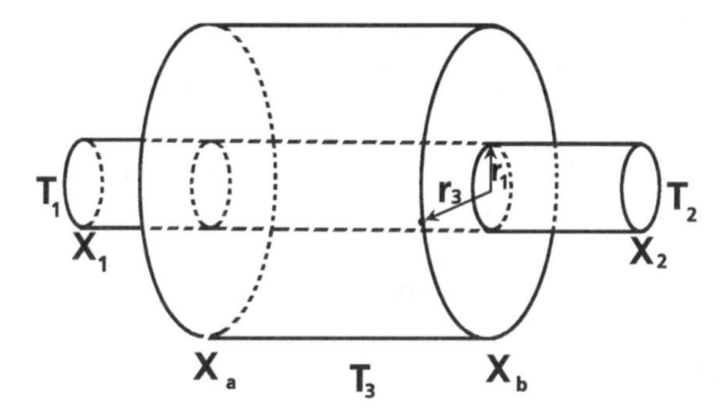

Fig. 1. Two coaxial right circular cylinders, of radii r_1 and r_3 $(r_1 < r_3)$, used to define an annular space that can be the site of a radial thermal or concentration field.

$$V = \pi\left(r_3^2 - r_1^2\right)\left(x_b - x_a\right) . \tag{1}$$

Let the annular-ring surfaces at $x = x_a$ and $x = x_b$ be of adiabatically insulating material; let the surface of the inner cylinder from x_1 to x_a and from x_b to x_2 be of adiabatically insulating material; and let all other surfaces be of diathermic material. Let the inner cylinder be closed by membranes of diathermic material at $x = x_1$ and $x = x_2$.

By means of suitable constant-temperature heat reservoirs, maintain the outer surface of the larger cylinder at temperature T_3, maintain the cross sectional area of the smaller cylinder at $x = x_1$ at temperature T_1, and maintain the cross sectional area at $x = x_2$ at temperature T_2.

Fill the inner cylinder with an arbitrary amount of an ideal gas. Fill the annular space with an amount n of a monatomic ideal gas. Let the composite system relax to a steady state in which there is a uniform pressure p throughout the annular space. For convenience, use the notation

$$\Delta r \equiv r_3 - r_1, \quad \langle r \rangle \equiv \tfrac{1}{2}(r_1 + r_3), \quad \Delta x \equiv x_b - x_a. \tag{2}$$

Radial Temperature Field

Let $T_2 = T_1 \neq T_3$, and let $\Delta T \equiv T_3 - T_1$. Let the temperature T_r (in the annular space) on the cylindrical surface defined by the radius $r\ (r_1 \le r \le r_3)$ be such that

$$T_r = T_1 + \left[(r - r_1) / \Delta r\right]\Delta T \ . \tag{3}$$

Then

$$dT_r = (\Delta T / \Delta r)dr, \quad dr = (\Delta r / \Delta T)dT_r \ . \tag{4}$$

The volume dV_r inside the annular space enclosed between the cylindrical surfaces defined by r and $r + dr$ is

$$dV_r = (\Delta x)2\pi r dr \ . \tag{5}$$

The amount dn_r of the monatomic ideal gas contained in the volume dV_r in the steady state is

$$dn_r = p dV_r / RT_r \ . \tag{6}$$

We note from Eq. (3) that

$$r = r_1 - T_1(\Delta r / \Delta T) + T_r(\Delta r / \Delta T) \ . \tag{7}$$

To determine n, we have

$$n = \int_{r=r_1}^{r=r_3} dn_r = \int_{r=r_1}^{r=r_3} p dV_r / RT_r$$

$$= \frac{pV}{R\theta}\frac{\Delta r}{\langle r \rangle}\left[\frac{\theta}{\Delta T} - \frac{T_1}{\Delta T} + \frac{r_1}{\Delta r}\right], \tag{8}$$

where

$$\theta \equiv \left(T_3 - T_1\right) / \ln\left(T_3 / T_1\right) \tag{9}$$

is the logarithmic mean of the temperatures T_1, T_3 (see Section 2).

In the steady state, the gas in the annular space satisfies the equation of state

$$pV = nR\Phi \ , \tag{10}$$

with

$$\Phi \equiv \theta \frac{\langle r \rangle}{\Delta r} \frac{\Delta T}{\theta - T_1 + \left(r_1 / \Delta r\right)\Delta T} \ . \tag{11}$$

The effective temperature Φ of the gas depends on thermal parameters $\left(T_1, T_3\right)$ and on geometric parameters $\left(r_1, \Delta r, \langle r \rangle\right)$.

Let us now evaluate the thermodynamic properties of the monatomic ideal gas in the steady-state thermal field with constant gradient in the radial direction.

$$\overline{U}_r = \overline{C}_v T_r \ , \tag{12}$$

$$U = \int_{r=r_1}^{r=r_3} \overline{U}_r dn_r = \int_{r=r_1}^{r=r_3} \overline{C}_v T_r dn_r = n\overline{C}_v \Phi \ , \tag{13}$$

$$\overline{S}_r = \overline{S}_1\left(T_1, p\right) + \overline{C}_p \ln\left(T_r / T_1\right) \ , \tag{14}$$

$$S = \int_{r=r_1}^{r=r_3} \overline{S}_r dn_r$$

$$= n\left(\overline{S}_1 - \overline{C}_p \ln T_1\right) + \frac{n\overline{C}_p \Phi}{\Delta T \langle r \rangle}\left[r_1 - T_1(\Delta r / \Delta T)\right] \ln\left(T_1 T_3\right)^{1/2} \ln\left(T_3 / T_1\right) +$$

$$\frac{n\overline{C}_p \Phi}{\Delta T \langle r \rangle} \frac{\Delta r}{\Delta T}\left(T_3 \ln T_3 - T_1 \ln T_1 - \Delta T\right) . \tag{15}$$

Adiabatic Relaxation

Instantaneously adiabatically insulate all the surfaces of the annular space, and let the gas inside the annular space relax to a state of equilibrium. For the change from the initial steady state to the final equilibrium state

$$\Delta U = 0 \ ; \tag{16}$$

so

$$U_f = U_i = n\overline{C}_v\Phi \ . \tag{17}$$

The temperature in the equilibrium state is thus Φ, and the pressure in the equilibrium state is p.

$$S_f = n\left[\overline{S}_1 + \overline{C}_p \ln\left(\Phi / T_1\right)\right] = n\left(\overline{S}_1 - \overline{C}_p \ln T_1\right) + n\overline{C}_p \ln \Phi \ , \tag{18}$$

and

$$\Gamma = \Delta S$$

$$= n\overline{C}_p\left\{\ln\Phi - \frac{\Phi}{\Delta T\langle r\rangle}\left[\left(r_1 - T_1\frac{\Delta r}{\Delta T}\right)\ln\left(T_1T_3\right)^{1/2}\ln\left(T_3 / T_1\right) + \right.\right.$$

$$\left.\left.\frac{\Delta r}{\Delta T}\left(T_3\ln T_3 - T_1\ln T_1 - \Delta T\right)\right]\right\} \geq 0 \ . \tag{19}$$

Let us consider a special case:

$$T_3 = 2T_1, \ \ r_3 = 3r_1, \ \ \langle r\rangle = 2r_1, \ \ \Delta T = T_1 \ . \tag{20}$$

Then

$$\theta = T_1 / \ln 2, \ \ \Phi = T_1 / \left(1 - \tfrac{1}{2}\ln 2\right) = 1.5304T_1 \ . \tag{21}$$

Let us specialize further to:

$$r_1 = 1, \ \ T_1 = 100K \ ; \tag{22}$$

then

$$\Gamma = \Delta S = 0.017946n\overline{C}_p > 0 \ . \tag{23}$$

Differentially Tailored Expansion

Let the x_b surface (adiabatically insulating) be movable; let the surroundings exert a pressure p_{ex} on that surface (extend the $r = r_3$ surface to accommodate the motion of the x_b surface). Let the exposed parts of the $r = r_1$ surface be adiabatically insulated. As the x_b surface moves, let the insulation on the $r = r_1$ surface move with it so that the exposed part is insulated but the interior part is not. The $r = r_1$ surface communicates with heat reservoirs of temperature T_1 at x_1 and x_2 $\left(T_2 = T_1\right)$. The $r = r_3$ surface communicates with a heat reservoir of temperature T_3.

For a change dV in the annular volume, the first law gives

$$dU = dQ - p_{ex}dV \ , \tag{24}$$

where $dQ = dQ_1 + dQ_2 + dQ_3$.

Let $p_{ex} \approx p = nR\Phi / V$, and let $dV = \sigma dx$, where σ is the area of the annular-ring surface. Then

$$dW = -pdV = -nR\Phi(dV / V) \ . \tag{25}$$

For a differentially tailored expansion at constant Φ, we have

$$\Delta U = 0 \ , \tag{26}$$

$$W = -nR\Phi \int_{V_i}^{V_f} dV / V = -nR\Phi \ln\left(V_f / V_i\right) \ , \tag{27}$$

$$Q = -W = nR\Phi \ln\left(V_f / V_i\right) \ , \tag{28}$$

$$\Delta S = -nR \ln\left(p_f / p_i\right) = nR \ln\left(V_f / V_i\right) = Q / \Phi \ . \tag{29}$$

As in Section 2, we could devise a Carnot-like cycle using the gas in nonequilibrium steady states as working fluid.

Two-Dimensional Thermal Field

Along the $r = r_1$ surface, between x_a and x_b, let the temperature T_{1x} be such that

$$T_{1x} = T_a + \frac{x - x_a}{\Delta x}\left(T_b - T_a\right) \quad \left(\mathrm{x_a} \le x \le x_b\right) \ . \tag{30}$$

At point x in the interval $\mathrm{x_a} \le x \le x_b$, let the temperature T_{rx} $\left(r_1 \le r \le r_3\right)$ in the radial direction be such that

$$T_{rx} = T_{1x} + \frac{r - r_1}{\Delta r}\left(T_3 - T_{1x}\right) \ . \tag{31}$$

The resulting thermal field has a gradient in the axial direction and a gradient in the radial direction.

The element of volume in the annular space, dV_{rx}, is now

$$dV_{rx} = 2\pi r dr dx \ . \tag{32}$$

Let there be an amount n of a monatomic ideal gas in the annular space. Seal off the inner cylinder at x_1, x_2 with membranes of diathermic material, and let the inner cylinder contain an arbitrary amount of an ideal gas. Let the annular-ring surfaces at x_a, x_b be adiabatically insulated.

Let the composite system settle into a steady state, with a uniform pressure p throughout the annular space.

In the steady state, the amount of gas dn_{rx} in the element of volume dV_{rx} is

$$dn_{rx} = pdV_{rx} / RT_{rx} . \tag{33}$$

The total amount n of gas in the annular space is, therefore,

$$n = \iint dn_{rx} = \iint pdV_{rx} / RT_{rx} . \tag{34}$$

The internal energy and the entropy of the gas in the annular space are

$$U = \iint \overline{U}_{rx} dn_{rx} , \tag{35}$$

$$S = \iint \overline{S}_{rx} dn_{rx} . \tag{36}$$

Catalogue of Processes

We can, if we wish, carry out a series of processes—each process starting in the steady state just described—and verify that $\Gamma_{\uparrow} \geq 0$ for each process. For example:

1) Instantaneously, adiabatically insulate all of the surfaces that bound the annular space; and let the gas in the annular space relax adiabatically to an equilibrium state.

2) Instantaneously, adiabatically insulate all exposed surfaces of the composite system (see Fig. 1); and let the gas in the annular space and the gas in the inner cylinder relax adiabatically to an equilibrium state with a common temperature.

3) Let there be a small opening in the wall of the inner cylinder at some point x between x_a and x_b, and let the opening be capable of being sealed shut by a sliding cover. Seal the opening. Fill the inner cylinder with an amount n_{in} of a monatomic ideal gas, and fill the annular space with an amount n_{an} of the same ideal gas. Establish axial and radial thermal fields as in Eqs. (30),(31). When the composite system reaches a steady state, let the pressure in the inner cylinder be p_{in} and the pressure in the annular space be p_{an}. Move the cover so as to open the hole between the inner cylinder and the annular space. We expect the condition of migrational equilibrium for flow of the gas between the two chambers to be

$$p_{in} = p_{an} , \tag{37}$$

and we can apply the Thomson-Gibbs Corollary to see if such is indeed the case. [Assume that the hole is large enough and that the pressures p_{in}, p_{an} are large enough so that there is no thermomolecular pressure effect—see Section 8].

4) Let the situation be the same as in 3), but let the gas in the inner cylinder be a different kind of monatomic ideal gas from that in the annular space. Select the

amounts n_{in} and n_{an} so that in the steady state $p_{in} = p_{an}$. Then open the connecting hole between the two chambers and let the two gases mix. We can verify that $\Gamma_{\uparrow} \geq 0$ for the mixing process.

5) Etc.

Concentration-Field Analogues

Some of the thermal-field processes that we have just discussed have concentration-field analogues. Let the entire apparatus be made of diathermic material. Fill the annular space with a rubber substance in which a given monatomic ideal gas is slightly soluble. Fill the inner cylinder with the gas in question, and by means of a movable piston (at x_2, say) maintain the pressure of the gas at p_1. Surround the apparatus with a box (made of diathermic material) in which the gas in question is maintained at pressure p_3 (by another movable piston), and place the box in a constant-temperature heat reservoir of temperature T. Let the $r = r_1$ and the $r = r_3$ surfaces of the annular space be of porous metal. The gas in question passes from the inner cylinder into the rubber substance, diffuses to the $r = r_3$ surface, and then passes into the surrounding box (for the case $p_1 > p_3$). After a steady state is established, we will have a radial concentration field inside the rubber substance.

We can then suddenly seal the porous-metal surfaces to make them impermeable to the gas, allowing the gas dissolved in the rubber substance to relax to an equilibrium state of uniform concentration. We can then verify that $\Gamma_{\uparrow} > 0$ for the concentration depolarization process inside the rubber substance. This depolarization process is the analogue of the thermal depolarization process in example 1) of the preceding section. And so on.

A Migrational-Equilibrium Problem

Take the apparatus shown in Fig. 1 and attach to the ports at x_1, x_2 constant-pressure mass reservoirs containing a given monatomic ideal gas (i.e. attach piston-and-cylinder devices with the internal radius r_{res} of each attached cylinder such that $r_{res} >> r_1$, and fill the attached devices with the gas in question). Let the mass reservoir at x_1 have pressure p_1, and let the reservoir at x_2 have pressure p_2. Let the wall of the $r = r_1$ cylinder between x_a and x_b be permeable to the gas [palladium is permeable to hydrogen, and soft glass is much more permeable to helium than is borosilicate glass, for example]. Let the annular space also contain the gas in question.

Place the entire apparatus, mass reservoirs and all, in a constant-temperature heat reservoir of temperature T; and let everything settle into a steady state. In that steady state the amount n_3 of gas in the annular space will not be changing, nor will the amount n_{ab} of gas in the $r = r_1$ cylinder between x_a and x_b be changing:

$$\dot{n}_3 = 0, \quad \dot{n}_{ab} = 0 ; \tag{38}$$

the amounts of gas in the mass reservoirs, however, will be changing at rates such that

$$\dot{n}_1 + \dot{n}_2 = 0 \; . \tag{39}$$

For the given apparatus, with given p_1, p_2, T, we desire to establish the properties of the gas in the annular space in the steady state (which is a state of migrational equilibrium for the gas in the annular space). Assume that $p_1 > p_2$ so that the direction of flow is from side 1 to side 2. Assume also that the solubility of the gas in the wall material of the $r = r_1$ cylinder between x_a and x_b follows Henry's law.

In the steady state, then, there is a gradient in the pressure of the flowing gas between x_a and x_b; consequently there will be a net inflow of gas to the annular space across the permeable wall close to x_a and a net outflow of gas close to x_b— the inflow and outflow balancing so as to satisfy Eq. (38). The gas inside the annular space thus shows a pattern of circulation, indicating non-uniformity of pressure inside the annular space. Let the mean pressure of the gas in the annular space be $\langle p_3 \rangle$, and let the mean value of the chemical potential of the gas in that space be $\langle \mu_3 \rangle$.

In the region between x_a and x_b in the inner cylinder, because of the pressure gradient, we shall deal with mean values of the thermodynamic properties, indicated by $\langle \ldots \rangle$. In particular, let $\langle \overline{K}_{ab} \rangle$ be the mean molar macroscopic kinetic energy of the flowing gas. [The macroscopic kinetic energy of the gas in the mass reservoirs and in the annular space may be neglected.]

Start in a steady state characterized by $T, p_1, p_2, \langle p_3 \rangle$, and instantaneously replace the mass reservoir at x_1 by a new mass reservoir of constant pressure $p_1 + dp_1$. Let the composite system settle into a new steady state characterized by $T, p_1 + dp_1, p_2, \langle p_3 \rangle + d\langle p_3 \rangle$. First let's carry through a straightforward analysis, and then let's subject the analysis to critical review. For the changes attributable to the process of settling into the new steady state, we have

$$\begin{aligned} dU = \overline{U}_1 dn_1 + \overline{U}_2 dn_2 + \left(\langle \overline{U}_{ab} \rangle + \langle \overline{K}_{ab} \rangle \right) dn_{ab} + \langle \overline{U}_3 \rangle dn_3 + \\ n_{ab} d\left(\langle \overline{U}_{ab} \rangle + \langle \overline{K}_{ab} \rangle \right) + n_3 d\langle \overline{U}_3 \rangle \\ = dQ - \left(p_1 + dp_1 \right) \overline{V}_1 dn_1 - p_2 \overline{V}_2 dn_2 \; . \end{aligned} \tag{40}$$

It follows then that

$$\begin{aligned} dQ = \overline{H}_1 dn_1 + \overline{H}_2 dn_2 + \left(\langle \overline{U}_{ab} \rangle + \langle \overline{K}_{ab} \rangle \right) dn_{ab} + \langle \overline{U}_3 \rangle dn_3 + \\ n_{ab} d\left(\langle \overline{U}_{ab} \rangle + \langle \overline{K}_{ab} \rangle \right) + n_3 d\langle \overline{U}_3 \rangle \; . \end{aligned} \tag{41}$$

Continuing our catalogue, we have

$$dS = \overline{S}_1 dn_1 + \overline{S}_2 dn_2 + \langle \overline{S}_{ab} \rangle dn_{ab} + \langle \overline{S}_3 \rangle dn_3 +$$

$$n_{ab} d\langle \overline{S}_{ab} \rangle + n_3 d\langle \overline{S}_3 \rangle \; , \tag{42}$$

$$d_{\uparrow} S_{sur} = -dQ / T \; , \tag{43}$$

$$\begin{aligned} d\Gamma_{\uparrow} &= dS + d_{\uparrow} S_{sur} \\ &= \overline{S}_1 dn_1 + \overline{S}_2 dn_2 + \langle \overline{S}_{ab} \rangle dn_{ab} + \langle \overline{S}_3 \rangle dn_3 + \\ &\quad n_{ab} d\langle \overline{S}_{ab} \rangle + n_3 d\langle \overline{S}_3 \rangle - \\ &\quad T^{-1} [\overline{H}_1 dn_1 + \overline{H}_2 dn_2 + \left(\langle \overline{U}_{ab} \rangle + \langle \overline{K}_{ab} \rangle \right) dn_{ab} + \langle \overline{U}_3 \rangle dn_3 + \\ &\quad n_{ab} d\left(\langle \overline{U}_{ab} \rangle + \langle \overline{K}_{ab} \rangle \right) + n_3 d\langle \overline{U}_3 \rangle] \\ &= -T^{-1} [\mu_1 dn_1 + \mu_2 dn_2 + \left(\langle \overline{A}_{ab} \rangle + \langle \overline{K}_{ab} \rangle \right) dn_{ab} + \langle \overline{A}_3 \rangle dn_3 + \\ &\quad n_{ab} d_T \left(\langle \overline{A}_{ab} \rangle + \langle \overline{K}_{ab} \rangle \right) + n_3 d_T \langle \overline{A}_3 \rangle] \; . \end{aligned} \tag{44}$$

[Although all the differentials refer to a condition of constant temperature T, the restriction is noted explicitly (d_T) for the differential of the Helmholtz function: $d_T \overline{A} = d_T(\overline{U} - T\overline{S}) = d\overline{U} - Td\overline{S}$.]

Proceeding with the analysis, we get

$$\begin{aligned} \frac{d\Gamma_{\uparrow}}{dn_1} &= -T^{-1} [\mu_1 + \mu_2 \frac{dn_2}{dn_1} + \left(\langle \overline{A}_{ab} \rangle + \langle \overline{K}_{ab} \rangle \right) \frac{dn_{ab}}{dn_1} + \langle \overline{A}_3 \rangle \frac{dn_3}{dn_1} + \\ &\quad n_{ab} \frac{d_T \langle \overline{A}_{ab} \rangle}{dn_{ab}} \frac{dn_{ab}}{dn_1} + n_3 \frac{d_T \langle \overline{A}_3 \rangle}{dn_3} \frac{dn_3}{dn_1} + n_{ab} \frac{d\langle \overline{K}_{ab} \rangle}{dn_{ab}} \frac{dn_{ab}}{dn_1}] \; . \end{aligned} \tag{45}$$

Let us pause here for a moment to establish a useful relation. For an amount of substance n for a pure species in a fixed volume V at temperature T and pressure p, we observe that

$$d\overline{A} = -\overline{S} dT - p d\overline{V} \; , \tag{46}$$

$$\left. d\overline{A} \right)_T = -p d\overline{V}, \quad n \left. \frac{d\overline{A}}{dn} \right)_T = -pn \frac{d\overline{V}}{dn} \; , \tag{47}$$

$$V = n\overline{V}, \quad 0 = dV = n d\overline{V} + \overline{V} dn \; . \tag{48}$$

So

$$-n\frac{d\overline{V}}{dn} = \overline{V} \tag{49}$$

and

$$n\frac{d\overline{A}}{dn}\bigg)_T = p\overline{V} \ . \tag{50}$$

Applying relation (50) to the terms in Eq. (45), we get

$$\frac{d\Gamma_\uparrow}{dn_1} = -T^{-1}[\mu_1 + \mu_2\frac{dn_2}{dn_1} + \left(\langle\mu_{ab}\rangle + \langle\overline{K}_{ab}\rangle\right)\frac{dn_{ab}}{dn_1} + \langle\mu_3\rangle\frac{dn_3}{dn_1} + n_{ab}\frac{d\langle\overline{K}_{ab}\rangle}{dn_{ab}}\frac{dn_{ab}}{dn_1}]. \tag{51}$$

Invoking the Thomson-Gibbs Corollary gives

$$0 = \mu_1 + \mu_2\frac{dn_2}{dn_1} + \left(\langle\mu_{ab}\rangle + \langle\overline{K}_{ab}\rangle\right)\frac{dn_{ab}}{dn_1} + \langle\mu_3\rangle\frac{dn_3}{dn_1} + n_{ab}\frac{d\langle\overline{K}_{ab}\rangle}{dn_{ab}}\frac{dn_{ab}}{dn_1}. \tag{52}$$

The mass-balance condition requires that

$$dn_1 + dn_2 + dn_3 + dn_{ab} = 0 \tag{53}$$

and

$$1 + \frac{dn_2}{dn_1} + \frac{dn_3}{dn_1} + \frac{dn_{ab}}{dn_1} = 0 \ . \tag{54}$$

Before proceeding further, let us look at two special cases—but before doing that, let's establish a useful result.

Let the pressure in the mass reservoir at x_1 be $p + dp$, and let the pressure in the mass reservoir at x_2 be p. In the steady state, the rate of flow of gas through the $r = r_1$ cylinder from x_1 to x_2, if the flow can be treated as Poiseuille flow [1], is $\dot{n}_2$, where

$$\dot{n}_2 = \frac{\left(\pi r_1^2\right)^2\left[(p + dp)^2 - p^2\right]}{16\pi\left(x_2 - x_1\right)\zeta RT}$$

$$= \frac{\left(\pi r_1^2\right)^2 2pdp}{16\pi\left(x_2 - x_1\right)\zeta RT} \tag{55}$$

to the first order of differentials and ζ is the coefficient of viscosity. Also, let us approximate the amount n of gas between x_1 and x_2 in the steady state by

$$n = \frac{p\pi r_1^2\left(x_2 - x_1\right)}{RT} . \tag{56}$$

Equation (56) allows us to write Eq. (55) as

$$\dot{n}_2 = \frac{\pi r_1^2 ndp}{8\pi\left(x_2 - x_1\right)^2 \zeta} . \tag{57}$$

It follows, then, that

$$\frac{d\dot{n}_2}{dn} = \frac{\pi r_1^2 dp}{8\pi\left(x_2 - x_1\right)^2 \zeta} \tag{58}$$

and that

$$\frac{d\dot{n}_2}{dn} \to 0 \text{ as } dp \to 0 . \tag{59}$$

Now, the mean velocity of the gas as it passes through the $r = r_1$ cylinder from x_1 to x_2 is

$$\langle \text{velocity} \rangle = \dot{n}_2 \langle \overline{V}_{12} \rangle / \pi r_1^2 ; \tag{60}$$

and the mean molar macroscopic kinetic energy of the gas, $\langle \overline{K}_{12} \rangle$, as it passes through the cylinder is (M being the molecular weight of the gas)

$$\langle \overline{K}_{12} \rangle = \tfrac{1}{2} M \left(\dot{n}_2 \langle \overline{V}_{12} \rangle / \pi r_1^2\right)^2 . \tag{61}$$

So [Eq. (59)]

$$\frac{d\langle \overline{K}_{12} \rangle}{dn} = M \frac{\dot{n}_2 \langle \overline{V}_{12} \rangle}{\pi r_1^2} \frac{\langle \overline{V}_{12} \rangle}{\pi r_1^2} \frac{d\dot{n}_2}{dn} \to 0 \text{ as } dp \to 0 . \tag{62}$$

With Eq. (62) in hand, we are ready for our two special cases. This would also be an apt time for that critical review promised above—so let's do that first.

Critical Review

In the change in pressure $p_1 \to p_1 + dp_1$, let $dp_1 > 0$; then the present situation is formally analogous to the first case of water flow into a standpipe, Section 9 (Fig. 9.2). The increase in the pressure on side 1 will result in an increase in the rate of outflow from mass reservoir 1. We expect, therefore, that the mass of gas in the annular space will increase and so will the mass of gas in the inner cylinder between x_a and x_b. At all times during the transition from the initial to the final steady state, it will be the case that

$$-\dot{n}_1(p_1 + dp_1) \geq \dot{n}_2(p_2). \tag{63}$$

By precisely the argument used in Section 9, following after Eq.(9.20), we conclude that

$$-dn_1 = dn_{ab} + dn_3, \quad dn_2 = 0. \tag{64}$$

So Eqs. (52)-(54) become

$$0 = \mu_1 + \left(\langle \mu_{ab} \rangle + \langle \overline{K}_{ab} \rangle\right)\frac{dn_{ab}}{dn_1} + \langle \mu_3 \rangle \frac{dn_3}{dn_1} + n_{ab}\frac{d\langle \overline{K}_{ab} \rangle}{dn_{ab}}\frac{dn_{ab}}{dn_1}, \tag{65}$$

$$dn_1 + dn_3 + dn_{ab} = 0, \tag{66}$$

$$1 + \frac{dn_3}{dn_1} + \frac{dn_{ab}}{dn_1} = 0. \tag{67}$$

Let us now look at some special cases.

Special Cases

i) Let the $r = r_1$ surface be impermeable to the gas (i.e. $n_3 =$ constant and $dn_3 = 0$). The migrational equilibrium condition in this case is for the gas in reservoir 1 to be in migrational equilibrium with the gas in reservoir 2 across the connecting cylinder stretching from x_1 to x_2. In the state of migrational equilibrium the gas will be at rest, so $\langle \overline{K}_{12} \rangle = 0$. Equation (65), in this case, takes the form

$$0 = \mu_1 + \langle \mu_{ab} \rangle \frac{dn_{ab}}{dn_1} + n_{ab}\frac{d\langle \overline{K}_{ab} \rangle}{dn_{ab}}\frac{dn_{ab}}{dn_1}. \tag{68}$$

For the case in hand $\left(\langle \overline{K}_{ab} \rangle = 0\right)$, Eq. (62) yields

$$\frac{d\langle \overline{K}_{ab}\rangle}{dn_{ab}} = 0 \; ; \tag{69}$$

so Eq. (68) reduces to

$$0 = \mu_1 + \langle \mu_{ab}\rangle \frac{dn_{ab}}{dn_1} \; . \tag{70}$$

Equation (67) leads to

$$1 + \frac{dn_{ab}}{dn_1} = 0 \tag{71}$$

and [Eq. (70)]

$$0 = \mu_1 - \langle \mu_{ab}\rangle \; . \tag{72}$$

We expect to find that, in the state of migrational equilibrium,

$$\mu_1 = \mu_2 = \langle \mu_{ab}\rangle ; \tag{73}$$

and Eq. (72) is consistent with our expectation.

ii) Let the $r = r_1$ surface be permeable to the gas again between x_a and x_b. Let $p_1 = p_2 = \langle p_{ab}\rangle$, with the gas in the inner cylinder being at rest. Then $\mu_1 = \mu_2 = \langle \mu_{ab}\rangle \equiv \mu_{in}$, and Eq. (65) becomes

$$0 = \mu_{in}\left(1 + \frac{dn_{ab}}{dn_1}\right) + \langle \mu_3\rangle \frac{dn_3}{dn_1} + n_{ab}\frac{d\langle \overline{K}_{ab}\rangle}{dn_{ab}}\frac{dn_{ab}}{dn_1} \; . \tag{74}$$

Combining Eq. (67) with Eqs. (69),(74) gives

$$0 = \left(\langle \mu_3\rangle - \mu_{in}\right)\frac{dn_3}{dn_1} \tag{75}$$

or

$$\langle \mu_3\rangle = \mu_{in} \tag{76}$$

just as we expect.

General Case

Now on to the general case. The combination of Eq. (65) and Eq. (67) gives

$$0 = \mu_1 + \left(\langle \mu_{ab}\rangle + \langle \overline{K}_{ab}\rangle\right)\frac{dn_{ab}}{dn_1} -$$

$$\langle \mu_3\rangle \left(1 + \frac{dn_{ab}}{dn_1}\right) + n_{ab}\frac{d\langle \overline{K}_{ab}\rangle}{dn_{ab}}\frac{dn_{ab}}{dn_1} \tag{77}$$

and

$$\langle \mu_3 \rangle = f_1 \mu_1 + f_2 \left[\langle \mu_{ab} \rangle + \langle \overline{K}_{ab} \rangle + n_{ab} \left(d \langle \overline{K}_{ab} \rangle / dn_{ab} \right) \right], \tag{78}$$

where

$$f_1 = \frac{1}{1 + \dfrac{dn_{ab}}{dn_1}}, \; f_2 = \frac{dn_{ab} / dn_1}{1 + \dfrac{dn_{ab}}{dn_1}}, \quad f_1 + f_2 = 1 \,. \tag{79}$$

With Eq. (79), we come to the end of our analysis.

REFERENCE

1. K. F. Herzfeld and H. M. Smallwood, in H. S. Taylor and S. Glasstone, editors, 3rd ed., *A Treatise on Physical Chemistry, Vol. 2, States of Matter* (Van Nostrand, Princeton, 1951) p. 105.

TWELVE

THE PRINCIPLE OF MINIMUM ENTROPY PRODUCTION VS. THE THOMSON-GIBBS COROLLARY

I display a heat-conduction case for which the principle of minimum entropy production fails, but for which the Thomson-Gibbs Corollary holds.

The Principle of Minimum Entropy Production

In Section 8, I showed that the Thomson-Gibbs Corollary was a complete solution, in the Dartmouth domain, to the problem of migrational equilibrium in a nonequilibrium spatial field. Another approach to determining the conditions of migrational equilibrium in nonequilibrium steady states is via the principle of minimum entropy production (minimum rate of entropy generation).

The work of Onsager [1,2], Prigogine [3], and others [4,5] has shown that there is a domain of near-equilibrium steady states in which the rate of entropy generation $\dot{\Gamma}$ can be displayed as a sum of terms, each term being a product of a current (J_i) and an affinity (X_i) (a flux and a force) [5]:

$$\dot{\Gamma} = \sum_i J_i X_i \ . \tag{1}$$

In the near-equilibrium domain of steady states, the currents are linear functions of the affinities:

$$J_k = \sum_i L_{ki} X_i \ , \tag{2}$$

where

$$L_{ki} \equiv \left(\partial J_k \ / \ \partial X_i \right)_{X'} \ , \tag{3}$$

the subscript X' meaning constant X_j for $j \neq i$ and the right-hand side of (3) being evaluated for the state in which each X_i is equal to zero (the reference equilibrium state [5]).

The condition of migrational equilibrium for a given quantity in the near-equilibrium domain of steady states is that the current, say J_k, associated with the quantity vanish, i.e. $J_k = 0$. The vanishing of the current J_k establishes a relation among the affinities X_i and the coefficients L_{ki} characteristic of the state of migrational equilibrium.

A statement of the principle of minimum entropy production adequate for our purposes is [5]:

If a system that is the site of $\nu + 1$ *steady (independent) currents converges (along an appropriate path) via a sequence of steady states on a state involving* ν *steady currents, then the state with* ν *currents is a state of minimum rate of entropy generation relative to the (steady) states with* $\nu + 1$ *currents.*

In the near-equilibrium domain of steady states, it is observed that

$$\left.\frac{\partial \dot{\Gamma}}{\partial J_k}\right)_{X'} = 0 \text{ implies } J_k = 0 \text{ and vice versa.} \tag{4}$$

For simple systems not involving the motion of electric charges in magnetic fields, the L_{ik} coefficients satisfy a symmetry condition, the Onsager reciprocal relations [1,2], in the near-equilibrium domain of steady states:

$$L_{ik} = L_{ki} \ . \tag{5}$$

For a simple system in that domain, then,

$$\left.\frac{\partial \dot{\Gamma}}{\partial X_k}\right)_{X'} = J_k + \sum_i L_{ik} X_i = J_k + \sum_i L_{ki} X_i = 2J_k \tag{6}$$

and

$$\left.\frac{\partial \dot{\Gamma}}{\partial J_k}\right)_{X'} = \frac{\left.\partial \dot{\Gamma} / \partial X_k\right)_{X'}}{\left.\partial J_k / \partial X_k\right)_{X'}} = 2J_k L_{kk}^{-1} \ ; \tag{7}$$

so the situation is as stated in relation (4).

The principle of minimum entropy production, Eq. (4), thus supplies a test,

$$\left.\partial \dot{\Gamma} / \partial J_k\right)_{X'} = 0,$$

for states of migrational equilibrium $\left(J_k = 0\right)$ in the given domain of steady states—see Book II for numerous applications of this basic idea.

Outside the near-equilibrium domain of steady states, where the currents are non-linear functions of the affinities, $\left.\partial \dot{\Gamma} / \partial J_k\right)_{X'} = 0$

no longer necessarily implies that $J_k = 0$, and so is not a reliable test for the conditions of migrational equilibrium—i.e. the principle of minimum entropy production breaks down [6,7]. Attempts to generalize the principle of minimum entropy production (or to replace it by something else) are matters of ongoing research [8-13].

Let us now look at a heat-conduction example in which the principle of minimum entropy production fails but in which the Thomson-Gibbs Corollary continues to hold.

Heat-Conduction Example, Principle of Minimum Entropy Production Fails

Take two thermally conducting wire segments of identical length, cross sectional area, and other properties; and connect the segments to heat reservoirs of temperatures T_1, T_{23}, T_4 in the manner indicated in Fig. 1.

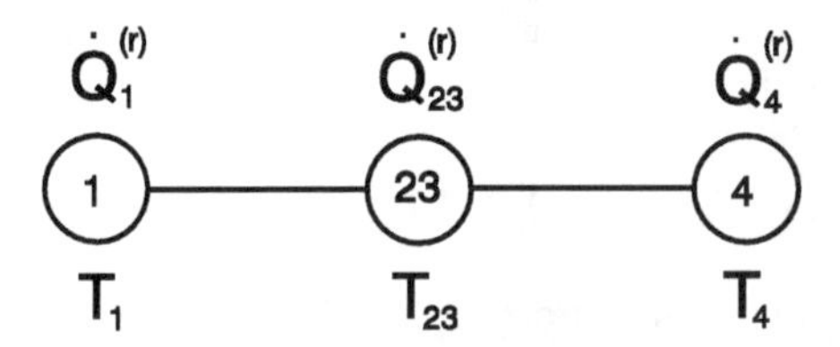

Fig. 1. Two identical wire segments serving as heat conductors between heat reservoirs 1 & 23 and 23 & 4. The segments are insulated against lateral heat losses.

Adiabatically insulate the side walls of the wire segments so that they exchange heat only with the heat reservoirs. Let the rate of influx of heat into a heat reservoir be $\dot{Q}^{(r)}$. Let the thermal conductivity of each segment be independent of temperature, and let the thermal conductivity, length, and cross sectional area of each be such that, in the steady state,

$$\dot{Q}_1^{(r)} = T_{23} - T_1, \quad \dot{Q}_{23}^{(r)} = T_1 + T_4 - 2T_{23}, \quad \dot{Q}_4^{(r)} = T_{23} - T_4 \,. \tag{8}$$

In the steady state, then, considering the wires as our system,

$$\dot{U} = -\left(\dot{Q}_1^{(r)} + \dot{Q}_{23}^{(r)} + \dot{Q}_4^{(r)}\right) = 0 \,, \tag{9}$$

$$\dot{\Gamma} = T_1^{-1}\dot{Q}_1^{(r)} + T_{23}^{-1}\dot{Q}_{23}^{(r)} + T_4^{-1}\dot{Q}_4^{(r)}$$

$$= \dot{Q}_1^{(r)}\left(T_1^{-1} - T_4^{-1}\right) + \dot{Q}_{23}^{(r)}\left(T_{23}^{-1} - T_4^{-1}\right) , \tag{10}$$

where Eq. (9) has been used in the second form of Eq. (10).

The principle of minimum entropy production would require that $\dot{\Gamma}$ be a minimum at $\dot{Q}_{23}^{(r)} = 0$ for conditions of constant T_1, T_4 (constant 1,4, for short); to see that such is not the case, it suffices to evaluate $\left.\partial\dot{\Gamma} / \partial T_{23}\right)_{1,4}$ at the point

$T_{23} = \frac{1}{2}(T_1 + T_4)$ [i.e. $\dot{Q}_{23}^{(r)} = 0$].

$$\left.\frac{\partial \dot{\Gamma}}{\partial T_{23}}\right)_{1,4} = T_1^{-1} + T_4^{-1} - 2T_{23}^{-1} \tag{11}$$

and

$$\left.\frac{\partial \dot{\Gamma}}{\partial T_{23}}\right)_{1,4} \left[\text{at } T_{23} = \tfrac{1}{2}(T_1 + T_4)\right] = \frac{(T_1 - T_4)^2}{T_1 T_4 (T_1 + T_4)} \quad . \tag{12}$$

The right-hand side of Eq. (12) is not zero for $T_1 \neq T_4$; so $\dot{\Gamma}$, at constant T_1, T_4, does not have a minimum at $\dot{Q}_{23}^{(r)} = 0$. In fact, the minimum of $\dot{\Gamma}$, for fixed T_1, T_4, occurs at the place

$$T_{23} = (T_1 T_4)^{1/2}, \text{ where } \dot{Q}_{23}^{(r)} \neq 0.$$

The principle of minimum entropy production thus fails for the case in hand.

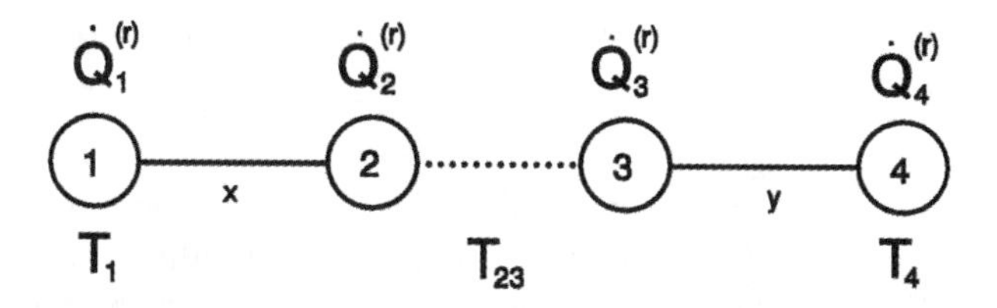

Fig. 2. Two identical wire segments serving as heat conductors between heat reservoirs 1 & 2 and 3 & 4; the segments are insulated against lateral heat losses. The dotted line indicates that heat reservoirs 2 & 3 are constrained to have at all times the same common temperature T_{23}.

Heat-Conduction Example, Thomson-Gibbs Corollary Holds

Take the same two wire segments as in the preceding section and connect them to heat reservoirs 1,2,3,4 as in Fig. 2. The dotted line between reservoirs 2 and 3 means that they are to be constrained to always have the same temperature—let that common temperature be called T_{23}. We are thus dealing with the same set of temperatures, T_1, T_{23}, T_4, as before. Let specific properties (i.e. properties per unit mass) of the wire material be represented by lower case letters: *u*, specific internal energy; *s*, specific entropy; etc. Let us impose a few additional restrictions on the properties of the wire segments, to make the necessary calculations more manageable.

Over the range of temperatures we shall be dealing with, let the density of the wire material ρ and the specific heat capacity *c* be constant. Let each wire segment be of length λ, cross sectional area σ, and mass m $(m = \rho\sigma\lambda)$. Lay off a coordinate x $(0 \le x \le \lambda)$ from reservoir 1 to reservoir 2, and a coordinate y $(0 \le y \le \lambda)$ from reservoir 3 to reservoir 4. Let there be a linear temperature distribution along each wire segment:

$$T_x = T_1 + \left(\tfrac{x}{\lambda}\right)\left(T_{23} - T_1\right), \quad T_y = T_{23} + \left(\tfrac{y}{\lambda}\right)\left(T_4 - T_{23}\right). \tag{13}$$

Then,

$$dT_x = \left[\left(T_{23} - T_1\right)/\lambda\right]dx, \quad dT_y = \left[\left(T_4 - T_{23}\right)/\lambda\right]dy, \tag{14}$$

$$dx = \left[\lambda/\left(T_{23} - T_1\right)\right]dT_x, \quad dy = \left[\lambda/\left(T_4 - T_{23}\right)\right]dT_y, \tag{15}$$

$$dm_x = \rho\sigma dx, \quad dm_y = \rho\sigma dy, \tag{16}$$

$$u_x = u_1 + c\left(T_x - T_1\right), \quad u_y = u_1 + c\left(T_y - T_1\right), \tag{17}$$

$$s_x = s_1 + c\ln\left(T_x/T_1\right), \quad s_y = s_1 + c\ln\left(T_y/T_1\right). \tag{18}$$

Start in the initial steady state with $\dot{Q}_2^{(r)} + \dot{Q}_3^{(r)} = 0$ $\left[\text{i.e. } T_{23} = \tfrac{1}{2}\left(T_1 + T_4\right)\right]$.

Then

$$U = \int_{x=0}^{x=\lambda} u_x dm_x + \int_{y=0}^{y=\lambda} u_y dm_y = \rho\sigma\left[\int_0^\lambda u_x dx + \int_0^\lambda u_y dy\right]$$

$$= m\left[\left(u_1 - cT_1\right) + \tfrac{1}{2}c\left(T_{23} + T_1\right)\right] + m\left[\left(u_1 - cT_1\right) + \tfrac{1}{2}c\left(T_4 + T_{23}\right)\right], \tag{19}$$

$$S = \int_{x=0}^{x=\lambda} s_x dm_x + \int_{y=0}^{y=\lambda} s_y dm_y = \rho\sigma\left[\int_0^\lambda s_x dx + \int_0^\lambda s_y dy\right]$$

$$= m\left[\left(s_1 - c\ln T_1\right) + c\left(\frac{T_{23}}{T_{23} - T_1}\ln T_{23} + \frac{-T_1}{T_{23} - T_1}\ln T_1 - 1\right)\right] +$$

$$m\left[\left(s_1 - c\ln T_1\right) + c\left(\frac{T_4}{T_4 - T_{23}}\ln T_4 + \frac{-T_{23}}{T_4 - T_{23}}\ln T_{23} - 1\right)\right]. \tag{20}$$

Now let

$$r_{123} \equiv T_{23}/\left(T_{23} - T_1\right), \quad r_1 \equiv -T_1/\left(T_{23} - T_1\right), \quad r_{123} + r_1 = 1, \tag{21}$$

$$r_{423} \equiv -T_{23} / \left(T_4 - T_{23}\right), \quad r_4 \equiv T_4 / \left(T_4 - T_{23}\right), \quad r_{423} + r_4 = 1. \tag{22}$$

Then,

$$\begin{aligned} S &= m\left[\left(s_1 - c\ln T_1\right) + c\ln T_{23}^{r_{123}} T_1^{r_1} - c\right] + \\ &\quad m\left[\left(s_1 - c\ln T_1\right) + c\ln T_4^{r_4} T_{23}^{r_{423}} - c\right] \\ &= 2m\left(s_1 - c\ln T_1\right) - 2mc + mc\ln T_1^{r_1} T_{23}^{r_{123}+r_{423}} T_4^{r_4} \ . \end{aligned} \tag{23}$$

Now instantaneously replace heat reservoirs 2 and 3 by new reservoirs of temperature T'_{23}, and let the system reach a new steady state. For the change from the initial steady state to the final steady state, we have

$$\begin{aligned} U_f &= m\left[\left(u_1 - cT_1\right) + \tfrac{1}{2}c\left(T'_{23} + T_1\right)\right] + \\ &\quad m\left[\left(u_1 - cT_1\right) + \tfrac{1}{2}c\left(T_4 + T'_{23}\right)\right] \end{aligned} \tag{24}$$

and U_i from Eq. (19); so

$$\begin{aligned} \Delta U &= \tfrac{1}{2}mc\left(T'_{23} - T_{23}\right) + \tfrac{1}{2}mc\left(T'_{23} - T_{23}\right) \\ &= Q_1 + Q_2 + \left(Q_3 + Q_4\right) \ . \end{aligned} \tag{25}$$

The total heat input $Q_1 + Q_2, Q_3 + Q_4$ to each of the wire segments is well defined:

$$\Delta U_x = Q_1 + Q_2, \ \ \Delta U_y = Q_3 + Q_4, \ \ \Delta U = \Delta U_x + \Delta U_y; \tag{26}$$

but the separate heat inputs Q_j $(j = 1, 2, 3, 4)$ are difficult to assess.

From Eq. (25) we see that

$$Q_1 + Q_2 = Q_3 + Q_4 = \tfrac{1}{2}mc\left(T'_{23} - T_{23}\right) \ . \tag{27}$$

Continuing the analysis, we have

$$S_f = 2m\left(s_1 - c\ln T_1\right) - 2mc + mc\ln T_1^{r'_1} T_{23}^{\prime\, r'_{123}+r'_{423}} T_4^{r'_4} \tag{28}$$

and S_i from Eq. (23); so

$$\Delta S = mc\ln T_1^{\Delta r_1} T_4^{\Delta r_4} T_{23}^{\prime\, r'_{123}+r'_{423}} T_{23}^{-\left(r_{123}+r_{423}\right)} \ . \tag{29}$$

Also,

$$\Delta_{\uparrow} S_{sur} = -T_1^{-1} Q_1 - T_{23}'^{-1}\left(Q_2 + Q_3\right) - T_4^{-1} Q_4 \ . \tag{30}$$

What we would like to do is to obtain a general expression for $\Gamma_{\uparrow} \equiv \Delta S + \Delta_{\uparrow} S_{sur}$ and show that

$$\left.\partial\Gamma_{\uparrow} / \partial T_{23}'\right)_{1,4} = 0 \text{ at } T_{23}' = \tfrac{1}{2}\left(T_1 + T_4\right).$$

It is difficult, however, to continue in general terms—the evaluation of the quantities Q_j $\left(j = 1,2,3,4\right)$ is something of a stumbling block. We can, however, gain insight into the problem by doing calculations on a special case.

Let $T_1 = 400K$, $T_{23} = 300K$, $T_4 = 200K$ be the initial steady state; then, in separate experiments instantaneously change

$$T_{23} \text{ to } T_{23}' = 310K,\ 301K,\ 299K,\ 290K$$

and let the system relax into the appropriate steady state. Determine $d\Gamma_{\uparrow} / \Delta T_{23}$ for each case.

For

$$T_{23}' = 310K, 301K, \text{let } Q_1 = \tfrac{1}{2} mc\left(T_{23}' - T_{23}\right),\ Q_2 = 0,$$
$$Q_3 = \tfrac{1}{2} mc\left(T_{23}' - T_{23}\right),\ Q_4 = 0.$$

For

$$T_{23}' = 299K, 290K, \text{let } Q_1 = 0,\ Q_2 = \tfrac{1}{2} mc\left(T_{23}' - T_{23}\right),$$
$$Q_3 = 0,\ Q_4 = \tfrac{1}{2} mc\left(T_{23}' - T_{23}\right).$$

We find, then, the pairings

T_{23}' / K	310	301	299	290
$\left(10^3\, K / mc\right) d\Gamma_{\uparrow} / \Delta T_{23}$	0.497	0.483	-0.770	-0.787

The pairings suggest that a sequence of positive numbers and a sequence of negative numbers both converge on the same state; so it must be the case that

$$\lim_{\Delta T_{23} \to 0-} d\Gamma_{\uparrow} / \Delta T_{23} = \lim_{\Delta T_{23} \to 0+} d\Gamma_{\uparrow} / \Delta T_{23} = 0 \ , \tag{31}$$

i.e. $d\Gamma_{\uparrow} / dT_{23}' = 0$ at $T_{23}' = \tfrac{1}{2}\left(T_1 + T_4\right)$, and the Thomson-Gibbs Corollary is satisfied.

Consider the case $T_{23}' = 310K$. It would be possible, in principle, for that segment of wire having initial temperatures T_x such that $310K \geq T_x \geq 300K$ to gain energy from reservoir 2 until the entire segment is at temperature $310K$ and then to gain additional energy from reservoir 1 to bring the segment into its final steady state configuration. In that case, Q_2 would equal

$\frac{1}{2}mc\left(T'_{23}-T_{23}\right)^2/\left(T_1-T_{23}\right)$; and a similar result would hold for the case $T'_{23}=301K$. In cases where $T'_{23}>T_{23}$, the fraction of the total heat input to the segment stretching between reservoirs 1 and 2 that could possibly come from reservoir 2 is $\left(T'_{23}-T_{23}\right)/\left(T_1-T_{23}\right)$; so as $T'_{23}-T_{23}\to 0$, the heat input behaves more and more like

$$Q_1=\tfrac{1}{2}mc\left(T'_{23}-T_{23}\right),\ Q_2=0.$$

There is an analogous consideration for cases where $T'_{23}<T_{23}$.

Regardless of how the energy exchanged between the appropriate reservoirs and the x and y segments of wire is apportioned among the reservoirs, as T'_{23} gets sufficiently close to T_{23}, the sequences of positive and negative values for $d\Gamma_{\uparrow}/\Delta T_{23}$ must satisfy relation (31).

Discussion

Nonequilibrium states of systems that satisfy the hypothesis of local equilibrium [3] evidently fall within the Dartmouth domain—such systems have well-defined fields of intensities, and the internal energy of the *equilibrium-based image* closely matches the internal energy of the system. The Thomson–Gibbs Corollary provides a complete solution of the problem of migrational equilibrium for systems having states that fall within the Dartmouth domain—and thus for systems satisfying the local equilibrium hypothesis. We saw above that there are situations involving nonequilibrium states of systems satisfying the local equilibrium hypothesis for which the principle of minimum entropy production fails. The Thomson–Gibbs Corollary is thus more general than the principle of minimum entropy production.

REFERENCES

1. L. Onsager, Phys. Rev. **37**, 405 (1931).
2. L. Onsager, Phys. Rev. **38**, 2265 (1931).
3. I. Prigogine, *Introduction to the Thermodynamics of Irreversible Processes*, 2nd ed. (Wiley, New York, 1961).
4. S. de Groot and P. Mazur, *Non-Equilibrium Thermodynamics* (North-Holland, Amsterdam, 1962).
5. Book II.
6. R. J. Tykodi, J. Chem. Phys. **57**, 37 (1972).
7. R. J. Tykodi, Physica, **72**, 341 (1974).
8. E. T. Jaynes, in R. D. Rosenkrantz, editor, *E. T. Jaynes: Papers on Probability, Statistics and Statistical Physics* (Reidel, Dordrect, 1983) pp. 401-424.
9. R. J. Donnelly, R. Herman, and I. Prigogine, editors, *Non-Equilibrium Thermodynamics, Variational Techniques, and Stability* (University of Chicago Press, Chicago, 1966).
10. C. Truesdell, *Rational Thermodynamics*, 2nd ed. (Springer, New York,

1984).
11. S. Sieniutycz and, P. Salamon, editors, *Nonequilibrium Theory and Extremum Principles* (Taylor & Francis, New York, 1990).
12. D. Jou, J. Casas-Vazquez, and G. Lebon, *Extended Irreversible Thermodynamics* (Springer, Berlin, 1993).
13. See the series of papers following-up on Ref. 5; references to all the papers in the series appear in the last member of the series: R. J. Tykodi, J. Non-Equilib. Thermodyn. **16**, 267 (1991). In the paper just cited, there are a few errors in subscripting practice: in Eqs. (20), interchange the subscripts i,k on the L and K coefficients; in Eqs. (21), interchange the subscripts i,k throughout.

THIRTEEN

BEYOND THE DARTMOUTH DOMAIN

For convenience, I recapitulate here the analysis of the thermodynamic properties of a monatomic ideal gas in a thermal field of constant gradient from Section 2. I then use the ideal gas model to discuss the problem of extending the thermodynamic formalism beyond the Dartmouth domain

Monatomic Ideal Gas in a Thermal Field of Constant Gradient, Dartmouth Domain

A model system that will prove extremely useful for discussing the topics dealt with in this section is that of a monatomic ideal gas in a thermal field of constant gradient. Let us first deal with the gas in the Dartmouth domain.

Let a right circular cylinder serve as a container with an interior volume V. Let the axis of the cylinder be of length λ, and let each of the circular end faces have inner area σ. Let the side wall of the cylinder be of rigid, adiabatically insulating material; and let the end faces be thin, rigid, and of diathermic material with negligible heat capacity.

Lay off a coordinate x along a line parallel to the axis of the cylinder; use a heat reservoir of temperature T_0 to maintain the face of the cylinder at $x = 0$ at the temperature T_0; likewise, maintain the face of the cylinder at $x = \lambda$ at the temperature T_λ. Fill the interior volume V with an amount n of an ideal monatomic gas, and allow the apparatus to reach a steady state. In the steady state, the pressure p is to be uniform throughout the cylinder (the mean free path between collisions for the gas particles is to be small compared to the diameter of the cylinder, i.e. there is to be no thermomolecular pressure effect [see Section 8]). Let the temperature on the cross sectional area at the coordinate value x be T_x, and assume that

$$T_x = T_0 + \left(\tfrac{x}{\lambda}\right)\left(T_\lambda - T_0\right) . \qquad (1)$$

See Fig 1.

Properties of the Ideal Gas Model

The molar heat capacities of the gas are $\overline{C}_v = (3/2)R$, $\overline{C}_p = (5/2)R$, where R is the universal gas constant; and we have the relations $\overline{C}_p - \overline{C}_v = R$, $\gamma \equiv \overline{C}_p / \overline{C}_v = 5/3$, and $pV = nRT$.

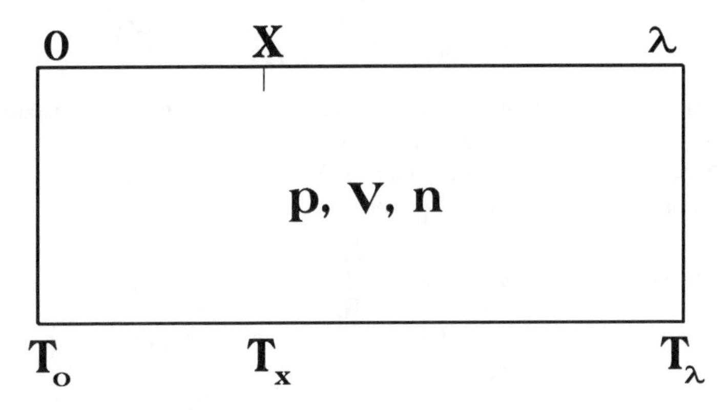

Fig. 1. Gas in a thermal field. An amount n of an ideal monatomic gas in a volume V at pressure p in a steady state; the steady-state thermal field is of constant gradient: $T_x = T_0 + \left(\frac{x}{\lambda}\right)\left(T_\lambda - T_0\right)$.

The System

Let us now carry out an analysis of the thermodynamic properties of the gas in the steady-state thermal field.

In the steady state, the amount of gas dn_x contained between planes at x and $x + dx$ in the cylinder, i.e. the amount of gas contained in the volume σdx at the indicated place, is

$$dn_x = p\sigma dx / RT_x \; ; \tag{2}$$

and the total amount of gas n in the cylinder is

$$n = \int_{x=0}^{x=\lambda} dn_x = \frac{p\sigma}{R}\int_0^\lambda \frac{dx}{T_x} \; . \tag{3}$$

It follows from Eq. (1) that

$$dx = \left[\lambda / \left(T_\lambda - T_0\right)\right] dT_x \; ; \tag{4}$$

so

$$n = \frac{p\sigma\lambda}{R\left(T_\lambda - T_0\right)}\int_{T_0}^{T_\lambda} \frac{dT_x}{T_x} = \frac{pV}{R\left(T_\lambda - T_0\right)} \ln\frac{T_\lambda}{T_0} \; , \tag{5}$$

since $\sigma\lambda = V$.

In the steady state, then, the gas follows the equation of state

$$pV = nR\frac{T_\lambda - T_0}{\ln\left(T_\lambda / T_0\right)} = nR\theta \; , \tag{6}$$

where

$$\theta \equiv \theta(T_\lambda, T_0) \equiv (T_\lambda - T_0) / \ln(T_\lambda / T_0) \tag{7}$$

is the logarithmic mean of the temperatures T_λ, T_0. Logarithmic means are of some use in discussions of heat transfer [1].

For two non-negative real numbers a, b $(0 \le a, b < \infty)$ the logarithmic mean of the numbers lies between the ingredient numbers:

$$\min(a,b) \le \theta(a,b) \le \max(a,b).$$

Let the order of placement of a,b in $\theta(a,b)$ be significant; then, from the definition [Eq. (7)], it follows that

$$\theta(a,b) = \theta(b,a) \; ; \tag{8}$$

so we are free to show the variables in either order. For non-negative real numbers, the logarithmic mean lies between the arithmetic and geometric means of those numbers:

$$\frac{a+b}{2} \ge \frac{a-b}{\ln(a/b)} \ge (ab)^{1/2} \; . \tag{9}$$

The particles of the gas have kinetic energy only, so the molar internal energy $\overline{U}$ of such particles at temperature T is

$$\overline{U} = (3/2)RT = \overline{C}_v T \; . \tag{10}$$

The internal energy U of the amount n of gas in the steady state with end temperatures T_0, T_λ is

$$\begin{aligned} U &= \int_{x=0}^{x=\lambda} \overline{U}_x dn_x = \int_{x=0}^{x=\lambda} \overline{C}_v T_x dn_x = \frac{\overline{C}_v p\sigma}{R} \int_0^\lambda dx \\ &= \frac{\overline{C}_v p\sigma\lambda}{R} = \frac{\overline{C}_v pV}{R} = n\overline{C}_v \theta \; . \end{aligned} \tag{11}$$

The molar entropy of the gas particles at temperature T_x, pressure p is

$$\overline{S}_x = \overline{S}_0 + \overline{C}_p \ln(T_x / T_0) \; , \tag{12}$$

where $\overline{S}_0 = \overline{S}_0(T_0, p)$—I show the variable-dependence of $\overline{S}_0$ explicitly only when needed. The total entropy S of the amount n of gas in the steady-state thermal field is, then,

$$S = \int_{x=0}^{x=\lambda} \overline{S}_x dn_x = \frac{p\sigma}{R} \int \left[\overline{S}_0 + \overline{C}_p \ln(T_x / T_0)\right](dx / T_x)$$

$$= \frac{pV}{R(T_\lambda - T_0)} \int_{T_0}^{T_\lambda} \left[\overline{S}_0 + \overline{C}_p \ln(T_x / T_0) \right] (dT_x / T_x)$$

$$= n\left[\overline{S}_0 + (\overline{C}_p / 2) \ln(T_\lambda / T_0) \right]. \qquad (13)$$

Thermally Depolarizing the System

Let us now take the system in the steady state with end temperatures T_0, T_λ and instantaneously slip a thin, rigid, adiabatically insulating membrane over each end of the cylinder so as to isolate the system from interactions with the surroundings. The system will relax, at constant U, V, n, to a state of equilibrium. Since the internal energy remains constant during the process, the internal energy at equilibrium will be [Eq. (11)]

$$U = n\overline{C}_v \theta \; ; \qquad (14)$$

and the uniform temperature at equilibrium will be $\theta(T_0, T_\lambda)$. Since n, V, θ are all the same as in the steady state, the equilibrium pressure will be p—the same as in the steady state.

For the equilibrium state, then,

$$S = n\left[\overline{S}_0 + \overline{C}_p \ln(\theta / T_0) \right]. \qquad (15)$$

The generated entropy Γ for the equilibration step after the isolation operation is [Eqs. (13),(15)] (in this case $\Gamma_\uparrow = \Gamma$)

$$\Gamma = \Delta S$$

$$= n\left[\overline{S}_0 + \overline{C}_p \ln(\theta / T_0) - \overline{S}_0 - (\overline{C}_p / 2) \ln(T_\lambda / T_0) \right]$$

$$= n\overline{C}_p \ln\left[\theta / (T_0 T_\lambda)^{1/2} \right]. \qquad (16)$$

In order that Γ be ≥ 0, it is required that

$$\theta(T_0, T_\lambda) \geq (T_0 T_\lambda)^{1/2}. \qquad (17)$$

The logarithmic mean of the two temperatures T_0, T_λ is never less than their geometric mean. We have thus used a model system to demonstrate an instance of the mathematical inequality [2-4]

$$\frac{a+b}{2} \geq \frac{a-b}{\ln(a/b)} \geq (ab)^{1/2} \qquad (18)$$

for any two non-negative real numbers $a, b \; (0 \leq a, b < \infty)$.

We shall return to the monatomic ideal gas model in later parts of this section.

BEYOND THE DARTMOUTH DOMAIN FOR SYSTEMS HAVING EQUILIBRIUM-BASED IMAGES

In Section 1, the Dartmouth version of the second law was stated as:

If the system and the surroundings in both the initial and final states have equilibrium-based images, then, in the Dartmouth domain, the relation

$$\Gamma_{\uparrow} \equiv \Delta S(system) + \Delta_{\uparrow} S(surroundings) \geq 0 \tag{19}$$

must be true. $\Gamma_{\uparrow}$ *is that part of* Γ *directly attributable to the process, and the changes* $\Delta_{\uparrow}$ *are for effects directly attributable to the process.*

A natural question is, Can relation (19) be extended beyond the Dartmouth domain for systems having *equilibrium-based images*? [For systems that do not have *equilibrium-based images*, i.e. that do not have well-defined fields of intensities, attempts at a thermodynamic form of analysis are, in my opinion, ill-advised.] In order to carry through such an extension, we would need to find a suitable entropy function that would behave as required outside the Dartmouth domain. Three possibilities readily come to mind:

i) Use the *equilibrium-based-image* entropy function outside the Dartmouth domain as well as inside.

ii) Use an entropy function of equilibrium form but based on the actual local molar internal energy and actual local molar volume rather than on *equilibrium-based-image* values.

iii) Construct an entropy function that depends explicitly on spatial derivatives of the intensities as well as on the intensities themselves.

I also explore a fourth possibility: try extending the definition of entropy onto planes tangent to the equilibrium entropy–energy–volume surface for the *equilibrium-based image*.

The *Equilibrium-based-Image* Entropy Outside the Dartmouth Domain

Whether the *equilibrium-based-image* entropy satisfies relation (19) outside the Dartmouth domain is a question of fact: we need to evaluate $\Gamma_{\uparrow}$ for model systems which have calculable properties outside the Dartmouth domain to see if relation (19) holds in all cases that we can investigate. If we find at least one counterexample, we can dismiss further consideration of possibility i).

To see some of the points at issue, let us look at the specialized model of a monatomic ideal gas in a thermal field of constant gradient.

Monatomic Ideal Gas in a Thermal Field of Constant Gradient

Let the apparatus be as described in the first part of this section. Lay off a coordinate x along a line parallel to the axis of the cylinder; use a heat reservoir of temperature T_0 to maintain the face of the cylinder at $x = 0$ at the temperature T_0; likewise, maintain the face of the cylinder at $x = \lambda$ at the temperature T_λ. Fill the interior volume V with an amount n of a monatomic ideal gas, and allow the apparatus to reach a steady state. In the steady state, let the pressure p be uniform throughout the cylinder. Let the temperature on the cross sectional area at the coordinate value x be T_x, and assume that

$$T_x = T_0 + \left(\tfrac{x}{\lambda}\right)\left(T_\lambda - T_0\right) . \tag{20}$$

See Fig. 1.

Let the thermal gradient be so severe that locally, at point x,

$$\overline{\mathbf{U}}_x \neq \overline{U}_x . \tag{21}$$

To keep things as simple as possible, let us assume that

$$\overline{\mathbf{V}}_x = \overline{V}_x = RT_x / p . \tag{22}$$

[The two conditions $\overline{\mathbf{U}}_x \neq \overline{U}_x$ and $\overline{\mathbf{V}}_x = \overline{V}_x$ may possibly be inconsistent, inequality in one perhaps necessarily requiring inequality in the other; but let us pursue our simplified version of the model.]

In the steady state, the amount of gas dn_x contained between planes at x and $x + dx$ in the cylinder, i.e. the amount of gas contained in the volume σdx at the indicated place, is

$$dn_x = p\sigma dx / RT_x ; \tag{23}$$

and the total amount of gas n in the cylinder is

$$n = \int_{x=0}^{x=\lambda} dn_x = \frac{p\sigma}{R}\int_0^\lambda \frac{dx}{T_x} . \tag{24}$$

In the steady state, the gas follows the equation of state [see the earlier analysis]

$$pV = nR\frac{T_\lambda - T_0}{\ln\left(T_\lambda / T_0\right)} = nR\theta , \tag{25}$$

where

$$\theta \equiv \theta\left(T_\lambda, T_0\right) \equiv \left(T_\lambda - T_0\right) / \ln\left(T_\lambda / T_0\right) \tag{26}$$

is the logarithmic mean of the temperatures T_λ, T_0.

Let $\overline{\mathbf{U}}_x$ depend on the intensities and on spatial derivatives of the intensities. Introduce the spatial derivative operator D_x^i:

$$D_x^i \equiv d^i / dx^i \ . \tag{27}$$

Write for $\overline{\mathbf{U}}_x$ in the present case

$$\overline{\mathbf{U}}_x = \overline{U}_x + \sum_i \int_0^{D_x^i T_x} \left.\frac{\partial \overline{\mathbf{U}}_x}{\partial D_x^i T_x}\right)_{\ldots} dD_x^i T_x \quad \mathrm{i} = 1,2,\ldots,\mathrm{r} \ . \tag{28}$$

Consider first-derivative dependence only:

$$\overline{\mathbf{U}}_x = \overline{U}_x + \int_0^{D_x T_x} \left.\frac{\partial \overline{\mathbf{U}}_x}{\partial D_x T_x}\right)_{T_x,p} dD_x T_x \ . \tag{29}$$

Let $\partial \overline{\mathbf{U}}_x / \partial D_x T_x)_{T_x,p}$ be independent of p. Then

$$\int_0^{D_x T_x} \left.\frac{\partial \overline{\mathbf{U}}_x}{\partial D_x T_x}\right)_{T_x,p} dD_x T_x = \mathbf{F}_x\left(D_x T_x, T_x\right) \tag{30}$$

and

$$\overline{\mathbf{U}}_x = \overline{U}_x + \mathbf{F}_x = \overline{C}_v T_x + \mathbf{F}_x \ . \tag{31}$$

The total internal energy of the gas is therefore

$$\mathbf{U} = \int_{x=0}^{x=\lambda} \overline{\mathbf{U}}_x dn_x = \int_{x=0}^{x=\lambda} (\overline{U}_x + \mathbf{F}_x) dn_x$$

$$= n\overline{C}_v \theta + \frac{pV}{R\left(T_\lambda - T_0\right)} \int_{T_0}^{T_\lambda} \mathbf{F}_x d \ln T_x \ . \tag{32}$$

Now let

$$\langle \mathbf{F} \rangle \equiv \frac{1}{\ln(T_\lambda / T_0)} \int_{T_0}^{T_\lambda} \mathbf{F}_x d \ln T_x \tag{33}$$

and

$$\theta^* \equiv \theta + \left(\langle \mathbf{F} \rangle / \overline{C}_v\right) ; \tag{34}$$

then Eq. (32) becomes

$$\mathbf{U} = n\overline{C}_v \theta + n\langle \mathbf{F} \rangle = n\overline{C}_v \theta^* \ . \tag{35}$$

The *equilibrium-based-image* entropy for the gas in the steady state is [Eq. (13)]

$$S=\int_{x=0}^{x=\lambda}\overline{S}_x dn_x=\int_{x=0}^{x=\lambda}\left[\overline{S}_0(T_0,p)+\overline{C}_p\ln(T_x/T_0)\right]dn_x$$

$$=n\left[\overline{S}_0+(\overline{C}_p/2)\ln(T_\lambda/T_0)\right]. \qquad (36)$$

Thermally Depolarizing the System

Let us now take the system in the steady state with end temperatures T_0, T_λ and instantaneously slip a thin, rigid, adiabatically insulating membrane over each end of the cylinder so as to isolate the system from interactions with the surroundings. The system will relax, at constant $\mathbf{U}, V, n$, to a state of equilibrium. Since the internal energy remains constant during the process, we have

$$\mathbf{U}_f=\mathbf{U}_i\ , \qquad (37)$$

$$n\overline{C}_v T_f=n\overline{C}_v\theta^*\ , \qquad (38)$$

$$T_f=\theta^*\ . \qquad (39)$$

The final uniform temperature for the system will be θ^*. The final pressure at equilibrium, p_f, will be related to the initial pressure, $p_i=p$, by

$$p_f/p_i=\theta^*/\theta\ . \qquad (40)$$

For the equilibrium state, then,

$$S_f=n\left[\overline{S}_0(T_0,p_f)+\overline{C}_p\ln(\theta^*/T_0)\right]. \qquad (41)$$

The generated entropy Γ for the equilibration step after the isolation operation is (in this case $\Gamma_\uparrow=\Gamma$)

$$\Gamma=\Delta S$$

$$=n\left[\overline{S}_0(T_0,p_f)+\overline{C}_p\ln(\theta^*/T_0)-\overline{S}_0(T_0,p_i)-(\overline{C}_p/2)\ln(T_\lambda/T_0)\right]$$

$$=n\left[-R\ln(\theta^*/\theta)+\overline{C}_p\ln\left\{\theta^*/(T_0T_\lambda)^{1/2}\right\}\right]$$

$$=n\left[\overline{C}_p\ln\left\{\theta/(T_0T_\lambda)^{1/2}\right\}+\overline{C}_v\ln(\theta^*/\theta)\right], \qquad (42)$$

where we have made use of the relation $R = \overline{C}_p - \overline{C}_v$.

We saw above that the first term inside the square brackets in Eq. (42) is non-negative. In the initial state,

$$\theta^* / \theta = \mathbf{U}_i / U_i \ ; \tag{43}$$

so if the effect of the severe thermal gradient were to make the internal energy of the system greater than that of the *equilibrium-based image*, then Γ would certainly be ≥ 0. If, on the other hand, the thermal gradient were to make $\mathbf{U}_i$ smaller than U_i, then, if $\mathbf{U}_i$ were not bounded from below, for a sufficiently small value of $\mathbf{U}_i / U_i$ the expression in relation (42) would cease to be non-negative and relation (19) would fail.

To be sure of the status of relation (42), we would have to be able to determine the sign of $\langle \mathbf{F} \rangle$ and, if that turns out to be negative, whether $\mathbf{U}_i$ is bounded from below.

Next, let us try to construct an entropy function that depends on spatial derivatives of the intensities as well as on the intensities themselves; and let us limit the dependence on spatial derivatives to first-derivative dependence only. For our monatomic ideal gas model, let us try

$$\overline{\mathbf{S}}_x = \overline{S}_x + \int_0^{D_x T_x} \left. \frac{\partial \overline{\mathbf{S}}_x}{\partial D_x T_x} \right)_{T_x, p} dD_x T_x$$

$$= \overline{S}_x + \mathbf{Z}_x \left(T_x, p, D_x T_x \right) . \tag{44}$$

Then, for the thermal depolarization process that we discussed above, we have

$$\mathbf{S}_i = \int_{x=0}^{x=\lambda} \overline{\mathbf{S}}_x dn_x = \int_{x=0}^{x=\lambda} \left(\overline{S}_x + \mathbf{Z}_x \right) dn_x$$

$$= n \left[\overline{S}_0 \left(T_0, p \right) + \left(\overline{C}_p / 2 \right) \ln \left(T_\lambda / T_0 \right) + \langle \mathbf{Z} \rangle \right] , \tag{45}$$

where

$$\langle \mathbf{Z} \rangle \equiv \frac{1}{\ln \left(T_\lambda / T_0 \right)} \int_{T_0}^{T_x} \mathbf{Z}_x d \ln T_x \ , \tag{46}$$

and

$$\mathbf{S}_f = n \left[\overline{S}_0 \left(T_0, p\theta^* / \theta \right) + \overline{C}_p \ln \left(\theta^* / T_0 \right) \right] . \tag{47}$$

So,

$$\Gamma = \Delta \mathbf{S}$$

$$= n \left[\overline{C}_v \ln (\theta^* / \theta) + \overline{C}_p \ln \left\{ \theta / \left(T_0 T_\lambda \right)^{1/2} \right\} - \langle \mathbf{Z} \rangle \right] . \tag{48}$$

A sufficient condition for $\Gamma \geq 0$ is that

$$-\langle \mathbf{Z} \rangle + \overline{C}_v \ln(\theta^* / \theta) \geq 0 \; . \tag{49}$$

It is quite likely that we are asking more of our simplified ideal gas model than it can rightfully deliver; but if we take relation (49) at face value, then we must require, if $\theta^* < \theta$, that the effect of the severe thermal gradient be to make the entropy of the system less than that of the *equilibrium-based image* (i.e. $\langle \mathbf{Z} \rangle < 0$). We must also require that θ^* be bounded from below:

$$\theta^* \geq \theta \exp\left(\langle \mathbf{Z} \rangle / \overline{C}_v \right) . \tag{50}$$

However, if θ^* cannot decrease below the bound shown in relation (50), then we have to investigate relation (42) once again to see if it may be always non-negative after all.

Onto the Tangent Plane

In dealing with nonequilibrium states outside of the Dartmouth domain, an idea worth considering is the extension of the entropy concept onto planes tangent to the equilibrium entropy–energy–volume surface of the system.

Consider the system we just dealt with. The local *equilibrium-based-image* values of $\overline{S}_x, \overline{U}_x, \overline{V}_x$ lie on the equilibrium $\overline{S}, \overline{U}, \overline{V}$ surface in the entropy–energy–volume thermodynamic space for the system under investigation. Consider the point $\overline{S}_x, \overline{U}_x, \overline{V}_x$ on the equilibrium surface for the monatomic ideal gas system. The lines whose slopes represent the derivatives

$$\partial \overline{S}_x / \partial \overline{U}_x \Big)_{\overline{V}_x} = 1 / T_x \text{ and } \partial \overline{S}_x / \partial \overline{V}_x \Big)_{\overline{U}_x} = p_x / T_x$$

lie in the plane tangent to the $\overline{S}, \overline{U}, \overline{V}$ surface at the point $\overline{S}_x, \overline{U}_x, \overline{V}_x$. Let us consider extending the definition of entropy onto the tangent plane via the relation

$$\overline{\mathbf{S}}_x = \overline{S}_x + \frac{1}{T_x}\left(\overline{\mathbf{U}}_x - \overline{U}_x \right) + \frac{p_x}{T_x}\left(\overline{\mathbf{V}}_x - \overline{V}_x \right) . \tag{51}$$

Relation (51) can also be exhibited in the forms

$$\overline{\mathbf{U}}_x - \overline{U}_x = T_x\left(\overline{\mathbf{S}}_x - \overline{S}_x \right) - p_x\left(\overline{\mathbf{V}}_x - \overline{V}_x \right) \tag{52}$$

and

$$\overline{\mathbf{U}}_x + p_x \overline{\mathbf{V}}_x - T_x \overline{\mathbf{S}}_x = \overline{U}_x + p_x \overline{V}_x - T_x \overline{S}_x . \tag{53}$$

In terms of the quantities $\mathbf{F}_x$ and $\mathbf{Z}_x$ introduced in Eqs. (30) and (44), Eq. (51) leads to

$$\mathbf{U} - U = n \langle \mathbf{F} \rangle \tag{54}$$

and

$$\mathbf{S} - S = n\langle \mathbf{Z} \rangle = n\langle \mathbf{F} / T \rangle \tag{55}$$

—see Eqs. (35) and (45).

Consequently, Eqs. (48) and (49) take the forms

$$\Gamma = \Delta\mathbf{S}$$
$$= n\left[\overline{C}_v \ln(\theta^* / \theta) + \overline{C}_p \ln\left\{\theta / \left(T_0 T_\lambda\right)^{1/2}\right\} - \langle \mathbf{F} / T \rangle\right], \tag{56}$$

$$-\langle \mathbf{F} / T \rangle + \overline{C}_v \ln(\theta^* / \theta) \geq 0. \tag{57}$$

We conclude once again that if $\langle \mathbf{F} \rangle < 0$ $(\text{i.e. } \theta^* < \theta)$, then θ^* must be bounded from below [Eq. (50)]:

$$\theta^* \geq \theta \exp\left(\langle \mathbf{F} / T \rangle / \overline{C}_v\right). \tag{58}$$

Another Suggestion

An obvious way of trying to extend the definition of entropy into the domain of nonequilibrium states that lie outside of the Dartmouth domain is to use the equilibrium formula

$$\overline{\mathbf{S}} = \overline{\mathbf{S}}\left(\overline{\mathbf{U}}, \overline{\mathbf{V}}\right), \tag{59}$$

where the independent variables $\overline{\mathbf{U}}, \overline{\mathbf{V}}$ are actual values of the system rather than values for the *equilibrium-based image* of the system. Consider the case of a monatomic ideal gas in a thermal field that has a gradient so severe that $U \neq \mathbf{U}$. As usual, let the intensities of the field be specified; and let the gas have a temperature T_x and a pressure p_x at point *x* in the thermal field. The gas has an *equilibrium-based image* whose local properties are, by construction, such that

$$d\overline{U}_x = T_x d\overline{S}_x - p_x d\overline{V}_x, \tag{60}$$

where $\overline{U}_x(T_x, p_x)$ and $\overline{S}_x(T_x, p_x)$ are equilibrium formulas and $\overline{V}_x(T_x, p_x)$ represents the value from the equilibrium equation of state—see Section 1. Our assumption is that because of the severe gradient in the thermal field

$$\overline{\mathbf{U}}_x \neq \overline{U}_x, \quad \overline{\mathbf{V}}_x \neq \overline{V}_x. \tag{61}$$

Now it is almost certain that for the case being considered

$$\left.\frac{\partial \overline{\mathbf{S}}_x}{\partial \overline{\mathbf{U}}_x}\right)_{\overline{\mathbf{V}}_x} \neq \frac{1}{T_x}. \tag{62}$$

Suppose that, on the contrary, equality holds in relation (62). Then we should have at each and every point of the field with the severe gradient

$$\left.\frac{\partial \overline{\mathbf{S}}_x}{\partial \overline{\mathbf{U}}_x}\right)_{\overline{\mathbf{V}}_x} = \frac{1}{T_x} = \left.\frac{\partial \overline{S}_x}{\partial \overline{U}_x}\right)_{\overline{V}_x} . \tag{63}$$

Since relation (63) would hold at an infinite number of places (in fact at all the places) in the thermal field, we should be forced to conclude that

$$\overline{\mathbf{U}}_x = \overline{U}_x, \quad \overline{\mathbf{V}}_x = \overline{V}_x , \tag{64}$$

contrary to our assumption in relation (61); and there would be a contradiction. We must conclude, therefore, that if we require relation (61) to hold with the entropy function of Eq. (59), then relation (62) must follow—which in itself is a disagreeable result, and we choose not to follow this line of investigation further.

Discussion

To know whether relations (42), (48), and (56) are such that $\Gamma \geq 0$ for our monatomic ideal gas model, we have to be able to evaluate the functions $\langle \mathbf{F} \rangle$ and $\langle \mathbf{Z} \rangle$ in sufficient detail to know the sign of each and the numerical value of each for a particular given condition. I do not know how to do that, so that task will have to fall to others to accomplish.

The idea of constructing an entropy function that depends explicitly on one or more nonequilibrium parameters characterizing a system in a nonequilibrium state has been used extensively by workers in the field of *extended thermodynamics of irreversible processes* [5,6]. There, however, the emphasis is on the rate of entropy generation; and the additional parameters used to characterize the entropy are, for choice, currents (fluxes) rather than spatial derivatives of intensities. The basic strategy used in *extended thermodynamics of irreversible processes* is to posit a particular form for the local differential of the generalized entropy function and to use that form in the expression for the local rate of entropy generation.

Suppose we try to use a similar strategy for the case of a monatomic ideal gas in a (one-dimensional) thermal field. Let the intensities have *given* values T_x, p_x at point x in the field. Let us start with the relations

$$\overline{\mathbf{S}}_x = \overline{\mathbf{S}}_x\left(\overline{\mathbf{U}}_x, \overline{\mathbf{V}}_x, D_x^i T_x, D_x^i p_x\right) , \tag{65}$$

$$d\overline{\mathbf{S}}_x = \frac{1}{T_x} d\overline{\mathbf{U}}_x + \frac{p_x}{T_x} d\overline{\mathbf{V}}_x + \sum_i \left.\frac{\partial \overline{\mathbf{S}}_x}{\partial D_x^i T_x}\right)_{\dots} dD_x^i T_x + \sum_i \left.\frac{\partial \overline{\mathbf{S}}_x}{\partial D_x^i p_x}\right)_{\dots} dD_x^i p_x , \tag{66}$$

where we are again dealing with spatial derivatives of the intensities as the extra arguments in the generalized entropy function.

To proceed further, let us again limit ourselves to first-derivative dependence only; then,

$$d\overline{\mathbf{S}}_x = \frac{1}{T_x} d\overline{\mathbf{U}}_x + \frac{p_x}{T_x} d\overline{\mathbf{V}}_x + \left.\frac{\partial \overline{\mathbf{S}}_x}{\partial D_x T_x}\right)_{T_x, p_x, D_x p_x} dD_x T_x +$$

$$\left.\frac{\partial \overline{\mathbf{S}}_x}{\partial D_x p_x}\right)_{T_x, p_x, D_x T_x} dD_x p_x. \tag{67}$$

I find nothing worthy of comment in the Maxwell-type relations following from Eq. (67).

Next, let us express each of $\overline{\mathbf{S}}_x, \overline{\mathbf{U}}_x, \overline{\mathbf{V}}_x$ in the form

$$\overline{\mathbf{X}}_x\left(T_x, p_x, D_x T_x, D_x p_x\right) = \overline{X}_x\left(T_x, p_x\right) +$$

$$\int_0^{D_x T_x} \left.\frac{\partial \overline{\mathbf{X}}_x}{\partial D_x T_x}\right)_{T_x, p_x, D_x p_x} dD_x T_x + \int_0^{D_x p_x} \left.\frac{\partial \overline{\mathbf{X}}_x}{\partial D_x p_x}\right)_{T_x, p_x, D_x T_x} dD_x p_x \tag{68}$$

for $\overline{\mathbf{X}}_x = \overline{\mathbf{S}}_x, \overline{\mathbf{U}}_x, \overline{\mathbf{V}}_x$. Upon substitution of the forms (68) in relation (67), we get

$$\left.\frac{\partial \overline{\mathbf{S}}_x}{\partial D_x T_x}\right)_{\overline{\mathbf{U}}_x, \overline{\mathbf{V}}_x, D_x p_x} = \left.\frac{\partial \overline{\mathbf{S}}_x}{\partial D_x T_x}\right)_{T_x, p_x, D_x p_x} - \frac{1}{T_x} \left.\frac{\partial \overline{\mathbf{U}}_x}{\partial D_x T_x}\right)_{T_x, p_x, D_x p_x} -$$

$$\frac{p_x}{T_x} \left.\frac{\partial \overline{\mathbf{V}}_x}{\partial D_x T_x}\right)_{T_x, p_x, D_x p_x}, \tag{69}$$

with an analogous expression for $\left.\partial \overline{\mathbf{S}}_x / \partial D_x p_x\right)_{\overline{\mathbf{U}}_x, \overline{\mathbf{V}}_x, D_x T_x}$.

The left-hand sides of relation (69) and its analogue are related to the amount by which $d\overline{\mathbf{S}}_x$ exceeds $\left(1/T_x\right) d\overline{\mathbf{U}}_x + \left(p_x / T_x\right) d\overline{\mathbf{V}}_x$; i.e. let

$$\delta \overline{\mathbf{S}}_x \equiv d\overline{\mathbf{S}}_x - \frac{1}{T_x} d\overline{\mathbf{U}}_x - \frac{p_x}{T_x} d\overline{\mathbf{V}}_x, \tag{70}$$

then

$$\left.\frac{\partial \overline{\mathbf{S}}_x}{\partial D_x T_x}\right)_{\overline{\mathbf{U}}_x, \overline{\mathbf{V}}_x, D_x p_x} = \left.\frac{\delta \overline{\mathbf{S}}_x}{\partial D_x T_x}\right)_{T_x, p_x, D_x p_x}, \tag{71}$$

with an analogous expression for $\left.\partial \overline{\mathbf{S}}_x / \partial D_x p_x\right)_{\overline{\mathbf{U}}_x,\overline{\mathbf{V}}_x,D_xT_x}$.

In the determinant shown below as relation (72), the independent variables are $\overline{\mathbf{U}}_x, \overline{\mathbf{V}}_x, D_xT_x, D_xp_x$. Let us require, in the way of stability conditions for the system, that the determinant of second derivatives from relation (67) be ≥ 0, that each principal minor of even order in the determinant be ≥ 0, and that each principal minor of odd order in the determinant be ≤ 0.

$$\begin{vmatrix} -\frac{1}{T_x^2}\frac{\partial T_x}{\partial \overline{\mathbf{U}}_x} & -\frac{1}{T_x^2}\frac{\partial T_x}{\partial \overline{\mathbf{V}}_x} & -\frac{1}{T_x^2}\frac{\partial T_x}{\partial D_xT_x} & -\frac{1}{T_x^2}\frac{\partial T_x}{\partial D_xp_x} \\ \frac{\partial(p_x/T_x)}{\partial \overline{\mathbf{U}}_x} & \frac{\partial(p_x/T_x)}{\partial \overline{\mathbf{V}}_x} & \frac{\partial(p_x/T_x)}{\partial D_xT_x} & \frac{\partial(p_x/T_x)}{\partial D_xp_x} \\ \frac{\partial^2 \overline{\mathbf{S}}_x}{\partial \overline{\mathbf{U}}_x \partial D_xT_x} & \frac{\partial^2 \overline{\mathbf{S}}_x}{\partial \overline{\mathbf{V}}_x \partial D_xT_x} & \frac{\partial^2 \overline{\mathbf{S}}_x}{\partial (D_xT_x)^2} & \frac{\partial^2 \overline{\mathbf{S}}_x}{\partial D_xp_x \partial D_xT_x} \\ \frac{\partial^2 \overline{\mathbf{S}}_x}{\partial \overline{\mathbf{U}}_x \partial D_xp_x} & \frac{\partial^2 \overline{\mathbf{S}}_x}{\partial \overline{\mathbf{V}}_x \partial D_xp_x} & \frac{\partial^2 \overline{\mathbf{S}}_x}{\partial D_xT_x \partial D_xp_x} & \frac{\partial^2 \overline{\mathbf{S}}_x}{\partial (D_xp_x)^2} \end{vmatrix} \geq 0 \tag{72}$$

Our requirements result in each of the entries on the main diagonal being ≤ 0:

$$\left.-\frac{1}{T_x^2}\frac{\partial T_x}{\partial \overline{\mathbf{U}}_x}\right)_{\overline{\mathbf{V}}_x,D_xT_x,D_xp_x} \leq 0\ , \tag{73}$$

$$\left.\frac{\partial(p_x/T_x)}{\partial \overline{\mathbf{V}}_x}\right)_{\overline{\mathbf{U}}_x,D_xT_x,D_xp_x} \leq 0\ , \tag{74}$$

$$\left.\frac{\partial^2 \overline{\mathbf{S}}_x}{\partial (D_xT_x)^2}\right)_{\overline{\mathbf{U}}_x,\overline{\mathbf{V}}_x,D_xp_x} \leq 0\ , \tag{75}$$

$$\left.\frac{\partial^2 \overline{\mathbf{S}}_x}{\partial (D_xp_x)^2}\right)_{\overline{\mathbf{U}}_x,\overline{\mathbf{V}}_x,D_xT_x} \leq 0\ . \tag{76}$$

Relation (73) tells us that one of the heat capacities of the system is non-negative, and relation (74) displays a connection between a kind of compressibility and a kind of thermal expansibility of the system.

Conclusion

For systems in nonequilibrium states possessing *equilibrium-based images*, the Dartmouth domain is defined by

$$\mathbf{U} \approx U \text{ or } \left|(\mathbf{U} - U)/U\right| << 1 \,. \tag{77}$$

Relation (19) is thought to hold everywhere inside the Dartmouth domain.

We discussed extending relation (19) outside the Dartmouth domain by finding a suitable entropy function. We examined the use of the entropy of the *equilibrium-based image* outside the Dartmouth domain and we tried extending the definition of entropy onto tangent planes of the equilibrium entropy-energy-volume surface; we also considered generalizing the entropy function to make it depend explicitly on spatial derivatives of the intensities. When we applied these ideas to a model of a monatomic ideal gas in a (one-dimensional) thermal field, we found that in order to get conclusive results the calculations had to be carried out in more detail than the author was capable of providing. Outside the Dartmouth domain, what the most useful definition of the entropy is, relative to relation (19), was left as an unresolved question.

Appendix

We should bear in mind the possibility that in the explicit dependence of $\overline{\mathbf{U}}_x$ or $\overline{\mathbf{S}}_x$ on spatial derivatives of the intensities—Eqs. (28) and (66) for the monatomic ideal gas model—the first spatial derivative having a significant effect on $\overline{\mathbf{U}}_x$ or $\overline{\mathbf{S}}_x$ may be not the first-order derivative but a higher-order derivative. Suppose it to be the case that spatial derivatives of the intensities of order $\geq k$ have a significant effect on $\overline{\mathbf{U}}_x$ or $\overline{\mathbf{S}}_x$; but that spatial derivatives of order $< k$ have no significant effect, and $k \geq 2$. Then systems in fields of intensities with distributions of the intensities such that spatial derivatives of the intensities of orders $\geq k$ vanish identically would belong to the Dartmouth domain regardless of the "strengths" of the spatial derivatives of the intensities of orders $< k$.

Consider again the monatomic ideal gas model, Eqs. (28) and (66). Suppose, for example, that $D_x T_x$ has no significant effect on $\overline{\mathbf{U}}_x$ or $\overline{\mathbf{S}}_x$, but that $D_x^2 T_x$ does. Then the thermodynamic behavior of the monatomic ideal gas in a thermal field of constant gradient (i.e., $D_x^2 T_x \equiv 0$) would be exactly the same as that of the *equilibrium-based image* of the gas—see Section 1; and all the states of the model would fall in the Dartmouth domain regardless of the severity of the gradient.

REFERENCES

1. F. Kreith and M. S. Bohm, *Principles of Heat Transfer*, 4th ed. (Harper & Row, New York, 1986).
2. R. J. Tykodi, Am. J. Phys. **64,** 644 (1996).

3. A. R. Plastino, A. Plastino, and H. G. Miller, Am. J. Phys. **65,** 1102 (1997).
4. L. R. Berrone and C. D. Galles, Am. J. Phys. **66,** 87 (1998).
5. D. Jou, J. Casas-Vazquez, and G. Lebon, *Extended Irreversible Thermodynamics* (Springer, Berlin, 1993).
6. B. C. Eu, *Kinetic Theory and Irreversible Thermodynamics* (Wiley, New York, 1992).

FOURTEEN

AFTERTHOUGHTS

I set up this section as a place where, after the contents and the structure of the first 13 sections of Book I were pretty firmly in place, I could deal with new considerations and with reconsiderations of previously established results without undoing the structure of the first 13 sections. I do have a few "afterthoughts" to add to the corpus of Book I.

Complex Situations, Compound Effects

The cases of migrational equilibrium in a nonequilibrium spatial field that we investigated thus far dealt with a single field—a thermal or a concentration field. A more complex situation would be one in which we studied the behavior of a system of interest in a compound nonequilibrium spatial field, a field formed by the superposition of a thermal field and a concentration field, say.

Consider the case of the concentration-field analogue of the thermal diffusion effect in gases, Section 10—see Fig. 10.3. Now put the left-hand side of the apparatus in a thermostat of temperature T_α and put the right-hand side in a thermostat of temperature T_β, in the manner of Fig. 10.2—the part of the MIX region between the thermostats to be insulated against lateral heat exchanges. Adjust the analysis in Section 10 to accommodate the compound thermal-concentration field.

Start with the system in a steady state in which there is a steady flow of component 3 from the terminal part at L through the MIX region and into the terminal part at R $\left(\dot{n}_L^{(3)} + \dot{n}_R^{(3)} = 0\right)$ and in which the pressures $p_0^{(1)}, p_0^{(2)}, p_\lambda^{(1)}, p_\lambda^{(2)}$ are such that components 1 and 2 are in states of migrational equilibrium. Let $\overline{K}_L^{(3)}, \overline{K}_R^{(3)}$ be the molar macroscopic kinetic energy associated with component 3 in terminal parts L,R.

To establish the conditions of migrational equilibrium for components 1 and 2 in the concentration field, carry out a differentially tailored series of steps involving one of the mass reservoirs containing component 1 or component 2, the net result of the series of steps being to move a small amount of the relevant component into or out of the mass reservoir, holding the temperatures and all the other mass-reservoir pressures constant. At the end of the differentially tailored process, the system is again in the original steady state, except for the fact that each mass reservoir may have experienced a differential change in its amount of substance. Mass-conservation considerations require that

$$dn_L^{(3)} + dn_R^{(3)} = 0,\ dn_0^{(1)} + dn_\lambda^{(1)} = 0,\ dn_0^{(2)} + dn_\lambda^{(2)} = 0. \tag{1}$$

In this case, $dn_L^{(3)}, dn_R^{(3)}$ are "undetermined quantities" to be evaluated later. The first law for the differentially tailored process gives

$$\sum\nolimits_{\omega,i} \text{``}\overline{U}_\omega^{(i)}\text{''} dn_\omega^{(i)} = dQ_\alpha + dQ_\beta - \sum\nolimits_{\omega,i} p_\omega^{(i)} \overline{V}_\omega^{(i)} dn_\omega^{(i)}, \tag{2}$$

where $\text{``}\overline{U}_\omega^{(i)}\text{''} \equiv \overline{U}_\omega^{(i)} + \overline{K}_\omega^{(i)}$, dQ_α and dQ_β are "undetermined quantities" to be evaluated later, and the indexes ω / i range over L,R/3 and $0, \lambda / 1, 2$. Then

$$\sum\nolimits_{\omega,i} \text{``}\overline{H}_\omega^{(i)}\text{''} dn_\omega^{(i)} = dQ_\alpha + dQ_\beta. \tag{3}$$

For the entropy changes attributable to the process, we have

$$dS = \sum\nolimits_{\omega,i} \overline{S}_\omega^{(i)} dn_\omega^{(i)}, \tag{4}$$

$$d_\uparrow S_{sur} = -T_\alpha^{-1} dQ_\alpha - T_\beta^{-1} dQ_\beta, \tag{5}$$

$$d\Gamma_\uparrow = dS + d_\uparrow S_{sur}$$

$$= \sum\nolimits_{\omega,i} \overline{S}_\omega^{(i)} dn_\omega^{(i)} - T_\alpha^{-1} dQ_\alpha - T_\beta^{-1} dQ_\beta. \tag{6}$$

Let us suppose that we carried through the differentially tailored process in such a way that $dn_0^{(1)} \neq 0$. Upon dividing Eq. (6) through by $dn_0^{(1)}$, introducing the notation

$$r^{(21)} \equiv dn_0^{(2)} / dn_0^{(1)},\ r^{(31)} \equiv dn_L^{(3)} / dn_0^{(1)}, \tag{7}$$

and making use of Eq. (1), we have

$$\frac{d\Gamma_\uparrow}{dn_0^{(1)}} = \overline{S}_0^{(1)} - \overline{S}_\lambda^{(1)} + r^{21}\left(\overline{S}_0^{(2)} - \overline{S}_\lambda^{(2)}\right) + r^{31}\left(\overline{S}_L^{(3)} - \overline{S}_R^{(3)}\right) -$$

$$T_\alpha^{-1} \frac{dQ_\alpha}{dn_0^{(1)}} - T_\beta^{-1} \frac{dQ_\beta}{dn_0^{(1)}}. \tag{8}$$

Note that dividing Eq. (3) through by $dn_0^{(1)}$ gives (remember that only component 3 has any macroscopic kinetic energy for the state in question)

$$\frac{dQ_\alpha}{dn_0^{(1)}} + \frac{dQ_\beta}{dn_0^{(1)}} = \overline{H}_0^{(1)} - \overline{H}_\lambda^{(1)} + r^{21}\left(\overline{H}_0^{(2)} - \overline{H}_\lambda^{(2)}\right) + r^{31}\left(``\overline{H}_L^{(3)}" - ``\overline{H}_R^{(3)}"\right). \tag{9}$$

Now set

$$dQ_\alpha \equiv \left(``\overline{H}_L" - \left[H_L\right]\right)dn_L + \left(\overline{H}_0^{(1)} - \left[H_0^{(1)}\right]\right)dn_0^{(1)} + \left(\overline{H}_0^{(2)} - \left[H_0^{(2)}\right]\right)dn_0^{(2)} \tag{10}$$

and

$$dQ_\beta \equiv \left(``\overline{H}_R" - \left[H_R\right]\right)dn_R + \left(\overline{H}_\lambda^{(1)} - \left[H_\lambda^{(1)}\right]\right)dn_\lambda^{(1)} + \left(\overline{H}_\lambda^{(2)} - \left[H_\lambda^{(2)}\right]\right)dn_\lambda^{(2)}, \tag{11}$$

where the $\left[H\right]$ quantities are molar nonequilibrium enthalpies of transport. It follows from Eqs. (9)-(11) that

$$\frac{dQ_\alpha}{dn_0^{(1)}} = \overline{H}_0^{(1)} + r^{21}\overline{H}_0^{(2)} + r^{31}``\overline{H}_L" - \left\{\left[H_0^{(1)}\right] + r^{21}\left[H_0^{(2)}\right] + r^{31}\left[H_L\right]\right\} \tag{12}$$

and

$$\frac{dQ_\beta}{dn_0^{(1)}} = -\left(\overline{H}_\lambda^{(1)} + r^{21}\overline{H}_\lambda^{(2)} + r^{31}``\overline{H}_R"\right) + \left\{\left[H_\lambda^{(1)}\right] + r^{21}\left[H_\lambda^{(2)}\right] + r^{31}\left[H_R\right]\right\}. \tag{13}$$

Introduce the notation

$$Z._{}^{(1)} + r^{21}Z._{}^{(2)} + r^{31}Z._{}^{(3)} \equiv Z._{}^{\dagger}. \tag{14}$$

Then Eqs. (12),(13) become

$$\frac{dQ_\alpha}{dn_0^{(1)}} = \overline{H}_\alpha^\dagger - \left[H_\alpha^\dagger\right] + r^{31}\overline{K}_L, \quad \frac{dQ_\beta}{dn_0^{(1)}} = -\overline{H}_\beta^\dagger + \left[H_\beta^\dagger\right] - r^{31}\overline{K}_R, \tag{15}$$

and we have

$$\frac{dQ_\alpha}{dn_0^{(1)}} + \frac{dQ_\beta}{dn_0^{(1)}} = \overline{H}_\alpha^\dagger - \overline{H}_\beta^\dagger + r^{31}\left(\overline{K}_L - \overline{K}_R\right) - \left\{\left[H_\alpha^\dagger\right] - \left[H_\beta^\dagger\right]\right\}. \quad (16)$$

In the condensed notation (14), Eq. (9) reads

$$\frac{dQ_\alpha}{dn_0^{(1)}} + \frac{dQ_\beta}{dn_0^{(1)}} = \overline{H}_\alpha^\dagger - \overline{H}_\beta^\dagger + r^{31}\left(\overline{K}_L - \overline{K}_R\right). \quad (17)$$

Upon comparing Eqs. (16) and (17), we see that

$$\left[H_\alpha^\dagger\right] = \left[H_\beta^\dagger\right] \equiv \left[H^\dagger\right]. \quad (18)$$

Returning to Eq. (8), we now find that

$$\frac{d\Gamma_\uparrow}{dn_0^{(1)}} = \overline{S}_\alpha^\dagger - \overline{S}_\beta^\dagger - T_\alpha^{-1}\left\{\overline{H}_\alpha^\dagger - \left[H_\alpha^\dagger\right] + r^{31}\overline{K}_L\right\} -$$

$$T_\beta^{-1}\left\{-\overline{H}_\beta^\dagger + \left[H_\beta^\dagger\right] - r^{31}\overline{K}_R\right\}. \quad (19)$$

We can write Eq. (19) in terms of the Planck function $\overline{Y} \equiv \overline{S} - T^{-1}\overline{H}$:

$$\frac{d\Gamma_\uparrow}{dn_0^{(1)}} = \overline{Y}_\alpha^\dagger - \overline{Y}_\beta^\dagger + T_\alpha^{-1}\left[H_\alpha^\dagger\right] - T_\alpha^{-1}r^{31}\overline{K}_L -$$

$$T_\beta^{-1}\left[H_\beta^\dagger\right] + T_\beta^{-1}r^{31}\overline{K}_R. \quad (20)$$

The Thomson-Gibbs Corollary applied to Eq. (20) gives for the condition of migrational equilibrium for component 1 in the compound thermal-concentration field

$$\overline{Y}_\alpha^\dagger + T_\alpha^{-1}\left[H_\alpha^\dagger\right] - T_\alpha^{-1}r^{31}\overline{K}_L = \overline{Y}_\beta^\dagger + T_\beta^{-1}\left[H_\beta^\dagger\right] - T_\beta^{-1}r^{31}\overline{K}_R. \quad (21)$$

Since in the state of migrational equilibrium for components 1 and 2, the macroscopic kinetic energies of those components are each equal to zero, we may, formally, write Eq. (21) as

$$\text{“}\overline{Y}_\alpha^\dagger\text{”} + T_\alpha^{-1}\left[H_\alpha^\dagger\right] = \text{“}\overline{Y}_\beta^\dagger\text{”} + T_\beta^{-1}\left[H_\beta^\dagger\right], \quad (22)$$

where “$\overline{Y}$” $\equiv \overline{S} - T^{-1}$“$\overline{H}$” [compare Eq. (10.32)].

I shall not carry the analysis any further. As a finishing touch, however, let's evaluate the "undetermined quantities" $dn_L^{(3)}, dn_R^{(3)}$ and dQ_α, dQ_β used in the above analysis. In a separate series of operations, induce a (slow) steady flow $\dot{n}_0^{(1)}$ of component 1 between its terminal parts at 0 and λ by varying $p_0^{(1)}$, keeping T_α, T_β and the pressures for the other terminal parts constant (constant T, p', for short); and evaluate the derivatives

$$\delta\dot{n}_L^{(3)} / \tilde{\delta}\dot{n}_0^{(1)}\Big)_{T,p'}, \quad \delta\dot{n}_R^{(3)} / \tilde{\delta}\dot{n}_0^{(1)}\Big)_{T,p'}$$

via a sequence of steady states such that $\dot{n}_0^{(1)} \to 0$ [see Eq. (1.14)]. Make the identifications

$$dn_L^{(3)} / dn_0^{(1)} = \delta\dot{n}_L^{(3)} / \tilde{\delta}\dot{n}_0^{(1)}\Big)_{T,p'} \quad \text{and}$$

$$dn_R^{(3)} / dn_0^{(1)} = \delta\dot{n}_R^{(3)} / \tilde{\delta}\dot{n}_0^{(1)}\Big)_{T,p'} \tag{23}$$

or

$$dn_L^{(3)} = \delta\dot{n}_L^{(3)} / \tilde{\delta}\dot{n}_0^{(1)}\Big)_{T,p'} dn_0^{(1)} \quad \text{and}$$

$$dn_R^{(3)} = \delta\dot{n}_R^{(3)} / \tilde{\delta}\dot{n}_0^{(1)}\Big)_{T,p'} dn_0^{(1)} \tag{24}$$

where appropriate. Similarly, evaluate the derivatives

$$\delta\dot{Q}_\alpha^{(r)} / \tilde{\delta}\dot{n}_0^{(1)}\Big)_{T,p'} \quad \text{and} \quad \delta\dot{Q}_\beta^{(r)} / \tilde{\delta}\dot{n}_0^{(1)}\Big)_{T,p'}$$

via a sequence of steady states such that $\dot{n}_0^{(1)} \to 0$ and make the identifications

$$dQ_\alpha / dn_0^{(1)} = -\delta\dot{Q}_\alpha^{(r)} / \tilde{\delta}\dot{n}_0^{(1)}\Big)_{T,p'} \quad \text{and}$$

$$dQ_\beta / dn_0^{(1)} = -\delta\dot{Q}_\beta^{(r)} / \tilde{\delta}\dot{n}_0^{(1)}\Big)_{T,p'} \tag{25}$$

where appropriate.

Transformational Equilibrium in a Chemical-Reaction Field: Two Consecutive Reversible First-Order Reactions

Consider the case of two consecutive reversible first-order reactions as in the following scheme, Eq. (26):

$$A_1\left(\xrightarrow{k_{12}}, \xleftarrow{k_{21}}\right)A_2\left(\xrightarrow{k_{23}}, \xleftarrow{k_{32}}\right)A_3. \tag{26}$$

Let the reactions take place in a stirred-flow reactor of constant volume in a thermostat of temperature T. Let the reactor chamber communicate via membranes, each permeable to one component alone, with three mass reservoirs, each containing one of the species A_1, A_2, A_3. Assume that, by means of suitable catalysts and anti-catalysts, the individual species can be rendered stable (no transformations) in their mass reservoirs, and that in the reactor the reactions of Eq. (26) take place at reasonable rates.

Use the subscript index to identify the species; thus for a gas-phase reaction scheme of type (26), let p_i, say, be the pressure of species $A_i(i = 1, 2, 3)$ in its mass reservoir.

Suppose that for a particular temperature T and particular reservoir pressures p_1, p_2, p_3 we succeed in establishing a steady state such that

$$\dot{n}_1 + \dot{n}_3 = 0, \quad \dot{n}_2 = 0, \tag{27}$$

where $\dot{n}_i$ is the molar rate of influx of species A_i into its mass reservoir. We may then say of species A_2 that it is in a state of *transformational equilibrium* with respect to the configurations A_1, A_3 to which it is free to transform, but into which (on a macroscopic scale) it does not go. The condition of transformational equilibrium is the analogue in a chemical-reaction field of the condition of migrational equilibrium in a spatial field free of chemical reactions.

To establish the conditions of transformational equilibrium for species A_2 in the reaction scheme (26), gas-phase reactions, start in the state of transformational equilibrium, Eq. (27), and move an amount dn_2 of A_2 in a differentially tailored fashion out of or into the A_2 mass reservoir. The net effect of the process may be changes in all the mass reservoirs:

$$dn_1 + dn_2 + dn_3 = 0, \quad dn_2 \neq 0, \tag{28}$$

where dn_1, dn_3 are "undetermined quantities" to be evaluated later.

For the process as described, then (neglect the macroscopic kinetic energy of the material in a mass reservoir),

$$d\Gamma_{\uparrow} = -T^{-1}\left(\mu_1 dn_1 + \mu_2 dn_2 + \mu_3 dn_3\right). \tag{29}$$

and

$$\frac{d\Gamma_{\uparrow}}{dn_2} = -T^{-1}\left(\mu_1 \frac{dn_1}{dn_2} + \mu_2 + \mu_3 \frac{dn_3}{dn_2}\right). \tag{30}$$

The application of the Thomson-Gibbs Corollary to Eq. (30) gives

$$\mu_1 \frac{dn_1}{dn_2} + \mu_2 + \mu_3 \frac{dn_3}{dn_2} = 0 \tag{31}$$

or

$$\left(\mu_1 - \mu_3\right)\frac{dn_1}{dn_2} + \mu_2 - \mu_3 = 0 \tag{32}$$

where Eq.(28) has been used to obtain Eq. (32) from Eq. (31).

To evaluate dn_1 / dn_2: in a separate experiment, start in a state of transformational equilibrium, and then change the pressure in the mass reservoir containing component 2 from p_2 to p_2' and establish a steady state such that

$$\dot{n}_1 + \dot{n}_2 + \dot{n}_3 = 0. \tag{33}$$

Then, keeping T, p_1, p_3 constant, evaluate the derivative

$$\left.\delta\dot{n}_1 / \tilde{\delta}\dot{n}_2\right)_{T, p_1, p_3}$$

via a sequence of steady states such that $\dot{n}_2 \rightarrow 0$ [see Eq. (1.14)]. Make the identification

$$dn_1 / dn_2 = \left.\delta\dot{n}_1 / \tilde{\delta}\dot{n}_2\right)_{T, p_1, p_3} \tag{34}$$

where appropriate.

Acoustic Pulses (Sound Waves) in Ideal Gases

I wish to say a few things about nonequilibrium states of ideal gas systems that are being traversed by acoustic pulses (longitudinal sound waves).

Consider an amount n of a monatomic ideal gas in a right circular cylinder of interior volume V. The cylinder has walls of diathermal material and resides in an air thermostat set at the temperature T. At equilibrium, the gas in the cylinder has uniform temperature T and uniform pressure p. A pressure sensor communicates with the gas at a point of the side wall. At equilibrium, the gaseous system follows the equation of state $pV = nRT$, where R is the universal gas constant. The left-hand face of the cylinder is rigid and movable (in a frictionless, gas-tight fashion); it is held in place by stops, against which it is pressed by the gas. The (inner) distance between the end faces of the cylinder is λ. Lay off a coordinate x along a line parallel to the axis of the cylinder. Let the right-hand face of the cylinder be (slightly) elastic though fixed in place. Couple each end face of the cylinder to a pulse generator—see Fig. 1.

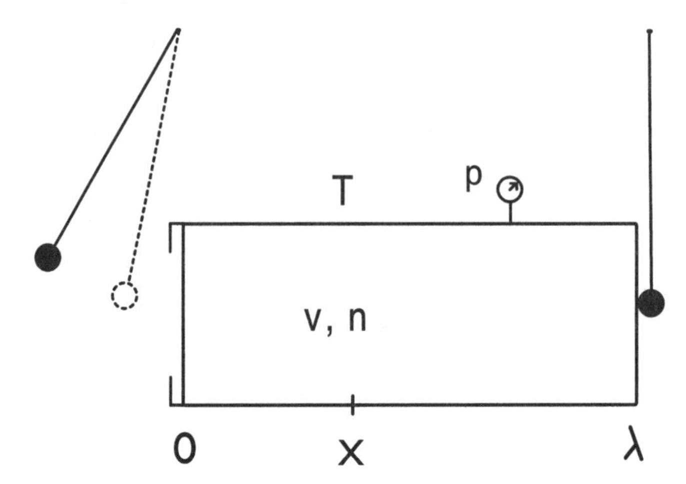

Fig. 1. Right circular cylinder of inner volume V containing amount n of a monatomic ideal gas. The cylinder rests in an air thermostat of temperature T. At equilibrium the pressure in the gas is uniform and is equal to p. Rod–and–bob pulse generators can be coupled to the left-hand and right-hand end faces of the cylinder.

The pulse generator coupled to a face of the cylinder is simply a (massless) rod with a metal bob of mass m on one end. The other end of the rod is affixed to a swivel joint located at such a place that the rod–and–bob combination swings like a pendulum from the swivel joint and can strike and rebound from the center of the cylinder face. The bob–end of the pulse generator interacts with a ratchet mechanism in such a way that the ratchet holds the bob in place before it is released to swing down and give an impulse to the gas sample and the ratchet catches the rebounding bob and holds it in place. Assume that the ratchet absorbs a negligible amount of energy as it fulfills its catching and holding functions.

To add an acoustic pulse to the gas sample: let the left-hand bob be at rest at a height h_L above a reference plane. Free the bob and let it swing down, strike the movable left-hand face of the cylinder, and rebound. Catch the rebounding bob with the ratchet just as it comes to rest at a height h_L' above the reference plane. The energy transferred to the gas, $\Delta\text{“}U\text{”} - Q$, in the form of an acoustic pulse is (assume that there are essentially no losses of the available potential energy due to friction, inelastic collision, etc.)

$$\Delta\text{“}U\text{”} - Q = -mg\left(h_L' - h_L\right) = -mg\Delta_L h \equiv \varepsilon > 0, \tag{35}$$

where g is the acceleration due to gravity, ε is the pulsed energy introduced into the gas in forming an acoustic dipole, “U” is the total energy of the gas (i.e. the internal energy U plus kinetic, gravitational, and other forms of energy), and Q is the heat supplied to the gas by the air thermostat.

Let the right-hand pulse generator be inactive, i.e. let it be decoupled from the right-hand cylinder face, and let the density of the gas before the introduction of the acoustic pulse be ρ_*. The introduction of the acoustic pulse into the gaseous system creates a moving acoustic dipole, one side of which (the leading edge) has a density greater than ρ_* and the other side of which (the trailing edge) has a density smaller than ρ_*—see Fig. 2.

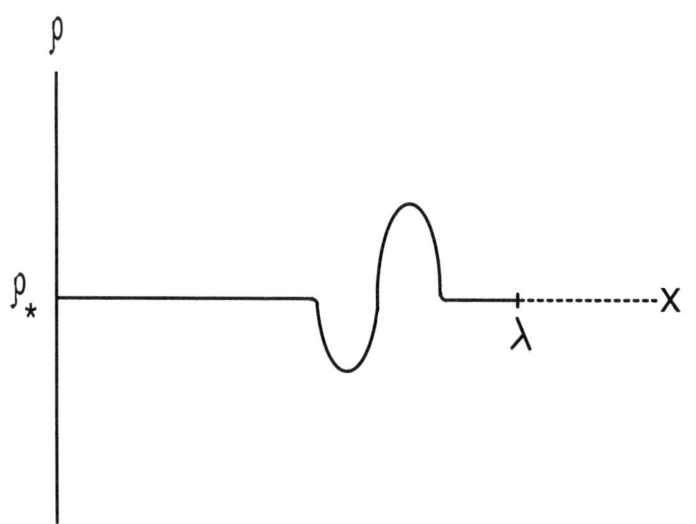

Fig. 2. Density profile of the gas in the cylinder at a particular time t_1 $\left(0 < t_1 < \lambda c_s^{-1}\right)$ after inserting the acoustic pulse into the gas.

The acoustic dipole travels through the gas at the speed of sound, c_s, in the gas:

$$c_s = \left(\gamma RT / M\right)^{1/2}, \tag{36}$$

where $\gamma \equiv \overline{C}_p / \overline{C}_v$ for the gas and M is the molecular weight of the gas. As the dipole sweeps over the pressure sensor, the readings of the sensor first go up, then go down, and then show the value $p_* = \rho_* RT / M$.

Assume that the dipole experiences only slight dissipative interactions in its motion through the gas, so that on a short-term basis (minutes) the dipole remains intact and carries the full pulse energy ε, but on a long-term basis (hours, days) the energy of the dipole is eventually dissipated and is incorporated into the random motions of the molecules of the system and the surroundings.

With the right-hand pulse generator decoupled from the right-hand face of the cylinder, the acoustic dipole travels back and forth through the gas, being reflected at each end face; and the density at a point x ($0 \le x \le \lambda$) inside the cylinder, $\rho(x,t)$, is a periodic function of the time, the period being $2\lambda / c_s$ (short-term basis):

$$\rho\left(x, t + k2\lambda c_s^{-1}\right) = \rho(x,t) \quad k = 1,2,3,\ldots. \tag{37}$$

What I wish to do next is to establish a steady flow of pulsed energy from the left generator through the gas to the right generator, so that the gas will be in a steady state with pulses (sound waves) passing through it. To see the basic idea, return to Fig. 1, and let the right generator be coupled to the right end face of the cylinder in the manner indicated, resting against the right face. When the acoustic dipole interacts with the right face, it will kick the right bob (also of mass *m*) from its lowest level h_R to some higher level h'_R. Let $\Delta_R h \equiv h'_R - h_R > 0$, and assume that matters are so arranged that $\Delta_L h + \Delta_R h = 0$; then in the steady state—once the gas is "saturated" with acoustic dipoles— there will be no *continuing* accumulation of pulsed energy in the gas, the gas merely serving as a conduit for pulsed energy from the left generator to the right generator. Let a train of pulses emanate from the left generator, the pulses all having the same $\Delta_L h$ and the time interval between successive pulses being always the same. (If necessary, mount a string of generators on a railway flatcar. Lay a track for the flatcar so it passes right by the left end face of the cylinder—do the same at the right end face. As the train of generators flashes by the cylinder, activate the generators so that they impinge on the left face as they shoot by.) With a little ingenuity, we can arrange for the time interval between successive pulses to be as small or as large as we please.

Let the frequency of pulses striking the left end face of the cylinder (the number of pulses per unit time) be $\dot{N}$. Then the number of acoustic dipoles inside the cylinder in the steady state is

$$\text{\# dipoles inside the cylinder} = \dot{N}\lambda / c_s. \tag{38}$$

Assume that each dipole contributes the same amount of pulsed energy to the system in the steady state. For that steady state, then, assume that

$$\text{“}U\text{”}(\text{steady state}) = U(T,V,n) + \varepsilon\left(\dot{N}\lambda c_s^{-1}\right), \tag{39}$$

$$S(\text{steady state}) = S(T,V,n). \tag{40}$$

Depolarizing the System

Start with the system in the steady state described above. Instantaneously decouple the left and right generators from the cylinder faces, and cover the cylinder surfaces with adiabatically insulating material so as to adiabatically insulate the system from its surroundings. If we wait long enough, the dipoles will eventually decay; and the gas will come to an equilibrium state characterized by the quantities V, n, T', p', with $T' > T$ and $p' > p_*$. In fact

$$p' / p_* = T' / T. \tag{41}$$

For the changes in going from the initial steady state to the final equilibrium state, we have

$$\begin{aligned} 0 = \Delta\text{“}U\text{”} &= U(T',V,n) - U(T,V,n) - \varepsilon\dot{N}\lambda c_s^{-1} \\ &= n\overline{C}_v(T' - T) - \varepsilon\dot{N}\lambda c_s^{-1}. \end{aligned} \tag{42}$$

So

$$T' = T + \left(\varepsilon\dot{N}\lambda / n\overline{C}_v c_s\right). \tag{43}$$

Also,

$$\begin{aligned} \Delta S &= S(T',V,n) - S(T,V,n) \\ &= n\overline{C}_v \ln T'/T = n\overline{C}_v \ln\left[1 + \left(\varepsilon\dot{N}\lambda / n\overline{C}_v c_s T\right)\right] > 0. \end{aligned} \tag{44}$$

Since for the adiabatic relaxation process, $\Gamma_{\uparrow} = \Gamma = \Delta S > 0$, we see that the process satisfies the Dartmouth condition.

Migrational Equilibrium in an Acoustic Field

Suppose we put a 2-component mixture of monatomic ideal gases in an acoustic field. Will the steady flow of acoustic dipoles through the gas mixture cause any separation of the two components? If we can establish the conditions for the migrational equilibrium of the components of the mixture, we can answer that question.

Take the apparatus of Fig. 1, and let the interior of the cylinder communicate via small holes at $x = 0$ and $x = \lambda$ (each hole covered by a membrane permeable to component 1 alone) with mass reservoirs of component 1 at T, p_ω $(\omega = 0, \lambda)$. Fill the cylinder with the 2-component gas mixture, the amounts of the two components being n_1, n_2 and the mole fraction of component 1 in the mixture being $X_1 = n_1 / (n_1 + n_2)$—see Fig 3. Establish a

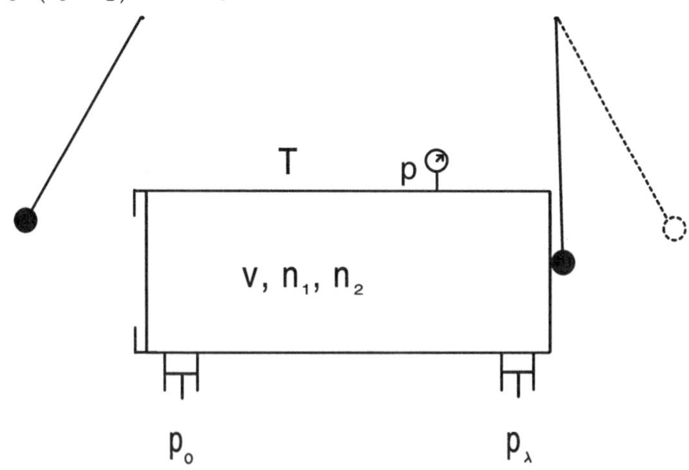

Fig. 3. Mass reservoirs for component 1 connected to the gas mixture through membranes permeable to component 1 alone at the 0 end and the λ end of the cylinder. The material in the mass reservoirs is at temperature T, pressure p_ω $(\omega = 0, \lambda)$.

steady-state acoustic field in the cylinder with a pulse insertion frequency of $\dot{N}$, each pulse starting from the same height h_L. Find a pair of pressures p_0, p_λ for the mass reservoirs for which component 1 is in a state of migrational equilibrium in the gaseous mixture. Then, keeping $T, p_\lambda, \dot{N}, h_L$, constant, transfer an amount $d\xi$ of component 1 in differentially tailored fashion from the mass reservoir at 0 to the reservoir at λ. For the process, the first law states that

$$d\text{“}U_{mix}\text{”} - \overline{U}_0 d\xi + \overline{U}_\lambda d\xi = dQ + p_0 \overline{V}_0 d\xi - p_\lambda \overline{V}_\lambda d\xi, \qquad (45)$$

or [Eq.(39)]

$$d\left(\varepsilon \dot{N} \lambda \langle c_s \rangle^{-1}\right) - \overline{U}_0 d\xi + \overline{U}_\lambda d\xi = dQ + p_0 \overline{V}_0 d\xi - p_\lambda \overline{V}_\lambda d\xi, \quad (46)$$

where dQ is an “undetermined quantity” to be evaluated later and I shall comment on $\langle c_s \rangle$ shortly.

Divide Eq.(46) through by $d\xi$ and rearrange terms:

$$d\left(\varepsilon \dot{N} \lambda \langle c_s \rangle^{-1}\right) / d\xi + \overline{H}_\lambda - \overline{H}_0 = dQ / d\xi, \qquad (47)$$

where $\overline{H}_\omega = \overline{U}_\omega + p_\omega \overline{V}_\omega$.

The speed of sound in a pure, ideal gas component k, $c_s^{(k)}$, is

$$c_s^{(k)} = \left(\gamma^{(k)} RT / M^{(k)}\right)^{1/2}.$$

What of the speed of sound in a mixture of monatomic ideal gases? Since γ has the same value, $\gamma = 5/3$, for each monatomic ideal gas, I shall take the speed of sound in the mixture, $\langle c_s \rangle$, to be

$$\langle c_s \rangle = \left(\gamma RT / \langle M \rangle\right)^{1/2}. \tag{48}$$

Let X_k be the mole fraction of component k in the mixture. Then

$$\langle M \rangle = \sum_k X_k M^{(k)}, \quad \text{with } \sum_k X_k = 1, \tag{49}$$

the molecular weight of component k being $M^{(k)}$.

For the change in entropy of the system, we have

$$dS = \left(\bar{S}_\lambda - \bar{S}_0\right) d\xi; \tag{50}$$

and for the change in the entropy of the surroundings attributable to the process,

$$d_\uparrow S_{sur} = -T^{-1} dQ. \tag{51}$$

So,

$$d\Gamma_\uparrow = \left(\bar{S}_\lambda - \bar{S}_0\right) d\xi - T^{-1} dQ \tag{52}$$

and

$$\frac{d\Gamma_\uparrow}{d\xi} = \left(\bar{S}_\lambda - \bar{S}_0\right) - \frac{1}{T}\frac{dQ}{d\xi}$$

$$= \left(\bar{S}_\lambda - \bar{S}_0\right) - \frac{1}{T}\left(\bar{H}_\lambda - \bar{H}_0\right) - \frac{1}{T}\frac{d\left(\varepsilon \dot{N} \lambda \langle c_s \rangle^{-1}\right)}{d\xi}$$

$$= \frac{1}{T}\left(\mu_0 - \mu_\lambda - \frac{d\left(\varepsilon \dot{N} \lambda \langle c_s \rangle^{-1}\right)}{d\xi}\right), \tag{53}$$

where $\mu_\omega = \bar{U}_\omega + p_\omega \bar{V}_\omega - T\bar{S}_\omega = \bar{H}_\omega - T\bar{S}_\omega \ (\omega = 0, \lambda)$.

On invoking the Thomson-Gibbs Corollary, we get

$$0 = \mu_0 - \mu_\lambda - \frac{d\left(\varepsilon \dot{N} \lambda \langle c_s \rangle^{-1}\right)}{d\xi} \tag{54}$$

as the condition of migrational equilibrium for the components of the monatomic ideal gas mixture in the acoustic field.

"Steady" on a Time-Average Basis

With the given constraints acting on the system (constant $T, V, n, \dot{N}, \lambda, h_L, p_0, p_\lambda$), the response of the system is apt to be an oscillating or fluctuating one. The pressure indicated on the pressure sensor attached to the cylinder will register the total pressure of the mixture at its point of interaction with the gas mixture, $p = p(t)$, which changes with time: as a dipole sweeps across the sensing area of the sensor, the indication of the sensor will rise and fall. Consider a point in the gas mixture at the $x = 0$ end of the cylinder, close to the semipermeable membrane that communicates with the mass reservoir of pressure p_0, with p_0 and p_λ chosen so as to satisfy Eq.(54). The partial pressure of component 1 at the point in question will most likely fluctuate or oscillate as the train of dipoles sweeps by, so there will be times at which

$$\mu_0(\text{at point in mixture}) \neq \mu_0(\text{in mass reservoir}) \tag{55}$$

and there will be some leakage of mass into and out of the mass reservoir. However, if we average the partial pressures of component 1 at the point in question over the time it takes for $\dot{N}\lambda\langle c_s\rangle^{-1}$ dipoles to sweep past the point, we should find that the leaks into and out of the mass reservoir just balance, and the time-average value of the chemical potential for component 1, $\hat{\mu}_0(\text{at point in mixture})$, satisfies the relation

$$\hat{\mu}_0(\text{at point in mixture}) = \mu_0(\text{in mass reservoir}), \tag{56}$$

where I have used the overhead caret to signify a time-average value (to show a time-average value for a complex expression, I shall also use the label "$\Big|_{AV}$").

An expression analogous to Eq.(56) should hold at the λ end of the cylinder. In terms of time-average values, we can write Eq.(54) as

$$0 = \hat{\mu}_0(\text{in mixture}) - \hat{\mu}_\lambda(\text{in mixture}) - \left.\frac{d\left(\varepsilon\dot{N}\lambda\langle c_s\rangle^{-1}\right)}{d\xi}\right|_{AV}, \tag{57}$$

To be able to proceed further, let us assume that the time-average total pressure is the same at both the 0 end and the λ end of the gaseous mixture. Then we can write Eq.(57) as

$$\ln \hat{X}_1^{(0)} / \hat{X}_1^{(\lambda)} = \frac{1}{RT}\left.\frac{d\left(\varepsilon\dot{N}\lambda\langle c_s\rangle^{-1}\right)}{d\xi}\right|_{AV}, \tag{58}$$

and we see that if there is coupling between the flux of energy pulses and the flow of component 1, i.e. if the right-hand side of Eq.(58) is not zero, then we can have some separation of the components of the gaseous mixture in the acoustic field.

Finally, let's evaluate the "undetermined quantity" dQ, in the form $dQ / d\xi$. Let the rate of influx of heat to the air thermostat be $\dot{Q}^{(r)}$. Start from a state of migrational equilibrium [Eq.(57)]. Keeping $T, \dot{N}, h_L, p_\lambda$ constant, select a value for p_0 that will induce a (slow) steady (in the time-average sense) flow of component 1 from the reservoir at 0 through the gas mixture to the reservoir at λ. We expect to find that upon keeping constant the quantities $T, \dot{N}, h_L, p_\lambda, p_0$, the system will behave in such a way, on a time-average basis, that

$$\hat{\dot{\xi}}_0 + \hat{\dot{\xi}}_\lambda = 0, \tag{59}$$

where $\dot{\xi}_\omega$ $(\omega = 0, \lambda)$ is the rate of influx of amount of component 1 into the mass reservoir at ω and the carets indicate time-average values. Evaluate the derivative

$$\delta\hat{\dot{Q}}^{(r)} / \tilde{\delta}\hat{\dot{\xi}}_0 \Big)_{T,...}$$

via a sequence of steady (on a time-average basis) states such that $\hat{\dot{\xi}}_0 \rightarrow 0$

[see Eq.(1.14)]. Make the identification

$$dQ / d\xi = \delta\hat{\dot{Q}}^{(r)} / \tilde{\delta}\hat{\dot{\xi}}_0 \Big)_{T,...} \tag{60}$$

where appropriate.

[There is additional discussion of systems in acoustic fields in an addendum, pages 468-474.]

BOOK II. THE OLD IDEA

The Rate of Entropy Generation

A reprint of the book *Thermodynamics of Steady States* by Ralph J. Tykodi, originally published by Macmillan in 1967. The original version contained answers to selected problems; but a pamphlet had been prepared with answers to all the problems, available to instructors using the book as a text. The text of the pamphlet containing answers to all the problems is reprinted here as an addendum to the book.

ERRATUM: page 257, line 15, for "$\left(fT_{\alpha}, p_{\alpha}, T_{\beta}, p_{\beta}, \langle T \rangle\right) = 0$" read "$f\left(T_{\alpha}, p_{\alpha}, T_{\beta} \cdot p_{\beta}, \langle T \rangle\right) = 0$."

Answers to Book II exercises start on page 398.

THERMODYNAMICS OF STEADY STATES

RALPH J. TYKODI
Emeritus Professor of Chemistry
University of Massachusetts Dartmouth

TO

Mary and Martin Kilpatrick

Preface

This book is both an original monograph and a textbook dealing with the application of thermodynamics to nonequilibrium situations; it represents an attempt to extend and to make rigorous a line of development initiated by William Thomson [1] and pursued by other investigators such as Eastman [2], Wagner [3], and (in a certain sense) Brønsted [4]. Each of the cited authors tried to treat nonequilibrium situations on a par with equilibrium ones, using (essentially) the same severely classical techniques for both types of phenomena. These classical approaches to the treatment of nonequilibrium situations failed to achieve the desired degree of rigor and generality, and a nonclassical method of treatment originating with Onsager [5] has come to the fore in the last few decades [5–9]—it today enjoys great popularity under the title *Thermodynamics of Irreversible Processes.*

Although I did not originally intend it so, this book has ended up being something of a synthesis of the lines of development initiated by William Thomson and by Lars Onsager. The treatment here is along decidedly classical lines (I maintain throughout a global rather than a local view of systems and processes), yet some of the conceptions introduced by Onsager do find a natural place in the development. Before reading this book the reader should be acquainted with the elements of classical thermodynamics.

I have divided the book into three parts: Principles, Applications, and Comments. I lay down the theoretical foundations in the part on Principles and comment on them in the part on Comments. The section on Applications contains analyses of systems I have found to be of interest. I have not attempted to make the coverage in any way exhaustive.

With respect to my earlier work in this area [10], this book shows an improvement in notation (I now indicate special quantities by using a typographical convention instead of German script), and it includes much new material appearing here in print for the first time. In keeping with the textbook function of this work, I have constructed a series of exercises for the reader; the exercises are partly of the theorem-proving type and partly of the data-processing type. In my overall style of presentation I have been influenced by the writings of Michael Polanyi [11]. When developing a logical chain of argument or when introducing routine assumptions, I march shoulder to shoulder with the reader: *if we combine Eqs. . . . , we get . . .* ; *let us assume that. . . .* When dealing with definitions, speculative assumptions, or matters of opinion, I step forward and assume full responsibility for the presentation: *I here find it convenient to introduce the quantities . . .* ; *I assume that the system . . .* ; *I feel that. . . .*

This book does for the class of nonequilibrium situations involving stationary states and steady-rate processes what any good book on ordinary thermodynamics does for equilibrium situations: it develops a series of relations of interest in their own right; it helps the experimentalist plan his experiments efficiently by making use of the necessary interconnections among experimentally determined quantities; it provides the experimenter with some consistency checks on his measurements; and it yields equations interrelating macroscopic quantities—equations that are of use to engineers and that serve as guides for more detailed kinetic theory or statistical mechanical analysis of the phenomena.

I wish to thank those persons who have encouraged me in my work on nonequilibrium thermodynamics: my former associates in the chemistry department of the Illinois Institute of Technology, foremost among whom has been Peter G. Lykos; other colleagues, in particular Ted A. Erikson, Charles K. Hersh, Isadore Hauser, Myron Tribus, Robert G. Parr, James C. M. Li, and L. T. Fan; many friends; and, above all, my wife Lois. I am indebted to the Illinois Institute of Technology for financial support that enabled me to spend part of a summer working on the manuscript.

Acknowledgements

I am indebted in the following ways for permission to reproduce previously published material: to John Wiley and Sons, Inc. (Interscience), for permission to quote from Brønsted's monograph on *Energetics*; to John Chipman and the American Chemical Society for permission to reproduce Figure 8–9; to the American Institute of Physics for permission to reproduce Figure 2–2; and to Ted A. Erikson for permission to include Figures 2–3 through 2–6.

References

1. W. Thomson, *Mathematical and Physical Papers*, Vol. I (Cambridge University Press, Cambridge, 1882), pp. 232–291.
2. E. Eastman, *J. Am. Chem. Soc.*, **48**, 1482 (1926); **49**, 794 (1927).

3. C. Wagner, *Ann. Phys.*, **3**, 629 (1929).
4. J. N. Brønsted, *Principles and Problems in Energetics* (Interscience Publishers, New York, 1955), Chapter 7.
5. L. Onsager, *Phys. Rev.*, **37**, 405 (1931); **38**, 2265 (1931).
6. I. Prigogine, *Etude Thermodynamique des Phenomenes Irreversibles* (Desoer, Liege, 1947).
7. S. de Groot, *Thermodynamics of Irreversible Processes* (North-Holland Publishing Company, Amsterdam, 1951).
8. K. Denbigh, *Thermodynamics of the Steady State* (Methuen and Company, Ltd., London, 1951).
9. I. Prigogine, *Introduction to the Thermodynamics of Irreversible Processes* (Charles C. Thomas, Springfield, Illinois, 1955).
10. R. J. Tykodi and T. A. Erikson, *J. Chem. Phys.*, **31**, 1506–1525 (1959); **33**, 40–49 (1960).
11. M. Polanyi, *Personal Knowledge* (University of Chicago Press, Chicago, 1958).

R. J. T.

Contents

PART II: APPLICATIONS

PART III: COMMENTS

APPENDIXES

List of Primary Symbols

[The page numbers cited below are holdovers from the 1967 edition. To get the correct page number for this edition, add 195 to the cited page number.]

(*Italic number following definition indicates the page symbol is first used*)

$(*)$	reference state, *20*.
$\delta/\delta\eta$	partial derivative with respect to η evaluated at the point $\eta = 0$, *27*.
.	(dot over letter)—time derivative, i.e. $\dot{Z} \equiv dZ/dt$.
–	(bar over letter)—property per mole; molar, partial molar, or mean molar property according to context.
$\langle\ \rangle$	average value.
$\langle\ \rangle_t$	time-average value.
[ ]	integral steady-state quantity.
$[\ ]^\partial$	differential steady-state quantity.
$\equiv$	is identically equal to, is by definition equal to.
$\approx$	is approximately equal to.
$\propto$	is proportional to.
a	place designation; activity, *21*; Helmholtz free energy density, *139*.
a_{ij}	element of the determinant A, *36*.
A	Helmholtz free energy, $A \equiv U-TS$, *6*; functional determinant, *36*.
A_{ij}	cofactor of the element a_{ij} in the determinant A, *36*.
$\mathscr{A}$	molar total Helmholtz free energy, i.e., ordinary molar Helmholtz free energy plus additional energy terms, *11*.
b	place designation.
B	interfacial area, *17*; reactant in a chemical reaction, *20*; cross sectional area, *131*; magnitude of the magnetic field vector **B**, *192*.
B	magnetic field vector, *188*.
$[c]$	Thomson-type steady-state function, *66*; Thomson coefficient, *106*.
C	concentration, *18*; heat capacity, *55*; nonreactive component, *148*.
$[C]$	integral molar steady-state heat capacity function, *65*.
$[C]^\partial$	differential molar steady-state heat capacity function, *52*.
D	diffusion coefficient, *18*; product in a chemical reaction, *20*.

E emf of a cell, *102*.
$\mathbf{f}$ dummy field vector, *188*.
F the Faraday; number of degrees of freedom, *29*.
$\mathscr{F}$ generalized potential difference, *69*.
g acceleration due to gravity, *93*; Gibbs free energy density, *138*.
$[\mathbf{g}]$ steady-state Gibbs free energy density function, *142*.
G Gibbs free energy, $G \equiv H - TS$, *6*.
$[G]$ integral molar steady-state Gibbs free energy function, *50*.
$[G]^{\partial}$ differential molar steady-state Gibbs free energy function, *65*.
$\mathscr{G}$ molar total Gibbs free energy, i.e., ordinary molar Gibbs free energy plus additional energy terms, *11*.
h running index; height above a reference plane, *93*; enthalpy density, *138*.
$[h]$ molar steady-state linkage function, *49*.
H enthalpy, *6*.
$[H]$ integral molar steady-state enthalpy function, *49*.
$[H]^{\partial}$ differential molar steady-state enthalpy function, *65*.
$\mathscr{H}$ molar total enthalpy, i.e., ordinary molar enthalpy plus additional energy terms, *11*.
i running index.
I electric current, *43*.
$\mathscr{I}$ moment of inertia, *191*.
j running index.
J molar flux, *18*; a current other than a heat current, *156*.
k running index; reaction rate constant, *22*.
$\mathbf{k}_T$ thermal diffusion ratio, *88*.
K equilibrium constant, *21*; phenomenological coefficient, *31*.
$\mathscr{K}$ molar kinetic energy of macroscopic motion, *41*.
L phenomenological coefficient, *31*.
m running index; molality, *94*.
M molecular weight, *40*.
$[M]$ integral steady-state function, *100*.
$[M]^{\partial}$ differential steady-state function, *100*.
n number of moles, *6*.
N torque, *149*.
$[N]$ integral steady-state coefficient, *12*.
$[N]^{\partial}$ differential steady-state coefficient, *12*.
P pressure.
q heat gained by the surroundings of a gradient part, *9*.
$\mathbf{q}$ separation factor, *88*.
Q heat supplied to the system, *6*.
$Q^{(r)}$ heat supplied to a reservoir, *9*.
$[Q]$ integral molar steady-state latent heat function, *65*.

$[Q]^{\partial}$ differential molar steady-state latent heat function, *52*.
$\mathscr{Q}$ activity quotient, *21*.
r composition ratio, *120*; radial distance, *190*.
$[r]$ coupling coefficient, *41*.
$[r]^{\partial}$ differential coupling coefficient, *42*.
R the gas constant; electrical resistance, *113*.
$[R]$ steady-state coupling coefficient, *41*.
s entropy associated with the surroundings of a given linkage, *47*; entropy density, *138*.
$[s]$ molar steady-state linkage function, *49*.
$[\mathbf{s}]$ steady-state entropy density function, *141*.
S entropy, *6*.
$S^{(r)}$ entropy of a reservoir, *9*.
$[S]$ integral molar steady-state entropy function, *49*.
$[S]^{\partial}$ differential molar steady-state entropy function, *52*.
$\mathbf{S}$ separation, *88*.
t transference number, *153*; time, *173*.
T temperature.
$\langle T \rangle$ effective average temperature of the surroundings of a given linkage, *47*.
u internal energy density, *139*.
U internal energy, *6*.
$U^{(r)}$ internal energy of a reservoir, *27*.
$\mathscr{U}$ molar total energy, i.e., internal energy plus kinetic energy plus gravitational energy plus . . ., *9*.
V volume.
W work supplied to the system, *6*.
W_0 work other than pressure-volume work supplied to the system, *12*.
X mole fraction, *42*.
Y generalized steady current, *27*.
z dummy intensive variable.
Z dummy variable.

α place designation.
$\boldsymbol{\alpha}$ thermal diffusion factor, *88*.
β place designation.
γ place designation; activity coefficient, *96*.
Γ Stefan–Boltzmann constant, *140*; thermomagnetic or galvanomagnetic coefficient, *193*.
δ_{jk} the Kronecker delta symbol, *37*.
$\in$ symbol for set inclusion, *31*.
ζ dummy variable; figure of merit, *123*.
η dummy variable; efficiency, *123*.

$[\eta]$ alternative measure of efficiency, *123*.
θ *measured-to-mean* mass flow ratio, *17*; relative activity, *100*; angular measure, *149*.
ϑ *measured-to-all tube* mass flow ratio, *17*; fraction, *114*; fractional parameter, *143*.
Θ rate of entropy production, *11*.
κ thermal conductivity, *131*.
λ distance parameter, *41*; wavelength, *139*.
Λ characteristic length, *41*.
$\mu^{(i)}$ chemical potential of the *i*th species, *6*.
$\bar{\mu}$ electrochemical potential, *109*.
ν stoichiometric coefficient, *20*; number of steady currents or affinities, *29*; number of components, *94*.
ξ degree of advancement of a chemical reaction or process, *5*.
π Peltier heat, *106*.
ρ running index.
σ reference thermostatic state, *29*.
$\boldsymbol{\sigma}$ Soret coefficient, *97*.
τ current ratio (transference number), *117*.
φ gravitational potential energy per gram, *40*.
χ dummy variable usually equal to α or β.
ψ dummy variable; electric potential, *43*.
ω dummy variable usually equal to α or β.
Ω generalized affinity, *27*.

THERMODYNAMICS OF STEADY STATES

PART I: PRINCIPLES

1

Introduction

In the last decade of his life, J. N. Brønsted grew increasingly concerned with the foundations and formulations of classical thermodynamics. His writings on the subject show a keen appreciation of the thermodynamic way of doing things, and his critical analyses of classical procedures will always be of value. In addition to his critical work, he also undertook a general reformulation of the language of thermodynamics in such a way as to stress *processes* rather than *states*; he called his formulation of the principles of thermodynamics by the name *energetics*. Shortly before his death in 1947, he summarized his work in this area in a short monograph entitled *Principles and Problems in Energetics.*

Although this book does not follow Brønsted in the formal part of his energetics, it owes much to his critical writings. It would be difficult to find a clearer statement of the scope and nature of classical thermodynamics than that appearing in the introductory chapter of Brønsted's monograph [1]:

> The subject of *energetics* describes ordinary macroscopic phenomena, and attempts from a general point of view to establish the laws governing the changes brought about by these phenomena. It deals with the forms and components of energy, and with its transformations. This field of knowledge is usually termed *thermodynamics*, but in spite of the great importance of thermal phenomena their position in relation to the whole field is hardly sufficiently outstanding to justify the traditional name. On the other hand, the concept of energy is a universal one and provides a natural name for this division of science. The field covered by energetics is, however, limited by the fact that purely mechanical phenomena, including motion, are completely described under the heading of mechanics, and therefore are not usually included in energetics.

It is important to bear in mind that energetics seeks to solve problems solely on the basis of macroscopic observations. In other words, it does not appeal to ideas involving the existence of discrete particles of matter or energy. The introduction of such ideas represents an excursion outside the bounds of pure energetics. This statement does not, of course, seek to question the importance of molecular theory and quantum theory for science as a whole or for the general problems with which energetics is concerned. However, it is of great importance for the clarity of scientific understanding that the different methods of attack should be kept strictly separated from one another. Several serious weaknesses in the existing system of traditional thermodynamics can be attributed to an insufficient appreciation of the value of this prescript.

Historically, the origin of energetics lies in the study of the transformations of energy. Carnot, Robert Mayer, Joule, Clausius, William Thomson, and other pioneers in energetics were concerned with the "motive power of heat" and the "equivalence of different forms of energy." "Heat" and "work" were the practical concepts whose relations were sought. Later developments (notably by Willard Gibbs) led to a system of functional energetics in which a material system is described by means of the functions that are characteristic of its state and that serve to determine its behavior under different conditions. These two tendencies in energetics differ somewhat in that the former emphasizes energetic transformations and the latter, energetic functions. However, both lines of development, in conformity with their historical origin, are based on the study (in the light of experience) of suitably chosen elementary processes, while the principles of energetics are the generalizations arrived at by a logical interpretation of this study.

The Brønsted admonition against inopportunely introducing microscopic considerations into purely thermodynamic discussions reflects a basic feature of a well-established thermodynamic tradition; indeed a thermodynamicist was once heard to remark that thermodynamics is the study of systems with macroscopically describable surroundings—it is the surroundings that are studied in order that the properties of the system may be inferred. This book is imbued with the classical spirit so aptly epitomized by Brønsted; it deals with the *necessary* relations among the macroscopic properties of well-defined systems. Once we have established the set of thermodynamic relations for a given class of phenomena, we usually desire (in this modern age) to "explain" the observed behavior of the quantities and coefficients introduced into the thermodynamic analysis in terms of the properties of atoms and molecules. Evidently, we cannot "explain" a set of experimental facts and necessary relations until we know what the facts and relations are. For the large class of nonequilibrium situations amenable to thermodynamic analysis, the gathering of experimental facts and the establishment of necessary relations among those facts is work for the present; the explanation via statistical mechanical analysis of the patterns revealed by the facts is largely work for the future.

Nonequilibrium Situations

Classical thermodynamics deals primarily with equilibrium (static) states and with quasi-static processes (sequences of equilibrium states). In order to successfully carry through a classical analysis, we must be able to distinguish between the *system* and the *surroundings*, and we must be able to keep track of given amounts of heat, work, and energy exchanged between system and surroundings. Now these requirements are broad enough to encompass not only classical thermodynamics but also a class of nonequilibrium situations involving steady states and quasi-steady processes (sequences of steady states). We can use for steady states procedures analogous to those we use for static states, and as a result we can rigorously derive truly thermodynamic relationships for many nonequilibrium situations. We must take care, however, always to relate derived concepts to the primitive concepts of heat, work, and energy. A common set of concepts and procedures is adequate for the treatment of equilibrium states, quasi-static processes, steady states, and quasi-steady processes. Excluded from this extended classical domain are situations involving rapidly varying values of the parameters of the system;* such situations properly fall in the domain of kinetic theory and statistical mechanics.

It is surprising how easily and naturally classical procedures can be adapted to the treatment of steady-state situations. This book, then, presents a general thermodynamic approach, entirely classical in spirit, to the problems posed by nonequilibrium situations; it analyzes in detail a large number of stationary states and steady-rate processes; and it concerns itself exclusively with the macroscopic features of the investigated phenomena.

The Porous Plug Experiment

For purposes of orientation consider that well-known example of an irreversible process: the flow of a gas across a porous plug (the Joule–Thomson experiment)—gas and plug being thermally insulated from the surroundings. Let the system of interest be marked out as in Figure 1–1, where α and β refer to the high-pressure and low-pressure sides, respectively. Let us assume that the gas is flowing across the plug at a steady rate of $\dot{\xi}$ moles per second, that the gradients in temperature and pressure can be localized in a region in the vicinity of the plug, and that sufficiently far from the plug the thermodynamic state of the gas is characterized by the variables (T_α, P_α) and

* I am referring here to situations that are not even quasi-steady—i.e., to situations for which the concept *macroscopic state of the system* is inapplicable, it being necessary to specify the instantaneous and rapidly varying values of the parameters at each and every point of the system.

(T_β, P_β). The system is thus divided into *terminal parts* that are homogeneous in intensive variables and a *gradient part* that contains gradients in intensive variables. Let us further assume that the gradient part remains stationary (i.e., pointwise invariant in time) during the steady-flow process. Now, if we take a definite mass of gas and the porous plug as our system, and if we apply

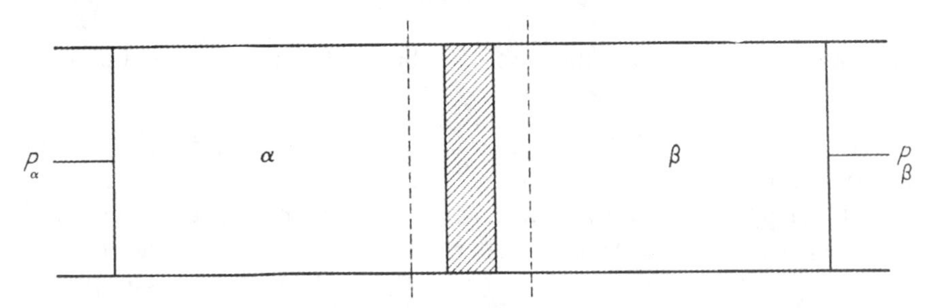

Figure 1–1. A mass of gas in adiabatic flow across a porous plug; $P_\alpha > P_\beta$.

the first law of thermodynamics to the adiabatic steady-flow process, then we obtain the relation*

$$\dot{U}(\text{system}) = \dot{W} = -P_\alpha \dot{V}_\alpha - P_\beta \dot{V}_\beta. \tag{1.1}$$

Since the flow is between terminal parts of the system, we have

$$\dot{U}(\text{system}) = \overline{U}_\alpha \dot{n}_\alpha + \overline{U}_\beta \dot{n}_\beta, \tag{1.2}$$

$$\dot{V}_\omega = \dot{n}_\omega \overline{V}_\omega(T_\omega, P_\omega) \qquad \omega = \alpha, \beta, \tag{1.3}$$

and

$$-\dot{n}_\alpha = \dot{n}_\beta \equiv \dot{\xi}, \tag{1.4}$$

where the overbar indicates a molar property of the gas. The gradient part, being stationary, does not contribute to $\dot{U}$(system). Putting Eqs. (1.1) to (1.4) together (and neglecting kinetic energy contributions to the internal energy), we get

$$(\overline{U}_\beta - \overline{U}_\alpha)\dot{\xi} = (P_\alpha \overline{V}_\alpha - P_\beta \overline{V}_\beta)\dot{\xi} \tag{1.5}$$

and

$$\overline{H}_\beta = \overline{H}_\alpha, \tag{1.6}$$

where $\overline{H} \equiv \overline{U} + P\overline{V}$. Equation (1.6) is the well-known constant enthalpy condition for the porous plug experiment. In deriving it, I made use of

* My choice of symbols for the standard thermodynamic functions is U, H, A, G, and S; Q and W represent heat and work supplied *to* the system; $\mu^{(i)}$ is the chemical potential of the ith species, and n stands for number of moles.

several features characteristic of my version of the thermodynamics of irreversible processes; a listing, therefore, of my primary assumptions is here in order.

Basic Premises

In addition to invoking the laws of ordinary thermodynamics, I assume that

(I) The system is capable of sustaining steady-rate operations under the described conditions.

(II) The system can always be divided into terminal parts and gradient parts, the gradient parts being stationary during steady-flow operations.

(III) Certain nonflow states can be treated as the limit of a sequence of steady-flow states.

I also systematically investigate the consequences of yet another assumption:

(IV) In a *proper* (ordinarily affine*) *sequence* of steady-flow states converging on a nonflow state, the limiting state is a state of minimum rate of entropy production relative to the neighboring flow states.

A fifth assumption appears in Chapter 3.

Assumption I plays the same role here that the analogous assumption plays in equilibrium thermodynamics: it is there always tacitly assumed that equilibrium states can be achieved under the given conditions; if a set of experiments yields conflicting results, we conclude that at some stage we failed to attain equilibrium. It is equally so with steady states: conflicting results imply failure to achieve steady states at some point in the procedure.

Assumption II is one of the distinguishing features of the present approach. The division of the system into terminal parts and gradient parts is especially fruitful when applied to steady states. In a steady state the gradient parts are stationary (pointwise invariant in time) so the change in any thermodynamic property of the system in a given time interval is found by simply summing up the changes in the terminal parts for the appropriate time interval.

Assumption III allows us to embed a nonflow state in a larger class of steady-flow states so as to have some maneuvering room for determining conditions of migrational equilibrium.† Much the same thing is done in equilibrium thermodynamics when variations are taken about an equilibrium state so as to determine the extremum properties of the equilibrium state.

Assumption IV holds when the reference nonflow state is a state of

* I explain the technical significance of this word in Chapter 3.

† A formal definition of the meaning ascribed to this phrase appears later in the chapter.

thermostatic equilibrium;* it is interesting to trace the consequences of this assumption for more general cases. If we could show that a properly generalized form of assumption IV (see the *special fields* section of the Appendix) is always true, the result would be of sufficient importance to merit calling it the Fourth Law of Thermodynamics.

Whereas the meaning of assumptions I and II is fairly clear, the full import of assumptions III and IV can only be appreciated inductively, by examination of a number of special cases. In the succeeding chapters we shall arrive at more precise formulations of assumptions III and IV, and in the concluding chapter we shall again take up assumptions III and IV in the light of the intervening analyses.

Brief Outline

Archimedes of Syracuse, upon having discovered the law of the lever, is reported to have said, "Give me a place to stand and I will move the world." The Archimedean lever with which we set in motion the machinery of thermodynamics is the closed system: the first law of thermodynamics can always be applied in an unambiguous fashion to closed systems. With a little ingenuity we can always manage to encapsulate the system or process of interest inside a closed system; then, starting from the first law of thermodynamics, we can proceed to winnow out the sought-after relations. A brief outline of the method I employ, together with a number of necessary preliminaries, is as follows.

In each case I define the *system* in such a way as to insure that only heat and work can be exchanged between the system and the surroundings; therefore, if we restrict our attention to steady-rate processes, we are always entitled to write

$$\dot{U}(\text{system}) = \dot{Q} + \dot{W} \tag{1.7}$$

for the rate at which the system gains internal energy. I normally divide the system into terminal parts and gradient parts, and I assume that the terminal parts of the system are spatially uniform.

Next, I find it useful to distinguish between *monothermal* and *polythermal* processes (and systems). A monothermal process is one in which the *entire* system is in heat communication with a *single* heat reservoir of temperature T during the process. A polythermal process (or system) is one in which the terminal parts of the system are separately in heat communication with heat reservoirs of temperatures T_i during the process. I allow for heat communication between the surroundings (exclusive of labeled thermostats) and the

* The rate of entropy production is zero in a state of thermostatic equilibrium, whereas in any neighboring steady-flow state the rate of entropy production must be positive definite.

gradient parts in the polythermal case, but I shall not now delve into the details of the analysis.* For a monothermal process then

$$\dot{Q} = -\dot{Q}^{(r)} = -T\dot{S}^{(r)}, \tag{1.8}$$

where $\dot{Q}^{(r)}$ represents the rate of influx of heat into the heat reservoir and $\dot{S}^{(r)}$ represents the rate of accumulation of entropy in the heat reservoir. Similarly, for a polythermal process,

$$\dot{Q} = -\dot{q} - \sum_{i} \dot{Q}_i^{(r)} = -\dot{q} - \sum_{i} T_i\dot{S}_i^{(r)}, \tag{1.9}$$

where $\dot{q}$ represents the heat gained per unit time by the surroundings (exclusive of labeled thermostats) via exchange with gradient parts, and $\dot{S}_i^{(r)}$ (e.g.) represents the rate of accumulation of entropy in the ith heat reservoir (which exchanges heat with the ith terminal part of the system).

In those cases for which there is a steady exchange of mass between terminal parts of the system, I assume that the gradient parts of the system are stationary, i.e., the state of the gradient parts is pointwise invariant in time during the steady-flow process. Under such conditions we can write

$$\dot{U}(\text{system}) = \sum_{i} \mathcal{U}_i \dot{n}_i, \tag{1.10}$$

where $\mathcal{U}_i$ is the total energy per mole (internal energy plus other kinds of energy—kinetic, gravitational, etc.) of the material in the ith terminal part of the system, and $\dot{n}_i$ represents the rate of influx of matter into the ith terminal part. The sum is taken over the terminal parts only, since the gradient parts (being stationary) do not contribute to $\dot{U}$(system) during the steady-flow process.† It is usually possible to arrive at a satisfactory description of any steady-rate process by combining Eqs. (1.7) and (1.10).

In dealing with systems in which there is no flow of a given type, I assume that the nonflow state can be taken as the limit of a sequence of steady-flow states. In addition, I find it necessary to assume that the system is capable of sustaining steady-rate operations under the described conditions. Sometimes I have to introduce new quantities to account for the fact that stationary states and steady-rate processes are, after all, different from static states and quasi-static processes. I introduce such quantities in as operational a manner as possible, using the equations of classical thermostatics as guides.

* The *monothermal, polythermal* language reflects the emphasis on the *surroundings* that was mentioned earlier. A system may be immersed in a thermostat and yet, due to going processes inside of it, it may not be uniform in temperature. We could not properly refer to such a system as isothermal. However, the surroundings of such a system would be characterized by a constant, well-defined temperature, and a going process in such a system could be referred to as monothermal.

† From the definition of a terminal part—spatially uniform in intensive properties—it follows that for a steady-flow process $d\mathcal{U}_i/dt = 0$.

The preceding outline forms the basis for treating two types of problems: (i) the steady migration of a given chemical species from one terminal part of the system to another, and (ii) the *migrational equilibrium* of a given chemical species in a spatial field. A chemical species is said to be in migrational equilibrium in a spatial field if there is no macroscopic tendency for the given species to migrate from one place in the field to another. Chemical reactions can be made to fall into classes (i) and (ii) if we consider migration with change of identity—there being a correlation between reactants supplied to or removed from one point of the spatial field and products removed from or supplied to another point. Examples of case (i) are the forced vaporization process and chemical reactions in monothermal fields.* Examples of case (ii) are such polythermal field phenomena as the thermomolecular pressure effect, the Soret effect, etc.† Of the two sorts of problems, those of type (ii) are much the more important. By introducing migrational equilibrium functions, I try to systematize the treatment of steady states in the same way that Gibbs systematized the treatment of static states in his paper on heterogeneous equilibrium.

Reference

1. J. N. Brønsted, *Principles and Problems in Energetics* (Interscience Publishers, New York, 1955), p. 1.

* See Chapter 2.
† See Chapters 5 and 8.

2

Monothermal Steady-Rate Processes

General Remarks

For a system that is the site of a monothermal steady-rate process (provided that we can split the system into terminal parts and gradient parts), we can write

$$\dot{U}(\text{system}) \equiv \sum \mathscr{U}_i \dot{n}_i = -T\dot{S}^{(r)} + \dot{W}, \tag{2.1}$$

$$\dot{H}(\text{system}) \equiv \sum \mathscr{H}_i \dot{n}_i = -T\dot{S}^{(r)} + \dot{W} + \sum P_i \dot{V}_i, \tag{2.2}$$

$$\dot{A}(\text{system}) \equiv \sum \mathscr{A}_i \dot{n}_i = -T\dot{S}^{(r)} - \sum T_i \bar{S}_i \dot{n}_i + \dot{W}, \tag{2.3}$$

$$\dot{G}(\text{system}) \equiv \sum \mathscr{G}_i \dot{n}_i = -T\dot{S}^{(r)} - \sum T_i \bar{S}_i \dot{n}_i + \dot{W} + \sum P_i \dot{V}_i, \tag{2.4}$$

where $\mathscr{U}_i$ represents the total energy per mole of material in the ith terminal part, $\bar{Z}_i$ represents a molar property of the material in the ith terminal part, $\mathscr{H}_i \equiv \mathscr{U}_i + P_i\bar{V}_i$, $\mathscr{A}_i \equiv \mathscr{U}_i - T_i\bar{S}_i$, $\mathscr{G}_i \equiv \mathscr{H}_i - T_i\bar{S}_i$, $\dot{S}^{(r)}$ represents the rate of accumulation of entropy in the surrounding heat reservoir of temperature T, and $\dot{n}_i$ and $\dot{V}_i$ represent the rate of accumulation of mass and volume, respectively, in the ith terminal part of the system. The sums in Eqs. (2.1) to (2.4) are over the terminal parts only since the gradient parts are stationary during the steady-flow process. The rate of entropy production Θ,

$$\Theta \equiv \dot{S}(\text{system}) + \dot{S}(\text{surroundings}), \tag{2.5}$$

under steady-flow conditions is

$$\Theta = \dot{S}^{(r)} + \sum \bar{S}_i \dot{n}_i. \tag{2.6}$$

In terms of the rate of entropy production Θ and the rate of performance of other-than-pressure-volume work $\dot{W}_0$,

$$\dot{W}_0 \equiv \dot{W} + \sum P_i \dot{V}_i, \tag{2.7}$$

Eqs. (2.1) to (2.4) take the well-known forms

$$\dot{U}(\text{system}) = -T\dot{S}^{(r)} + \dot{W}, \tag{2.8}$$

$$\dot{H}(\text{system}) = -T\dot{S}^{(r)} + \dot{W}_0, \tag{2.9}$$

$$-\dot{A}(\text{system}) = T\Theta - \dot{W} + \sum (T_i - T)\bar{S}_i \dot{n}_i, \tag{2.10}$$

$$-\dot{G}(\text{system}) = T\Theta - \dot{W}_0 + \sum (T_i - T)\bar{S}_i \dot{n}_i. \tag{2.11}$$

For most cases of interest it will be true that $T_i = T$ (however, see Ex. 2–4).

At this point I find it convenient to define coefficients $[N]$ and $[N]^\partial$ such that

$$[N] \equiv \frac{\Theta}{R\dot{n}_k^{\,2}}, \tag{2.12}$$

$$[N]^\partial \equiv [N] + \dot{n}_k\left(\frac{\partial [N]}{\partial \dot{n}_k}\right)_T, \tag{2.13}$$

where R is the gas constant, and $\dot{n}_k$ is a suitably chosen rate parameter. The relationship between the quantities $[N]$ and $[N]^\partial$ is like that between integral molar and differential molar properties for gas-solid adsorption systems [1]: in general, an integral molar property $\bar{Z}$ of the adsorbed molecules is related to a differential molar property Z^∂ via a relation of the form

$$Z^\partial = \bar{Z} + n_a\left(\frac{\partial \bar{Z}}{\partial n_a}\right)_T, \tag{2.14}$$

where n_a represents the number of moles of gas adsorbed on the solid [1]. The similarity in form between Eqs. (2.13) and (2.14) should now be evident.

Equations (2.11) and (2.12) yield

$$-\sum \frac{\mathcal{G}_i \dot{n}_i}{\dot{n}_k} + \frac{\dot{W}_0}{\dot{n}_k} = [N]RT\dot{n}_k + \sum (T_i - T)\bar{S}_i \frac{\dot{n}_i}{\dot{n}_k}. \tag{2.15}$$

If we write $\Delta\mathcal{G}$ for the first sum in Eq. (2.15), we have (for $T_i = T$)

$$-\Delta\mathcal{G} + \frac{\dot{W}_0}{\dot{n}_k} = [N]RT\dot{n}_k. \tag{2.16}$$

Equation (2.16) has the form of Ohm's law: potential difference = resistance × current, with the $[N]RT$ combination playing the role of a thermodynamic resistance. Occasionally it is useful to consider a *specific* rate parameter (i.e., rate parameter per unit area, per unit volume, etc.). Under such circumstances it is convenient to define a corresponding specific coefficient: if the specific rate parameter is $\dot{n}_k/Z$, then the specific coefficient $[N_Z]$ is $[N_Z] \equiv Z[N]$.

Forced Vaporization

Consider the monothermal steady vaporization process (steady vaporization of a liquid *without boiling*) indicated schematically in Figure 2–1. The terminal parts of the system consist of liquid at temperature T and pressure P_0, and vapor at temperature T and pressure P. For convenience let us assume

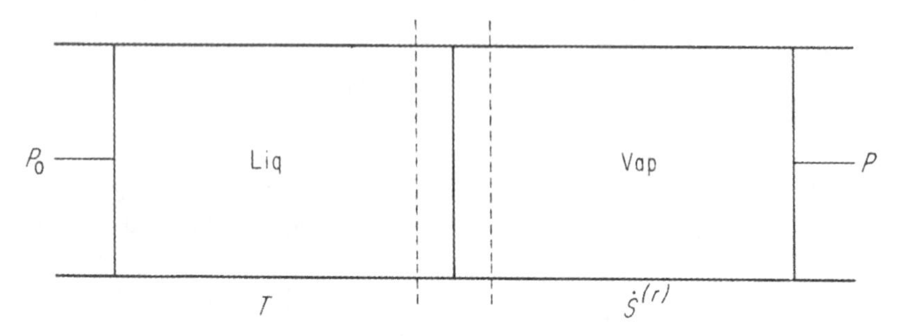

Figure 2–1. Monothermal steady vaporization of a liquid; P_0 = saturated vapor pressure at temperature T; $P \leqslant P_0$.

that the liquid-gas interface is fixed in space and that liquid is brought up to and vapor removed from the interface.

Under steady-flow conditions ($\dot{n}_{\text{liq}} + \dot{n}_{\text{gas}} = 0$) with $\dot{W}_0 = 0$, Eq. (2.16) takes the form

$$-(\mathscr{G}_{\text{gas}} - \mathscr{G}_{\text{liq}}) = [N]RT\dot{n}_{\text{gas}}. \tag{2.17}$$

If we neglect kinetic energy terms, assume that the vapor can be treated as an ideal gas, and write simply $\dot{\xi}$ for $\dot{n}_{\text{gas}}$, then we can rearrange Eq. (2.17) to give

$$\ln P = \ln P_0 - \dot{\xi}[N]. \tag{2.18}$$

More than thirty years ago Alty [2, 3] made some measurements on the rate of vaporization of water and of carbon tetrachloride in an attempt to find out something about the accommodation coefficient [4] at the liquid–vapor interface. His experimental arrangement was such that he could measure the rate of steady vaporization $\dot{\xi}$, the temperature T of the surrounding thermostat, the approximate temperature T_s at the interface, and the pressure P in the gas phase at a moderate distance from the interface. We have at our disposal, then, the variables $\dot{\xi}$, T, T_s, P; and we can form differential coefficients such as $(\partial \ln P/\partial T_s)_T$ and $RT^2(\partial \ln P/\partial T)_{\dot{\xi}}$. From Eq. (2.18) we see that

$$\left(\frac{\partial \ln P}{\partial T_s}\right)_T = -[N]^{\partial}\left(\frac{\partial \dot{\xi}}{\partial T_s}\right)_T, \tag{2.19}$$

and

$$RT^2\left(\frac{\partial \ln P}{\partial T}\right)_{\dot{\xi}} = \Delta\bar{H}_{\text{vap}} - \dot{\xi}RT^2\left(\frac{\partial [N]}{\partial T}\right)_{\dot{\xi}}. \tag{2.20}$$

Figure 2–2 exhibits Alty's data in the form of a ln P versus $\dot{\xi}$ plot; for small rates of vaporization the coefficient $[N]$ is independent of the rate parameter $\dot{\xi}$. The carbon tetrachloride data stand out as peculiar: beyond a

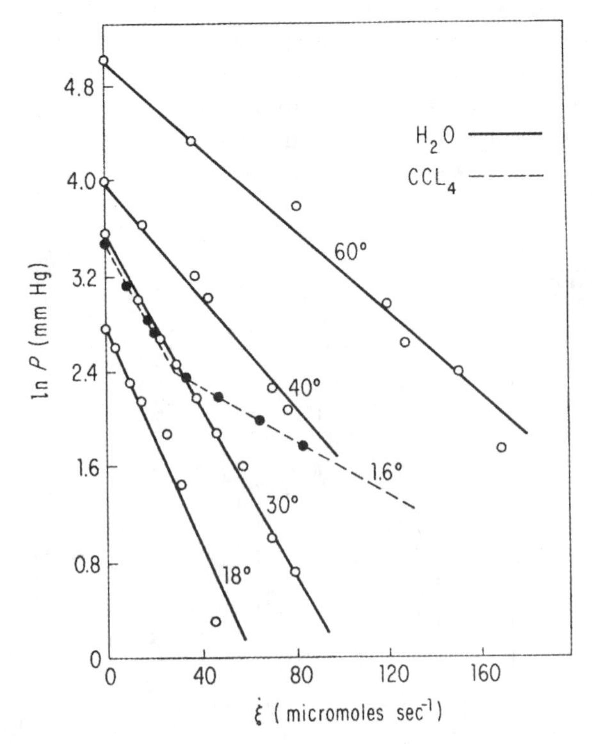

Figure 2–2. Plot of ln P versus $\dot{\xi}$ for the data of Alty. (From the *J. Chem. Phys.*, **31**, 1521 (1959), by permission.)

certain flow rate the slope of the ln P versus $\dot{\xi}$ plot becomes less steep, indicating a *decrease* in flow resistance with *increasing* rate of flow.

Exercise 2–1. Alty investigated the vaporization of water from a certain glass tube at the monothermal temperatures 18, 40, and 60°C. An analysis of Alty's data [5, 6] reveals that over the range 18 to 60°C for flow rates in the range 0 to 40 micromoles per second, the $[N]$ coefficient for the given tube can be roughly represented by the relation

$$[N] = -0.166 + \frac{61.0}{T},$$

where $[N]$ is expressed in seconds per micromole.

Compute the approximate value of the term $-\dot{\xi}RT^2(\partial[N]/\partial T)_{\dot{\xi}}$ for the conditions $\dot{\xi} = 10$ micromoles per second and $T = 298°K$; compare the magnitude of the computed term to that of $\Delta\bar{H}_{vap}$ at 298°K.

T. A. Erikson [7] has made some interesting studies of the steady vaporization process. In a series of measurements he had five glass tubes with differing diameters* hooked up to a common pressure manifold—the entire apparatus being immersed in a thermostat—and he measured the rate of vaporization from each tube for a given liquid at various settings of the

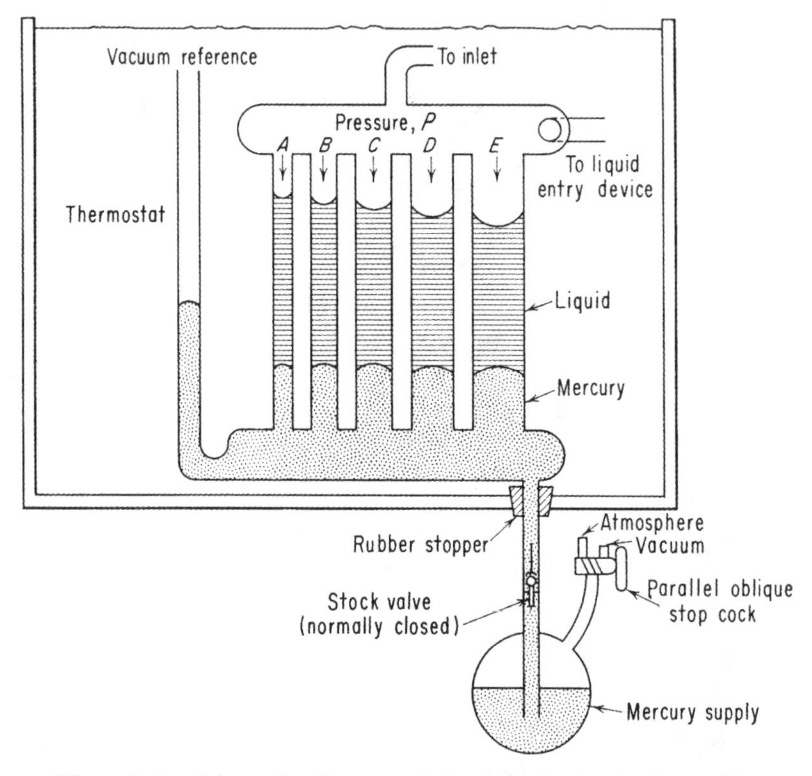

Figure 2–3. Schematic diagram of the forced vaporization equipment.

thermostat temperature T and the manifold pressure P (see Figure 2–3). Erikson presents his data in an interesting fashion: his observations consist of a set of rates of vaporization (for a given liquid at a given temperature) for each tube corresponding to various values of the pressure in the common

* The diameters ranged from 2.8 mm (tube A) to 19.0 mm (tube E).

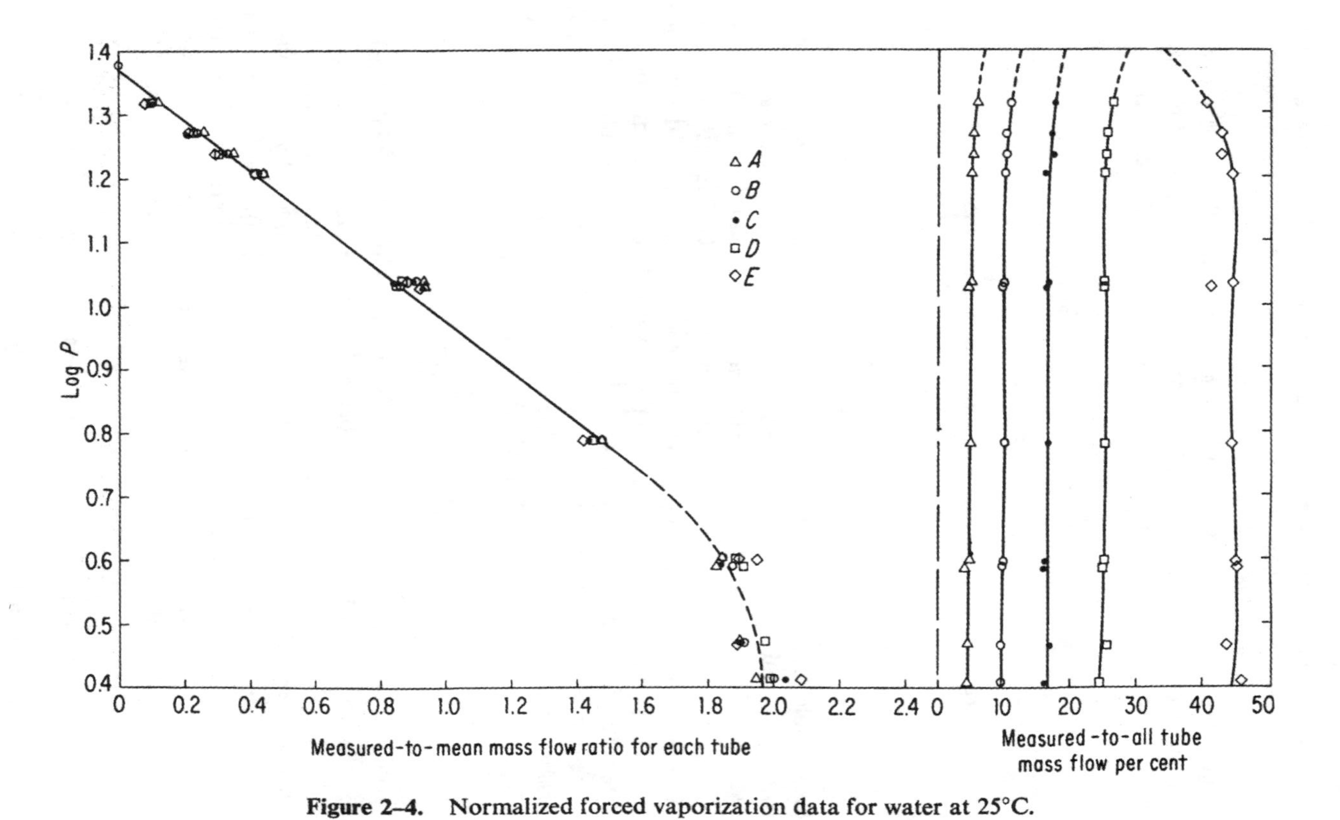

Figure 2–4. Normalized forced vaporization data for water at 25°C.

manifold; he finds it convenient to average the measured rates of vaporization for each tube and to express the rate of vaporization for each tube as a fraction of the average rate of vaporization for that tube, i.e., if at a given temperature there are m different settings of the manifold pressure, and if $\dot{\xi}_j^{(i)}$ is the steady rate of vaporization from the ith tube at the jth setting of the pressure, then Erikson defines a reduced flow rate $\theta_j^{(i)}$ such that

$$\theta_j^{(i)} \equiv \frac{\dot{\xi}_j^{(i)}}{\frac{1}{m}\sum_{k=1}^{m} \dot{\xi}_k^{(i)}} \equiv \frac{\dot{\xi}_j^{(i)}}{\langle \dot{\xi}^{(i)} \rangle}.$$

He also introduces another reduced variable $\vartheta_j^{(i)}$ such that

$$\vartheta_j^{(i)} \equiv \frac{\dot{\xi}_j^{(i)}}{\sum_{i=1}^{5} \dot{\xi}_j^{(i)}},$$

this variable representing the fraction of the total mass vaporized that comes from a given tube at a fixed pressure setting. Erikson calls θ the *measured-to-mean* mass flow ratio, and he calls ϑ the *measured-to-all tube* mass flow ratio.

Figure 2–4 shows Erikson's results for water at 25°C. The usefulness of the θ variable is at once apparent: the data from all the tubes fall very close to a single curve. The behavior of the ϑ variable close to equilibrium is quite interesting: here the relative contribution of tube E (the tube of largest diameter) to the total mass flow varies significantly with the manifold pressure. Erikson calculates the coefficient $[N^{(i)}]$ for the ith tube from a plot of $\log_{10} P$ versus $\theta^{(i)}$ (Figure 2–4) via the relation

$$[N^{(i)}] = \frac{-2.3 \text{ slope}}{\langle \dot{\xi}^{(i)} \rangle} \tag{2.21}$$

for the linear part of the plot. He then defines a specific coefficient $[N_B^{(i)}]$ such that $[N_B^{(i)}] \equiv B^{(i)}[N^{(i)}]$ with $B^{(i)}$ being the area [8] of the liquid meniscus in the ith tube. The $[N_B]$ coefficient defined in this manner depends on the tube diameter. Erikson eliminates this apparatus dependency by extrapolating his results to a tube of infinite diameter* (Figure 2–5). The limiting value $[N_B^{(\infty)}]$ is then a well-defined, apparatus-independent property of the vaporization process for a given liquid at a given temperature. Figure 2–6 shows the temperature dependence of the $[N_B^{(\infty)}]$ coefficient for water; we expect $[N_B^{(\infty)}]$ to approach zero at the critical point.

* In a small diameter tube the meniscus is essentially a spherical cap having an area twice as large as the cross-sectional area of the tube. In a large diameter tube the meniscus is essentially flat and its area coincides with the cross-sectional area of the tube. The ratio of the meniscus area to the cross-sectional area of the tube, $B^{(i)}/B_t$, thus varies from 2 to 1 as the tube diameter goes from 0 to ∞.

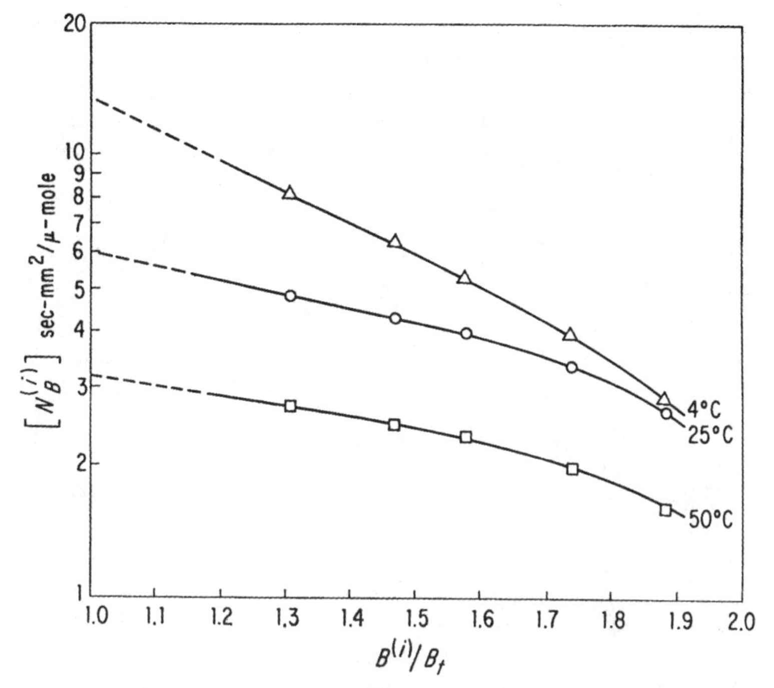

Figure 2–5. Dependence of $[N_B^{(i)}]$ on tube diameter and extrapolation to infinite diameter.

Exercise 2–2. For Erikson's experimental arrangement show that if each separate tube has a linear dependence of $\ln P$ on ξ, i.e., if $\ln P = \ln P_0 - \xi^{(i)}[N^{(i)}]$ with constant $[N^{(i)}]$, then the data from all the tubes fall on a single straight line when plotted in the form $\ln P$ versus $\theta^{(i)}$.

Exercise 2–3. Consider the situation (steady diffusion of a gas through a metal diaphragm) indicated schematically in Figure 2–7 and apply Eq. (2.16) to this situation. Assume approximate equilibrium at the metal–gas interface, a linear concentration gradient in the metal diaphragm, and the applicability of Fick's law of diffusion:

$$J = -D\frac{\partial C}{\partial x},$$

where J is the molar flux (moles sec^{-1} area^{-1}), D is the diffusion coefficient, and C is the molar concentration of the gas dissolved in the metal. Show that given the preceding assumptions

$$D \approx \frac{\Delta x}{B[N]\langle C\rangle},$$

where B is the metal–gas interfacial area and $\langle C\rangle$ is the average concentration of the dissolved gas.

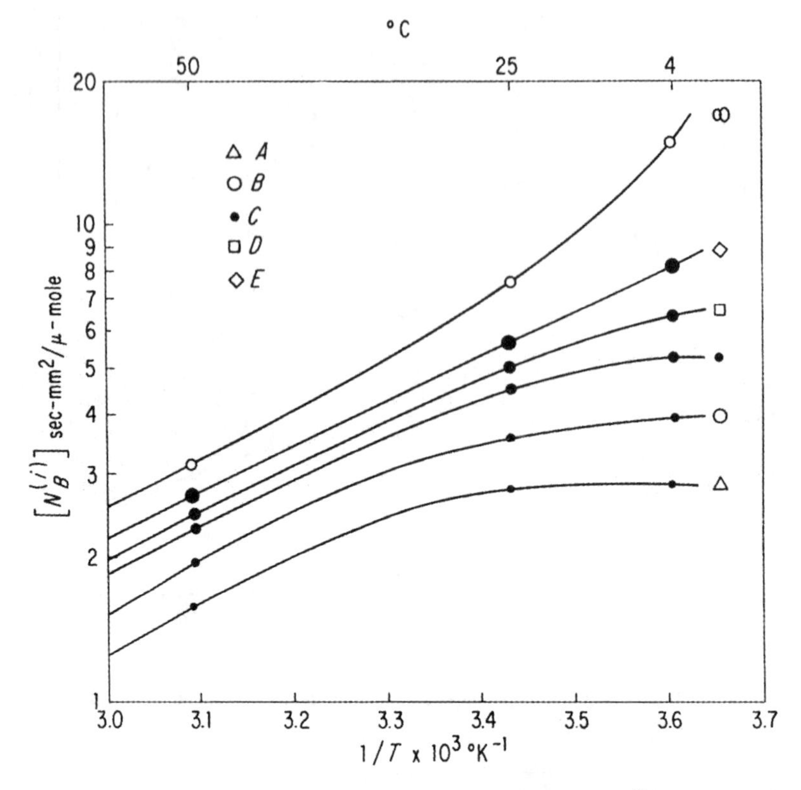

Figure 2–6. Temperature dependence of $[N_B^{(i)}]$.

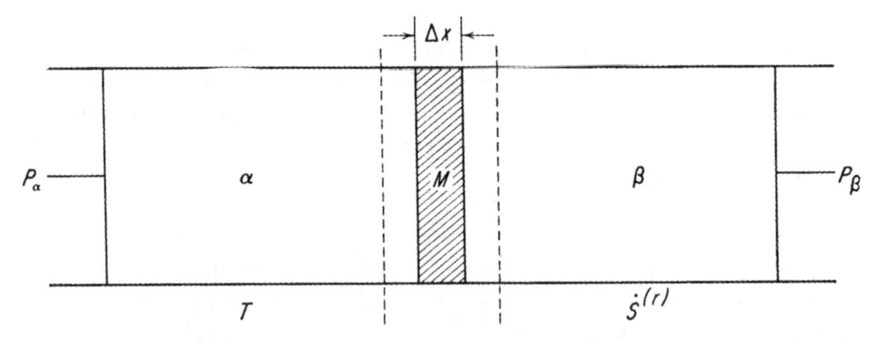

Figure 2–7. Monothermal steady diffusion of a gas through a metal diaphragm.

Exercise 2–4. Suppose that a pure liquid and a pure solid in equilibrium at a temperature $T*$ are heated to a temperature $T > T*$ (the pressure being kept constant)—forced melting of a one-component system. Assume that the liquid takes on the temperature T and that the solid (i) takes on the temperature T, or (ii) remains at the temperature $T*$ during the forced melting process. Show that for both (i) and (ii) $\mu_{\text{solid}} - \mu_{\text{liquid}} > 0$ and that

$$\text{(i)} \qquad \frac{T*}{T} \ln \frac{T}{T*} \approx \frac{\dot{n}_{\text{liq}} [N]_{\text{mel}} R}{\Delta \bar{S}(*)_{\text{mel}}},$$

$$\text{(ii)} \qquad \frac{\Delta T}{T} \approx \frac{\dot{n}_{\text{liq}} [N]_{\text{mel}} R}{\Delta \bar{S}(*)_{\text{mel}}},$$

and hence that for either case

$$\frac{\Delta T}{T} \propto \dot{n}_{\text{liq}},$$

where $\Delta T \equiv T - T*$, and $\Delta \bar{S}(*)_{\text{mel}}$ is the molar entropy of fusion at the temperature $T*$.
Hint: For case (ii) make use of Eq. (2.15).

Chemical Reaction

Consider the reversible chemical reaction

$$\sum \nu_i B_i = \sum \nu_j D_j \tag{2.22}$$

(the reactants B_i being transformed to products D_j with stoichiometric coefficients ν_i and ν_j, respectively) taking place in a steady-flow gradient reactor of volume V in a thermostat of temperature T. Let the reactants be individually supplied to and the products individually removed from the reaction vessel via semipermeable membranes communicating with terminal parts containing the appropriate species alone. For chemical reactions, the coefficient $[N]$ is best defined in terms of the number of reaction measures reacting in unit time $\dot{\xi}$; thus, if the reaction in Eq. (2.22) is proceeding from left to right, then*

$$\dot{\xi} = \frac{-\dot{n}_i}{\nu_i} = \cdots = \frac{\dot{n}_j}{\nu_j} = \cdots, \tag{2.23}$$

and

$$[N] \equiv \frac{\Theta}{R\dot{\xi}^2} = \frac{\Theta \nu_i^2}{R\dot{n}_i^2} = \cdots. \tag{2.24}$$

* Observe that the changes in state discussed in the porous plug experiment and in the forced vaporization experiment can be written in the form of Eq. (2.22); hence the variable $\dot{\xi}$ introduced into those discussions represents but a special case of the use of Eq. (2.23).

I want to point out here two general relations pertaining to reversible chemical reactions taking place in a steady monothermal fashion:

$$\Delta H \equiv \sum \nu_j \bar{H}_j - \sum \nu_i \bar{H}_i = \frac{-T\dot{S}^{(r)}}{\dot{\xi}} = \frac{\dot{Q}}{\dot{\xi}}, \tag{2.25}$$

$$-\Delta G \equiv \sum \nu_i \mu_i - \sum \nu_j \mu_j = \frac{T\Theta}{\dot{\xi}} = [N]RT\dot{\xi} = [N]RT\frac{\dot{n}_k}{\nu_k}, \tag{2.26}$$

where I have neglected kinetic energy terms, have considered only pressure-volume work ($\dot{W}_0 = 0$), have assumed that $T_i = T$, and in the final part of Eq. (2.26) have described the process in terms of the rate parameter $\dot{n}_k$ (the rate of influx of mass into the kth terminal part containing the product species D_k); I compute the change in any thermodynamic property, ΔG for example, relative to the *terminal parts of the system*. Equation (2.25) is an especially clear statement of the physical significance of the enthalpy change ΔH accompanying a chemical reaction.

The Ohm's law type of formulation, Eq. (2.16), for monothermal steady-rate processes normally yields an $[N]$ coefficient that is independent of the rate parameter (for moderate flow rates) under those conditions for which the system is allowed to adjust itself to the promptings of the environment—i.e., the gradient part has a definite structure and a number of natural "back emf's" are induced by the steady-flow process. If we destroy the natural gradient structure of the system (by stirring, e.g.), then the overall $[N]$ coefficient may or may not be independent of the rate parameter. Although processes with a constant $[N]$ coefficient are the most interesting, we can treat *any* monothermal steady-rate process according to the formalism of Eq. (2.16). If we find that the integral coefficient $[N]$ depends markedly on the rate parameter, we can discuss the process in terms of the differential coefficient $[N]^{\partial}$.

If we wish to relate Eq. (2.26) to the formulas of chemical kinetics, we must move down the sequence *gradient reactor, stirred-flow reactor, batch reactor*. Consider first the stirred-flow reactor (same setup as for the steady-flow gradient reactor, but the contents of the reaction vessel are stirred). Under steady-flow conditions an equation of the form (2.26) holds. If we assume that the change in Gibbs free energy ΔG(out) for the reaction is approximately the same as the free energy change ΔG(in) computed from the chemical potentials *inside* the stirred reactor, we can then write

$$\ln \mathcal{Q} = \ln K - \frac{[N_V]}{\nu_k}\left(\frac{\dot{n}_k}{V}\right) \tag{2.27}$$

where V is the volume of the reactor, $[N_V] \equiv V[N]$, K is the equilibrium constant for the reaction, and $\mathcal{Q} \equiv \Pi a_j^{\nu_j}/\Pi a_i^{\nu_i}$, the activities in the activity quotient being computed for the state of affairs holding inside the stirred reactor.

Consider a further step in the sequence: let the reaction be taking place in a batch reactor, and assume that the changing conditions inside the batch reactor can be treated as equivalent to a sequence of steady-flow states in a stirred-flow reactor—i.e., assume that Eq. (2.27) can be applied to a batch reactor provided that $\dot{C}_k$ (the rate of change of the concentration of the D_k species in the batch reactor) is substituted for the quantity $\dot{n}_k/V$. In the batch reactor, then

$$\ln \mathcal{Q} = \ln K - \frac{[N_V]}{\nu_k}\dot{C}_k. \tag{2.28}$$

Exercise 2–5. Show that for a first-order reversible reaction $B \rightleftharpoons D$, close to equilibrium,

$$[N_V] \approx \frac{1}{k_b C_D^{(\mathrm{eq})}},$$

where k_b is the backward reaction rate constant, and $C_D^{(\mathrm{eq})}$ is the equilibrium concentration of the D species.

If we apply Eq. (2.28) to a number of well-known kinetic situations, we find that the $[N_V]$ coefficient depends (rather strongly) on the rate parameter. Such a state of affairs can result from any of a number of possibilities:

(i) In the stirred-flow reactor ΔG(in) may differ significantly from ΔG(out).

(ii) The states inside a batch reactor may not be comparable to steady-flow states in a stirred-flow reactor.

(iii) In accord with an electrical analogy, it may be the case that capacitance and inductance effects are important.

(iv) It may be that the $[N_V]$ coefficient is independent of the rate parameter in a steady-flow gradient reactor but is not independent of the rate parameter in a stirred-flow reactor or in a batch reactor due to loss of the natural gradients ("back emf's") in the system.

The investigations of Prigogine et al. [9] appear to favor position (iv). Kirkwood and Crawford [10], by injecting some elements of relaxation theory into their description of chemical reactions, have developed an approach somewhat similar to that of position (iii).

Other Monothermal Steady-Rate Processes

I can think of several other (potentially) steady-rate processes that would lead to interesting experimental investigations. Consider the steady monothermal vaporization of water from a saturated solution of a sparingly soluble salt. As the vaporization proceeds, small crystals of the salt will nucleate and grow in a region bordering on the liquid–vapor interface. A steady process is possible provided that the small crystals being precipitated do not clutter up the liquid meniscus and thus interfere with the vaporization process. If a steady process can be achieved, we can analyze the situation by the methods of this chapter, can extract from the analysis an overall $[N]$ coefficient, and can resolve the overall $[N]$ coefficient into a part due to the vaporization process and a part due to the precipitation process.

Consider next a three-phase, 2-component system consisting of solid solvent, solid solute, and solution. Such a collection of phases can coexist in equilibrium at the eutectic temperature $T*$. If we place the three-phase system in a thermostat at some temperature $T > T*$, then the solid phases will pass into solution until at least one of them vanishes. If we assume that dissolution of the solid phases can be made to take place at a steady rate with all terminal parts at temperature T and that the solution phase has a composition approximately corresponding to that of a solution saturated with respect to the solute at temperature T, then we can carry through the standard analysis and can evaluate an $[N]$ coefficient for the process. Here again we can probably break up the overall $[N]$ coefficient into more fundamental pieces. This case is rather interesting in that the progress of the dissolution process can be followed by dilatometry.

A final thought is that it would be interesting to study a chemical reaction proceeding at a steady rate, far from equilibrium, and involving terminal parts and a gradient part. By way of example, consider a steady hydrogen–oxygen flame burning in a water thermostat (Figure 2–8). Let the hydrogen and oxygen be led from storage tanks through volume flow meters (wet-test meters, say), being saturated with water vapor as they pass through the meters; assume that the hydrogen and oxygen each enter the mixing chamber at temperature T and partial pressure $P' \equiv P - P_{H_2O}$, where P_{H_2O} is the saturated vapor pressure of water at temperature T. Consider the case where oxygen is in excess, and let the subscript x stand for properties of the excess oxygen bubbling out of the thermostat into the atmosphere. If we write P_{atm} for atmospheric pressure, then let $P'_{atm} \equiv P_{atm} - P_{H_2O}$.

For the reaction

$$2H_2(g, P', T) + O_2(g, P', T) = 2H_2O(liq, P_{atm}, T) \qquad (2.29)$$

conducted in a steady monothermal fashion with oxygen in excess, it follows that

$$\dot{\xi} = -\tfrac{1}{2}\dot{n}_{H_2}; \qquad \dot{n}_x \equiv -\dot{n}_{O_2} + \tfrac{1}{2}\dot{n}_{H_2}; \qquad \dot{n}_{H_2O} = -\dot{n}_{H_2}. \qquad (2.30)$$

Equations (2.11) and (2.24) yield, in the present instance,

$$-(\dot{n}_{H_2}\mu_{H_2} + \dot{n}_{O_2}\mu_{O_2} + \dot{n}_{H_2O}\mu_{H_2O} + \dot{n}_x\mu_x) = [N]RT\dot{\xi}^2. \qquad (2.31)$$

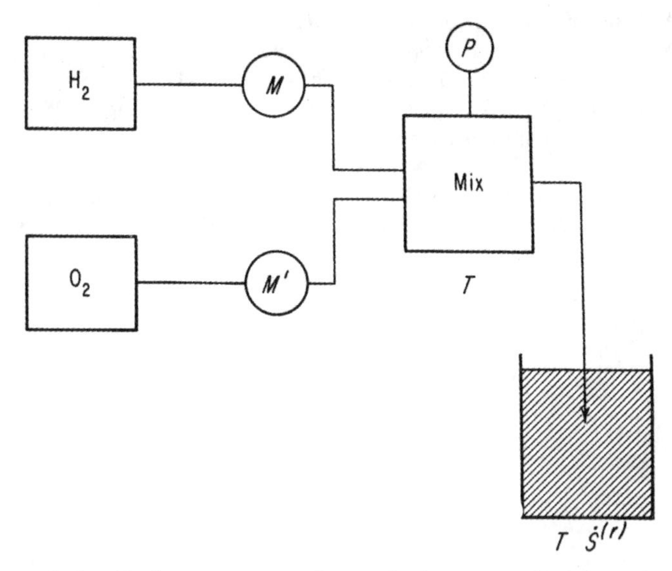

Figure 2–8. Hydrogen–oxygen flame: hydrogen and oxygen from reservoir cylinders pass through flow meters M and M' (wet-test meters, say) into a mixing chamber where the temperature and pressure of the gaseous mixture are T and P, respectively. From the mixing chamber the gaseous mixture goes to a flame at the end of a nozzle immersed in a water thermostat of temperature T.

Equations (2.30) and (2.31) combine to give

$$-\left\{-2\mu_{\mathrm{H_2}} - \mu_{\mathrm{O_2}} + 2\mu_{\mathrm{H_2O}} - \frac{\dot{n}_x}{\dot{\xi}}(\mu_{\mathrm{O_2}} - \mu_x)\right\} = [N]RT\dot{\xi}, \tag{2.32}$$

$$-\Delta G + \frac{\dot{n}_x}{\dot{\xi}}(\mu_{\mathrm{O_2}} - \mu_x) = [N]RT\dot{\xi}. \tag{2.33}$$

If we use expressions derived from the perfect gas law in Eq. (2.33), we then get

$$3 \ln P' + \frac{\dot{n}_x}{\dot{\xi}} \ln \frac{P'}{P'_{\mathrm{atm}}} = [N]\dot{\xi} + \frac{\Delta G^{(\mathrm{ref})}}{RT}, \tag{2.34}$$

where $\Delta G^{(\mathrm{ref})}$ is defined by $\Delta G \equiv \Delta G^{(\mathrm{ref})} + RT \ln(1/P_{\mathrm{H_2}}^2 P_{\mathrm{O_2}})$, and I have made use of the condition $P_{\mathrm{H_2}} = P_{\mathrm{O_2}} = P'$. It would certainly be interesting to evaluate the $[N]$ coefficient for this situation and to see how it varied with changing experimental conditions.

References

1. R. J. Tykodi, *J. Chem. Phys.*, **22**, 1647 (1954).
2. T. Alty, *Proc. Roy. Soc.* (London), **A131**, 554 (1931).

3. T. Alty and F. Nicoll, *Can. J. Research*, **4**, 547 (1931).
4. T. A. Erikson, *J. Phys. Chem.*, **64**, 820 (1960).
5. R. J. Tykodi and T. A. Erikson, *J. Chem. Phys.*, **31**, 1521 (1959).
6. T. A. Erikson, M.S. Thesis, Illinois Institute of Technology (1959).
7. T. A. Erikson, unpublished results.
8. T. A. Erikson, *J. Phys. Chem.*, **69**, 1809 (1965).
9. I. Prigogine, P. Outer, and C. Herbo, *J. Phys. Chem.*, **52**, 321 (1948).
10. J. G. Kirkwood and B. Crawford, Jr., *J. Phys. Chem.*, **56**, 1041 (1952).

3

The Rate of Entropy Production

For the purposes of this text, I classify steady-state situations as either *ordinary* or *special.* Special situations are those involving external magnetic or centrifugal fields; ordinary situations (the vast majority of those of interest) are simply situations *not* involving magnetic or centrifugal fields. The body of the text deals with ordinary situations, and I formulate Eq. (3.5) in a way especially suited to such situations; I discuss the change in formulation needed to deal with special situations in a section of the Appendix.

General Remarks

When a system is the site of one or more steady-rate processes, there are steady rates of accumulation of energy, entropy, volume, mass, chemical reaction measures, electric charge, etc., in terminal parts of the system or in reservoirs adjoined to the system. In steady closed-loop operation there can be a steady accumulation of work in a work reservoir—the rate of accumulation of work in the reservoir being normally factorable into a well-defined "current" and a "potential difference." I choose a set of work-delivering currents and steady rates of accumulation of thermodynamic quantity as a *canonical set*: in equilibrium thermodynamics the fundamental relation

$$dU = T\,dS - P\,dV + \sum \mu^{(j)}\,dn^{(j)} + \cdots \tag{3.1}$$

gives rise to a preferred set of independent variables for both the energy function and the entropy function; thus $U = U(S, V, n^{(j)}, \ldots)$ and $S = S(U, V, n^{(j)}, \ldots)$. In analogy to the equilibrium case, the canonical set consists of quantities of the type $\dot{S}^{(r)}$ or $\dot{U}^{(r)}$, $\dot{V}$, $\dot{n}^{(j)}, \ldots$, where the superscript r refers to an adjoined reservoir. In a heat reservoir of the classical ideal type, $\dot{U}^{(r)} = \dot{Q}^{(r)}$, and I find the choice between $\dot{S}^{(r)}$ and $\dot{Q}^{(r)}$ to be arbitrary and largely a matter of convenience. I refer to the quantities of the canonical set as currents and symbolize them by Y_i, thus $\{Y_i\} = \{\dot{S}^{(r)}$ or $\dot{Q}^{(r)}, \dot{V}, \dot{n}^{(j)}, \ldots\}$. The currents are thus time derivatives of extensive thermodynamic quantities.

By combining the first and second laws of thermodynamics we can always express the rate of entropy production $\Theta \equiv \dot{S}(\text{system}) + \dot{S}(\text{surroundings})$ for a steady-state situation in the form

$$\Theta = \sum_i Y_i \Omega_i, \tag{3.2}$$

with the Y_i forming a subset of *independent* currents chosen from the canonical set, and the Ω_i being *affinities* conjugate to the (independent) currents Y_i. I define the affinities Ω_i via Eq. (3.2), and I note that they are normally combinations of intensive variables.*

Assumption IV

In discussing problems of migrational equilibrium we shall be analyzing the behavior of pertinent parameters of the system for a characteristic limiting operation: we shall be interested in how the values of the pertinent parameters change as a given current or affinity approaches zero along a prescribed path via a sequence of steady states; i.e., symbolically, we shall study the limit of $\{Z(\eta, \ldots) - Z(0, \ldots)\}/\eta$ as $\eta \to 0$, with Z being the pertinent parameter and η being the given current or affinity. These limits, of course, are nothing but partial derivatives evaluated at the point $\eta = 0$ with certain things being held constant:

$$\lim_{\eta \to 0} \frac{Z(\eta, \ldots) - Z(0, \ldots)}{\eta} \equiv \left(\frac{\partial Z}{\partial \eta}\right)_{\ldots}\bigg|_{\eta = 0,\, z_\omega = z_\omega(*)}, \tag{3.3}$$

where the derivative is evaluated for a *specific steady state*—one for which $\eta = 0$ and for which the intensive variables z_ω of the system have the definite values $z_\omega(*)$. Now, since such derivatives appear over and over again in the text, and since the right-hand side of Eq. (3.3) is rather cumbersome, I propose the following notational convention: let

$$\left(\frac{\delta Z}{\delta \eta}\right)_{\ldots} \equiv \left(\frac{\partial Z}{\partial \eta}\right)_{\ldots}\bigg|_{\eta = 0,\, z_\omega = z_\omega(*)}. \tag{3.4}$$

* By choosing a canonical set of currents and by defining my affinities via Eq. (3.2), I circumvent Truesdell's objections [1, 2] to other formulations.

I can now state assumption IV in precise mathematical form (however, see the *special fields* section of the Appendix):

$$\left(\frac{\delta \Theta}{\tilde{\delta} Y_k}\right)_{\Omega'} = 0, \qquad \left(\frac{\delta^2 \Theta}{\tilde{\delta} Y_k^{\,2}}\right)_{\Omega'} > 0, \tag{3.5}$$

where the subscript Ω' indicates that the affinities conjugate to the non-vanishing currents are to be kept constant; i.e., constant Ω_i for each $i \neq k$. I call the sequence of steady states (constant Ω_i for $i \neq k$) used in the limiting operation of Eq. (3.5) an *affine sequence*, and it is this sort of sequence that I had in mind in my earlier (Chapter 1) verbal statement of assumption IV. Perhaps it will help if I put Eq. (3.5) into words:

ASSUMPTION IV: If a system that is the site of $\nu + 1$ steady (independent) currents converges via a sequence of steady states on a state involving ν steady currents in such a way that the affinities conjugate to the ν non-vanishing currents are maintained constant, then the state with ν currents is a state of minimum rate of entropy production relative to the (steady) states with $\nu + 1$ currents.

From Eq. (3.2) we see that

$$\begin{aligned}\left(\frac{\delta \Theta}{\tilde{\delta} Y_k}\right)_{\Omega'} &= \sum_i Y_i \left(\frac{\delta \Omega_i}{\tilde{\delta} Y_k}\right)_{\Omega'} + \sum_i \Omega_i \left(\frac{\delta Y_i}{\tilde{\delta} Y_k}\right)_{\Omega'} \\ &= \sum_i \Omega_i \left(\frac{\delta Y_i}{\tilde{\delta} Y_k}\right)_{\Omega'},\end{aligned} \tag{3.6}$$

since in the limiting state $Y_k = 0$ and $(\delta\Omega_i/\tilde{\delta} Y_k)_{\Omega'} = 0$ for $i \neq k$; hence we can also write assumption IV as

$$0 = \left(\frac{\delta \Theta}{\tilde{\delta} Y_k}\right)_{\Omega'} = \sum_i \Omega_i \left(\frac{\delta Y_i}{\tilde{\delta} Y_k}\right)_{\Omega'}. \tag{3.7}$$

For just one current Y_1, with the limiting state being one of thermostatic equilibrium, Eq. (3.7) indicates that the conjugate affinity Ω_1 vanishes at equilibrium; i.e., $\Omega_1(0) = 0$. By way of example consider the monothermal steady vaporization process described in Chapter 2. It follows from Eqs. (2.12), (2.17), and (2.18) that

$$\Theta = \frac{\dot{n}_{\text{gas}}(\mu_{\text{liq}} - \mu_{\text{gas}})}{T} = Y_1 \Omega_1 \tag{3.8}$$

and that

$$\left(\frac{\delta \Theta}{\tilde{\delta} \dot{n}_{\text{gas}}}\right)_T = \frac{\mu_{\text{liq}} - \mu_{\text{gas}}}{T}. \tag{3.9}$$

Thus Eq. (3.7) implies that at thermostatic equilibrium $\mu_{\text{liq}} = \mu_{\text{gas}}$.

Degrees of Freedom

In order to specify the equilibrium thermodynamic state of the two-phase system H_2O(liq) + H_2O(gas) we only need to know the temperature or the vapor pressure since the system has but 1 degree of freedom. To specify the state of steady vaporization for the same two-phase system, we require considerably more information: we need to know the thermostat temperature, the pressure in the gas phase, and the diameter and nature (glass, metal, etc.) of the tube in which the process is taking place. In general, to describe a steady-state system it is necessary to describe the apparatus, the chemicals (i.e., the contents of the apparatus), and the precise way in which the apparatus interacts energetically with the surroundings. In his forced vaporization work, Erikson (Chapter 2) was able to overcome the apparatus dependence of his steady-state quantities by a clever extrapolation procedure. The apparatus dependence of steady-state quantities should always be kept in mind; and, where possible, explicit extrapolation or limiting procedures should be worked out for obtaining apparatus-independent results.

Continuing our examination of Erikson's experiments on the forced vaporization of water, we see that for a given tube the steady state is completely characterized by the thermostat temperature T and the gas-phase pressure P, thus $\Theta = \Theta(T, P)$, and $\dot{\xi} = \dot{\xi}(T, P)$. Since $(\partial\dot{\xi}/\partial P)_T \neq 0$, we can invert the $\dot{\xi}$ relation to get $P = P(T, \dot{\xi})$ and $\Theta = \Theta(T, \dot{\xi})$. We can thus express the rate of entropy production in terms of a parameter T characteristic of a related state of thermostatic equilibrium and in terms of the steady current $\dot{\xi}$ traversing the system. I should like to extend this idea to other systems. For a given system we can consider any steady state to have been generated from a reference thermostatic (equilibrium) state σ with variables $T_\sigma, P_\sigma, \mu_\sigma^{(i)}, \ldots,$ a thermostatic state of F_σ degrees of freedom (in the phase rule sense). Now the primary variables pertaining to a given steady-state system are the intensive variables in each of its terminal parts. I wish to take the set of primary variables and to perform a transformation such that I get an equivalent set of variables dealing with a reference thermostatic state σ and with ν independent currents (or affinities); the Jacobian of the transformation must, of course, have certain desirable properties.* In terms of the new set of variables we can say that a steady state characterized by ν independent currents has $F_\sigma + \nu$ degrees of freedom and that the rate of entropy production Θ, being a function of the (steady) state of the system, requires for its complete determination the specification of $F_\sigma + \nu$ independent variables. In describing a given steady-state system, then, the most convenient set of variables will usually be a set

* I discuss the inversion problem and the properties of the Jacobian in a section of Chapter 15.

of ν independent currents (or affinities) and F_σ parameters characteristic of the reference thermostatic state; thus

$$\Theta = \Theta(Y_j, \sigma) = \Theta(\Omega_j, \sigma), \tag{3.10}$$

$$Y_i = Y_i(\Omega_j, \sigma), \tag{3.11}$$

$$\Omega_i = \Omega_i(Y_j, \sigma), \tag{3.12}$$

where I have rather inelegantly used the symbol σ to stand for the set of variables $T_\sigma, P_\sigma, \mu_\sigma^{(i)}, \ldots$.

The expression $\Theta = \sum_i Y_i \Omega_i$ for the rate of entropy production is symmetric in the currents and affinities; also, we see from Eqs. (3.10) to (3.12) that there is an easy interconvertibility in the roles of currents and affinities. These observations suggest that a *duality principle* may hold for steady-state situations. I here introduce my fifth, and final, basic assumption:

(V) A duality principle holds for steady-state situations: a new theorem results upon the interchange of the roles of currents and affinities in any given theorem, provided that the original theorem deals *solely* with currents, affinities, and the rate of entropy production (in any combination) and makes use of no additional quantities.

An example may help to clarify the preceding material. Consider the (one-dimensional) steady monothermal diffusion [3, 4] of a gas through a metal diaphragm* as indicated schematically in Figure 2–7. In the steady state $\dot{n}_\alpha + \dot{n}_\beta = 0$, and

$$\Theta = \frac{\dot{n}_\beta(\mu_\alpha - \mu_\beta)}{T} = Y_1 \Omega_1. \tag{3.13}$$

The steady state is completely described by the three variables T, P_α, P_β. We may consider this steady state to have originated from any of several thermostatic states. (i) We could have started from the state $[T, P_\alpha = \langle P \rangle, P_\beta = \langle P \rangle]$, where $\langle P \rangle \equiv \frac{1}{2}(P_\alpha + P_\beta)$, and could have reached the steady state in question by letting P_α and P_β approach $P_\alpha(\dot{n})$ and $P_\beta(\dot{n})$, respectively, while keeping $\langle P \rangle$ constant. (ii) We could have started from the state $[T, P_\alpha = P_\alpha(\dot{n}), P_\beta = P_\alpha(\dot{n})]$ and could have reached the final (steady) state by letting P_β approach $P_\beta(\dot{n})$ while keeping P_α constant—in this case the affinity Ω_1 is a measure of the deviation of the final state from the reference thermostatic state. (iii) . . . Each thermostatic reference state has 2 degrees of freedom (T, P) and each steady-flow state has one current $(\dot{n}_\beta)$; hence the number of degrees of freedom of the steady state is $F_\sigma + \nu = 2 + 1 = 3$, and $\Theta = \Theta(\dot{n}_\beta, T, P)$.

* See Ex. 2–3.

From the discussion in Chapter 2, treating the gas as ideal, we get (for a certain range of flow rates)

$$\Theta = \dot{n}_\beta R \ln \frac{P_\alpha}{P_\beta}, \tag{3.14}$$

$$\Theta = R[N(T, P)]\dot{n}_\beta^{\,2}, \tag{3.15}$$

$$\Omega_1 = R \ln \frac{P_\alpha}{P_\beta} = R[N(T, P)]\dot{n}_\beta, \tag{3.16}$$

where the "resistance factor" $[N]$ depends on the temperature and on the concentration of dissolved gas in the membrane via the reference state variables T, P.

Reprise

We can make explicit mention of a thermostatic reference state σ in the limiting procedures based on Eq. (3.4); thus, for example, $(\delta/\bar{\delta}\eta)_{\ldots,\sigma}$ is an operation performed while holding constant F_σ variables characteristic of the reference state, and $(\delta/\bar{\delta}Y_k)_{\Omega',\sigma}$ is an operation performed via a uniquely specified sequence of steady states: the final state is characterized by F_σ parameters related to the reference thermostatic state, by $\nu - 1$ affinities Ω_i $(i \neq k)$, and by the condition $Y_k = 0$, a total of $F_\sigma + (\nu - 1) + 1$ constraints in all. Hence the final state is fully characterized, and the neighboring states participating in the operation $(\delta/\bar{\delta}Y_k)_{\Omega',\sigma}$ form a determinate set. Of the many possible sequences of steady states implied by the notation $(\delta/\bar{\delta}Y_k)_{\Omega'}$, the operation $(\delta/\bar{\delta}Y_k)_{\Omega',\sigma}$ singles out *one* such sequence and evaluates the derivative $(\partial/\partial Y_k)_{\Omega',\sigma}$ along that sequence. Both forms, the less explicit one $(\delta/\bar{\delta}Y_k)_{\Omega'}$ and the more explicit one $(\delta/\bar{\delta}Y_k)_{\Omega',\sigma}$, have their uses and both will appear in the material that follows.

Linear Current-Affinity Relationships

For a given steady-state situation with a number (ν) of independent currents Y_i chosen from the canonical set and with the rate of entropy production expressed in terms of the Y_i—i.e., $Y_i \in \{\dot{S}^{(r)}$ or $\dot{Q}^{(r)}, \dot{V}, \dot{n}^{(j)}, \ldots\}$ and $\Theta = \sum_i Y_i\Omega_i$—we can assume that the currents may be expressed as linear functions of the affinities, and vice versa, in a neighborhood of a thermostatic reference state σ. Let us assume that

$$Y_i = \sum_j L_{ij}(\sigma)\Omega_j, \tag{3.17}$$

$$\Omega_i = \sum_j K_{ij}(\sigma) Y_j, \tag{3.18}$$

given that

$$Y_i \in \{\dot{S}^{(r)} \text{ or } \dot{Q}^{(r)}, \dot{V}, \dot{n}^{(j)}, \ldots\}, \tag{3.19}$$

$$\Theta = \sum_i Y_i \Omega_i \geqslant 0, \tag{3.20}$$

where the Y_i are independent currents. For a fixed reference state σ and constant coefficients $L_{ij}(\sigma)$ and $K_{ij}(\sigma)$, I shall show that

$$0 = \left(\frac{\delta \Theta}{\delta Y_k}\right)_{\Omega', \sigma} \leftrightarrow L_{ik}(\sigma) = L_{ki}(\sigma) \quad [\text{and } K_{ik}(\sigma) = K_{ki}(\sigma)]; \tag{3.21}$$

all summations will be over the index i.

We see from Eqs. (3.17) and (3.20) that

$$\begin{aligned} \left(\frac{\delta \Theta}{\delta Y_k}\right)_{\Omega', \sigma} &= \Omega_k + \sum_{i \neq k} \Omega_i \left(\frac{\delta Y_i}{\delta Y_k}\right)_{\Omega', \sigma} \\ &= \Omega_k + \sum_{i \neq k} \Omega_i \frac{L_{ik}}{L_{kk}}. \end{aligned} \tag{3.22}$$

We also have the condition

$$0 = Y_k = L_{kk}\Omega_k + \sum_{i \neq k} L_{ki}\Omega_i \tag{3.23}$$

with $L_{kk} > 0$ by virtue of the positive definite character of the rate of entropy production (Eq. 3.20). Upon eliminating Ω_k between Eqs. (3.22) and (3.23) we get

$$\left(\frac{\delta \Theta}{\delta Y_k}\right)_{\Omega', \sigma} = \sum_{i \neq k} \frac{(L_{ik} - L_{ki})\Omega_i}{L_{kk}}. \tag{3.24}$$

Now, if $L_{ik} = L_{ki}$, then $(\delta\Theta/\delta Y_k)_{\Omega',\sigma}$ must equal zero. Conversely, if $(\delta\Theta/\delta Y_k)_{\Omega',\sigma} = 0$, then L_{ik} must equal L_{ki}; this is so because Eq. (3.24) holds for a whole set of values Ω_i $(i \neq k)$, and if L_{ik} were not equal to L_{ki} then Eq. (3.24) with $(\delta\Theta/\delta Y_k)_{\Omega',\sigma} = 0$ would be a demonstration of the linear dependence of the affinities Ω_i. Since we assumed that we were dealing with a set of *independent* affinities, we are forced to conclude that $L_{ik} = L_{ki}$. If the matrix of coefficients L_{ij} is symmetric, then the matrix of coefficients K_{ij} (the reciprocal matrix) is symmetric also.* Thus for ordinary steady states in the region where linear current-affinity relations are expected to hold, assumption IV holds if and only if the Onsager reciprocal relations [5] hold. Note that the coefficients used here are global, not local, ones. I assume that assumption IV holds for any ordinary steady-state situation whatsoever, and I do not limit it to the region of linear current-affinity relationships (see the Appendix).

(As an aside observe that Eqs. (3.16) and (3.18) reveal that $[N] = K_{11}/R$.)

* The dual (assumption V) of Eq. (3.5) is the relation $(\delta\Theta/\delta\Omega_k)_{Y'} = 0$; this dual form of Eq. (3.5) leads to a relation analogous to Eq. (3.21):

$$0 = \left(\frac{\delta \Theta}{\delta \Omega_k}\right)_{Y', \sigma} \leftrightarrow K_{ik} = K_{ki} \quad (\text{and } L_{ik} = L_{ki}).$$

Observations

Consider the following two sets of relations, sets that I refer to as *petit principles of entropy production* and *grand principles of entropy production.*

Petit *Principles of Entropy Production*

$$\left(\frac{\delta \Theta}{\delta Y_k}\right)_{\Omega', \sigma} = 0, \tag{3.25}$$

$$\left(\frac{\delta^2 \Theta}{\delta Y_k^2}\right)_{\Omega', \sigma} > 0, \tag{3.26}$$

$$\left(\frac{\delta \Omega_k}{\delta Y_k}\right)_{Y', \sigma} \geqslant \left(\frac{\delta \Omega_k}{\delta Y_k}\right)_{\Omega', \sigma} > 0, \tag{3.27}$$

$$0 = \left(\frac{\partial Y_k}{\partial \Omega_j}\right)_{\Omega', \sigma} - \left(\frac{\partial Y_j}{\partial \Omega_k}\right)_{\Omega', \sigma} + \sum_i \Omega_i \left\{\left(\frac{\partial^2 Y_i}{\partial \Omega_j \, \partial \Omega_k}\right)_{\Omega', \sigma} - \left(\frac{\partial^2 Y_i}{\partial \Omega_k^2}\right)_{\Omega', \sigma} \left(\frac{\partial Y_k}{\partial \Omega_j}\right)_{\Omega', \sigma} \left(\frac{\partial Y_k}{\partial \Omega_k}\right)_{\Omega', \sigma}^{-1}\right\} \tag{3.28}$$

at the point where Eqs. (3.25) and (3.26) hold, i.e., at the point where $Y_k = 0$ and $\Omega_i = \Omega_i(*)$ for $i \neq k$. Each of Eqs. (3.25) to (3.28) has a dual. I feel that the *petit* principles of entropy production are of unlimited validity for ordinary steady-state situations.

Grand *Principles of Entropy Production*

$$\left(\frac{\partial^2 \Theta}{\partial Y_k^2}\right)_{\Omega', \sigma} > 0, \tag{3.29}$$

$$\left(\frac{\partial \Omega_k}{\partial Y_k}\right)_{Y', \sigma} \geqslant \left(\frac{\partial \Omega_k}{\partial Y_k}\right)_{\Omega', \sigma} > 0, \tag{3.30}$$

$$\left(\frac{\partial Y_k}{\partial \Omega_j}\right)_{\Omega', \sigma} - \left(\frac{\partial Y_j}{\partial \Omega_k}\right)_{\Omega', \sigma} = 0. \tag{3.31}$$

Each of Eqs. (3.29) to (3.31) has a dual. I feel that the *grand* principles of entropy production have a wide (but perhaps limited) range of validity for ordinary steady-state situations. In the linear current-affinity region Eqs. (3.29) to (3.31) are indistinguishable from Eqs. (3.26) to (3.28).*

Relations (3.29) to (3.31) have a strong intuitive appeal: Eq. (3.29) is of the form of a *haste-makes-waste* principle, the more we "push" a steady

* The *petit* principles of entropy production claim to be valid only at the point $Y_k = 0$, whereas the *grand* principles claim to be valid for *any* value of Y_k; this distinction is immaterial in the linear current-affinity region.

process the greater we make the rate of entropy production; Eq. (3.30) is analogous to the conditions of stability for thermostatic systems; and Eq. (3.31) is of the form of a general reciprocity relation.

Exercise 3–1. Prove that relations (3.29) to (3.31) hold for linear current-affinity relations in ordinary steady-state situations.

Exercise 3–2. Show that assumption IV and Eq. (3.29) lead to the relation (and its dual)

$$Y_k\left(\frac{\partial \Theta}{\partial Y_k}\right)_{\Omega',\sigma} \geqslant 0. \tag{3.32}$$

Explorations

At this point I wish to explore some of the interconnections among the *petit* principles and among the *grand* principles of entropy production. My first observation is that $(\delta^2\Theta/\delta Y_k^2)_{\Omega',\sigma} > 0$ implies that $(\delta\Omega_k/\delta Y_k)_{\Omega',\sigma} > 0$. By straightforward differentiation of Eq. (3.20) we see that

$$\left(\frac{\partial^2\Theta}{\partial Y_k^2}\right)_{\Omega',\sigma} = -\left(\frac{\partial\Omega_k}{\partial Y_k}\right)^3_{\Omega',\sigma}\left(\frac{\partial^2 Y_k}{\partial\Omega_k^2}\right)_{\Omega',\sigma}\left[Y_k + \sum_i \Omega_i\left(\frac{\partial Y_i}{\partial\Omega_k}\right)_{\Omega',\sigma}\right] + \left(\frac{\partial\Omega_k}{\partial Y_k}\right)_{\Omega',\sigma}\left[2 + \left(\frac{\partial\Omega_k}{\partial Y_k}\right)_{\Omega',\sigma}\sum_i \Omega_i\left(\frac{\partial^2 Y_i}{\partial\Omega_k^2}\right)_{\Omega',\sigma}\right] \tag{3.33}$$

and that

$$\left(\frac{\delta^2\Theta}{\delta Y_k^2}\right)_{\Omega',\sigma} = \left(\frac{\delta\Omega_k}{\delta Y_k}\right)_{\Omega',\sigma}\left[2 + \left(\frac{\delta\Omega_k}{\delta Y_k}\right)_{\Omega',\sigma}\sum_i \Omega_i\left(\frac{\partial^2 Y_i}{\partial\Omega_k^2}\right)_{\Omega',\sigma}\Bigg|_{Y_k=0}\right]. \tag{3.34}$$

In the linear current-affinity region in the neighborhood of the thermostatic reference state σ the relation $0 < L_{kk}^{-1} = (\delta\Omega_k/\delta Y_k)_{\Omega',\sigma}$ holds due to the positive definite character of Θ; thus there exists a region for which $(\delta\Omega_k/\delta Y_k)_{\Omega',\sigma} > 0$. Now let us vary the affinities Ω_i $(i \neq k)$ so as to move far away from the reference state σ (out of the linear current-affinity region). Since we are assuming that $(\delta\Omega_k/\delta Y_k)_{\Omega',\sigma}$ is a continuous function of the affinities Ω_i $(i \neq k)$, it can only change to a negative value by passing through the value zero; but a zero value (see Eq. 3.34) would violate the condition $(\delta^2\Theta/\delta Y_k^2)_{\Omega',\sigma} > 0$. Hence $(\delta^2\Theta/\delta Y_k^2)_{\Omega',\sigma} > 0$ implies that $(\delta\Omega_k/\delta Y_k)_{\Omega',\sigma} > 0$. An analogous treatment of Eq. (3.33) shows that $(\partial^2\Theta/\partial Y_k^2)_{\Omega',\sigma} > 0$ implies that $(\partial \Omega_k/\partial Y_k)_{\Omega',\sigma} > 0$. Since assumption IV thus implies that $(\delta\Omega_k/\tilde{\delta} Y_k)_{\Omega',\sigma} > 0$, we can make use of the observation that

$$\left(\frac{\delta\Theta}{\tilde{\delta} Y_k}\right)_{\Omega',\sigma} = \left(\frac{\delta\Omega_k}{\tilde{\delta} Y_k}\right)_{\Omega',\sigma}\left(\frac{\partial\Theta}{\partial\Omega_k}\right)_{\Omega',\sigma}\Bigg|_{Y_k=0} \tag{3.35}$$

to express assumption IV in the alternative form

$$0 = \left(\frac{\partial\Theta}{\partial\Omega_k}\right)_{\Omega',\sigma}\Bigg|_{Y_k=0} = \sum_i \Omega_i\left(\frac{\partial Y_i}{\partial\Omega_k}\right)_{\Omega',\sigma}\Bigg|_{Y_k=0}. \tag{3.36}$$

Equation (3.36) and its dual are of use in those cases (rare for most experimental situations) where we have explicit expressions of the form (3.11) or (3.12).

Let us now derive Eq. (3.28) from Eq. (3.36). Equation (3.36) says that

$$0 = \sum_i \Omega_i\left(\frac{\partial Y_i}{\partial\Omega_k}\right)_{\Omega',\sigma} \qquad \text{for } Y_k = 0. \tag{3.37}$$

Let us differentiate Eq. (3.37) with respect to one of the other affinities Ω_j keeping Y_k constant:

$$0 = \left(\frac{\partial Y_j}{\partial\Omega_k}\right)_{\Omega',\sigma} + \left(\frac{\partial Y_k}{\partial\Omega_k}\right)_{\Omega',\sigma}\left(\frac{\partial\Omega_k}{\partial\Omega_j}\right)_{\Omega'',Y_k,\sigma}$$
$$+ \sum_i \Omega_i\left\{\left(\frac{\partial^2 Y_i}{\partial\Omega_j\,\partial\Omega_k}\right)_{\Omega',\sigma} + \left(\frac{\partial^2 Y_i}{\partial\Omega_k^2}\right)_{\Omega',\sigma}\left(\frac{\partial\Omega_k}{\partial\Omega_j}\right)_{\Omega'',Y_k,\sigma}\right\}, \tag{3.38}$$

where the subscript Ω'' means constant Ω_i for $i \neq j$ or k. The condition $dY_k = 0$ leads to the relation

$$\left(\frac{\partial\Omega_k}{\partial\Omega_j}\right)_{\Omega'',Y_k,\sigma} = -\left(\frac{\partial Y_k}{\partial\Omega_j}\right)_{\Omega',\sigma}\left(\frac{\partial Y_k}{\partial\Omega_k}\right)_{\Omega',\sigma}^{-1}. \tag{3.39}$$

Upon combining Eqs. (3.38) and (3.39) we get Eq. (3.28). Equation (3.28) is thus a direct consequence of assumption IV; in the linear current-affinity region for ordinary steady-state situations Eq. (3.28) reduces to the Onsager reciprocal relations.

Equation (3.37) is equivalent to two simultaneous equations. In deriving Eq. (3.28) I made use of derivatives of the basic simultaneous equations rather than of the basic equations themselves. For those cases where the currents can be displayed as a Taylor's series expansion in the affinities, with a finite number of terms in the expansion, it turns out that the basic simultaneous equations imply the stronger relation (3.31) as well as the relation (3.28). There are, however, some problems connected with a Taylor's series expansion of this type, and I investigate these matters in a section of the Appendix.

Let us next find the relationship between $(\partial \Omega_k/\partial Y_k)_{\Omega',\sigma}$ and $(\partial \Omega_k/\partial Y_k)_{Y',\sigma}$. Let the determinant

$$\begin{vmatrix} \left(\frac{\partial \Omega_1}{\partial Y_1}\right)_{Y',\sigma} & \left(\frac{\partial \Omega_1}{\partial Y_2}\right)_{Y',\sigma} & \cdots & \left(\frac{\partial \Omega_1}{\partial Y_\nu}\right)_{Y',\sigma} \\ \left(\frac{\partial \Omega_2}{\partial Y_1}\right)_{Y',\sigma} & \left(\frac{\partial \Omega_2}{\partial Y_2}\right)_{Y',\sigma} & \cdots & \left(\frac{\partial \Omega_2}{\partial Y_\nu}\right)_{Y',\sigma} \\ \vdots & \vdots & & \vdots \\ \left(\frac{\partial \Omega_\nu}{\partial Y_1}\right)_{Y',\sigma} & \left(\frac{\partial \Omega_\nu}{\partial Y_2}\right)_{Y',\sigma} & \cdots & \left(\frac{\partial \Omega_\nu}{\partial Y_\nu}\right)_{Y',\sigma} \end{vmatrix}$$

be called A, let the elements be called a_{ij}, and let the cofactor of element a_{ij} be called A_{ij}. Also let $\Omega_k = \Omega_k(Y_i, \sigma)$. Then

$$\left(\frac{\partial \Omega_k}{\partial Y_k}\right)_{\Omega',\sigma} = \left(\frac{\partial \Omega_k}{\partial Y_k}\right)_{Y',\sigma} + \sum_{\substack{i=1 \\ i \neq k}}^{\nu} \left(\frac{\partial \Omega_k}{\partial Y_i}\right)_{Y',\sigma} \left(\frac{\partial Y_i}{\partial Y_k}\right)_{\Omega',\sigma}. \tag{3.40}$$

From the condition Ω_j = constant $(j \neq k)$, we see that

$$\sum_{i=1}^{\nu} \left(\frac{\partial \Omega_j}{\partial Y_i}\right)_{Y',\sigma} \left(\frac{\partial Y_i}{\partial Y_k}\right)_{\Omega',\sigma} = 0 \qquad \begin{matrix} j = 1, 2, \ldots, \nu \\ j \neq k. \end{matrix} \tag{3.41}$$

Relations (3.41) are a set of $\nu - 1$ linear relations in the quantities $(\partial Y_i/\partial Y_k)_{\Omega',\sigma}$ $(i \neq k)$:

$$\sum_{\substack{i=1 \\ i \neq k}}^{\nu} a_{ji} \left(\frac{\partial Y_i}{\partial Y_k}\right)_{\Omega',\sigma} = -a_{jk} \qquad \begin{matrix} j = 1, 2, \ldots, \nu \\ j \neq k. \end{matrix} \tag{3.42}$$

These relations can be solved simultaneously for the quantities $(\partial Y_i/\partial Y_k)_{\Omega',\sigma}$:

$$\left(\frac{\partial Y_i}{\partial Y_k}\right)_{\Omega',\sigma} = \frac{A_{ki}}{A_{kk}}. \tag{3.43}$$

Since

$$\sum_{\substack{i=1 \\ i \neq k}}^{\nu} \frac{a_{ki} A_{ki}}{A_{kk}} = \frac{A - a_{kk} A_{kk}}{A_{kk}}, \tag{3.44}$$

we see (Eq. 3.40) that

$$\left(\frac{\partial \Omega_k}{\partial Y_k}\right)_{\Omega',\sigma} = \left(\frac{\partial \Omega_k}{\partial Y_k}\right)_{Y',\sigma} + \frac{A - a_{kk} A_{kk}}{A_{kk}}. \tag{3.45}$$

In the linear current-affinity case, $a_{ij} = K_{ij}$, $A > 0$, $A_{kk} > 0$, $a_{kk} > 0$, and [6] $a_{kk} A_{kk} > A$ if the matrix of coefficients is symmetric, whereas $a_{kk} A_{kk} < A$ if the matrix of coefficients is antisymmetric (i.e., $K_{ki} = -K_{ik}$; see the *special fields* section of the Appendix). The preceding argument can, of course, be carried through in dual form to obtain the relationship between $(\partial Y_k/\partial \Omega_k)_{Y',\sigma}$ and $(\partial Y_k/\partial \Omega_k)_{\Omega',\sigma}$.

For formal completeness I list some relations of general validity (but of limited practical importance):

$$\sum_i \left(\frac{\partial Y_k}{\partial \Omega_i}\right)_{\Omega',\sigma} \left(\frac{\partial \Omega_i}{\partial Y_j}\right)_{Y',\sigma} = \delta_{jk}, \tag{3.46}$$

$$Y_i = \sum_j \left\{ \left(\frac{\partial \Theta}{\partial Y_j}\right)_{Y',\sigma} - \Omega_j \right\} \left(\frac{\partial Y_j}{\partial \Omega_i}\right)_{\Omega',\sigma}, \tag{3.47}$$

$$\Omega_i = \sum_j \left\{ \left(\frac{\partial \Theta}{\partial \Omega_j}\right)_{\Omega',\sigma} - Y_j \right\} \left(\frac{\partial \Omega_j}{\partial Y_i}\right)_{Y',\sigma}, \tag{3.48}$$

where δ_{jk} is the Kronecker delta; i.e., $\delta_{jk} = 0$ if $j \neq k$, and $\delta_{jk} = 1$ if $j = k$.

Exercise 3–3. Show that for linear current-affinity relationships Eqs. (3.47) and (3.48) reduce to Eqs. (3.17) and (3.18).

A final word in clarification of assumption V may be in order. Equations such as (3.16), (3.17), and (3.18) do not possess duals because they contain the "additional quantities" $R[N]$, L_{ij}, and K_{ij}.

Equations of State

The role played by explicit current-affinity relations (Eqs. 3.11, 3.12, 3.17, 3.18, etc.) in the thermodynamics of steady states is analogous to the role played by equations of state in equilibrium thermostatics. In thermostatics it is essential that we have a firm belief in the *existence* of equations of state, but we do not need an *explicit* equation of state in order to establish such basic thermodynamic results as the conditions of equilibrium, the conditions of stability, the phase rule, etc. Explicit equations of state are undeniably useful, but we may erect an elaborate thermodynamic structure upon the mere premise that equations of state exist. In an analogous fashion we merely need a firm belief in the existence of current-affinity relations in order to develop the thermodynamics of steady states. An overemphasis on equations of state in thermostatics sometimes leads students of the subject to conclude (erroneously) that classical thermodynamics applies only to ideal gases. I fear that a corresponding overemphasis on explicit current-affinity relations for steady-state situations may leave the student with the (erroneous) impression that the thermodynamics of steady states is valid only in the linear current-affinity region.

The resolution of the rate of entropy production Θ into currents and affinities generates for us a useful and suggestive language in terms of which we can efficiently state the theorems of the thermodynamics of steady states.

The problem of determining the sign and the magnitude of a given current as a function of the pertinent parameters for a given steady-state situation, however, is more a problem of general physics than of thermodynamics. To insist on displaying the currents as power series in the affinities is to generate needless complications. The functional forms are all wrong: a heat flux depends on grad T and not on grad $(1/T)$, a diffusion flux depends on grad C and not on grad ln C, and so on. Of course, in an infinitesimal neighborhood of a given point, grad $(1/T)$ is proportional to grad T and grad ln C is proportional to grad C. It is to this simple fact that we owe a distressing overemphasis on the linear current-affinity region.

I shall have more to say on these matters in succeeding chapters of the text.

References

1. B. Coleman and C. Truesdell, *J. Chem. Phys.*, **33**, 28 (1960).
2. C. Truesdell, *J. Chem. Phys.*, **37**, 2336 (1962).
3. R. J. Tykodi and T. A. Erikson, *J. Chem. Phys.*, **31**, 1521 (1959).
4. L. S. Darken and R. W. Gurry, *Physical Chemistry of Metals* (McGraw-Hill Book Co., New York, 1953), p. 440.
5. L. Onsager, *Phys. Rev.*, **37**, 405 (1931); **38**, 2265 (1931).
6. H. Cramer, *Mathematical Methods of Statistics* (Princeton University Press, Princeton, 1946), Chapter 11.

4

Migrational Equilibrium in Monothermal Fields

General Considerations and Conditions of Thermostatic Equilibrium

Consider a system in a monothermal field with all terminal parts at temperature T; for a state that is a steady-flow state involving the steady migration of a given chemical substance into the kth terminal part of the system, we observe that

$$\Theta = -\frac{\dot{G}(\text{system})}{T} + \frac{\dot{W}_0}{T}. \tag{4.1}$$

We find the condition for the migrational equilibrium of the given chemical substance in the monothermal field by applying the operation $(\delta/\bar{\delta}\dot{n}_k)_{\Omega'}$ to Eq. (4.1); for monothermal fields let us take the operation $(\delta/\bar{\delta}\dot{n}_k)_{\Omega'}$ in the form $(\delta/\bar{\delta}\dot{n}_k)_{T,\ldots}$ with the temperature and an appropriate set of free-energy differences being held constant. The application of the operation $(\delta/\bar{\delta}\dot{n}_k)_{T,\ldots}$ to Eq. (4.1) leads to

$$\left(\frac{\delta\Theta}{\bar{\delta}\dot{n}_k}\right)_{T,\ldots} = \frac{1}{T}\left\{-\left(\frac{\delta\dot{G}(\text{system})}{\bar{\delta}\dot{n}_k}\right)_{T,\ldots} + \left(\frac{\delta\dot{W}_0}{\bar{\delta}\dot{n}_k}\right)_{T,\ldots}\right\}. \tag{4.2}$$

The application of assumption IV to Eq. (4.2) leads to

$$\sum\left(\frac{\delta(\mathcal{G}_i\dot{n}_i)}{\bar{\delta}\dot{n}_k}\right)_{T,\ldots} \equiv \left(\frac{\delta\dot{G}(\text{system})}{\bar{\delta}\dot{n}_k}\right)_{T,\ldots} = \left(\frac{\delta\dot{W}_0}{\bar{\delta}\dot{n}_k}\right)_{T,\ldots}, \tag{4.3}$$

the sum being over the terminal parts of the system. For a steady exchange of mass between two terminal parts (α, β) of a system with only pressure-volume work being exchanged between system and surroundings (i.e., $\dot{W}_0 = 0$), it follows that $\dot{n}_\alpha + \dot{n}_\beta = 0$, and that

$$\Theta = \frac{\dot{n}_\alpha(\mathscr{G}_\beta - \mathscr{G}_\alpha)}{T} = Y_1\Omega_1, \tag{4.4}$$

$$0 = \left(\frac{\delta\Theta}{\delta\dot{n}_\alpha}\right)_T = \frac{\mathscr{G}_\beta - \mathscr{G}_\alpha}{T} = \Omega_1, \tag{4.5}$$

and, hence, that

$$\mathscr{G}_\beta = \mathscr{G}_\alpha, \tag{4.6}$$

provided that there are no other independent currents traversing the system. Equation (4.6) yields many of the well-known relations pertaining to thermostatic equilibrium. Thus, if α and β refer to two different points in a gravitational field, and φ_α and φ_β refer to the potential energy per gram of some fluid of molecular weight M at the points α and β, then $\mathscr{G} = \mu + M\varphi$, and

$$\mu_\beta + M\varphi_\beta = \mu_\alpha + M\varphi_\alpha. \tag{4.7}$$

Concentration Field

Consider next the problem of migrational equilibrium in a monothermal concentration field—the situation indicated schematically in Figure 4–1. A spatial region marked MIX contains two fluid components, 1 and 2. The MIX region communicates with terminal parts of the system at a and b via membranes permeable to component 1 alone and with terminal parts at α and

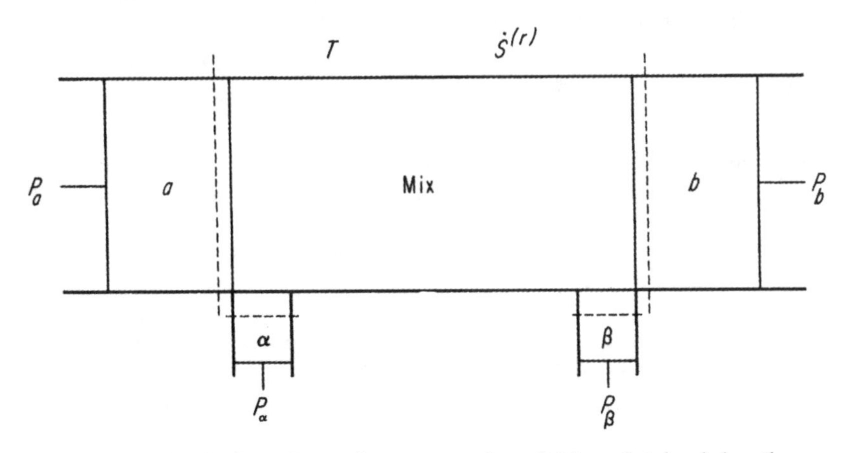

Figure 4–1. Monothermal concentration field maintained by the steady flow of component 1.

β via membranes permeable to component 2 alone. Let us induce a steady flow of component 1 through the MIX region and seek the conditions for the migrational equilibrium of component 2 in the concentration field set up by the flow of component 1. The situation we are considering is such that $\dot{W}_0 = 0$ and all terminal parts are at temperature T. If we establish a steady flow of both components 1 and 2 through the MIX region, then

$$\begin{aligned}\Theta &= -\frac{1}{T}(\mathcal{G}_a\dot{n}_a + \mathcal{G}_b\dot{n}_b + \mu_\alpha\dot{n}_\alpha + \mu_\beta\dot{n}_\beta) \\ &= \dot{n}_a\frac{\mathcal{G}_b - \mathcal{G}_a}{T} + \dot{n}_\alpha\frac{\mu_\beta - \mu_\alpha}{T} \\ &= Y_1\Omega_1 + Y_2\Omega_2, \end{aligned} \tag{4.8}$$

where $\dot{n}_a + \dot{n}_b = 0$, $\dot{n}_\alpha + \dot{n}_\beta = 0$, and $\mathcal{G} \equiv \mu + \mathcal{K}'$ with $\mathcal{K}$ being the molar kinetic energy of macroscopic motion. I have neglected kinetic energy terms associated with component 2. The operation $(\delta/\bar{\delta}\dot{n}_\alpha)_{T,\Delta\mathcal{G}^{(1)}}$, where $\Delta\mathcal{G}^{(1)} \equiv \mathcal{G}_b - \mathcal{G}_a$ is of the form $(\delta/\bar{\delta}Y_2)_{\Omega'}$. It follows, then, that

$$T\left(\frac{\delta\Theta}{\bar{\delta}\dot{n}_\alpha}\right)_{T,\Delta\mathcal{G}^{(1)}} = (\mathcal{G}_b - \mathcal{G}_a)\left(\frac{\delta\dot{n}_a}{\bar{\delta}\dot{n}_\alpha}\right)_{T,\Delta\mathcal{G}^{(1)}} + \mu_\beta - \mu_\alpha, \tag{4.9}$$

and, by the application of assumption IV, that

$$\mu_\alpha - \mu_\beta = -[R_{\alpha\beta}](\mathcal{G}_a - \mathcal{G}_b) \tag{4.10}$$

in the state of migrational equilibrium with respect to component 2; $[R_{\alpha\beta}] \equiv (\delta\dot{n}_a/\bar{\delta}\dot{n}_\alpha)_{T,\Delta\mathcal{G}^{(1)}}$. If the flow of the two fluids is coupled ($[R_{\alpha\beta}] \neq 0$), then the flow of component 1 sets up a gradient in the chemical potential of component 2. We can use Eq. (4.10) to test the validity of the relation $(\delta\Theta/\bar{\delta}Y_2)_{\Omega'} = 0$ for the situation considered: we can compare the $[R]$ quantity as computed from Eq. (4.10) to the value computed according to the definition

$$\lim_{\dot{n}_\alpha \to 0}\frac{\Delta\dot{n}_a}{\dot{n}_\alpha} \equiv [R_{\alpha\beta}] \qquad \text{constant } T, \Delta\mathcal{G}^{(1)}. \tag{4.11}$$

By introducing a parameter λ to measure distance in the direction of decreasing $\mathcal{G}$ in the (one-dimensional) concentration field, we can write Eq. (4.10) as

$$\frac{\mu_\beta - \mu_\alpha}{\lambda_\beta - \lambda_\alpha} \equiv \frac{\Delta\mu^{(2)}}{\Delta\lambda} = \frac{\Lambda[R_{\alpha\beta}]}{\Delta\lambda}\left(\frac{\Delta\mathcal{G}}{\Lambda}\right), \tag{4.12}$$

where Λ is a characteristic length (the distance between the semi-permeable membranes at a and b) for the field and $\Delta\mathcal{G} \equiv \mathcal{G}_{\text{source}} - \mathcal{G}_{\text{sink}}$. Upon the introduction of the abbreviation $[r_{\alpha\beta}] \equiv \Lambda[R_{\alpha\beta}]/\Delta\lambda$, we obtain

$$\frac{\Delta\mu^{(2)}}{\Delta\lambda} = [r_{\alpha\beta}]\frac{\Delta\mathcal{G}}{\Lambda}. \tag{4.13}$$

Thus far we have considered two points in the concentration field; we can also compute effects at a given point in the field. For a given point λ, the relations

$$\operatorname{grad} \mu_\lambda^{(2)} \equiv \left(\frac{\partial \mu_\lambda^{(2)}}{\partial \lambda}\right)_{T,\langle 2\rangle}, \tag{4.14}$$

$$[r_\lambda] \equiv \frac{\operatorname{grad} \mu_\lambda^{(2)}}{\Delta\mathscr{G}/\Lambda}, \tag{4.15}$$

are convenient; the subscript $\langle 2\rangle$ means that the *average concentration* (expressed in mole fraction or molality units) of component 2 in the concentration field is to be kept constant. From Eqs. (4.13) and (4.15) we see that

$$[r_{\alpha\beta}] = \frac{1}{\Delta\lambda}\int_\beta^\alpha [r_\lambda]\,d\lambda. \tag{4.16}$$

The coupling coefficients $[r]$ thus far introduced refer to a fixed $\Delta\mathscr{G}/\Lambda$ and are in the nature of integral quantities with respect to $\Delta\mathscr{G}/\Lambda$. For convenience we can also introduce differential coupling coefficients via the definitions

$$[r_{\alpha\beta}]^\partial \equiv \frac{1}{\Delta\lambda}\left(\frac{\partial\,\Delta\mu^{(2)}}{\partial(\Delta\mathscr{G}/\Lambda)}\right)_{T,\langle 2\rangle} \tag{4.17}$$

and

$$[r_\lambda]^\partial \equiv \left(\frac{\partial \operatorname{grad} \mu_\lambda^{(2)}}{\partial(\Delta\mathscr{G}/\Lambda)}\right)_{T,\langle 2\rangle}. \tag{4.18}$$

An equation similar to (4.16) holds between the differential coupling coefficients.

The application of the Gibbs–Duhem relation to the points of the concentration field yields

$$\bar{S}_\lambda \operatorname{grad} T_\lambda - \bar{V}_\lambda \operatorname{grad} P_\lambda + X_\lambda^{(1)} \operatorname{grad} \mu_\lambda^{(1)} + X_\lambda^{(2)} \operatorname{grad} \mu_\lambda^{(2)} = 0, \tag{4.19}$$

where $\bar{Z}_\lambda$ is a mean molar property of the 2-component system evaluated at a point λ in the concentration field, and $X_\lambda^{(i)}$ is the mole fraction of component i at the point λ. Now if $(\Delta\mathscr{G}/\Lambda) \approx -\operatorname{grad} \mu_\lambda^{(1)}$, then

$$[r_\lambda] \approx -\frac{\operatorname{grad} \mu_\lambda^{(2)}}{\operatorname{grad} \mu_\lambda^{(1)}} = \frac{X_\lambda^{(1)}}{X_\lambda^{(2)}} + \frac{\bar{S}_\lambda}{X_\lambda^{(2)}}\left(\frac{\operatorname{grad} T_\lambda}{\operatorname{grad} \mu_\lambda^{(1)}}\right) - \frac{\bar{V}_\lambda}{X_\lambda^{(2)}}\left(\frac{\operatorname{grad} P_\lambda}{\operatorname{grad} \mu_\lambda^{(1)}}\right). \tag{4.20}$$

Now quite clearly there will be gradients in temperature and pressure inside the MIX region due to the flow process and, consequently, $[r_\lambda] \neq (X_\lambda^{(1)}/X_\lambda^{(2)})$. I feel, however, that there is a good chance that the relation $[r_\lambda] \approx \langle X^{(1)}\rangle/\langle X^{(2)}\rangle$ will be close to the truth. We sorely need experimental investigations of concentration fields so as to be able to demonstrate the dependence of the various coupling coefficients on suitable parameters of the system (T, $\langle 2\rangle$, $\Delta\mathscr{G}/\Lambda$, etc.).

Electrokinetic Relations

If we separate two portions of a fluid by a porous membrane and if we place an inert electrode on either side of the membrane, then we find that the flow of fluid through the membrane gives rise to an electrical potential difference across the electrodes and, conversely, that a potential difference applied to the electrodes gives rise to a fluid flow or to a pressure difference across the membrane. The various phenomena that I have just described give rise to relations called *electrokinetic effects* [1, 2]. The situation is indicated schematically in Figure 4–2.

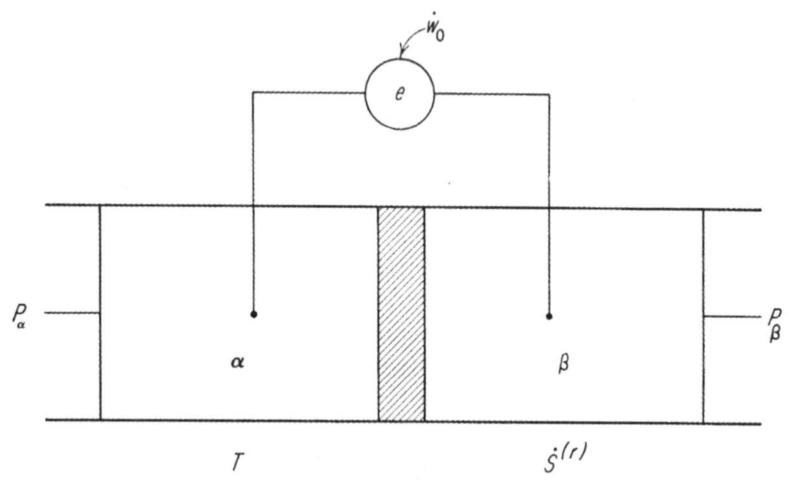

Figure 4–2. Electrokinetic apparatus, schematic; the electric engine at e receives work at the rate $\dot{W}_0$ and drives a current I around the circuit; $P_\alpha \geqslant P_\beta$.

Consider a state in which there is a steady flow of fluid across the membrane and a steady flow of electric current through the auxiliary circuit:

$$\Theta = -\frac{1}{T}(\dot{n}_\alpha\mu_\alpha + \dot{n}_\beta\mu_\beta) + \frac{\dot{W}_0}{T}. \tag{4.21}$$

Let us treat the fluid as incompressible, replacing $\dot{n}_\beta(\mu_\beta - \mu_\alpha)$ by

$$-\dot{V}_\beta(P_\alpha - P_\beta) \equiv -\dot{V}\Delta P,$$

and let us replace $\dot{W}_0$ by $I(\psi' - \psi'')$, where I is the negative electric current expressed in conventional units (amperes, say) and $\Delta\psi \equiv \psi' - \psi''$ is the difference in electrical potential between the binding posts of the electric engine at e. It is necessary to decide upon a sign convention for the direction of flow of negative electric current so as to be able to label (′ or ″) the binding posts of the electric engine properly.* Equation (4.21) thus takes the form

$$\Theta = \dot{V}\left(\frac{\Delta P}{T}\right) + I\left(\frac{\Delta\psi}{T}\right) = Y_1\Omega_1 + Y_2\Omega_2. \tag{4.22}$$

The application of the operation $(\delta/\tilde{\delta}I)_{T,\Delta P}$, which is of the form $(\delta/\tilde{\delta}\,Y_2)_{\Omega'}$, to Eq. (4.22) and the use of assumption IV lead to the relation

$$\left(\frac{\Delta\psi}{\Delta P}\right)_{T,I=0} = -\left(\frac{\delta\dot{V}}{\tilde{\delta}I}\right)_{T,\Delta P}. \tag{4.23}$$

If the right-hand side of Eq. (4.23) is independent of ΔP (as it is in the region of linear current-affinity relations [1]), then we can evaluate it by measuring the volume flow induced by unit electric current in the state for which $\Delta P = 0$; thus, for such circumstances,

$$\left(\frac{\Delta\psi}{\Delta P}\right)_{T,I=0} = -\left(\frac{\dot{V}}{I}\right)_{T,\Delta P=0}. \tag{4.24}$$

In a similar fashion, the operation $(\delta/\tilde{\delta}\dot{V})_{T,\Delta\psi}$ performed on Eq. (4.22) leads to the relations

$$\left(\frac{\Delta P}{\Delta\psi}\right)_{T,\dot{V}=0} = -\left(\frac{\delta I}{\tilde{\delta}\dot{V}}\right)_{T,\Delta\psi} \tag{4.25}$$

and (in the linear current-affinity region)

$$\left(\frac{\Delta P}{\Delta\psi}\right)_{T,\dot{V}=0} = -\left(\frac{I}{\dot{V}}\right)_{T,\Delta\psi=0}. \tag{4.26}$$

Exercise 4–1. Consider the linear current-affinity relations

$$\dot{V} = L_{11}\left(\frac{\Delta P}{T}\right) + L_{12}\left(\frac{\Delta\psi}{T}\right) \tag{4.27}$$

$$I = L_{21}\left(\frac{\Delta P}{T}\right) + L_{22}\left(\frac{\Delta\psi}{T}\right) \tag{4.28}$$

relative to the thermostatic reference state $(T, P_\beta, \Delta\psi = 0)$, treating the coefficients $L_{ik}(T, P_\beta)$ as constant for a given temperature T and reference pressure P_β. Show that

$$\left(\frac{\delta\dot{V}}{\tilde{\delta}I}\right)_{T,P_\beta,\Delta P} = \lim_{I\to 0}\left(\frac{(\partial\dot{V}/\partial\,\Delta\psi)_{T,P_\beta,\Delta P}}{(\partial I/\partial\,\Delta\psi)_{T,P_\beta,\Delta P}}\right)$$

is independent of ΔP and is in fact equal to $(\dot{V}/I)_{T,P_\beta,\Delta P=0}$.

* If it were necessary for us to be explicit about the sign convention, we could, for example, take the spontaneous direction of negative electric current flow for a small positive ΔP as the positive flow direction.

Exercise 4–2. Rather than struggle with the sign convention implicit in Eqs. (4.23) to (4.26), let us shift to absolute values. Thus, e.g.,

$$\left|\left(\frac{\Delta\psi}{\Delta P}\right)_{T,I=0}\right| = \left|\left(\frac{\dot{V}}{I}\right)_{T,\Delta P=0}\right|. \tag{4.29}$$

D. R. Briggs [3] made some measurements of the electrical potential difference developed across a cellulose plug and also across a glass capillary tube when conductivity water was forced through the plug or capillary at a steady rate. His data are listed in Table 4–1.

Table 4–1

Streaming Potential Data of Briggs [3]

ΔP (cm Hg)	$\Delta\psi$ (mv)	*Rate of Flow* (cm^3 sec^{-1})	
55.2	141	0.481	
47.9	122	0.424	
38.7	100	0.343	
28.2	75	0.238	cellulose
17.8	48.6	0.149	
7.5	21.5	0.060	
47.5	118.0	0.424	
38.4	98.0	0.338	
30.5	4370	1.56	
22.0	3100	1.21	capillary
14.4	2140	0.89	

Compute the ratio $(\Delta\psi/\Delta P)_{T,I=0}$ for each set of data. A nearly constant value for the ratio implies that Eqs. (4.27) and (4.28) are adequate representations of the data. From the computed values of the ratio $(\Delta\psi/\Delta P)_{T,I=0}$ and Eq. (4.29) calculate the volume flow (cm^3 sec^{-1}) that would accompany an electric current flow of 1 mamp for the case $\Delta P = 0$; make the calculation for both the cellulose and the capillary cases, paying especial attention to the choice of a consistent set of physical units in which to express the results. Also calculate for the case $\Delta P = 0$ the electric current that would be required to induce a volume flow of 1 cm^3 sec^{-1} across the cellulose plug; do the same for the capillary tube.

References

1. G. N. Lewis and M. Randall, *Thermodynamics*, 2nd edition, edited by K. Pitzer and G. Brewer (McGraw-Hill Book Co., New York, 1961), p. 458.
2. D. MacInnes, *The Principles of Electrochemistry* (Reinhold Publishing Co., New York, 1939), Chapter 23.
3. D. R. Briggs, *J. Phys. Chem.*, **32**, 641 (1928).

5

Migrational Equilibrium in Polythermal Fields: The Thermomolecular Pressure Effect

Let us turn now to the study of systems containing temperature gradients. In such systems we maintain a steady flow of heat in the system by controlling the action of the surroundings, and we look for the conditions of migrational equilibrium for given chemical species in such an environment. The necessary ideas are best illustrated by taking an example: consider the case of the thermomolecular pressure effect.

Thermomolecular Pressure Effect

Consider the system shown schematically in Figure 5–1: a 1-component gaseous substance is distributed through two chambers and a connecting capillary linkage, each chamber being in heat communication with a thermostat. If under conditions of no mass flow we find that $P_\alpha \neq P_\beta$ when $T_\alpha \neq T_\beta$, we say that a thermomolecular pressure effect exists. A study of the functional relatedness of the variables T_α, P_α, T_β, P_β (for states of no mass flow) for various gases and various kinds of capillary tubes is important for fields such as precision gas thermometry [1] and low temperature gas-solid adsorption [2]. In analyzing this situation, I divide the system into terminal parts (α, β) and a gradient part, the gradient part being stationary during steady-flow operations. Under conditions of steady heat flow alone, the relation

$$\begin{aligned}\dot{U}(\text{system}) &= -\dot{Q}_\alpha^{(r)} - \dot{Q}_\beta^{(r)} - \dot{q}(\alpha\beta) \\ &= -T_\alpha\dot{S}_\alpha^{(r)} - T_\beta\dot{S}_\beta^{(r)} - \dot{q}(\alpha\beta) = 0 \end{aligned} \tag{5.1}$$

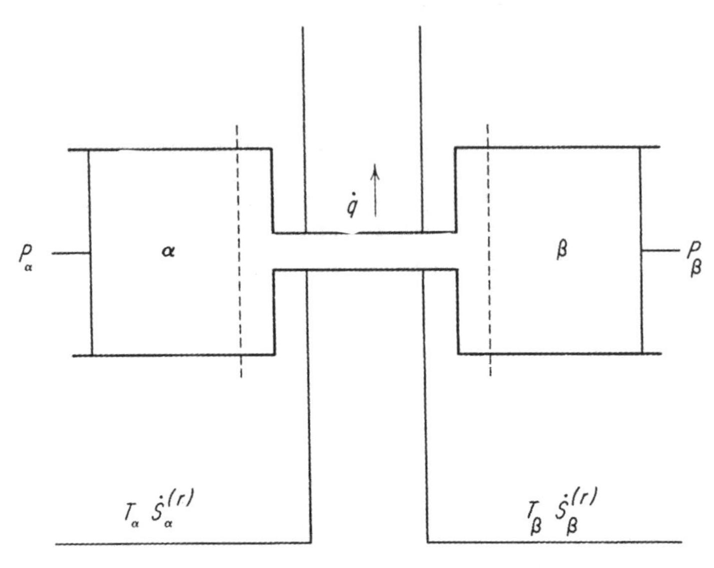

Figure 5–1. Thermomolecular pressure effect: two thermal states of a gas connected by a capillary linkage.

holds, where $\dot{Q}_\omega^{(r)}$ represents the rate of influx of heat into the heat reservoir of temperature T_ω, $\dot{S}_\omega^{(r)}$ represents the rate of accumulation of entropy in the heat reservoir at ω, and $\dot{q}(\alpha\beta)$ represents the heat gained by the surroundings (exclusive of the thermostats at α and β) via exchange with the gradient part. I call the section of the gradient part that falls outside the two thermostats the *link* or the *linkage*. The heat quantity $\dot{q}(\alpha\beta)$ represents the rate at which heat leaks across the lateral surface of the link to parts of the surroundings other than the thermostats at α and β. In order that the process described in Eq. (5.1) take place at a steady rate, it is necessary that the parts of the surroundings involved in the heat transfer $\dot{q}(\alpha\beta)$ be at fixed temperatures; i.e.,

$$\dot{q}(\alpha\beta) = \sum \dot{Q}_x^{(r)} = \sum T_x\dot{S}_x^{(r)} = \langle T(\alpha\beta)\rangle\dot{s}(\alpha\beta), \tag{5.2}$$

$$\dot{s}(\alpha\beta) \equiv \sum \dot{S}_x^{(r)}. \tag{5.3}$$

$$\langle T(\alpha\beta)\rangle \equiv \frac{\sum T_x\dot{S}_x^{(r)}}{\sum \dot{S}_x^{(r)}} = \frac{\dot{q}(\alpha\beta)}{\dot{s}(\alpha\beta)}. \tag{5.4}$$

The quantity $\langle T(\alpha\beta)\rangle$ is an effective average temperature for the surroundings to which heat is being leaked at the rate $\dot{q}(\alpha\beta)$. Two cases that frequently occur are (i) $\langle T(\alpha\beta)\rangle$ = ambient temperature (room temperature), and (ii) $\langle T(\alpha\beta)\rangle$ = the average over an infinite number of heat reservoirs ranging in temperature from T_α to T_β (a series of *Thomson reservoirs*). If both $\dot{q}(\alpha\beta) = 0$ and $\dot{s}(\alpha\beta) = 0$, I define $\langle T(\alpha\beta)\rangle$ as the limit of $\dot{q}(\alpha\beta)/\dot{s}(\alpha\beta)$ as $\dot{q}(\alpha\beta) \rightarrow 0$ for those cases where the limit is physically evident.*

In my analysis of polythermal fields, then, I give the various parts of the surroundings unequal emphasis: I give essential temperatures and heat reservoirs individual emphasis, whereas I lump together less essential ones and treat them in terms of "averaged" or "effective" parameters. In addition, when there is no danger of ambiguity, I shorten the notation $\dot{q}(\alpha\beta)$, $\langle T(\alpha\beta)\rangle$, etc., to $\dot{q}$, $\langle T\rangle$, etc.

Consider again the system shown in Figure 5–1. Induce a steady flow of matter between the terminal parts of the system such that $\dot{n}_\alpha + \dot{n}_\beta = 0$ and neglect kinetic energy terms; we then get,

$$\dot{n}_\alpha \overline{U}_\alpha + \dot{n}_\beta \overline{U}_\beta = -\dot{Q}_\alpha^{(r)} - \dot{Q}_\beta^{(r)} - \dot{q} - P_\alpha \dot{V}_\alpha - P_\beta \dot{V}_\beta, \tag{5.5}$$

$$\dot{n}_\alpha \overline{H}_\alpha + \dot{n}_\beta \overline{H}_\beta + \dot{Q}_\alpha^{(r)} + \dot{Q}_\beta^{(r)} + \dot{q} = 0, \tag{5.6}$$

$$\begin{aligned} \Theta &= \dot{n}_\alpha \overline{S}_\alpha + \dot{n}_\beta \overline{S}_\beta + \frac{\dot{Q}_\alpha^{(r)}}{T_\alpha} + \frac{\dot{Q}_\beta^{(r)}}{T_\beta} + \frac{\dot{q}}{\langle T\rangle} \\ &= \dot{n}_\alpha\left(\frac{\overline{H}_\beta - \overline{H}_\alpha}{T_\beta} + \overline{S}_\alpha - \overline{S}_\beta\right) + \dot{Q}_\alpha^{(r)}\left(\frac{1}{T_\alpha} - \frac{1}{T_\beta}\right) + \dot{q}\left(\frac{1}{\langle T\rangle} - \frac{1}{T_\beta}\right) \\ &= Y_1\Omega_1 + Y_2\Omega_2 + Y_3\Omega_3, \end{aligned} \tag{5.7}$$

where $\overline{Z}_\omega$ represents a molar property of the gas in the ω terminal part, $\dot{n}_\omega$ and $\dot{V}_\omega$ represent the rate of influx of mass and volume, respectively, to the ω terminal part, and $\overline{H}_\omega \equiv \overline{U}_\omega + P_\omega \overline{V}_\omega$. I have made use of Eq. (5.6) in arriving at the second line of Eq. (5.7). The gradient part of the system, being stationary during the steady-flow process, does not contribute to the left-hand side of Eq. (5.5). To find the condition for migrational equilibrium in the polythermal field we investigate the behavior of Eqs. (5.6) and (5.7) under the operation $(\delta/\bar{\delta}\dot{n}_\alpha)_{T_\alpha, T_\beta, \langle T\rangle}$, which is of the form $(\delta/\bar{\delta} Y_1)_{\Omega'}$:

$$\overline{H}_\alpha - \overline{H}_\beta + \left(\frac{\delta(\dot{Q}_\alpha^{(r)} + \dot{Q}_\beta^{(r)} + \dot{q})}{\bar{\delta}\dot{n}_\alpha}\right)_{T_\alpha, T_\beta, \langle T\rangle} = 0, \tag{5.8}$$

* If, for example, the link is efficiently insulated so that $\dot{q}(\alpha\beta) = 0$ and $\dot{s}(\alpha\beta) = 0$, and if removal of the insulation would lead to heat exchange with a room of temperature T_R, then it is clear that removal of the insulation followed by step-by-step restoration would lead to $\lim \dot{q}(\alpha\beta)/\dot{s}(\alpha\beta) = T_R$, regardless of the exact way in which the insulation was applied to the link.

$$\left(\frac{\delta\Theta}{\bar{\delta}\dot{n}_\alpha}\right)_{\Omega'} = \bar{S}_\alpha - \bar{S}_\beta + \left(\frac{\delta(\dot{S}_\alpha^{(r)} + \dot{S}_\beta^{(r)} + \dot{s})}{\bar{\delta}\dot{n}_\alpha}\right)_{T_\alpha, T_\beta, \langle T\rangle}$$
$$= \frac{\bar{H}_\beta - \bar{H}_\alpha}{T_\beta} + \bar{S}_\alpha - \bar{S}_\beta + \left(\frac{\delta\dot{Q}_\alpha^{(r)}}{\bar{\delta}\dot{n}_\alpha}\right)_{T_\alpha, T_\beta, \langle T\rangle}\left(\frac{1}{T_\alpha} - \frac{1}{T_\beta}\right) + \left(\frac{\delta\dot{q}}{\bar{\delta}\dot{n}_\alpha}\right)_{T_\alpha, T_\beta, \langle T\rangle}\left(\frac{1}{\langle T\rangle} - \frac{1}{T_\beta}\right). \tag{5.9}$$

In equilibrium thermostatics a heat effect in a heat reservoir surrounding a system that is undergoing a reversible isothermal isopiestic change in state can be related to an enthalpy change in the system. Consider, for example, the reversible isothermal vaporization of a liquid: the amount of heat $-dQ^{(r)}$ lost by the reservoir for the vaporization of dn moles of liquid is related to the enthalpy change $\bar{H}_{\mathrm{gas}} - \bar{H}_{\mathrm{liq}}$ by the relation $-dQ^{(r)} = dn(\bar{H}_{\mathrm{gas}} - \bar{H}_{\mathrm{liq}})$ or $dQ^{(r)}/dn = \bar{H}_{\mathrm{liq}} - \bar{H}_{\mathrm{gas}}$. I use this observation as a guide in setting up a number of useful steady-state definitions:*

$$\left(\frac{\delta\dot{Q}_\omega^{(r)}}{\bar{\delta}\dot{n}_\omega}\right)_{T_\alpha, T_\beta, \langle T\rangle} \equiv [H_\omega] - \bar{H}_\omega, \tag{5.10}$$

$$\left(\frac{\delta\dot{S}_\omega^{(r)}}{\bar{\delta}\dot{n}_\omega}\right)_{T_\alpha, T_\beta, \langle T\rangle} \equiv [S_\omega] - \bar{S}_\omega, \tag{5.11}$$

$$\left(\frac{\delta\dot{q}}{\bar{\delta}\dot{n}_\omega}\right)_{T_\alpha, T_\beta, \langle T\rangle} \equiv -[h(\chi\omega)], \tag{5.12}$$

$$\left(\frac{\delta\dot{s}}{\bar{\delta}\dot{n}_\omega}\right)_{T_\alpha, T_\beta, \langle T\rangle} \equiv -[s(\chi\omega)]. \tag{5.13}$$

Equations (5.10) to (5.13) are definitions of the steady-state square-bracketed quantities.† The quantity $[H_\omega]$, e.g., is an "effective" molar enthalpy for the gas at the point where it crosses the boundary of the heat reservoir. The application of the definitions (5.10) to (5.13) to Eqs. (5.8) and (5.9) yields the relations

$$[H_\alpha] - [H_\beta] - [h(\beta\alpha)] = 0, \tag{5.14}$$

* I am here using the less explicit limiting operation $(\delta/\bar{\delta} Y_i)_{\Omega'}$; see Chapter 3. I could use an explicit reference state σ and the explicit operation $(\delta/\bar{\delta} Y_i)_{\Omega',\sigma}$, adapting the notation accordingly: thus, e.g., I could write Eq. (5.10) as

$$\left(\frac{\delta\dot{Q}_\omega^{(r)}}{\bar{\delta}\dot{n}_\omega}\right)_{\Omega',\sigma} \equiv [H_\omega]_\sigma - \bar{H}_\omega.$$

Inasmuch as the eye easily tires of the rococo richness of the symbolism, I deem it more merciful to deal (where possible) with the less explicit form $(\delta/\bar{\delta} Y_i)_{\Omega'}$.

† Note that according to Eqs. (5.12) and (5.13) $[h(\beta\alpha)] = -[h(\alpha\beta)]$ and $[s(\beta\alpha)] = -[s(\alpha\beta)]$.

$$\left(\frac{\delta\Theta}{\bar{\delta}\dot{n}_\alpha}\right)_{\Omega'} = [S_\alpha] - [S_\beta] - [s(\beta\alpha)]$$

$$= \frac{\bar{H}_\beta - [H_\beta]}{T_\beta} + \bar{S}_\alpha - \bar{S}_\beta + \frac{[H_\alpha] - \bar{H}_\alpha}{T_\alpha} - \frac{[h(\beta\alpha)]}{\langle T\rangle}. \quad (5.15)$$

In addition, since $\dot{Q}_\omega^{(r)} = T_\omega \dot{S}_\omega^{(r)}$ and $\dot{q} = \langle T\rangle\dot{s}$, it follows from definitions (5.10) to (5.13) that

$$[H_\omega] = \mu_\omega + T_\omega[S_\omega], \quad (5.16)$$

$$[h(\chi\omega)] = \langle T\rangle[s(\chi\omega)], \quad (5.17)$$

where $\mu_\omega \equiv \bar{H}_\omega - T_\omega\bar{S}_\omega$. Upon combining other equations with it, we can write Eq. (5.14) in the alternative forms

$$\mu_\alpha + T_\alpha[S_\alpha] - \mu_\beta - T_\beta[S_\beta] - [h(\beta\alpha)] = 0, \quad (5.18)$$

$$\mu_\alpha + [S_\alpha]\{T_\alpha - \langle T\rangle\} - \mu_\beta - [S_\beta]\{T_\beta - \langle T\rangle\} = -\langle T\rangle\left(\frac{\delta\Theta}{\bar{\delta}\dot{n}_\alpha}\right)_{\Omega'}, \quad (5.19)$$

$$\frac{[G_\beta] - [G_\alpha]}{\langle T\rangle} = \left(\frac{\delta\Theta}{\bar{\delta}\dot{n}_\alpha}\right)_{\Omega'}, \quad (5.20)$$

where

$$[G_\omega] \equiv [H_\omega] - \langle T\rangle[S_\omega]. \quad (5.21)$$

In definition (5.21) I have again been guided by equilibrium thermostatic considerations. Assumption IV implies that

$$[S_\alpha] - [S_\beta] - [s(\beta\alpha)] = 0 \quad (5.22)$$

and that

$$[G_\beta] - [G_\alpha] = 0. \quad (5.23)$$

Now compare Eqs. (4.2), (4.5), (4.9), and (5.20). Observe that in each case the condition of migrational equilibrium, based on assumption IV, takes the form of a conservation condition for the spatial field in question: $\mathscr{G}$ = constant, $\mu + [R]\mathscr{G}$ = constant, $[G]$ = constant. These results are the generalization of Gibbs' fundamental result for isothermal heterogeneous equilibrium: $\mu^{(i)}$ = constant for all phases in which i is an actual component. This treatment of migrational equilibrium under steady-state conditions is thus as closely analogous to Gibbs' classical procedures as I can make it.

Consider now the special case where the connecting linkage between terminal parts α and β is efficiently insulated so that $\dot{q} = 0$. Under such conditions $[h(\beta\alpha)]$ and $[s(\beta\alpha)]$ are zero and the quantity $\langle T\rangle$ need not enter into any of the calculations. If the physical situation, however, permits us to define $\langle T\rangle$ as $\lim(\dot{q}/\dot{s})$ (see the earlier footnote on this subject), setting it equal to T_R, say, then we can define a $[G]$ function for this case also, and Eqs. (5.20), (5.21), and (5.23) will continue to hold. It is of course possible to have $\dot{q} \neq 0$ and yet $[h(\beta\alpha)] \equiv (\delta\dot{q}/\bar{\delta}\dot{n}_\alpha)_{T_\alpha, T_\beta, \langle T\rangle} = 0$. Under such circumstances $\Delta[H] = 0$ and, according to assumption IV, $\Delta[G] = 0$ and $\Delta[S] = 0$ also.

I have defined the steady-state bracketed quantities in such a way as to make them appear to be pseudo-properties of the gas under consideration, but they in fact depend on the nature of the linkage between the terminal parts of the system, on how that linkage is coupled energetically to the surroundings, and on the path used in the basic limiting operation $(\delta/\bar{\delta}\dot{n}_{\omega})$.* It is now necessary for me to discuss the properties of the linkage between the terminal parts of the system and the effect of those properties on the bracketed quantities thus far introduced.

Owing to the interchange of heat between the system and the surroundings in the state of no mass flow, we can distinguish three situations: (i) full-flux linkage—heat is exchanged only with the constant temperature reservoirs at T_{α} and T_{β}, i.e., $\dot{q} = 0$; (ii) zero-flux linkage—heat is exchanged only across the lateral surface of the linkage, i.e., $\dot{Q}_{\omega}^{(r)} = 0$ (an infinite number of heat reservoirs are coupled to the linkage); (iii) partial-flux linkage—heat is exchanged in such a way that $\dot{Q}_{\omega}^{(r)} \neq 0$ and $\dot{q} \neq 0$.

Full-Flux Linkage

We can approximate full-flux linkages in the laboratory by means of efficient thermal lagging. For a full-flux linkage $[h(\chi\omega)] = 0$, $[s(\chi\omega)] = 0$, and $\langle T \rangle$, although not needed anywhere in the analysis, can sometimes be set equal to a physically obvious reference temperature T_R.

Partial-Flux Linkage

Partial-flux linkages represent a common laboratory situation where there is *some* lateral leakage of heat from the link to the surroundings. The most common situation is that for which heat leaks to the room of temperature T_R. Under such circumstances $\langle T \rangle = T_R$, and in general, neither of the lower-case bracketed quantities are equal to zero.

Zero-Flux Linkage

The zero-flux linkage with its infinite set of heat reservoirs is a theoretical construct often used to maintain a fixed temperature gradient during steady-flow operations. For the zero-flux linkage $[H_{\omega}] = \bar{H}_{\omega}$, $[S_{\omega}] = \bar{S}_{\omega}$, $[G_{\omega}] = \bar{H}_{\omega} - \langle T \rangle \bar{S}_{\omega}$, $[h(\chi\omega)] = \bar{H}_{\omega} - \bar{H}_{\chi}$, $[s(\chi\omega)] = \bar{S}_{\omega} - \bar{S}_{\chi}$, and $\langle T \rangle = \Delta \bar{H}/\Delta \bar{S}$. Now it is often the case that

$$\bar{H}_{\alpha} - \bar{H}_{\beta} = \int_{T_{\beta}}^{T_{\alpha}} \frac{d\bar{H}}{dT}\, dT = \left\langle \frac{d\bar{H}}{dT} \right\rangle (T_{\alpha} - T_{\beta}) \tag{5.24}$$

* See Chapter 6 for a discussion of this point.

and that

$$\bar{S}_\alpha - \bar{S}_\beta = \int_{T_\beta}^{T_\alpha} \frac{d\bar{H}}{dT}\, d\ln T = \left\langle\frac{d\bar{H}}{dT}\right\rangle \ln\frac{T_\alpha}{T_\beta}. \tag{5.25}$$

Under such circumstances $\langle T\rangle = (T_\alpha - T_\beta)/\ln(T_\alpha/T_\beta)$; i.e., $\langle T\rangle$ is just the well-known log-mean temperature used in heat conductivity studies [3].

More Definitions

Consider again the system described in Figure 5–1. Under migrational equilibrium conditions $[G_\alpha] = [G_\beta]$; thus thermomolecular pressure phenomena* are characterized by the constancy of the function $[G]$. I could express the differential thermomolecular pressure effect in terms of the $[G]$ function:

$$\bar{V}\left(\frac{\partial P}{\partial T}\right)_{[G]} = \bar{S} - \left(\frac{\partial\{[S](T - \langle T\rangle)\}}{\partial T}\right)_{[G]} \tag{5.26}$$

in a manner analogous to the usual treatment of the differential Joule–Thomson effect; however, it seems better to me to proceed in a different fashion. I now introduce a new set of steady-state functions that are related to the previous set of functions in a manner analogous to that in which differential thermostatic adsorption quantities are related to integral thermostatic adsorption quantities:†

$$[S_\omega]^\partial \equiv -\left(\frac{\partial\mu_\omega}{\partial T_\omega}\right)_{\chi,\dot{n}_\omega=0}, \qquad \langle[S]^\partial\rangle \equiv -\frac{\mu_\omega - \mu_\chi}{T_\omega - T_\chi}, \tag{5.27}$$

$$[C_{\omega\omega}]^\partial \equiv T_\omega\left(\frac{\partial[S_\omega]^\partial}{\partial T_\omega}\right)_{\chi,\dot{n}_\omega=0}, \qquad [C_{\chi\omega}]^\partial \equiv T_\chi\left(\frac{\partial[S_\chi]^\partial}{\partial T_\omega}\right)_{\chi,\dot{n}_\omega=0}, \tag{5.28}$$

$$[Q_\omega]^\partial \equiv T_\omega(\bar{S}_\omega - [S_\omega]^\partial), \tag{5.29}$$

where the subscript χ on the differential coefficients means that the thermodynamic state at χ is to be kept constant; i.e., all the intensive variables at $\chi - T_\chi$, P_χ, etc., are to be kept constant. It is usually quite clear from the context when states of migrational equilibrium are being considered; hence I normally drop the explicit subscript notation $\dot{n}_\omega = 0$. From the definition of $[S_\omega]^\partial$ we see that

$$\bar{V}_\omega\left(\frac{\partial P_\omega}{\partial T_\omega}\right)_\chi = \bar{S}_\omega - [S_\omega]^\partial; \tag{5.30}$$

* When $\dot{n}_\omega = 0$ and $P_\alpha \neq P_\beta$ for the case $T_\alpha \neq T_\beta$, we say that a thermomolecular pressure effect exists.

† See Chapter 2 for the pertinent discussion.

thus (for $\dot{n}_\omega = 0$) the two conditions $P_\alpha \neq P_\beta$ for $T_\alpha \neq T_\beta$ and $[S_\omega]^\partial \neq \bar{S}_\omega$ mutually imply one another.*

In discussing the next series of operations I find it necessary (in certain integrals) for the sake of clarity to distinguish between the names of thermodynamic states (α, β) and the names of terminal parts (a, b). The thermodynamic state α is characterized by the set of parameters T_α, P_α, etc.; the terminal part at a will usually be in state α ($T_a = T_\alpha$, $P_a = P_\alpha$, etc.), but such need not always be the case (however, the ambiguity in notation is not serious, except in the case of integrals, and I normally use the same symbol (α, say) both for the name of the state and the name of the terminal part). Consider a series of operations in which the terminal parts of the system start out in the same thermodynamic state (α, say), then the condition of one of the terminal parts (b, say) is changed in such a manner as always to maintain the migrational equilibrium condition while keeping the state of the a terminal part fixed. For such a series of operations we can write

$$\begin{aligned} \mu_b - \mu_a = \mu_\beta - \mu_\alpha &= -\int_{T_b = T_\alpha}^{T_b = T_\beta} [S_b]^\partial \, dT_b = T_\alpha[S_\alpha] - T_\beta[S_\beta] - [h(\beta\alpha)] \\ &= -\langle [S]^\partial \rangle (T_\beta - T_\alpha), \end{aligned} \tag{5.31}$$

$$\left\langle \frac{\bar{S}_\beta - [S_\beta]^\partial}{R} \right\rangle_\alpha = \frac{\ln (P_\beta^{(2)}/P_\beta^{(1)})}{\ln (T_\beta^{(2)}/T_\beta^{(1)})} \text{ (ideal gas)}, \tag{5.32}$$

where $\langle(\bar{S}_\beta - [S_\beta]^\partial)/R\rangle_\alpha$ represents the average value of $(\bar{S}_b - [S_b]^\partial)/R$ as terminal part b passes between states 1 and 2—the state (α) of terminal part a remaining constant during the process. Evidently we can reach the state of affairs for which terminal part a is in state α and part b is in state β, either by starting out with the entire system in state α and carrying b to state β in a proper fashion or by starting out in state β and operating on part a. If we bring these considerations to bear upon Eq. (5.31), we see that

$$\mu_\beta - \mu_\alpha = -\int_{T_\alpha}^{T_\beta} [S_b]^\partial \, dT_b = \int_{T_\beta}^{T_\alpha} [S_a]^\partial \, dT_a. \tag{5.33}$$

The application of the operation $(\partial/\partial T_\beta)_\alpha$ to the second equality of Eq. (5.33) yields the relation†

$$-[S_b]^\partial = -[S_a(\beta)]^\partial, \tag{5.34}$$

* In terms of the notation that I introduce in the next paragraph, we have

$$P_\alpha - P_\beta = \int_{T_a = T_\beta}^{T_a = T_\alpha} \frac{\bar{S}_a - [S_a]^\partial}{\bar{V}_a} \, dT_a.$$

† In the final integral of Eq. (5.33) the value of the integrand at the lower limit is $[S_a(\beta)]^\partial$ since the integration starts from the state in which both terminal parts are in state β. (In terms of the "ambiguous" notation, $[S_b]^\partial \equiv [S_\beta]^\partial$.)

where $[S_a(\beta)]^\partial$ represents the value of $[S_a]^\partial$ for the state in which $T_a = T_b = T_\beta$, $P_a = P_b = P_\beta$, etc. Now since

$$[S_\beta]^\partial = [S_b(\beta)]^\partial + \int_{T_a=T_\beta}^{T_a=T_\alpha} \left(\frac{\partial [S_b]^\partial}{\partial T_a}\right)_\beta dT_a$$

$$= [S_b(\beta)]^\partial + \int_{T_\beta}^{T_\alpha} \frac{[C_{ba}]^\partial}{T_\beta} dT_a, \qquad (5.35)$$

we see that

$$[S_a(\beta)]^\partial - [S_b(\beta)]^\partial = \int_{T_\beta}^{T_\alpha} \frac{[C_{ba}]^\partial}{T_\beta} dT_a. \qquad (5.36)$$

The left-hand side of Eq. (5.36) is independent of T_α, whereas the right-hand side is a function of T_α; consistency requires that

$$[C_{\beta\alpha}]^\partial = 0 \qquad (5.37)$$

and hence that $[S_a(\beta)]^\partial = [S_b(\beta)]^\partial$. Equation (5.37) is also a direct consequence of Eq. (5.34).

Treatment of Experimental Results

Consider again the situation indicated in Figure 5–1. If we confine our experiments to mass-static measurements (measurements under migrational equilibrium conditions), then the best that we can do is to produce an experimental equation of correlation* $P_\alpha = P_\alpha(T_\alpha, T_\beta, P_\beta, \langle T \rangle)$ for a given gas and linkage; from the equation of correlation we can then compute differential quantities such as $[S_\omega]^\partial$, $[C_{\omega\omega}]^\partial$, etc., or average quantities like those exhibited in Eqs. (5.27) and (5.32). From mass-static measurements alone we cannot say much about the integral quantities $[S_\omega]$, $[C_{\omega\omega}]$, etc. The preceding thermodynamic considerations impose a constraint on the equation of correlation. We see from Eqs. (5.34) and (5.37) that the quantity $[S_\omega]^\partial$ is a function of the ω state variables only (for a given gas and a given linkage), regardless of the nature of the linkage (full, partial, or zero flux); i.e., $[S_\omega]^\partial = [S_\omega]^\partial(T_\omega, P_\omega)$.† It follows from Eq. (5.30), then, that $(\partial P_\omega/\partial T_\omega)_\chi$ must be expressible as a function of the ω state variables only (for a given gas and linkage); this result limits the possible forms of the equation of correlation: in fact the equation of correlation must have the symmetric form $f(T_\alpha, P_\alpha) = f(T_\beta, P_\beta)$.

* It is customary to assume that $(\partial P_\alpha/\partial\langle T \rangle)_{\beta,T\alpha} = 0$; however I know of no experimental information bearing on this point. This assumption has far-reaching consequences, and I shall, for purposes of identification, refer to it as *assumption Q*.

† If assumption Q did not hold, the quantity $[S_\omega]^\partial$ would depend on $\langle T \rangle$ as well as on T_ω and P_ω.

For a given gas, linkage, and temperatures T_α, T_β, I call the plot of P_α versus P_β or of $(P_\alpha - P_\beta)$ versus P_ω the *bithermal relation* or the *bitherm.* In the case of a perfect gas and a Knudsen linkage (molecules pass through the capillary linkage without exchanging energy with its walls or with other molecules; a Knudsen linkage is thus one for which $[h(\chi\omega)] = 0$), the equation of correlation is especially simple, namely $P_\alpha/P_\beta = (T_\alpha/T_\beta)^{1/2}$, and the bitherms are straight lines.

Exercise 5–1. Show that the Knudsen equation $P_\omega/P_\chi = (T_\omega/T_\chi)^{1/2}$ and the ideal gas relation together imply that (i) $\bar{S}_\omega - [S_\omega]^\partial = R/2$, (ii) $[Q_\omega]^\partial = RT_\omega/2$, (iii) $[C_{\omega\omega}]^\partial = \bar{C}_P - (R/2)$, and (iv) $[C_{\chi\omega}]^\partial = 0$. Result (iii) is suggestive: it seems to say that the gas loses 1 degree of freedom while passing through the capillary tube.

Exercise 5–2. (i) Show that for a given bitherm—the (T_ω, T_χ) bitherm, say—involving an ideal gas with an arbitrary linkage the relation

$$\frac{\bar{S}_\omega - \langle [S]^\partial \rangle}{R} = \frac{\bar{C}_P}{R} - \frac{T_\chi}{T_\chi - T_\omega}\left(\frac{\bar{C}_P}{R}\ln\frac{T_\chi}{T_\omega} - \ln\frac{P_\chi}{P_\omega}\right) \tag{5.38}$$

holds (assume $\bar{C}_P$ independent of T). (ii) Note that

$$\bar{S}_\omega - \langle [S]^\partial \rangle - (\bar{S}_\chi - \langle [S]^\partial \rangle) = \bar{S}_\omega - \bar{S}_\chi;$$

show that the right-hand side of Eq. (5.38) is consistent with this observation.

Exercise 5–3. Liang [4] proposed the following equation for treating the thermomolecular pressure effect between states 1 and 2:

$$\frac{P_1}{P_2} = \frac{\boldsymbol{\alpha}(x/\mathbf{f})^2 + \boldsymbol{\beta}(x/\mathbf{f}) + (T_1/T_2)^{1/2}}{\boldsymbol{\alpha}(x/\mathbf{f})^2 + \boldsymbol{\beta}(x/\mathbf{f}) + 1} \tag{5.39}$$

where $\boldsymbol{\alpha}$ and $\boldsymbol{\beta}$ are empirical constants, $\mathbf{f}$ (the pressure-shifting factor) is a constant characteristic of the gas under study, and $x \equiv P_2 d$ with d being the diameter of the tube. Show that Eq. (5.39) is illegitimate as an equation of correlation.

Hint: The derivative $(\partial P_\omega/\partial T_\omega)_\chi$ must be expressible as a function of the variables T_ω and P_ω only for a proper equation of correlation; show that such is not the case for Eq. (5.39). Note also that Eq. (5.39) cannot be put into the symmetric form $f(1) = f(2)$. There have been several attempts to improve upon Liang's equation [5–8]. It sometimes happens that an empirical equation violates the laws of thermodynamics. Such an empirical relation, although it cannot have any *general* validity, may nevertheless prove useful for recording and correlating data over a restricted range of conditions.

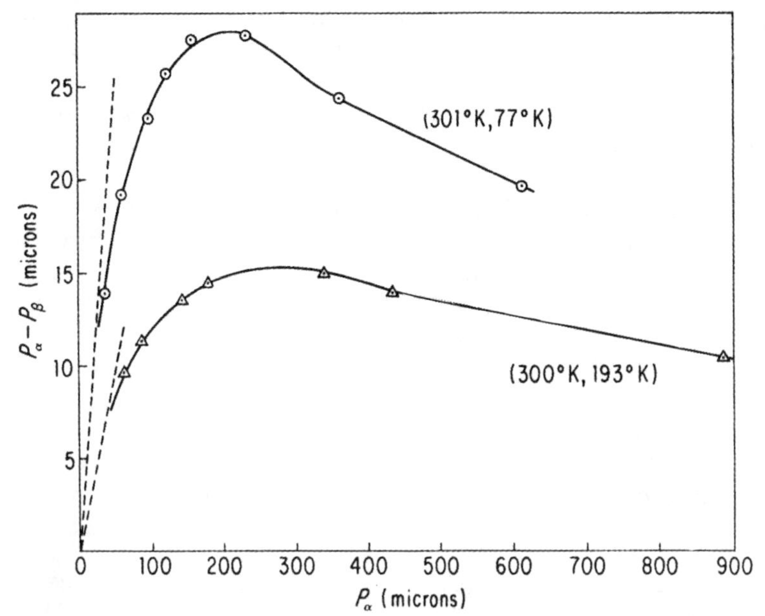

Figure 5–2. Experimental bitherms for argon with a Tru-bore capillary linkage 0.0510 cm in diameter (data of Los and Ferguson).

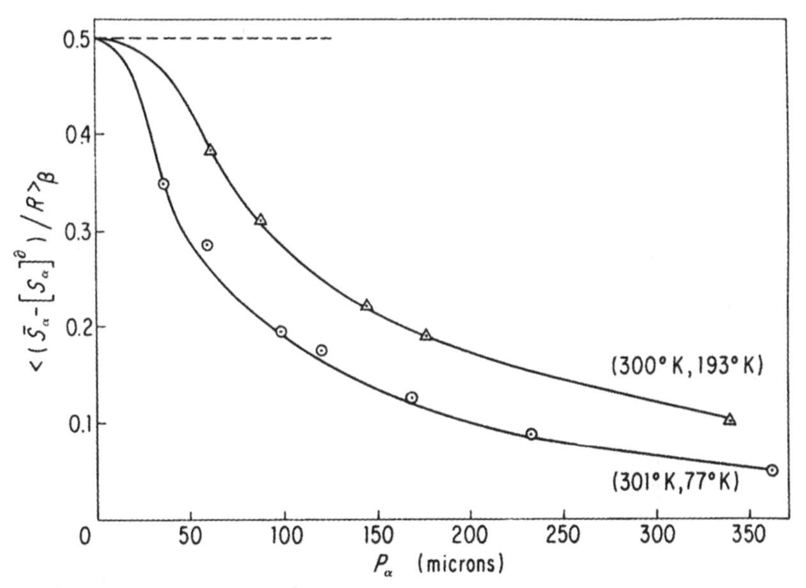

Figure 5–3. Plot of $\langle(\bar{S}_\alpha - [S_\alpha]^\partial)/R\rangle_\beta$ versus P_α as computed from the argon data, averages being taken over the ranges 301–77°K and 300–193°K; $T_\alpha > T_\beta$ in each case.

In terms of experimental data, if we wished to compute, e.g., $\bar{S}_\alpha - [S_\alpha]^\theta$ for a given gas and capillary linkage at some point (P_α, P_β) on the (T_α, T_β) bitherm, say the (298°K, 90°K) bitherm, then we should like to have data for the bitherms (303°K, 90°K), (298°K, 90°K), (293°K, 90°K) so as to be able to use an equation of the form (5.32) with $T_\alpha^{(1)} = 293$°K and $T_\alpha^{(2)} = 303$°K and to associate the resultant quantity $\langle(\bar{S}_\alpha - [S_\alpha]^\theta)/R\rangle_\beta$ with the point (P_α, P_β) on the (298°K, 90°K) bitherm. Families of related bitherms are almost nonexistent in the chemical literature; however, Erikson and I did make some crude calculations for argon and helium [9, 10]. To illustrate some of the conceptions thus far developed, I have taken the data of Los and Ferguson [2] pertaining to argon and a Tru-bore capillary linkage 0.0510 cm in diameter and have made a few sample calculations. Results are displayed in Figures 5–2 to 5–4, where the dashed lines indicate the behavior expected from the Knudsen limiting equation $P_\alpha/P_\beta = (T_\alpha/T_\beta)^{1/2}$.

We see in Figure 5–2 that the experimental data tend toward the Knudsen relation at low pressures and toward a uniform pressure distribution at high pressures; at intermediate pressures the difference $P_\alpha - P_\beta$ reaches a maximum

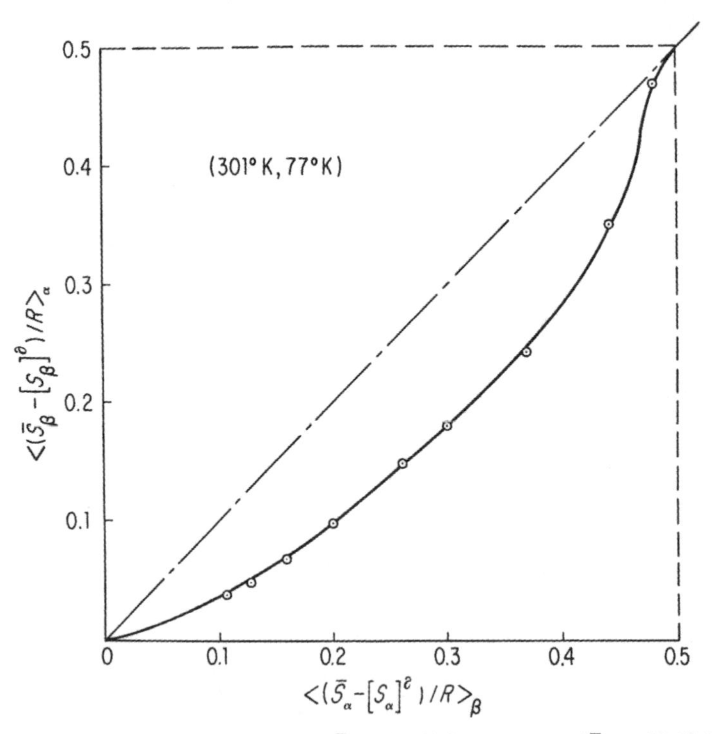

Figure 5–4. Relative plot of $\langle(\bar{S}_\beta - [S_\beta]^\theta)/R\rangle_\alpha$ versus $\langle(\bar{S}_\alpha - [S_\alpha]^\theta)/R\rangle_\beta$ for the argon case, the pairings being made according to the pressure pairings on the (301°K, 77°K) bitherm.

value, the size and position of which are characteristic of the temperatures T_α, T_β, the nature of the gas, and the diameter of the capillary linkage. I wish I could have displayed the quantities $(\bar{S}_\alpha - [S_\alpha]^{\theta})/R$ and $(\bar{S}_\beta - [S_\beta]^{\theta})/R$ as functions of P_ω for the (301°K, 77°K) bitherm, but the data are insufficient for this purpose; I have, instead, used equations of the form (5.32) to compute average values $\langle(\bar{S}_\omega - [S_\omega]^{\theta})/R\rangle_\chi$ over the complete temperature ranges 301–77°K and 300–193°K, and in Figure 5–3 I plot these average quantities versus the pressure P_α at the high temperature side of the capillary linkage. By combining the data for the two bitherms I have constructed the approximate (193°K, 77°K) bitherm, and in Figure 5–4 I have taken the average values $\langle(\bar{S}_\omega - [S_\omega]^{\theta})/R\rangle_\chi$ computed for the temperature ranges 301–193°K and 77–193°K and have made a direct intercomparison via the pressure pairings on the (301°K, 77°K) bitherm. Figure 5–4 indicates that for this case the difference $(\bar{S} - [S]^{\theta})/R$ is larger (at intermediate pressures) at the high temperature end of the capillary than at the low temperature end.

Exercise 5–4. In Table 5–1 I list the data of Los and Ferguson for argon in another Tru-bore capillary linkage.

Table 5–1

Thermomolecular Pressure Effect for Argon in a Tru-bore Capillary Linkage 0.1018 cm **in Diameter** T_α = 301.2°K; T_β = 77.3°K

(Data of Los and Ferguson[2])

$\frac{1}{2}(P_\alpha + P_\beta)$(mm)	$(P_\alpha - P_\beta)(\mu)$
0.0295	9.9
0.0491	12.2
0.0873	13.3
0.1067	13.5
0.1612	12.5
0.2196	11.4
0.3495	9.1
0.6023	6.3
0.8841	4.7
1.4595	3.2
1.9158	2.5
1.9694	2.37

(i) Calculate the quantity $(\bar{S}_\alpha - \langle[S]^{\theta}\rangle)/R$ for each entry in the table and make a plot of $(\bar{S}_\alpha - \langle[S]^{\theta}\rangle)/R$ versus P_β; take $\bar{C}_P$ for argon equal to $5R/2$. (ii) Calculate the quantity $\langle(\bar{S}_\alpha - [S_\alpha]^{\theta})/R\rangle_\beta$, averaging over the entire range 301–77°K, for each entry in the table and make a plot of

$$\langle(\bar{S}_\alpha - [S_\alpha]^{\theta})/R\rangle_\beta$$

versus P_β.

(iii) Write down an analytic expression for the difference

$$[(\bar{S}_\alpha - \langle [S]^\theta \rangle)/R] - \langle (\bar{S}_\alpha - [S_\alpha]^\theta)/R \rangle_\beta$$

and evaluate the difference for the two limiting cases $P_\alpha/P_\beta = (T_\alpha/T_\beta)^{1/2}$ and $P_\alpha = P_\beta$.

Conditions of Stability

The conditions of stability for thermostatic equilibrium states are relations such as $\bar{C}_V > 0$ and $(\partial \bar{V}/\partial P)_T < 0$. There exist analogous conditions of stability for steady states also. For the system shown in Figure 5–1 it is intuitively evident that the following relations must hold for steady states:

$$\left(\frac{\partial \dot{n}_\alpha}{\partial P_\alpha}\right)_{\beta,\langle T\rangle,T_\alpha} < 0, \qquad \left(\frac{\partial \dot{Q}_\alpha^{(r)}}{\partial T_\alpha}\right)_{\beta,\langle T\rangle,\dot{n}_\alpha=0} < 0. \tag{5.40}$$

Thus, if we keep fixed the thermodynamic state at β and keep fixed the temperatures T_α, $\langle T\rangle$, then upon increasing the pressure P_α we can only decrease the rate of influx of matter ($\dot{n}_\alpha$) into the α terminal part. Similarly, if we maintain a condition of no mass flow, keeping fixed the thermodynamic state at β and the temperature $\langle T\rangle$, then upon increasing the temperature T_α we can only decrease the rate of influx of heat ($\dot{Q}_\alpha^{(r)}$) into the heat reservoir at α. Whereas relations such as Eq. (5.40) are intuitively evident, the considerations mentioned in Chapter 3 enable us systematically to produce conditions of stability, some of which are not intuitively obvious. In Chapter 3 I showed that there was a neighborhood of a given thermostatic reference state σ for which the relation $(\partial \Omega_k/\partial Y_k)_{\Omega',\sigma} > 0$ held; I now wish to apply this condition to the currents and affinities described in Eq. (5.7) (taking state β as the thermostatic reference state):

$$\begin{aligned}\left(\frac{\partial \Omega_1}{\partial Y_1}\right)_{\Omega',\beta} &= \left(\frac{\partial\{T_\beta^{-1}(\bar{H}_\beta - \bar{H}_\alpha) + \bar{S}_\alpha - \bar{S}_\beta\}}{\partial \dot{n}_\alpha}\right)_{\beta,\langle T\rangle,T_\alpha} \\ &= -\frac{1}{T_\beta}\left(\frac{\partial P_\alpha}{\partial \dot{n}_\alpha}\right)_{\beta,\langle T\rangle,T_\alpha}\left\{\bar{V}_\alpha - T_\alpha\left(\frac{\partial \bar{V}_\alpha}{\partial T_\alpha}\right)_{P_\alpha} + T_\beta\left(\frac{\partial \bar{V}_\alpha}{\partial T_\alpha}\right)_{P_\alpha}\right\},\end{aligned} \tag{5.41}$$

$$\left(\frac{\partial \Omega_2}{\partial Y_2}\right)_{\Omega',\beta} = \left(\frac{\partial (T_\alpha^{-1} - T_\beta^{-1})}{\partial \dot{Q}_\alpha^{(r)}}\right)_{\beta,\langle T\rangle,\Omega_1} = -\frac{1}{T_\alpha^{2}}\left(\frac{\partial T_\alpha}{\partial \dot{Q}_\alpha^{(r)}}\right)_{\beta,\langle T\rangle,\Omega_1}, \tag{5.42}$$

$$\left(\frac{\partial \Omega_3}{\partial Y_3}\right)_{\Omega',\beta} = \left(\frac{\partial (\langle T\rangle^{-1} - T_\beta^{-1})}{\partial \dot{q}}\right)_{\alpha,\beta} = -\frac{1}{\langle T\rangle^2}\left(\frac{\partial \langle T\rangle}{\partial \dot{q}}\right)_{\alpha,\beta}. \tag{5.43}$$

The condition $(\partial \Omega_k/\partial Y_k)_{\Omega',\beta} > 0$ requires, then, that

$$\left\{\bar{V}_\alpha - T_\alpha\left(\frac{\partial \bar{V}_\alpha}{\partial T_\alpha}\right)_{P_\alpha} + T_\beta\left(\frac{\partial \bar{V}_\alpha}{\partial T_\alpha}\right)_{P_\alpha}\right\}\left(\frac{\partial P_\alpha}{\partial \dot{n}_\alpha}\right)_{\beta,\langle T\rangle,T_\alpha} < 0, \tag{5.44}$$

$$\left(\frac{\partial T_\alpha}{\partial \dot{Q}_\alpha^{(r)}}\right)_{\beta,\langle T\rangle,\Omega_1} < 0, \tag{5.45}$$

$$\left(\frac{\partial \langle T\rangle}{\partial \dot{q}}\right)_{\alpha,\beta} < 0. \tag{5.46}$$

Equations (5.44) to (5.46), when evaluated at the points $Y_k = 0$, represent applications of the *petit* principles of entropy production; when evaluated for arbitrary points Y_k, they represent applications of the *grand* principles of entropy production. Relation (5.46) is intuitively reasonable: we expect that increasing the quantity $\langle T\rangle$ while keeping the thermodynamic states at α and β constant will result in a *decrease* in the rate of heat leakage ($\dot{q}$) *to* the surroundings. Relation (5.45), although similar to relation (5.40), is difficult to appreciate intuitively because of the fact that Ω_1 is to be held constant; i.e., the temperature and pressure at α must vary in such a way as to maintain constant the combination $\bar{S}_\alpha - \bar{H}_\alpha T_\beta^{-1}$.

Relation (5.44) presents us with a serious problem. If the factor marked off by the braces { } is positive for all physically compatible pairs of states (α, β) then relation (5.44) implies that $(\partial P_\alpha/\partial \dot{n}_\alpha)_{\beta,\langle T\rangle,T_\alpha} < 0$, which is very reassuring. If, on the other hand, the factor { } were ever to become negative for some physically compatible pairs of states (α, β), then it would be impossible to satisfy, in a physically reasonable way, the condition $(\partial \Omega_1/\partial Y_1)_{\Omega',\beta} > 0$ for such states, and as a consequence some of the discussions in Chapter 3 pertaining to entropy production would be of limited validity. Now it is clear that the factor { } will be positive for states such that $T_\beta > T_\alpha$ and also for states such that $[\bar{V}_\alpha - T_\alpha(\partial \bar{V}_\alpha/\partial T_\alpha)_{P_\alpha}] \geqslant 0$. The only states for which { } might conceivably be negative are states for which $T_\beta < T_\alpha$ and at the same time $[\bar{V}_\alpha - T_\alpha(\partial \bar{V}_\alpha/\partial T_\alpha)_{P_\alpha}] < 0$. The quantity $[\bar{V}_\alpha - T_\alpha(\partial \bar{V}_\alpha/\partial T_\alpha)_{P_\alpha}]$ satisfies the relation

$$\bar{V}_\alpha - T_\alpha\left(\frac{\partial \bar{V}_\alpha}{\partial T_\alpha}\right)_{P_\alpha} = -\left(\frac{\partial T_\alpha}{\partial P_\alpha}\right)_{\bar{H}_\alpha} \bar{C}_P(\alpha) \tag{5.47}$$

with $(\partial T_\alpha/\partial P_\alpha)_{\bar{H}_\alpha}$ being the Joule–Thomson coefficient for the gas. We wish to investigate, then, that region of the P,T plane for which the Joule–Thomson coefficient is positive. The part of the P,T plane for which the Joule–Thomson coefficient is positive is a bounded region, bounded by the inversion curve and the temperature axis [11]. It is in the region inside the inversion curve, then, that we should look if we hope to find states such that { } < 0. Since $\{T_\beta(\partial \bar{V}_\alpha/\partial T_\alpha)_{P_\alpha} - (\partial T_\alpha/\partial P_\alpha)_{\bar{H}_\alpha}\bar{C}_P(\alpha)\}$ varies *roughly* as

$$[T_\beta(R/P_\alpha) - (\partial T_\alpha/\partial P_\alpha)_{\bar{H}_\alpha}\bar{C}_P(\alpha)],$$

the best combination for getting { } < 0 would seem to be a small value of T_β coupled to a large value of P_α, with α being a point inside the inversion curve.

If we are considering the *petit* principle of entropy production $(\delta\Omega_1/\tilde{\delta} Y_1)_{\Omega',\beta} > 0$, then the states α and β that appear in the quantity { } must be states that can coexist in a condition of migrational equilibrium; hence we cannot indiscriminately vary the quantities T_β and P_α. If we take P_α higher than the critical pressure P_c for the gas, then we must take T_β greater than the critical temperature T_c; if we take $P_\alpha < P_c$, then we must take T_β greater than or equal to the appropriate temperature on the vapor pressure (liquid or solid) curve. There is no use in our letting T_β approach 0°K, either: as $T \to 0$ the vapor pressure of a condensed phase approaches zero also; hence if we let $T_\beta \to 0$ we should have $(\partial \bar{V}_\alpha/\partial T_\alpha)_{P_\alpha} \to (R/P_\alpha) \to \infty$, and the product $T_\beta(\partial \bar{V}_\alpha/\partial T_\alpha)_{P_\alpha}$ would become indeterminate. Although it is not evident that { } *will* always be positive when we deal with the *petit* principle of entropy production, it is at least plausible that it *may* be so.* On the other hand, the *grand* principle of entropy production $(\partial\Omega_1/\partial Y_1)_{\Omega',\beta} > 0$ purports to hold for all values of Y_1, not just the value $Y_1 = 0$, and it seems likely that for some pairs of states (α, β) with $Y_1 \neq 0$ the quantity { } may very well be negative. The *grand* principle would thus seem to have a limited range of validity. I have more to say on this matter, however, in Chapter 15.

I introduced in Chapter 3 the additional *grand* principle of entropy production $(\partial Y_i/\partial\Omega_k)_{\Omega',\sigma} = (\partial Y_k/\partial\Omega_i)_{\Omega',\sigma}$ for ordinary steady-state situations. For the system shown in Figure 5–1 these reciprocity relations take the following forms:

$$T_\beta\left(\frac{\partial \dot{Q}_\alpha^{(T)}}{\partial P_\alpha}\right)_{\beta,\langle T\rangle,T_\alpha} = T_\alpha^2\left\{\bar{V}_\alpha - T_\alpha\left(\frac{\partial \bar{V}_\alpha}{\partial T_\alpha}\right)_{P_\alpha} + T_\beta\left(\frac{\partial \bar{V}_\alpha}{\partial T_\alpha}\right)_{P_\alpha}\right\}\left(\frac{\partial \dot{n}_\alpha}{\partial T_\alpha}\right)_{\beta,\langle T\rangle,\Omega_1}, \tag{5.48}$$

$$T_\beta\left(\frac{\partial \dot{q}}{\partial P_\alpha}\right)_{\beta,\langle T\rangle,T_\alpha} = \langle T\rangle^2\left\{\bar{V}_\alpha - T_\alpha\left(\frac{\partial \bar{V}_\alpha}{\partial T_\alpha}\right)_{P_\alpha} + T_\beta\left(\frac{\partial \bar{V}_\alpha}{\partial T_\alpha}\right)_{P_\alpha}\right\}\left(\frac{\partial \dot{n}_\alpha}{\partial \langle T\rangle}\right)_{\beta,T_\alpha,\Omega_1}, \tag{5.49}$$

$$T_\alpha^2\left(\frac{\partial \dot{q}}{\partial T_\alpha}\right)_{\beta,\langle T\rangle,\Omega_1} = \langle T\rangle^2\left(\frac{\partial \dot{Q}_\alpha^{(T)}}{\partial \langle T\rangle}\right)_{\beta,T_\alpha,\Omega_1}. \tag{5.50}$$

The condition of constant Ω_1 makes these relations difficult to interpret; I have nothing useful to say about them.

* An examination of the data of Roebuck and Osterberg pertaining to the Joule–Thomson coefficient for argon [12] and for nitrogen [13] reveals that everywhere inside the inversion curve, for each of the two gases, the quantity $[T_\beta(R/P_\alpha) - (\partial T_\alpha/\partial P_\alpha)_{H_\alpha}\bar{C}_P(\alpha)]$ is positive for pairs of states (α, β) compatible with the migrational equilibrium constraint.

Linkage Classes

Thus far we have considered the linkage between terminal parts in its heat-transfer function (full, partial, or zero flux) only. We can, however, consider the linkage between two terminal parts in a polythermal field from another point of view. In a certain sense a linkage is that device (apparatus, piece of hardware) that permits certain classes of thermodynamic states (α, β) to be in contact with one another under the migrational equilibrium constraint. Consider the system shown in Figure 5–1. For a given state $\alpha(T_\alpha, P_\alpha)$ the linkage establishes a class of states $\{\beta(i)\}$ $(T_{\beta(1)}, P_{\beta(1)}; T_{\beta(2)}, P_{\beta(2)}; \ldots; T_{\beta(i)}, P_{\beta(i)}; \ldots)$ each of which can coexist in a condition of migrational equilibrium across the given linkage with state α; there is such a class $\{\beta(i)\}$ for each state $\alpha(j)$. Now the cataloging of the classes $\{\beta(i)\}$ that can coexist in migrational equilibrium with states $\alpha(j)$ for a given gas and linkage is the thermodynamic way of describing the relevant properties of the linkage —the information is recorded neatly in the equation of correlation $(fT_\alpha, P_\alpha, T_\beta, P_\beta, \langle T \rangle) = 0$. The physical properties of the linkage are, of course, reflected in the structure of the linkage classes $\alpha(j)|\{\beta(i)\}$.

This way of looking at linkages gives rise to an interesting speculation. Consider Figure 5–1 again (full-flux linkage): let terminal parts a and b start out in state α; then, maintaining migrational equilibrium, change the state of b in a well-defined series of operations to some final state β. As terminal part b moves from state α to state β, it generates a linkage class $\{\beta(i)\}$ consistent with state α at terminal part a. Suppose now, with terminal part a in state α and part b in state β, it becomes possible to probe the interior of the linkage at some arbitrary point λ and to determine the values T_λ, P_λ. Now does each arbitrary interior point (T_λ, P_λ) of the linkage belong to the linkage class $\{\beta(i)\}$ generated by b in going from state α to state β? That is, for a given T_λ, P_λ can we find a point $\beta(k)$ of the set $\{\beta(i)\}$ such that $T_\lambda = T_{\beta(k)}$ and $P_\lambda = P_{\beta(k)}$?

Although the preceding question is a fundamental one, I see no simple way of answering it.

References

1. W. H. Keesom, *Helium* (Elsevier Publishing Co., Amsterdam, 1942), Section 2.74.
2. J. M. Los and R. R. Ferguson, *Trans. Faraday Soc.*, **48**, 730 (1952).
3. R. C. L. Bosworth, *Heat Transfer Phenomena* (John Wiley and Sons, New York, 1952), p. 129.
4. S. Liang, *J. Phys. Chem.*, **57**, 910 (1953).
5. M. Bennett and F. Tompkins, *Trans. Faraday Soc.*, **53**, 185 (1957).
6. H. Podgurski and F. Davis, *J. Phys. Chem.*, **65**, 1343 (1961).

7. E. A. Mason, R. B. Evans, III, and G. M. Watson, *J. Chem. Phys.*, **38**, 1808 (1963).
8. T. Takaishi and Y. Sensui, *Trans. Faraday Soc.*, **59**, 2503 (1963).
9. T. A. Erikson, M.S. Thesis, Illinois Institute of Technology (1959).
10. R. J. Tykodi and T. A. Erikson, *J. Chem. Phys.*, **31**, 1510 (1959).
11. M. Saha and B. Srivastava, *A Treatise on Heat*, fourth edition (The Indian Press Private, Ltd., Allahabad, 1958), Chapter 12.
12. J. R. Roebuck and H. Osterberg, *Phys. Rev.*, **46**, 785 (1934).
13. J. R. Roebuck and H. Osterberg, *Phys. Rev.*, **48**, 450 (1935).

6

Migrational Equilibrium in Polythermal Fields: General Relations

In this chapter I consider a number of more abstract relations and definitions pertaining to the thermomolecular pressure effect and to other steady-state bithermal systems.

Definitions and Theorems

I collect here a number of useful definitions aimed primarily at the thermomolecular pressure effect but useful for other bithermal phenomena as well. In order not to have to write similar formulas twice, once with subscript α and again with subscript β, I use the noncommittal subscripts ω and χ—each formula then yields two relations obtained by replacing ω,χ with either α,β or β,α throughout.*

$$[S_\omega]^\vartheta \equiv -\left(\frac{\partial \mu_\omega}{\partial T_\omega}\right)_{\chi,\dot{n}_\omega = 0}, \qquad \langle[S]^\vartheta\rangle \equiv -\frac{\mu_\omega - \mu_\chi}{T_\omega - T_\chi}, \tag{6.1}$$

where the subscript χ on the differential coefficient means that the thermodynamic state at χ is to be kept constant; i.e., all the intensive variables at χ

* I have included, for the sake of symmetry, a few quantities that are not strictly definitions but rather theorems derived from other definitions.

(T_χ, P_χ, etc.) are to be kept constant, and the subscript $\dot{n}_\omega = 0$ shows that the coefficient refers to states of migrational equilibrium. Since it is usually quite clear from the context when states of migrational equilibrium are being considered, I do not normally show the subscript $\dot{n}_\omega = 0$ explicitly; after all, we do not always tack on a subscript *eq* in classical thermostatics to show that equilibrium states are intended.

$$[H_\omega] \equiv \mu_\omega + T_\omega[S_\omega], \quad [H_\omega]^\partial \equiv \mu_\omega + T_\omega[S_\omega]^\partial, \tag{6.2}$$

$$[G_\omega] \equiv [H_\omega] - \langle T\rangle[S_\omega], \quad [G_\omega]^\partial \equiv [H_\omega]^\partial - \langle T\rangle[S_\omega]^\partial, \tag{6.3}$$

$$[C_{\omega\omega}] \equiv T_\omega\left(\frac{\partial[S_\omega]}{\partial T_\omega}\right)_\chi, \quad [C_{\omega\omega}]^\partial \equiv T_\omega\left(\frac{\partial[S_\omega]^\partial}{\partial T_\omega}\right)_\chi, \tag{6.4}$$

$$[C_{\omega\chi}] \equiv T_\omega\left(\frac{\partial[S_\omega]}{\partial T_\chi}\right)_\omega, \quad [C_{\omega\chi}]^\partial \equiv T_\omega\left(\frac{\partial[S_\omega]^\partial}{\partial T_\chi}\right)_\omega, \tag{6.5}$$

$$[Q_\omega] \equiv T_\omega(\bar{S}_\omega - [S_\omega]), \quad [Q_\omega]^\partial \equiv T_\omega(\bar{S}_\omega - [S_\omega]^\partial), \tag{6.6}$$

$$[h(\chi\omega)] \equiv \langle T\rangle[s(\chi\omega)], \quad [h(\chi\omega)] \equiv -[h(\omega\chi)], \quad [s(\chi\omega)] \equiv -[s(\omega\chi)]. \tag{6.7}$$

I have, of course, been guided in these definitions by the analogous equations of classical thermostatics. From Eqs. (6.1) to (6.7) and the results of Chapter 5 we can readily deduce the following theorems:

$$\langle [S]^\partial \rangle = \frac{T_\chi[S_\chi] - T_\omega[S_\omega] + [h(\chi\omega)]}{T_\chi - T_\omega}, \tag{6.8}$$

$$[S_\omega]^\partial = [S_\omega] + [C_{\omega\omega}] - [C_{\chi\omega}] - \left(\frac{\partial[h(\chi\omega)]}{\partial T_\omega}\right)_\chi, \tag{6.9}$$

$$\left(\frac{\partial[H_\omega]}{\partial T_\omega}\right)_\chi = [C_{\chi\omega}] + \left(\frac{\partial[h(\chi\omega)]}{\partial T_\omega}\right)_\chi, \quad \left(\frac{\partial[H_\omega]}{\partial T_\chi}\right)_\omega = [C_{\omega\chi}], \tag{6.10}$$

$$\left(\frac{\partial[H_\omega]^\partial}{\partial T_\omega}\right)_\chi = [C_{\omega\omega}]^\partial, \quad \left(\frac{\partial[H_\omega]^\partial}{\partial T_\chi}\right)_\omega = [C_{\omega\chi}]^\partial, \tag{6.11}$$

$$\left(\frac{\partial[G_\omega]}{\partial T_\omega}\right)_\chi = [C_{\chi\omega}]\left(1 - \frac{\langle T\rangle}{T_\chi}\right) - [S_\chi]\left(\frac{\partial\langle T\rangle}{\partial T_\omega}\right)_\chi, \tag{6.12}$$

$$\left(\frac{\partial[G_\omega]}{\partial T_\chi}\right)_\omega = [C_{\omega\chi}]\left(1 - \frac{\langle T\rangle}{T_\omega}\right) - [S_\omega]\left(\frac{\partial\langle T\rangle}{\partial T_\chi}\right)_\omega, \tag{6.13}$$

$$\left(\frac{\partial[G_\omega]^\partial}{\partial T_\omega}\right)_\chi = [C_{\omega\omega}]^\partial\left(1 - \frac{\langle T\rangle}{T_\omega}\right) - [S_\omega]^\partial\left(\frac{\partial\langle T\rangle}{\partial T_\omega}\right)_\chi, \tag{6.14}$$

$$\left(\frac{\partial[G_\omega]^\partial}{\partial T_\chi}\right)_\omega = [C_{\omega\chi}]^\partial\left(1 - \frac{\langle T\rangle}{T_\omega}\right) - [S_\omega]^\partial\left(\frac{\partial\langle T\rangle}{\partial T_\chi}\right)_\omega, \tag{6.15}$$

$$[Q_\omega] = \bar{H}_\omega - [H_\omega], \quad [Q_\omega]^\partial = \bar{H}_\omega - [H_\omega]^\partial. \tag{6.16}$$

Note that apparently $(\partial[H_\omega]/\partial T_\omega)_\chi \neq [C_{\omega\omega}]$, $(\partial[G_\omega]/\partial T_\omega)_\chi \neq -[S_\omega]$, and $(\partial[G_\omega]^\partial/\partial T_\omega)_\chi \neq -[S_\omega]^\partial$.

Exercise 6–1. Establish the correctness of the relations (6.8) to (6.16).

Exercise 6–2. Investigate the consequences of the assumptions that

$(\partial[H_\omega]/\partial T_\omega)_x = [C_{\omega\omega}]$, $(\partial[G_\omega]/\partial T_\omega)_x = -[S_\omega]$, $(\partial[G_\omega]^\theta/\partial T_\omega)_x = -[S_\omega]^\theta$.

Consider yet one more definition (this type of definition will be useful in discussing the thermocouple later on):

$$\left(\frac{\delta(\dot{Q}_\alpha^{(r)} + \dot{Q}_\beta^{(r)} + \dot{q})}{\delta\dot{n}_\alpha}\right)_{T_\alpha, T_\beta, \langle T\rangle} \equiv -\int_{T=T_\alpha}^{T=T_\beta} [c]\, dT. \tag{6.17}$$

We see from Eq. (5.8) that

$$\left(\frac{\delta(\dot{Q}_\alpha^{(r)} + \dot{Q}_\beta^{(r)} + \dot{q})}{\delta\dot{n}_\alpha}\right)_{T_\alpha, T_\beta, \langle T\rangle} = -(\bar{H}_\alpha - \bar{H}_\beta). \tag{6.18}$$

Thus

$$[c] = \frac{d\bar{H}}{dT} \tag{6.19}$$

and, for a perfect gas,

$$[c] = \bar{C}_P. \tag{6.20}$$

Note that

$$[c_\omega] = \left(\frac{\partial \bar{H}_\omega}{\partial T_\omega}\right)_x = \bar{S}_\omega - [S_\omega]^\theta + T_\omega\left(\frac{\partial \bar{S}_\omega}{\partial T_\omega}\right)_x. \tag{6.21}$$

In discussing this next series of operations, it is necessary (in certain integrals) for the sake of clarity to distinguish between the names of thermodynamic states (α, β) and the names of terminal parts (a, b).*

Consider a series of operations in which the terminal parts of the system start out in the same thermodynamic state (α say); then the condition of one of the terminal parts (b say) is changed in such a manner as to maintain migrational equilibrium while keeping the state of the a terminal part fixed. For such a series of operations we can write

$$\begin{aligned}[S_\beta] &= [S_b(\alpha)] + \int_{T_b=T_\alpha}^{T_b=T_\beta} \left(\frac{\partial[S_b]}{\partial T_b}\right)_\alpha dT_b \\ &= [S_b(\alpha)] + \int_{T_\alpha}^{T_\beta} \frac{[C_{bb}]}{T_b}\, dT_b,\end{aligned} \tag{6.22}$$

and from similar considerations,

$$\begin{aligned}[S_\beta] &= [S_b(\beta)] + \int_{T_a=T_\beta}^{T_a=T_\alpha} \left(\frac{\partial[S_b]}{\partial T_a}\right)_\beta dT_a \\ &= [S_b(\beta)] + \int_{T_\beta}^{T_\alpha} \frac{[C_{ba}]}{T_\beta}\, dT_a.\end{aligned} \tag{6.23}$$

* See the discussion in Chapter 5.

The quantity $[S_b(\alpha)]$, e.g., represents the value of $[S_b]$ for the state in which $T_a = T_b = T_\alpha$, $P_a = P_b = P_\alpha$, etc.

The equithermal state, the state, e.g., for which $T_a = T_b = T_\beta$, $P_a = P_b = P_\beta, \ldots$, is an interesting one; for such an equithermal state the following relations hold (provided that assumption IV and assumption Q hold):

$$T_\beta\{[S_a(\beta)] - [S_b(\beta)]\} = [h(ba|\beta)], \tag{6.24}$$

$$[S_a(\beta)] - [S_b(\beta)] = [s(ba|\beta)], \tag{6.25}$$

$$\{T_\beta - \langle T(ab|\beta)\rangle\}\{[S_a(\beta)] - [S_b(\beta)]\} = 0, \tag{6.26}$$

where $[h(ba|\beta)]$, $[s(ba|\beta)]$, and $\langle T(ab|\beta)\rangle$ are the appropriate values of $[h(\beta\alpha)]$, $[s(\beta\alpha)]$, and $\langle T\rangle$ for the given equithermal state.

Exercise 6–3. Establish the validity of relations (6.24) to (6.26).

In the case of full- and partial-flux linkages $\langle T(ab|\beta)\rangle \neq T_\beta$ (in general); so, by Eqs. (6.24) to (6.26), it follows that $[S_a(\beta)] = [S_b(\beta)]$, $[s(ba|\beta)] = 0$, and $[h(ba|\beta)] = 0$. The same set of relations holds for zero-flux linkages in spite of the fact that $\langle T(ab|\beta)\rangle = T_\beta$ for such linkages. From Eq. (5.11) we see that in the zero-flux case* $[S_a(\beta)] = \bar{S}_a(\beta)|_{\text{in}}$ and $[S_b(\beta)] = \bar{S}_b(\beta)|_{\text{in}}$. If we place the thermostats symmetrically about the linkage, then $\bar{S}_a(\beta)|_{\text{in}} = \bar{S}_b(\beta)|_{\text{in}}$.

The general relation (6.9)

$$[S_\beta]^\theta = [S_\beta] + T_\beta\left(\frac{\partial[S_\beta]}{\partial T_\beta}\right)_\alpha - T_\alpha\left(\frac{\partial[S_\alpha]}{\partial T_\beta}\right)_\alpha - \left(\frac{\partial[h(\alpha\beta)]}{\partial T_\beta}\right)_\alpha$$

and Eq. (5.34) together yield yet another useful result:

$$\begin{aligned}
[S_\beta]^\theta &= [S_b(\beta)]^\theta \\
&= [S_b(\beta)] + T_\beta \lim_{T_b\to T_\beta}\left(\frac{[S_b] - [S_b(\beta)]}{T_b - T_\beta}\right)_{a=\beta} \\
&\quad - T_a \lim_{T_b\to T_\beta}\left(\frac{[S_a] - [S_a(\beta)]}{T_b - T_\beta}\right)_{a=\beta} - \left(\frac{\partial[h(\alpha\beta)]}{\partial T_\beta}\right)_{a=\beta} \\
&= [S_b(\beta)] + T_\beta \lim_{T_b\to T_\beta}\left(\frac{[S_b] - [S_a]}{T_b - T_\beta}\right)_{a=\beta} - \left(\frac{\partial[h(\alpha\beta)]}{\partial T_\beta}\right)_{a=\beta} \\
&= [S_b(\beta)] + T_\beta\left(\frac{\partial[s(\alpha\beta)]}{\partial T_\beta}\right)_{a=\beta} - \left(\frac{\partial[h(\alpha\beta)]}{\partial T_\beta}\right)_{a=\beta} \\
&= [S_b(\beta)] + \{T_\beta - \langle T(ab|\beta)\rangle\}\left(\frac{\partial[s(\alpha\beta)]}{\partial T_\beta}\right)_{a=\beta},
\end{aligned} \tag{6.27}$$

* In the zero-flux case the thermostat at ω does not participate in a heat conduction process (i.e., $\dot{Q}_\omega^{(r)} = 0$ for $\dot{n}_\omega = 0$); hence the limiting operation $(\delta\dot{S}_\omega^{(r)}/\delta\dot{n}_\omega)_{T_\alpha,T_\beta,\langle T\rangle}$ represents the isothermal reversible change in molar entropy in taking the gas from the point at which it crosses the boundary of the thermostat at ω to a point adjacent to the piston face at the other end of the ω terminal part; i.e.,

$$(\delta\dot{S}_\omega^{(r)}/\delta\dot{n}_\omega)_{T_\alpha,T_\beta,\langle T\rangle} = \bar{S}_\omega|_{\text{in}} - \bar{S}_\omega|_{\text{piston}}.$$

where I have used assumption IV in the form of Eq. (5.22). We see that for full-* or zero-flux† linkages Eq. (6.27) easily reduces to

$$[S_\beta]^\partial = [S_b(\beta)]. \tag{6.28}$$

Since full- and zero-flux linkages are but the two extreme forms of partial-flux linkage, it would be most surprising if Eq. (6.28) did not hold for partial-flux linkages. To establish Eq. (6.28) for partial-flux linkages, however, we must be able to show that $0 = (\partial[s(\alpha\beta)]/\partial T_\beta)_{\alpha=\beta}$. Although I cannot give a direct proof of this relation, it is clear that it must hold, otherwise Eq. (6.27) would be inconsistent: the left-hand side of Eq. (6.27) is a function of the variables T_β, P_β only, whereas the right-hand side, if $(\partial[s(\alpha\beta)]/\partial T_\beta)_{\alpha=\beta}$ were not equal to zero, would be a function of the variables T_β, P_β, and $\langle T(ab|\beta)\rangle$. The entire argument, however, depends on our acceptance of assumption Q. We conclude, then, that Eq. (6.28) holds generally, regardless of the nature of the linkage, whenever assumption IV and assumption Q hold, and it follows that (for the system of Figure 5–1)

$$\left(\frac{\partial P_\beta}{\partial T_\beta}\right)_\alpha = \frac{\bar{S}_\beta - [S_\beta]^\partial}{\bar{V}_\beta} = \frac{\bar{S}_\beta - [S_b(\beta)]}{\bar{V}_\beta} = \frac{[Q_b(\beta)]}{T_\beta \bar{V}_\beta}, \tag{6.29}$$

where $[Q_b(\beta)] \equiv T_\beta(\bar{S}_\beta - [S_b(\beta)])$. Equation (6.29) shows that for the case considered a polythermal ($T_\alpha \neq T_\beta$) temperature coefficient is determined by a set of equithermal ($T_\alpha = T_\beta$) relations.

Exercise 6–4. According to the duality principle mentioned in Chapter 3, the dual of assumption IV is

$$0 = \left(\frac{\delta\Theta}{\bar{\delta}\Omega_k}\right)_{Y'}. \tag{6.30}$$

Show that for the system of Figure 5–1, full-flux linkage, Eq. (6.28) can be derived from Eq. (6.30).

Hint: Use Eq. (6.30) in the form $0 = (\delta\Theta/\bar{\delta}\Omega_2)_{Y_1,\beta}$ (see Eq. 5.7) and pass to the appropriate limit.

---

* $\dfrac{\partial[s(\alpha\beta)]}{\partial T_\beta} = 0.$

† $T_\beta = \langle T(ab|\beta)\rangle.$

Alternative Considerations

When a heat reservoir of temperature T encompasses a region constituting an open system in such a way that masses crossing the boundaries of the region enter from or leave to adjacent regions of the same temperature T, and all mass transfers take place in a reversible fashion, then

$$\dot{S}^{(r)} + \dot{S}(\text{region}) + \dot{S}(\text{out}) - \dot{S}(\text{in}) = 0, \tag{6.31}$$

where, in the steady state, $\dot{S}(\text{region}) = \sum_\rho \dot{S}_\rho$ [summation over all phases ρ of the region], $\dot{S}_\rho = \sum_i \dot{n}_\rho^{(i)} \bar{S}_\rho^{(i)}$ [summation over the components of phase ρ], and $\dot{S}(\text{in})$, e.g., is the convective flow of entropy into the region per unit time; i.e., $\dot{S}(\text{in}) = \dot{n}\bar{S}(\text{in})$. In the limit of vanishing transfer of mass, Eq. (6.31) takes the form

$$\frac{\delta \dot{S}^{(r)}}{\delta \dot{n}} + \sum_\rho \frac{\delta \dot{S}_\rho}{\delta \dot{n}} + \frac{\delta \dot{S}(\text{out})}{\delta \dot{n}} - \frac{\delta \dot{S}(\text{in})}{\delta \dot{n}} = 0. \tag{6.32}$$

In the limit of vanishing mass flow we can maintain the *form* of Eq. (6.32) for *any* heat reservoir by making use of the following conventions: Represent the value of $\dot{S}^{(r)}$ when the rate parameter has the value $\dot{n}$ by $\dot{S}^{(r)}(\dot{n})$; then, if $\dot{S}^{(r)}(0) = 0$, set $\delta\dot{S}(\text{in})/\delta\dot{n} = \bar{S}(\text{in})$; and if $\dot{S}^{(r)}(0) \neq 0$ [the heat reservoir is participating in a heat conduction process], set $(\delta\dot{S}(\text{in})/\delta\dot{n})_{\Omega'} \equiv [S(\text{in})]$. This definition of $[S]$ is entirely equivalent to that used in Eq. (5.11).

Now for a system composed of a number of terminal parts (running index j), a number of heat reservoirs (running index k), and a number of linkages of arbitrary type, it follows that in the steady state

$$\dot{U}(\text{system}) = \sum_j \dot{U}_j = -\sum_k T_k \dot{S}_k^{(r)} - \sum \dot{q}(kh) + \dot{W}, \tag{6.33}$$

$$\dot{H}(\text{system}) = \sum_j \dot{H}_j = -\sum_k T_k \dot{S}_k^{(r)} - \sum \dot{q}(kh) + \dot{W}_0, \tag{6.34}$$

$$\dot{G}(\text{system}) = \sum_j \dot{G}_j = -\sum_k T_k \dot{S}_k^{(r)} - \sum_j \sum_\rho T_\rho \dot{S}_\rho - \sum \dot{q}(kh) + \dot{W}_0, \tag{6.35}$$

$$\Theta = \sum_j \dot{S}_j + \sum_k \dot{S}_k^{(r)} + \sum \dot{s}(kh), \tag{6.36}$$

$$\dot{W}_0 = \sum_x \dot{W}_{0x} = \sum_x Y_{0x} \mathscr{F}_{0x}, \tag{6.37}$$

where $\mathscr{F}_{0x}$ represents the generalized "potential difference" appropriate to work of the type W_{0x}. We can recast Eq. (6.36) into other useful forms: let one of the thermostat temperatures (T_m say) be chosen as a reference temperature; then, on combining Eq. (6.34) or (6.35) with Eq. (6.36), we can write

$$\Theta = \sum_x Y_{0x}\frac{\mathscr{F}_{0x}}{T_m} - \sum_j \frac{\dot{H}_j - T_m\dot{S}_j}{T_m} + \sum_{\substack{k\\k\neq m}} \dot{Q}_k^{(r)}\left(\frac{1}{T_k} - \frac{1}{T_m}\right) + \sum \dot{q}(kh)\left(\frac{1}{\langle T(kh)\rangle} - \frac{1}{T_m}\right)$$

$$= \sum_x Y_{0x}\frac{\mathscr{F}_{0x}}{T_m} - \frac{1}{T_m}\left\{\sum_j\sum_\rho\sum_i \mathscr{G}_\rho^{(i)}\dot{n}_\rho^{(i)} + \sum_j\sum_\rho\sum_i (T_\rho - T_m)\bar{S}_\rho^{(i)}\dot{n}_\rho^{(i)}\right\} + \sum_{\substack{k\\k\neq m}} \dot{Q}_k^{(r)}\left(\frac{1}{T_k} - \frac{1}{T_m}\right) + \sum \dot{q}(kh)\left(\frac{1}{\langle T(kh)\rangle} - \frac{1}{T_m}\right). \quad (6.38)$$

We can eliminate the dependent currents in Eq. (6.38) via the steady-state equations of constraint and via any stoichiometric relations pertaining to the given situation; the resultant equation will be in the form $\Theta = \sum_h Y_h\Omega_h$.

For a state of vanishing mass flow (with respect to a rate parameter $\dot{n}$) we have

$$\sum_j\sum_\rho\sum_i \mathscr{G}_\rho^{(i)}\left(\frac{\delta\dot{n}_\rho^{(i)}}{\delta\dot{n}}\right)_{\Omega'} + \sum_k T_k\left\{\left(\frac{\delta\dot{S}_k(\text{in})}{\delta\dot{n}}\right)_{\Omega'} - \left(\frac{\delta\dot{S}_k(\text{out})}{\delta\dot{n}}\right)_{\Omega'}\right\} + \sum\left(\frac{\delta\dot{q}(kh)}{\delta\dot{n}}\right)_{\Omega'} = \left(\frac{\delta\dot{W}_0}{\delta\dot{n}}\right)_{\Omega'}, \quad (6.39)$$

$$0 = \left(\frac{\delta\Theta}{\delta\dot{n}}\right)_{\Omega'} = \sum_k\left\{\left(\frac{\delta\dot{S}_k(\text{in})}{\delta\dot{n}}\right)_{\Omega'} - \left(\frac{\delta\dot{S}_k(\text{out})}{\delta\dot{n}}\right)_{\Omega'}\right\} + \sum\left(\frac{\delta\dot{s}(kh)}{\delta\dot{n}}\right)_{\Omega'}, \quad (6.40)$$

where I have made use of Eq. (6.32) with attendant conventions (the heat reservoirs either communicate with terminal parts or they encompass regions through which mass flows but in which mass does not accumulate). Equations (6.39) and (6.40), taken together, are equivalent to the equation resulting from the application of the operation $(\delta/\delta\dot{n})_{\Omega'}$ to Eq. (6.38).

The equations of this section are basic ones for the analysis of several polythermal situations.

Exercise 6–5. Show that for the thermomolecular pressure case (Figure 5–1) Eq. (6.39) yields Eq. (5.18) directly.

Assumption IV

Assumption IV has some far-reaching consequences for polythermal field effects and is especially important in the treatment of thermocouples and thermocells. For the case considered in the preceding section (Eqs. 6.36 and

6.40) the terms of Eq. (6.40) combine in such a way as to leave quantities of the type $[S_k]$ and $[s(kh)]$, one set of such quantities for each link. If there are m (series-connected) links in the system and $[S_m(\text{out})]$ represents the $[S]$ quantity at the place where mass flows out of the link,* then we can write Eq. (6.40) as

$$0 = \sum_m [S_m(\text{out})] - [S_m(\text{in})] - [s_m(\text{in–out})]. \tag{6.41}$$

The systems investigated are normally constructed so that the separate linkages are independent of one another; therefore assumption IV requires that for *each* (series-connected) linkage in the system

$$[S_m(\text{out})] - [S_m(\text{in})] - [s_m(\text{in–out})] = 0. \tag{6.42}$$

Similar considerations imply that for each (series-connected) linkage in the system

$$[H_m(\text{out})] - [H_m(\text{in})] - [h_m(\text{in–out})] = 0 \tag{6.43}$$

and

$$[G_m(\text{out})] - [G_m(\text{in})] = 0. \tag{6.44}$$

Exercise 6–6. Linkages

(i) Series-Connected Linkages

Consider two series-connected full-flux linkages connecting three terminal parts (α, β, γ) with α and γ serving as mass source and sink, respectively—see Figure 6–1—(for definiteness let the effect be the thermomolecular pressure

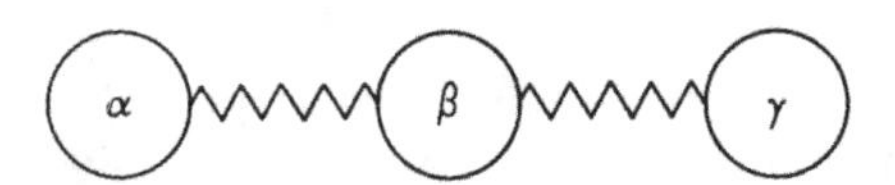

Figure 6–1. Two series-connected full-flux linkages.

effect). Let the steady-flow condition be that $\dot{n}_\alpha + \dot{n}_\gamma = 0$. Show that in the state of migrational equilibrium

$$[H_\alpha] - [H_\gamma] = T_\beta\{[S_\alpha] - [S_\gamma]\}. \tag{6.45}$$

Equation (6.45) is of the form of Eq. (5.14) with $[h(\gamma\alpha)] = T_\beta\{[S_\alpha] - [S_\gamma]\}$; it follows then that $\langle T(\alpha\gamma)\rangle = T_\beta$ and that in a certain sense $\Delta[H] = \int T\, d[S]$, where the integral is a path integral over the flow path.

(ii) Branched Linkages

Consider a branched partial-flux linkage connecting three terminal parts (α, β, γ) as in Figure 6–2 (let the effect be the thermomolecular pressure effect

* If the m link butts on to the k thermostat, then $(\delta\dot{S}_k(\text{in})/\delta\dot{n})_{\Omega'} \equiv [S_m(\text{out})]$.

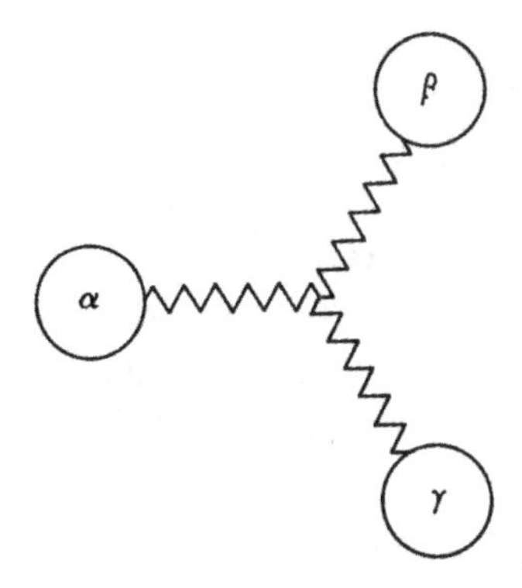

Figure 6–2. A branched, partial-flux linkage.

for definiteness). Let the steady-flow condition be that $\dot{n}_\alpha + \dot{n}_\beta + \dot{n}_\gamma = 0$. Show that for the state of migrational equilibrium the relations

$$[R_{\alpha\chi}][H_\alpha] + [R_{\beta\chi}][H_\beta] + [R_{\gamma\chi}][H_\gamma] = [h(\psi\omega\chi)], \tag{6.46}$$

$$[R_{\alpha\chi}][S_\alpha] + [R_{\beta\chi}][S_\beta] + [R_{\gamma\chi}][S_\gamma] = [s(\psi\omega\chi)], \tag{6.47}$$

$$[R_{\alpha\chi}] + [R_{\beta\chi}] + [R_{\gamma\chi}] = 0, \tag{6.48}$$

hold with $[R_{\omega\chi}] \equiv (\delta\dot{n}_\omega/\delta\dot{n}_\chi)_{\Omega'}$; one of the mass flows $\dot{n}_\alpha$, $\dot{n}_\beta$, or $\dot{n}_\gamma$ has been chosen as rate parameter and has been called $\dot{n}_\chi$.

By making use of the relation

$$[h(\psi\omega\chi)] = \langle T(\alpha\beta\gamma)\rangle[s(\psi\omega\chi)] \tag{6.49}$$

devise a definition for quantities $[G_\omega]$ such that

$$[R_{\alpha\chi}][G_\alpha] + [R_{\beta\chi}][G_\beta] + [R_{\gamma\chi}][G_\gamma] = 0. \tag{6.50}$$

Exercise 6–7. For the situation described in Ex. 6–6(i), any steady state is completely specified when the variables T_α, P_α, T_β, P_γ, T_γ are known for a given gas and apparatus; the steady state is thus characterized by 5 degrees of freedom (see Chapter 3). Similarly, the situation described in Ex. 6–6(ii) is characterized by 7 degrees of freedom: T_α, P_α, T_β, P_β, T_γ, P_γ, $\langle T(\alpha\beta\gamma)\rangle$. Take the state γ as reference thermostatic state for the two situations, and show that in each case the number of degrees of freedom is correctly given by the expression $F_\gamma + \nu$ (see Chapter 3).

Assumption III

Enough material is now at hand so that I can make further remarks concerning assumption III. Assumption III implies that the properties of the state of migrational equilibrium are independent of the exact path used in the $\delta/\bar{\delta} Y_k$ operation. Thus in the case of the thermomolecular pressure effect (Figure 5–1, full-flux linkage) we can, say, hold T_α, P_β, T_β constant and can approach the state of migrational equilibrium by varying P_α [the symbolic operation is $(\delta/\bar{\delta}\dot{n}_\alpha)_{\beta,T_\alpha}$], or we can hold P_α, P_β, T_β constant and can vary T_α [the symbolic operation in this case is $(\delta/\bar{\delta}\dot{n}_\alpha)_{\beta,P_\alpha}$]. Assumption III implies that the same *overall* result is obtained in either case:

$$\bar{H}_\alpha - \bar{H}_\beta + \frac{\delta \dot{Q}_\alpha^{(r)}}{\bar{\delta}\dot{n}_\alpha} + \frac{\delta \dot{Q}_\beta^{(r)}}{\bar{\delta}\dot{n}_\alpha} = 0, \tag{6.51}$$

the requirement being that the quantity $\delta(\dot{Q}_\alpha^{(r)} + \dot{Q}_\beta^{(r)})/\bar{\delta}\dot{n}_\alpha$ be independent of the path used in arriving at the state of migrational equilibrium. For the case being considered we shall see that $\delta(\dot{Q}_\alpha^{(r)} + \dot{Q}_\beta^{(r)})/\bar{\delta}\dot{n}_\alpha$ yields the same result for the two paths but that $(\delta\dot{Q}_\omega^{(r)}/\bar{\delta}\dot{n}_\alpha)_{\beta,T_\alpha} \neq (\delta\dot{Q}_\omega^{(r)}/\bar{\delta}\dot{n}_\alpha)_{\beta,P_\alpha}$.

With respect to the thermomolecular pressure effect any steady state involving a given gas and full-flux linkage is completely specified by the four variables T_α, P_α, T_β, P_β; therefore we can say that $\dot{Q}_\omega^{(r)} = \dot{Q}_\omega^{(r)}(T_\alpha, P_\alpha, T_\beta, P_\beta)$, and that $\dot{n}_\alpha = \dot{n}_\alpha(T_\alpha, P_\alpha, T_\beta, P_\beta)$, and hence that $\dot{Q}_\omega^{(r)} = \dot{Q}_\omega^{(r)}(T_\alpha, P_\beta, T_\beta, \dot{n}_\alpha)$. It is easy to see, then, that

$$\left(\frac{\delta \dot{Q}_\omega^{(r)}}{\bar{\delta}\dot{n}_\alpha}\right)_{\beta,P_\alpha} = \left(\frac{\delta \dot{Q}_\omega^{(r)}}{\bar{\delta}\dot{n}_\alpha}\right)_{\beta,T_\alpha} + \left(\frac{\delta T_\alpha}{\bar{\delta}\dot{n}_\alpha}\right)_{\beta,P_\alpha}\left(\frac{\partial \dot{Q}_\omega^{(r)}}{\partial T_\alpha}\right)_{\beta,\dot{n}_\alpha=0} \tag{6.52}$$

and that in general $(\delta\dot{Q}_\omega^{(r)}/\bar{\delta}\dot{n}_\alpha)_{\beta,P_\alpha} \neq (\delta\dot{Q}_\omega^{(r)}/\bar{\delta}\dot{n}_\alpha)_{\beta,T_\alpha}$. On the other hand, we see from Eq. (6.52) that

$$\left(\frac{\delta(\dot{Q}_\alpha^{(r)} + \dot{Q}_\beta^{(r)})}{\bar{\delta}\dot{n}_\alpha}\right)_{\beta,P_\alpha} = \left(\frac{\delta(\dot{Q}_\alpha^{(r)} + \dot{Q}_\beta^{(r)})}{\bar{\delta}\dot{n}_\alpha}\right)_{\beta,T_\alpha} + \left(\frac{\delta T_\alpha}{\bar{\delta}\dot{n}_\alpha}\right)_{\beta,P_\alpha}\left(\frac{\partial(\dot{Q}_\alpha^{(r)} + \dot{Q}_\beta^{(r)})}{\partial T_\alpha}\right)_{\beta,\dot{n}_\alpha=0}. \tag{6.53}$$

When $\dot{n}_\alpha = 0$, it follows that* $\dot{Q}_\alpha^{(r)} + \dot{Q}_\beta^{(r)} = 0$ and that

$$\left(\frac{\partial(\dot{Q}_\alpha^{(r)} + \dot{Q}_\beta^{(r)})}{\partial T_\alpha}\right)_{\beta,\dot{n}_\alpha=0} = 0;$$

hence Eq. (6.53) reduces to

$$\left(\frac{\delta(\dot{Q}_\alpha^{(r)} + \dot{Q}_\beta^{(r)})}{\bar{\delta}\dot{n}_\alpha}\right)_{\beta,P_\alpha} = \left(\frac{\delta(\dot{Q}_\alpha^{(r)} + \dot{Q}_\beta^{(r)})}{\bar{\delta}\dot{n}_\alpha}\right)_{\beta,T_\alpha}. \tag{6.54}$$

It is clear, then, that overall equations such as (6.51) are independent of the path used in the limiting operation but that separate terms of the type $\delta Y_i/\bar{\delta} Y_k$ do depend on the path. As far as assumption III is concerned, there is no reason to favor one path over another, and any well-defined sequence of steady states is allowable in the limiting operation. If we desire to resolve

* Remember that we are dealing with the full-flux case; see Eqs. (5.1) and (5.6) with $\dot{q} = 0$ and $\dot{n}_\alpha = 0$.

the separate terms $\delta Y_i/\tilde{\delta} Y_k$ into an ordinary part $\bar{Y}$ and an extraordinary part $[Y]$, then we get in general a different extraordinary part $[Y]_{\text{path}}$ for each different limiting path. Note, however, the following two points: (i) The differential quantities $[Z]^{\partial}$ do not depend on the limiting operation $(\delta/\tilde{\delta}\dot{n})_{\text{path}}$; hence they are functions of the terminal states only and show no path dependence whatsoever. (ii) Although the integral quantities $[Z]$ do have a path dependence based on the definitional relations (5.10) to (5.13), the *form* of each relation involving integral quantities is invariant—i.e., independent of the exact path used in the basic definitional relations.* Assumption IV supplies the motivation for singling out a special path (or a class of special paths) for consideration, and I calculate explicitly in the body of the text only extraordinary parts $[Y]_{\Omega'}$ associated with affine sequences. In using affine sequences for the limiting operation $\delta/\tilde{\delta} Y_k$ it is always possible to proceed in two stages: we can first apply the operation $(\delta/\tilde{\delta} Y_k)_{\Omega'}$ to the First Law equation $\dot{U}(\text{system}) = \dot{Q} + \dot{W}$ (or to an equation derived from it), and we can then combine the resultant equation with the equation $(\delta\,\Theta/\tilde{\delta} Y_k)_{\Omega'} = 0$ to get a final equation. The first equation depends only on the laws of ordinary thermodynamics and on assumptions I to III; the final equation depends in addition on assumption IV.

* Consider, e.g., two different limiting paths (′) and (″) used to define the quantities $[H']$, $[H'']$, etc. In spite of differences in path, it will still be true that $[H'] = \mu + T[S']$, $[H''] = \mu + T[S'']$, $[H_\alpha'] - [H_\beta'] = [h(\beta\alpha)']$, $[H_\alpha''] - [H_\beta''] = [h(\beta\alpha)'']$, etc. Remember, however, that there is in general only *one* path for which $(\delta\,\Theta/\tilde{\delta} Y_k) = 0$ if the final state $(Y_k = 0)$ is not one of thermostatic equilibrium.

7

Thermodynamics: The Well-Tempered Science

Macroscopic thermodynamics is a science that matches well the human temperament—R. T. Cox has summed it all up very well [1]:

> Classical thermodynamics is an admirable theory, unpretentious and consoling to common sense. The quantities with which it deals lie in the scale of ordinary measurement, and the laws on which it is based are verified in commonplace experience. As a logical system it is marvelous in the economy of its means. Its concepts and principles are never superfluous, its limits are clearly stated, and within these limits it is, humanly speaking, perfect, with nothing left out and nothing left over.

That thermodynamics has an *extra*-scientific appeal for many of its practitioners is clear from the steady stream of books and expository articles devoted to the subject each year. It seems as though each successful practitioner of the subject reformulates the meaning of the basic postulates of thermodynamics in terms of his own experience and then tries to tell the world about his own personal way of looking at the subject.

Thermodynamics, then, is very much a part of the human scene, and it continues to be a congenial, viable science in spite of the fact that technically it should have been swallowed up by statistical mechanics a long time ago; thermodynamics has so many features that appeal to human needs that it seems likely to go on leading an independent existence for a long time to come. The congeniality of thermodynamics is truly amazing: although, as I think of it, it finds its natural expression in the picturesque language of engineering—heat baths, piston-and-cylinder arrangements, etc.—it has been treated from points of view as diverse as those of Planck [2], Brønsted [3], Tisza [4], Landsberg [5], Hatsopoulos and Keenan [6], and Tribus [7]. This infectious congeniality of thermodynamics will, of course, drive its practitioners to extend its domain to the widest possible class of phenomena—hence the spate of books on nonequilibrium thermodynamics.

The Utility of Thermodynamics

Thermodynamics is par excellence a language for describing experiment. One of its useful language functions is the efficient storage and encoding of experimental information; the experimental information can then be processed (reflected into a number of different perspectives) so as to reveal different facets of the phenomena since some ways of looking at a given set of data are more useful than others. Thus, for example, pH scales and activity coefficients enable us to describe a wide range of phenomena via a comfortable range of numbers, and we may break down Gibbs free energy changes into enthalpy changes and entropy changes. I made use of this storage and encoding function of thermodynamics, e.g., in defining the coefficients $[N]$ and $[S]^{\theta}$; these coefficients merely represent efficient ways of recording data for the forced vaporization process and for, e.g., the thermomolecular pressure effect; the processing of the data then yields quantities such as $[C_{\omega\omega}]^{\theta}$ and $[Q_{\omega}]^{\theta}$.

In addition to merely describing experiments, thermodynamics also makes some positive statements about the systems experimented upon: it says that energy is conserved, that isolated systems tend toward states of equilibrium, and that systems in internal equilibrium show a characteristic form of behavior in the vicinity of 0°K. This factual content of the laws of thermodynamics results in some necessary relations among the quantities introduced in the descriptive stage of thermodynamics; the language of thermodynamics is thus *cross-referenced* (by Maxwell's relations, e.g.), and it is possible to determine some quantities indirectly instead of directly. Thus, e.g., by measuring vapor pressure as a function of temperature we can determine indirectly the heat of vaporization. The cross-referencing also enables us to check the consistency of our measurements; the separate partial pressures of the components in a multicomponent mixture, e.g., must satisfy the Gibbs–Duhem relation. In dealing with steady states, I introduced enough quantities so as to be able to *describe* the steady phenomena in sufficient detail. The laws of thermostatics (equilibrium thermodynamics) supply a certain amount of cross-referencing for the steady-state language. They imply, e.g., that $[C_{\chi\omega}]^{\theta} = 0$; the result, then, is a weakly cross-referenced thermodynamic language that is adequate for dealing with steady states.*

If we make a positive statement about the rate of entropy production in the system (assumption IV), we introduce an additional amount of cross-referencing into the steady-state language, and we make things easier for the experimenter by allowing him to determine some quantities indirectly instead of directly† and by giving him some consistency checks for his data.‡ Note that assumption IV is in no sense *necessary* for the successful establishment of a thermodynamic language for treating steady states; when assumption IV holds, it merely makes life that much easier for the experimenter.

Historical Conspectus

I now wish to make some remarks about the historical growth of thermodynamics. I distinguish three phases in the evolution of thermodynamics: (i) There was first an area of interest. In the period 1800–1850 there was a widespread interest in the workings of the steam engine, in problems of efficiency, and in the interrelations among heat, work, and energy. (ii) In the second stage a suitable language (the work of men such as Thomson, Clausius, Gibbs, and Planck) was developed for describing and dealing with the phenomena in the area of interest. The quantities introduced by the language were studied to determine how they varied with the experimental conditions. (iii) Finally, the quantities introduced and studied in stage (ii) were given an atomic interpretation through the invention of statistical mechanics (Maxwell, Boltzmann, Gibbs, Fowler, etc.).

I should like to see the thermodynamics of steady states develop in the same orderly fashion. That there is a class of nonequilibrium phenomena—an area of interest—to which we would like to apply the concepts and techniques of thermodynamics is shown by the early work of Thomson on the thermocouple and by the continuing interest we have in such nonequilibrium phenomena as the Soret effect, the thermomolecular pressure effect, thermoösmosis, and concentration cells with transference. The area of interest is certainly there; we need next an adequate language for describing the phenomena and for exhibiting the necessary relations among the phenomena. We need to study the phenomena in terms of our language so as to learn how the various quantities and coefficients depend on the experimental variables. After we have learned what to expect experimentally, we need to explain the experimental behavior in terms of atoms and molecules; i.e., we need to invent a statistical mechanics of nonequilibrium states.

* Up to this point I am considering only assumptions I to III; I am not yet considering assumption IV.

† See, e.g., Eqs. (4.9) to (4.11) and Eq. (6.29).

‡ See, e.g., Ex. 5-3 and Eq. (8.45).

Consider the following case in point. The term *activity coefficient* was introduced into the study of electrolytic solutions by Noyes and Bray [8] in 1911. In the period 1911 to 1923, activity coefficients were accepted as a useful way of recording data and were intensively studied. A great deal was learned empirically about the dependence of activity coefficients of electrolytes on fractional powers of the molality and on the ionic strength of the solution; this empirical information is summarized very well in the 1923 text of Lewis and Randall [9]. Then in 1923 Debye and Hückel [10] succeeded in rationalizing some of the experimental information by giving the limiting behavior of the solution an atomistic interpretation. I take this case as a paradigm for the thermodynamics of steady states: we need to learn about the experimental behavior of the quantities $[R]$, $[N]$, $[S]^{\partial}$, $[Q_b(\beta)]$, $[C_{\omega\omega}]$, etc., so as to be able to attempt an explanation of that behavior in terms of atoms and molecules; as I see it, the learning comes first and the explanation comes later.

Head versus Hand

It is a commonplace observation that the manner of carrying out and reporting on an experiment is influenced by the theoretical framework into which the experimenter expects to fit the results. The interlocking of theory and practice gives rise to procedures that we can characterize by considering two polar sequences: (i) *doing*, *describing*, *explaining*, and (ii) *speculating*, *testing*, *readjusting*. In sequence (i) the activity starts in the laboratory, and the acts of describing and explaining are conditioned by the laboratory procedures. The resulting language tends to be operational (couched in terms of idealized laboratory operations), and the task of theory is to build a bridge between the relations that it considers fundamental (the laws of mechanics, e.g.) and the laboratory-oriented quantities (heat, energy, and entropy, e.g.) of the descriptive language. In sequence (ii) the theoretical conception is primary, and it is the task of the experimenter to devise some way of testing the consequences of the theory and then to modify, if necessary, the theoretical conception in the light of the experimental results. The laboratory work accompanying sequence (ii) is sometimes cramped or hampered by the unoperational nature of the theoretical constructs being considered. That sequences (i) and (ii) are but the extreme poles of a continuous spectrum of investigative procedures is clear from the facts that in sequence (i) we usually have some definite reason for undertaking the experiment, some interest—often enough of a theoretical nature—extraneous to the actual laboratory manipulations, and that in sequence (ii) our starting theoretical conceptions are conditioned by our previous knowledge and experience, including our knowledge of earlier experimental results. The gist of the matter is that in sequence (i) the experimenter arranges things to suit himself, and the theoretician is constrained to operate within the framework set out by the experimenter, whereas in sequence (ii) the theoretician has complete freedom, and it is the experimenter who is constrained to try to actualize the theoretical conceptions within the confines of the laboratory.

I feel that classical thermodynamics leans more toward sequence (i) than toward sequence (ii). I also feel that the developments initiated by Onsager [11] and pursued by such men as Prigogine [12] and de Groot [13] lean more toward sequence (ii) than toward sequence (i) and result in a rather peculiar thermodynamic perspective. Considerations dealing with the regression of fluctuations in an isolated system are rather far removed from ordinary laboratory procedures. An obsessive concern with the explicit dependence of currents on affinities and other parameters of the system is rather uncharacteristic of thermodynamics: in the case of chemical equilibrium, for example, we do not need to know the exact kinetic mechanism (the precise forward and backward rate expressions) in order to find the thermodynamic conditions of equilibrium and the expression for the equilibrium constant. I feel that just as ordinary thermodynamics places its main emphasis on the conditions of equilibrium so the thermodynamics of steady states should place its *main* emphasis on the conditions of migrational equilibrium in given spatial fields rather than on problems of "matter and motion" (items that are more a part of general physics than of anything else). Furthermore, the fundamental experimental system is the container plus the contents plus the interaction of container and contents with the surroundings. The experimenter would prefer a global language that reflects the laboratory realities. Now the experimenter never *measures* directly what happens at a single point in space, and a language couched in terms of local properties and gradients makes his life that much more difficult and gives him little or no guidance in dealing with the effect of the container on the process being studied and in deciding on ways to minimize that effect.

Apologia

What it amounts to is that in a roundabout way I have said that I like thermodynamics, that I have my own personal way of looking at the subject, that I favor a laboratory-oriented language based on global concepts, that I find currently popular treatments of nonequilibrium thermodynamics more *kinetic* than *thermodynamic*, and that I have constructed a nonequilibrium thermodynamics more to my taste.

References

1. R. T. Cox, *Statistical Mechanics of Irreversible Change* (The Johns Hopkins Press, Baltimore, 1955), p. 3.
2. M. Planck, *Treatise on Thermodynamics* (Dover Publications, New York, 1945).

3. J. N. Brønsted, *Principles and Problems in Energetics* (Interscience Publishers, New York, 1955).
4. L. Tisza, *Ann. Phys.*, **13**, 1 (1961).
5. P. T. Landsberg, *Thermodynamics* (Interscience Publishers, New York, 1961).
6. G. Hatsopoulos and J. H. Keenan, *Principles of General Thermodynamics* (John Wiley and Sons, New York, 1965).
7. M. Tribus, *Thermostatics and Thermodynamics* (D. Van Nostrand Company, Princeton, 1961).
8. A. A. Noyes and W. C. Bray, *J. Am. Chem. Soc.*, **33**, 1643 (1911).
9. G. N. Lewis and M. Randall, *Thermodynamics and the Free Energy of Chemical Substances* (McGraw-Hill Book Company, New York, 1923).
10. P. Debye and E. Hückel, *Physik. Z.*, **24**, 185 (1923).
11. L. Onsager, *Phys. Rev.*, **37**, 405 (1931); **38**, 2265 (1931).
12. I. Prigogine, *Introduction to the Thermodynamics of Irreversible Processes*, second edition (John Wiley and Sons, New York, 1961).
13. S. de Groot and P. Mazur, *Non-Equilibrium Thermodynamics* (North-Holland Publishing Company, Amsterdam, 1962).

PART II. APPLICATIONS

8

Migrational Equilibrium in Polythermal Fields: Applications

Thermoösmosis ($\dot{W}_0 = 0$)

If we connect two different thermal states of a gaseous system by means of a rubber diaphragm instead of by a capillary linkage, the situation is closely analogous to the thermomolecular pressure case but we now speak of *thermoösmotic* effects. Figure 8–1 is a schematic rendering of the situation; let us assume that the gas–membrane interfaces at α and β are at the appropriate thermostat temperatures and that the temperature gradient is localized inside the rubber membrane. The standard procedure—the initiation of a steady flow of mass between terminal parts and the application of the operation $(\delta/\delta\dot{n}_\alpha)_{\beta,T_\alpha,\langle T\rangle}$—leads to the relation

$$\mu_\alpha^{(g)} - \mu_\beta^{(g)} + T_\alpha[S_\alpha^{(M)}] - T_\beta[S_\beta^{(M)}] - [h_M(\beta\alpha)] = 0 \tag{8.1}$$

as the condition for migrational equilibrium; the superscripts g and M refer to the gas phase and the membrane phase, respectively. Since in the state of migrational equilibrium $\mu_\omega^{(g)} = \mu_\omega^{(M)} = \mu_\omega$, it follows that we may write Eq. (8.1) as

$$[H_\alpha^{(M)}] - [H_\beta^{(M)}] = [h_M(\beta\alpha)] \tag{8.2}$$

or

$$[G_\alpha^{(M)}] - [G_\beta^{(M)}] = 0, \tag{8.3}$$

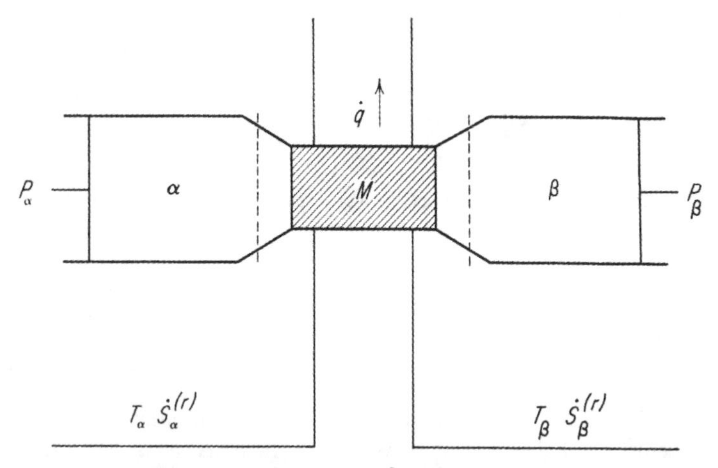

Figure 8–1. Two thermal states (α, β) of a gas connected by a rubber membrane (M); arbitrary linkage.

where

$$[H_\omega^{(M)}] \equiv \mu_\omega^{(M)} + T_\omega[S_\omega^{(M)}] \quad \text{and} \quad [G_\omega^{(M)}] \equiv [H_\omega^{(M)}] - \langle T(\alpha\beta)\rangle[S_\omega^{(M)}].$$

We can also write

$$\bar{V}_\omega^{(g)}\left(\frac{\partial P_\omega}{\partial T_\omega}\right)_\chi = \bar{S}_\omega^{(g)} - [S_\omega^{(M)}]^\partial \tag{8.4}$$

in a familiar fashion; thus the thermoösmotic effect ($P_\alpha \neq P_\beta$ for $T_\alpha \neq T_\beta$) depends on $[S_\omega^{(M)}]^\partial \equiv -(\partial\mu_\omega/\partial T_\omega)_\chi$ being different from $\bar{S}_\omega^{(g)}$.

If we treat the gas as ideal, Eqs. (8.4) and (5.29) enable us to write

$$\left(\frac{\partial \ln P_\omega}{\partial T_\omega}\right)_\chi \approx \frac{T_\omega(\bar{S}_\omega^{(g)} - [S_\omega^{(M)}]^\partial)}{RT_\omega^{\;2}} = \frac{[Q_\omega]^\partial}{RT_\omega^{\;2}}, \tag{8.5}$$

where

$$[Q_\omega]^\partial = \bar{H}_\omega^{(g)} - [H_\omega^{(M)}]^\partial = \bar{H}_\omega^{(g)} - \bar{H}_\omega^{(M)} + \bar{H}_\omega^{(M)} - [H_\omega^{(M)}]^\partial = \Delta\bar{H}_\omega + [Q_\omega^{(M)}]^\partial.$$

The quantity $[Q_\omega]^\partial$ thus contains an interphase and an intraphase effect.

By referring back to the remarks just preceding Eq. (5.31) we see that we can integrate Eq. (8.5) to obtain

$$\ln\frac{P_\beta}{P_\alpha} = -\frac{\langle[Q]^\partial\rangle}{R}\left(\frac{1}{T_\beta} - \frac{1}{T_\alpha}\right). \tag{8.6}$$

Denbigh [1], in a series of measurements involving a number of gases, a rubber membrane, and fixed temperatures T_1, T_2, found that the quantity $\langle[Q]^\partial\rangle$ was approximately constant (gave the same value, approximately, for all pairings (P_1, P_2) on the bitherm segment) for the bitherm segments that he investigated; the results obtained by Bearman [2] were similar.

Exercise 8–1. Show that Eq. (8.6) is just a disguised version of Eq. (5.38) with

$$\langle[Q]^{\partial}\rangle = T_{\omega}(\bar{S}_{\omega}^{(g)} - \langle[S]^{\partial}\rangle) - T_{\omega}\bar{C}_{P}^{(g)} + \frac{T_{\omega}T_{\chi}\bar{C}_{P}^{(g)}}{T_{\chi} - T_{\omega}} \ln \frac{T_{\chi}}{T_{\omega}}. \tag{8.7}$$

Exercise 8–2. If integration from state α to state β is to yield the same results as integration from β to α, then Eq. (8.7) must be invariant with respect to interchange of the subscripts ω and χ. Hence, show that if $\langle[Q]^{\partial}\rangle$ is independent of position on the bitherm then (i) $\bar{S}_{\omega}^{(g)} - \langle[S]^{\partial}\rangle$ must be independent of P_{ω}, and that (ii) $T_{\omega}(\bar{S}_{\omega}^{(g)} - \langle[S]^{\partial}\rangle - \bar{C}_{P}^{(g)})$ must equal $T_{\chi}(\bar{S}_{\chi}^{(g)} - \langle[S]^{\partial}\rangle - \bar{C}_{P}^{(g)})$.

Exercise 8–3. The apparatus of Denbigh and Raumann [1] consisted of a rubber membrane held in place by porous bronze discs. An arrangement of thermocouples allowed the temperature of each rubber-bronze disc junction to be determined. Temperatures and pressures characteristic of each face of the rubber membrane were measured. A sample of the data appears in the following table.

Table 8–1

Sample Collection of Data from the Thermoösmotic Measurements of Denbigh and Raumann [1]

Gas	T_1	T_2	P_1(cm Hg)	P_2(cm Hg)
	306.0	315.0	42.07	45.96
CO_2	289.8	301.9	22.09	25.07
	296.2	312.4	28.23	33.28
	299.1	315.9	66.66	68.24
N_2	298.6	316.0	66.88	68.50
	302.3	321.0	71.92	73.67
	300.4	316.8	54.82	54.23
H_2	299.7	316.5	62.51	61.74
	304.2	321.8	71.61	71.07

Use Eq. (8.6) to calculate the value of $[Q]^{\partial}$ for each set of data. Denbigh and Raumann took as appropriate values of the heat of evaporation $\Delta\bar{H}$ of the gas from the membrane the numbers (calories) -2800, $+100$, and $+800$ for the gases CO_2, N_2, and H_2, respectively. Average the $[Q]^{\partial}$ values for each gas and determine the average value of $[Q^{(M)}]^{\partial}$ for each gas from the relation $[Q]^{\partial} = \Delta\bar{H} + [Q^{(M)}]^{\partial}$.

Thermal Diffusion in Gases

I take up next the thermal diffusion of gases. The thermal diffusion effect in gases was predicted theoretically by Enskog and by Chapman [3] before it was ever measured experimentally; as a result the language used for describing the effect is slanted toward the kinetic theory derivations and uses of the effect,* rather than toward an efficient thermodynamic encoding of the experimental information. Consider a 2-component gaseous mixture in an apparatus consisting of two bulbs connected by a tube as in Figure 8–2. Let the heavier gas be referred to as component 1. Now, keep the two bulbs at different temperatures and examine the resulting state of migrational equilibrium with respect to component 1. If the mole fraction of component 1 is different in the two bulbs when the temperatures of the bulbs are different, then we say that a thermal diffusion effect exists.

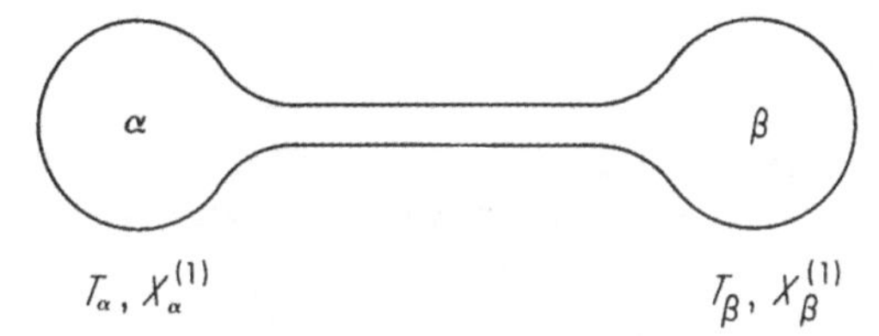

Figure 8–2. Thermal diffusion effect in gases; schematic.

We can carry out an ordinary thermodynamic analysis of the thermal diffusion phenomenon just as we did in the case of the thermomolecular pressure effect. Consider the situation shown in Figure 8–3. Establish a steady flow of component 1 between the terminal parts α(1) and β(1) and determine the conditions of migrational equilibrium with respect to component 1. Do the same for component 2. The result of these operations will be equations such as

$$\mu_\alpha^{(i)} + T_\alpha[S_\alpha^{(i)}] - \mu_\beta^{(i)} - T_\beta[S_\beta^{(i)}] - [h_i(\beta\alpha)] = 0 \qquad i = 1, 2 \tag{8.8}$$

and

$$[G_\alpha^{(i)}] = [G_\beta^{(i)}] \qquad i = 1, 2 \tag{8.9}$$

for the state of migrational equilibrium. In dealing with a multicomponent mixture, it is sometimes convenient to specify the thermodynamic state in terms of the temperature T_ω and the chemical potentials $\mu_\omega^{(j)}$ of each of the separate components, thus I write

$$[S_\omega^{(i)}]^\partial \equiv -\left(\frac{\partial \mu_\omega^{(i)}}{\partial T_\omega}\right)_{T_\chi, \mu_\chi^{(j)}} \equiv -\left(\frac{\partial \mu_\omega^{(i)}}{\partial T_\omega}\right)_\chi. \tag{8.10}$$

* The kinetic theory of gases uses the thermal diffusion effect as a way of learning about the law of force between molecules.

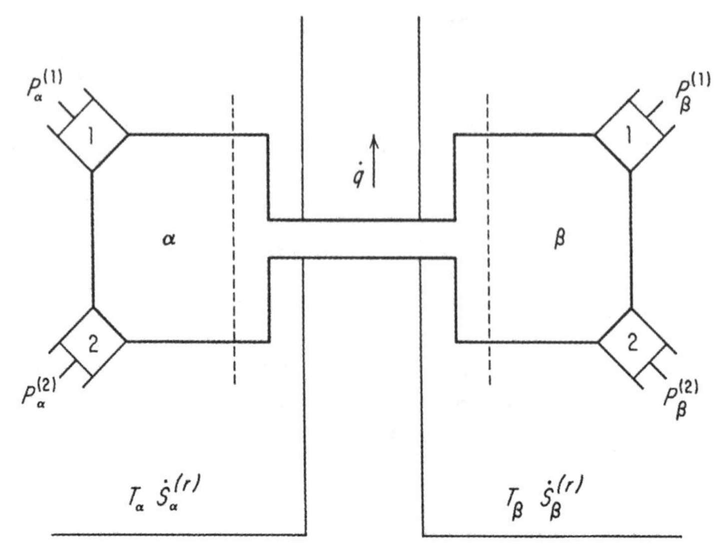

Figure 8–3. Gaseous thermal diffusion. Pure components 1 and 2 are in terminal parts in contact with the gaseous mixture through semipermeable membranes at α and β.

Now for the 2-component gaseous mixture at α, under conditions of migrational equilibrium, we have the relations

$$S_\alpha^{(m)}\,dT_\alpha - V_\alpha^{(m)}\,dP_\alpha + n_\alpha^{(1)}\,d\mu_\alpha^{(1)} + n_\alpha^{(2)}\,d\mu_\alpha^{(2)} = 0, \tag{8.11}$$

$$S_\alpha^{(m)} - V_\alpha^{(m)}\left(\frac{\partial P_\alpha}{\partial T_\alpha}\right)_\beta - n_\alpha^{(1)}[S_\alpha^{(1)}]^\partial - n_\alpha^{(2)}[S_\alpha^{(2)}]^\partial = 0, \tag{8.12}$$

$$\begin{aligned}\left(\frac{\partial \mu_\alpha^{(1)}}{\partial T_\alpha}\right)_\beta &\equiv -[S_\alpha^{(1)}]^\partial \\ &= -\bar{S}_\alpha^{(1)} + \bar{V}_\alpha^{(1)}\left(\frac{\partial P_\alpha}{\partial T_\alpha}\right)_\beta + \left(\frac{\partial \mu_\alpha^{(1)}}{\partial X_\alpha^{(1)}}\right)_{T_\alpha,P_\alpha}\left(\frac{\partial X_\alpha^{(1)}}{\partial T_\alpha}\right)_\beta,\end{aligned} \tag{8.13}$$

where the superscript m refers to properties of the gaseous mixture, P_α is the pressure of the gaseous mixture at α (i.e., $P_\alpha \approx P_\alpha^{(1)} + P_\alpha^{(2)}$), $X_\alpha^{(1)}$ is the mole fraction of component 1 in the gaseous mixture at α, and the superior bar indicates an appropriate molar, partial molar, or mean molar property.

Equation (8.11) is just the Gibbs–Duhem relation for the gaseous mixture at α; Eq. (8.12) is the result of applying the operation $(\partial/\partial T_\alpha)_\beta$ to Eq. (8.11); and Eq. (8.13) is the result of setting $\mu_\alpha^{(1)} = \mu_\alpha^{(1)}(T_\alpha, P_\alpha, X_\alpha^{(1)})$. It follows from Eq. (8.13) and the considerations of Chapter 5 that

$$\left(\frac{\partial X_\alpha^{(1)}}{\partial T_\alpha}\right)_\beta + \frac{\bar{V}_\alpha^{(1)}(\partial P_\alpha/\partial T_\alpha)_\beta}{(\partial \mu_\alpha^{(1)}/\partial X_\alpha^{(1)})_{T_\alpha,P_\alpha}} = \frac{\bar{S}_\alpha^{(1)} - [S_\alpha^{(1)}]^\partial}{(\partial \mu_\alpha^{(1)}/\partial X_\alpha^{(1)})_{T_\alpha,P_\alpha}} = f(T_\alpha, P_\alpha, X_\alpha^{(1)}) = f(\alpha); \tag{8.14}$$

and thus bithermal relations for the thermal diffusion effect are subject to the same sort of constraint as are the analogous relations for the thermomolecular pressure effect.*

The quantity $(\partial P_\alpha/\partial T_\alpha)_\beta$ is essentially a thermomolecular pressure effect for the gaseous mixture; we can make it vanish by choosing a wide enough tube or a high enough pressure. In the following relations I assume that matters have been so arranged that $(\partial P_\alpha/\partial T_\alpha)_\beta \approx 0$ and that the gases can be treated as ideal; in addition I make use of the fact that an extensive property $Z_\alpha^{(m)}$ of the mixture can be written in terms of the partial molar properties of the components; i.e., $Z_\alpha^{(m)} = n_\alpha^{(1)}\bar{Z}_\alpha^{(1)} + n_\alpha^{(2)}\bar{Z}_\alpha^{(2)}$ or $\bar{Z}_\alpha^{(m)} = X_\alpha^{(1)}\bar{Z}_\alpha^{(1)} + X_\alpha^{(2)}\bar{Z}_\alpha^{(2)}$:

$$\left(\frac{\partial \ln X_\alpha^{(1)}}{\partial \ln T_\alpha}\right)_\beta = \frac{[Q_\alpha^{(1)}]^\partial}{RT_\alpha}, \tag{8.15}$$

$$X_\alpha^{(1)}[Q_\alpha^{(1)}]^\partial + X_\alpha^{(2)}[Q_\alpha^{(2)}]^\partial = 0, \tag{8.16}$$

where

$$[Q_\alpha^{(i)}]^\partial \equiv T_\alpha(\bar{S}_\alpha^{(i)} - [S_\alpha^{(i)}]^\partial).$$

Exercise 8–4. Derive Eqs. (8.15) and (8.16) from Eqs. (8.12) and (8.13).

In discussions of thermal diffusion phenomena motivated by kinetic theory considerations [3, 4] the following quantities frequently occur (assume that $T_\alpha > T_\beta$):

the separation $\mathbf{S} \equiv X_\beta^{(1)} - X_\alpha^{(1)}$, (8.17)

the separation factor $\mathbf{q} \equiv (X_\beta^{(1)}/X_\alpha^{(1)})/(X_\beta^{(2)}/X_\alpha^{(2)})$, (8.18)

the thermal diffusion ratio $\mathbf{k}_T \equiv -(\partial X_\alpha^{(1)}/\partial \ln T_\alpha)_\beta$, (8.19)

the thermal diffusion factor $\boldsymbol{\alpha} \equiv -(1/X_\alpha^{(2)})(\partial \ln X_\alpha^{(1)}/\partial \ln T_\alpha)_\beta$. (8.20)

We see from Eqs. (8.15) and (8.16) and from the definitions (8.17) to (8.20) that the following relations hold:

$$\boldsymbol{\alpha} = \frac{\mathbf{k}_T}{X_\alpha^{(1)}X_\alpha^{(2)}}, \tag{8.21}$$

$$\langle \mathbf{k}_T \rangle = \frac{X_\beta^{(1)} - X_\alpha^{(1)}}{\ln(T_\alpha/T_\beta)} = \frac{\mathbf{S}}{\ln(T_\alpha/T_\beta)} = -\left\langle \frac{X^{(1)}[Q^{(1)}]^\partial}{RT} \right\rangle = \left\langle \frac{X^{(2)}[Q^{(2)}]^\partial}{RT} \right\rangle, \tag{8.22}$$

$$\langle \boldsymbol{\alpha} \rangle = \frac{\ln\{(X_\beta^{(1)}/X_\beta^{(2)})/(X_\alpha^{(1)}/X_\alpha^{(2)})\}}{\ln(T_\alpha/T_\beta)} = \frac{\ln \mathbf{q}}{\ln(T_\alpha/T_\beta)} = -\left\langle \frac{[Q^{(1)}]^\partial}{X^{(2)}RT} \right\rangle = \left\langle \frac{[Q^{(2)}]^\partial}{X^{(1)}RT} \right\rangle. \tag{8.23}$$

* The analog of assumption Q (see the note in Chapter 5) in this case is that

$$\left(\frac{\partial X_\alpha^{(1)}}{\partial \langle T \rangle}\right)_{\beta, T_\alpha} = 0.$$

For the case $X_\alpha^{(1)} \to 0$, we have $[Q_\alpha^{(2)}]^\partial \to 0$ and $\langle \boldsymbol{\alpha} \rangle \to -\langle [Q^{(1)}]^\partial/RT \rangle$; for the case $X_\alpha^{(2)} \to 0$, we have $[Q_\alpha^{(1)}]^\partial \to 0$ and $\langle \boldsymbol{\alpha} \rangle \to \langle [Q^{(2)}]^\partial/RT \rangle$. Note also that according to Eq. (8.15)

$$\frac{\ln (X_\beta^{(1)}/X_\alpha^{(1)})}{\ln (T_\alpha/T_\beta)} = -\left\langle \frac{[Q^{(1)}]^\partial}{RT} \right\rangle = \left\langle \frac{X^{(2)}[Q^{(2)}]^\partial}{X^{(1)}RT} \right\rangle. \tag{8.24}$$

Harrison Brown [4] suggested that average quantities such as $\langle \boldsymbol{\alpha} \rangle$ be associated with a unique reference temperature T_r given by

$$T_r \equiv \frac{T_\alpha T_\beta}{T_\alpha - T_\beta} \ln \frac{T_\alpha}{T_\beta}. \tag{8.25}$$

We can thus compute reference quantities $\langle [Q^{(i)}]^\partial \rangle_r$ from the relation

$$\langle [Q^{(i)}]^\partial \rangle_r \equiv RT_r \left\langle \frac{[Q^{(i)}]^\partial}{RT} \right\rangle; \tag{8.26}$$

and it is natural for us to assume that these reference quantities obey the relation

$$\langle X^{(1)} \rangle \langle [Q^{(1)}]^\partial \rangle_r + \langle X^{(2)} \rangle \langle [Q^{(2)}]^\partial \rangle_r = 0, \tag{8.27}$$

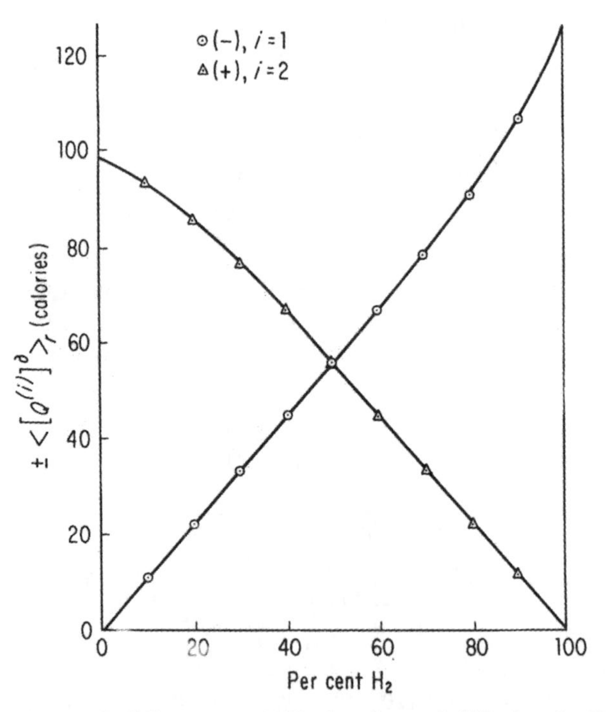

Figure 8-4. $\langle [Q^{(i)}]^\partial \rangle_r$ versus $\%H_2$ for thermal diffusion in H_2–D_2 mixtures (data of Heath, Ibbs, and Wild).

where $\langle X^{(i)} \rangle$ is defined by some such relation as

$$\langle X^{(i)} \rangle \equiv \frac{1}{T_\alpha - T_\beta} \int_{T_\beta}^{T_\alpha} X_\alpha^{(i)}\, dT_\alpha,$$

the state (β) at b being held constant. I assume that $\langle X^{(i)} \rangle$ can be approximated by the relation $\langle X^{(i)} \rangle \approx \frac{1}{2}(X_\alpha^{(i)} + X_\beta^{(i)})$.*

I have taken the thermal diffusion data of Heath, Ibbs, and Wild [5] pertaining to mixtures of hydrogen and deuterium and have made some illustrative calculations which I display in Table 8–2 and in Figures 8–4 and 8–5.

Table 8–2

Thermal Diffusion Results for H_2–D_2 Mixtures with $T_\alpha = 373°K$, $T_\beta = 288°K$, and $T_r = 326°K$
(Data of Heath, Ibbs, and Wild [5]; $RT_r = 648$ cal)

$\%H_2$ ($100\langle X^{(2)}\rangle$)	$\%Sep$ (100**S**)	$\langle \mathbf{k}_T \rangle$	$\langle \boldsymbol{\alpha} \rangle$	$-\langle [Q^{(1)}]^\theta \rangle_r$ (*cal*)	$\langle [Q^{(2)}]^\theta \rangle_r$ (*cal*)
10	0.374	0.0145	0.161	10.4	93.8
20	0.685	0.0265	0.165	21.5	85.6
30	0.920	0.0356	0.171	33.0	76.7
40	1.071	0.0416	0.173	45.0	67.3
50	1.112	0.0432	0.173	56.0	56.0
60	1.071	0.0416	0.173	67.3	45.0
70	0.938	0.0362	0.172	78.0	33.4
80	0.726	0.0281	0.176	91.0	22.8
90	0.420	0.0166	0.184	107.5	12.0

The quantities $\langle [Q^{(i)}]^\theta \rangle_r$ are, in this case, small compared to RT_r; this is but another way of saying that the thermal diffusion effect in H_2–D_2 mixtures is small. Figure 8–5 shows that at the extreme ends of the mole fraction scale the quantities $|\langle [Q^{(i)}]^\theta / RT \rangle|$ and $\langle \boldsymbol{\alpha} \rangle$ approach one another.

Exercise 8–5. Given that $X_\beta^{(1)} - X_\alpha^{(1)} = \mathbf{S}$, $X_\alpha^{(i)} + X_\beta^{(i)} = 2\langle X^{(i)} \rangle$, and $\ln\{(1+z)/(1-z)\} = 2\{z + (z^3/3) + (z^5/5) + \cdots\}$, show that for the case $(\mathbf{S}/2\langle X^{(i)} \rangle) \ll 1$ the following relations hold:

$$\ln \frac{X_\beta^{(1)}}{X_\alpha^{(1)}} \approx \frac{\mathbf{S}}{\langle X^{(1)} \rangle}, \quad \ln \frac{X_\beta^{(2)}}{X_\alpha^{(2)}} \approx -\frac{\mathbf{S}}{\langle X^{(2)} \rangle},$$

$$\ln \mathbf{q} \approx \frac{\mathbf{S}}{\langle X^{(1)} \rangle \langle X^{(2)} \rangle}.$$

* When the separation **S** is small compared to the average mole fractions—i.e., when $(\mathbf{S}/2\langle X^{(1)} \rangle) \ll 1$ and $(\mathbf{S}/2\langle X^{(2)} \rangle) \ll 1$—we have the useful approximate relation

$$\langle \boldsymbol{\alpha} \rangle \approx \frac{\langle \mathbf{k}_T \rangle}{\langle X^{(1)} \rangle \langle X^{(2)} \rangle}.$$

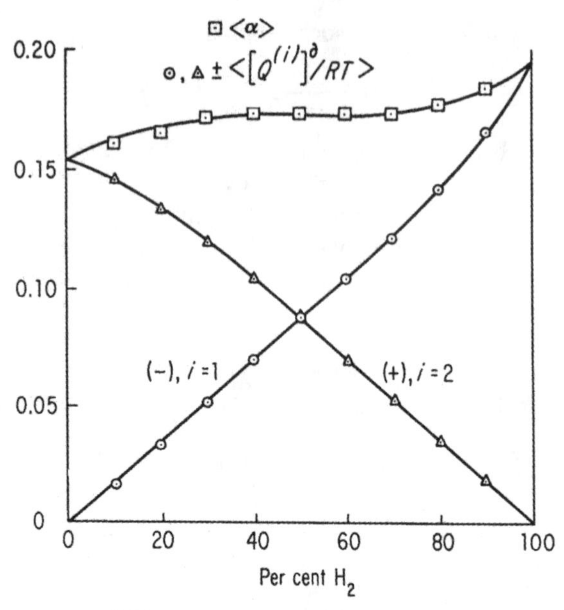

Figure 8-5. $\langle [Q^{(i)}]^{\partial}/RT \rangle$ and $\langle \boldsymbol{\alpha} \rangle$ versus $\% H_2$ for thermal diffusion in H_2–D_2 mixtures (data of Heath, Ibbs, and Wild).

Use these approximate relations to calculate the quantities $\langle \boldsymbol{\alpha} \rangle$, $\langle [Q^{(1)}]^{\partial}/RT \rangle$, and $\langle [Q^{(2)}]^{\partial}/RT \rangle$ from the data in Table 8-3, listing the results of Atkins, Bastick, and Ibbs [6] for mixtures of helium and krypton; plot the results in the manner of Figure 8-5.

Table 8-3

Thermal Diffusion Results for He–Kr Mixtures with $T_\alpha = 373°K$, $T_\beta = 288°K$, and $T_r = 326°K$
(Data of Atkins, Bastick, and Ibbs [6])

%He ($100\langle X^{(2)} \rangle$)	%Sep (100S)	$\langle \mathbf{k}_T \rangle$
30	1.75	0.0677
40	2.20	0.0852
50	2.58	0.1000
60	2.79	0.1080
70	2.76	0.1068

Exercise 8-6. Show that Eq. (8.15) can be integrated to yield

$$\ln \frac{X_\beta^{(1)}}{X_\alpha^{(1)}} = -\frac{\langle [Q^{(1)}]^{\partial} \rangle}{R} \left(\frac{1}{T_\beta} - \frac{1}{T_\alpha} \right)$$

and show that

$$\langle [Q^{(1)}]^{\partial} \rangle = \langle [Q^{(1)}]^{\partial} \rangle_r .$$

The Clusius–Dickel Separating Column

Just as the conditions of liquid–vapor equilibrium are the basis for practically useful distilling columns, so the thermal diffusion effect is the basis for a practically useful separating column designed by Clusius and Dickel [7] for separating gaseous mixtures. In essence the Clusius–Dickel column is a long cylindrical tube with a wire down the middle. We mount the tube in a vertical position, introduce the gaseous mixture, heat the wire to a high temperature ($\sim$800°C) by passing an electric current through it, cool the walls of the tube so as to maintain a large temperature difference between the wire and the wall, and wait for migrational equilibrium to be established along the column. The thermal diffusion effect tends to concentrate the heavier component along the wall of the tube, and the gravitational field induces a slow convective circulation of the gas, whereby the lighter component moves upward in the center of the tube and the heavier component moves downward along the wall of the tube. When the entire system comes to a state of migrational equilibrium with respect to each of the components, there is usually a significant difference in composition for points near the top of the tube relative to points near the bottom of the tube. The separation factor $\mathbf{q}$ for the gaseous mixture depends on the length of the column, the diameter of the tube, the difference in molecular weight for the component gases, the temperature difference between the heated wire and the wall of the tube, and on just how the motion of each component along the tube couples with the flow of heat from the wire to the wall. The Clusius–Dickel column has been used for the separation of gaseous isotopes, and kinetic theory analyses of the workings of the column have been attempted [7, 8].

For an analysis of the workings of the Clusius–Dickel column by steady-state thermodynamic methods, I arrange things in the manner indicated in Figure 8–6. I place the gaseous mixture into a region MIX in contact with two heat reservoirs a and b ($T_a > T_b$). On the b-side of the MIX region I set up two terminal parts, α and β, communicating with the MIX region via semipermeable membranes. With respect to the Clusius–Dickel column, T_a

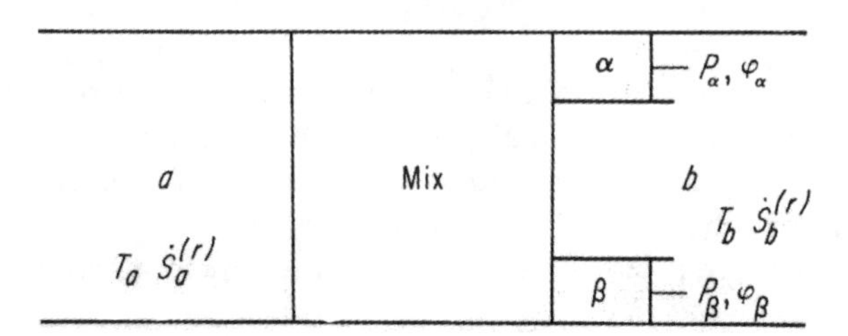

Figure 8–6. Schematic treatment of the Clusius–Dickel separating column.

corresponds to the temperature of the wire, T_b corresponds to the temperature of the wall, the distance between terminal parts α and β corresponds to the length of the column, and the distance between the reservoirs a and b corresponds to the radius of the tube. After filling the MIX region with the gaseous mixture, I introduce one of the components (component 1, say) into the terminal parts at α and β, adjusting the semipermeable membranes accordingly. I then set up a steady flow of component 1 through the MIX region from one terminal part to the other in such a way that $\dot{n}_\alpha + \dot{n}_\beta = 0$:

$$\dot{n}_\alpha\{(\bar{H}_\alpha + \varphi_\alpha) - (\bar{H}_\beta + \varphi_\beta)\} + \dot{Q}_a^{(r)} + \dot{Q}_b^{(r)} = 0, \tag{8.28}$$

$$\begin{aligned}\Theta &= \dot{n}_\alpha(\bar{S}_\alpha - \bar{S}_\beta) + \frac{\dot{Q}_a^{(r)}}{T_a} + \frac{\dot{Q}_b^{(r)}}{T_b} \\ &= \dot{n}_\alpha\left(\frac{\bar{H}_\beta + \varphi_\beta}{T_b} - \frac{\bar{H}_\alpha + \varphi_\alpha}{T_b} + \bar{S}_\alpha - \bar{S}_\beta\right) + \dot{Q}_a^{(r)}\left(\frac{1}{T_a} - \frac{1}{T_b}\right) \\ &= \dot{n}_\alpha \frac{\mu_\beta + \varphi_\beta - \mu_\alpha - \varphi_\alpha}{T_b} + \dot{Q}_a^{(r)}\left(\frac{1}{T_a} - \frac{1}{T_b}\right) \\ &= Y_1\Omega_1 + Y_2\Omega_2,\end{aligned} \tag{8.29}$$

where φ_ω is the molar gravitational energy of the material in the ω terminal part. Applying the operation $(\delta/\bar{\delta}\dot{n}_\alpha)_{T_a,T_b}$ to Eq. (8.28) and making use of assumption IV, we get, as the condition for the migrational equilibrium of component 1,

$$0 = \frac{\mu_\beta + \varphi_\beta - \mu_\alpha - \varphi_\alpha}{T_b} + \left(\frac{\delta\dot{Q}_a^{(r)}}{\bar{\delta}\dot{n}_\alpha}\right)_{T_a,T_b}\left(\frac{1}{T_a} - \frac{1}{T_b}\right)$$

or

$$0 = \mu_\beta + \varphi_\beta - \mu_\alpha - \varphi_\alpha + \left(\frac{\delta\dot{S}_a^{(r)}}{\bar{\delta}\dot{n}_\alpha}\right)_{T_a,T_b}(T_b - T_a). \tag{8.30}$$

Upon explicitly introducing the fact that it is component 1 we have been dealing with, we have

$$RT_b \ln\frac{X_\beta^{(1)}}{X_\alpha^{(1)}} = M^{(1)}g(h_\alpha - h_\beta) + \left(\frac{\delta\dot{S}_a^{(r)}}{\bar{\delta}\dot{n}_\alpha^{(1)}}\right)_{T_a,T_b}(T_a - T_b), \tag{8.31}$$

with $X_\omega^{(1)}$ being the mole fraction of component 1 in the gaseous mixture at a point adjacent to the ω terminal part, $M^{(1)}$ being the molecular weight of component 1, g being the acceleration due to gravity, and h_ω being the height of the ω terminal part above some reference plane.

If, now, we repeat the entire procedure with respect to component 2, we obtain

$$RT_b \ln\frac{X_\beta^{(2)}}{X_\alpha^{(2)}} = M^{(2)}g(h_\alpha - h_\beta) + \left(\frac{\delta\dot{S}_a^{(r)}}{\bar{\delta}\dot{n}_\alpha^{(2)}}\right)_{T_a,T_b}(T_a - T_b) \tag{8.32}$$

and

$$RT_b \ln \frac{(X_\beta^{(1)}/X_\alpha^{(1)})}{(X_\beta^{(2)}/X_\alpha^{(2)})} = g(M^{(1)} - M^{(2)})(h_\alpha - h_\beta) + (T_a - T_b)\left\{\left(\frac{\delta \dot{S}_a^{(r)}}{\tilde{\delta}\dot{n}_\alpha^{(1)}}\right)_{T_a,T_b} - \left(\frac{\delta \dot{S}_a^{(r)}}{\tilde{\delta}\dot{n}_\alpha^{(2)}}\right)_{T_a,T_b}\right\}. \quad (8.33)$$

Thus

$$RT_b \ln \mathbf{q} = g\,\Delta M\,\Delta h + \Delta T\,\Delta\left(\frac{\delta \dot{S}_a^{(r)}}{\tilde{\delta}\dot{n}_\alpha}\right)_{T_a,T_b}. \quad (8.34)$$

We see, then, that the separation factor $\mathbf{q}$ depends on the difference in molecular weight ΔM for the components, on the length Δh of the column, on the temperature difference ΔT between the wire and the wall, and on the quantity $\Delta(\delta \dot{S}_a^{(r)}/\tilde{\delta}\dot{n}_\alpha)_{T_a,T_b}$, which is a complicated function of the dimensions of the tube and of the exact way in which the motion of the gaseous mixture along the tube couples with the flow of heat from the wire to the wall [8].

Clusius and Dickel [7] used a column 20 meters long with $\Delta T \approx 700°\text{K}$ to effect a separation of the isotopes of chlorine. They filled the column with a mixture of $H^{35}Cl$ and $H^{37}Cl$. When the state of migrational equilibrium was reached, their analysis showed that the gas at the bottom of the tube was 75.7% $H^{35}Cl$, whereas the gas at the top of the tube was 99.6% $H^{35}Cl$. They had thus achieved a separation factor of ~80. From Eq. (8.34) we see that in this case

$$600 \ln 80 = 980 \times 2 \times 2000(4.18 \times 10^7)^{-1} + 700\,\Delta\left(\frac{\delta \dot{S}_a^{(r)}}{\tilde{\delta}\dot{n}_\alpha}\right)_{T_a,T_b} \approx 0.1 + 700\,\Delta\left(\frac{\delta \dot{S}_a^{(r)}}{\tilde{\delta}\dot{n}_\alpha}\right)_{T_a,T_b}, \quad (8.35)$$

and the simple gravitational effect is negligible compared to the thermal diffusion-based effect.

Thermal Diffusion in Solutions—Soret Effect

Consider two different thermal states of a ν-component solution connected by a narrow tube linkage (Figure 8–7). Let one of the components be designated as the solvent and be indicated by the index 1, and let the composition of the solution be expressed by the molalities* of the other $\nu - 1$ components relative to component 1 or in terms of the mole fractions $X_\omega^{(k)}$ of the same $\nu - 1$ components. If in Figure 8–7 $m_\alpha^{(k)} \neq m_\beta^{(k)}$ for $T_\alpha \neq T_\beta$, we say that the solution shows a *Soret effect* with respect to component k. (For convenience let $m_\omega^{(1)} \equiv 1000/M_1$, with M_1 being the gram-molecular weight of component 1.)

* The molality $m^{(k)}$ of component k is the number of moles of component k per kilogram of component 1.

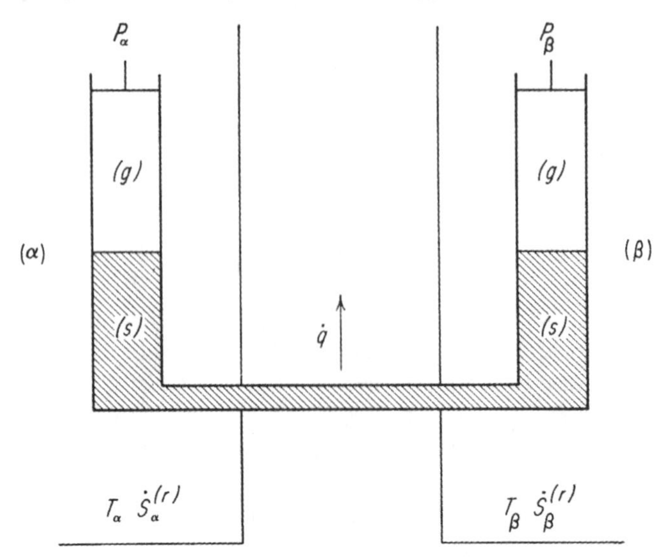

Figure 8–7. Two thermal states of a ν-component solution connected by a narrow-tube arbitrary linkage.

Now if the solvent is volatile while the other $\nu - 1$ components are not, then by letting P_ω differ from the solvent vapor pressure $P_\omega^{(1)}$ appropriate to the state of migrational equilibrium we can initiate a steady flow of component 1 between terminal parts and can converge on the state of migrational equilibrium via a sequence of steady-flow states. Equation (6.39) with $(\delta/\bar{\delta}\dot{n})_{\Omega'} = (\delta/\bar{\delta}\dot{n}_\alpha^{(1)})_{\Omega'}$ and $\dot{W}_0 = 0$—together with the relation $\mu_\omega^{(g)} = \mu_\omega^{(1)}$—yields

$$\mu_\alpha^{(1)} - \mu_\beta^{(1)} + T_\alpha[S_\alpha^{(1)}] - T_\beta[S_\beta^{(1)}] - [h_1(\beta\alpha)] = 0, \tag{8.36}$$

$$[H_\alpha^{(1)}] - [H_\beta^{(1)}] = [h_1(\beta\alpha)], \tag{8.37}$$

$$[G_\alpha^{(1)}] - [G_\beta^{(1)}] = 0. \tag{8.38}$$

For the steady flow of any single component k between terminal parts, Eq. (6.39) taken with respect to $(\delta/\bar{\delta}\dot{n}_\alpha^{(k)})_{\Omega'}$ yields

$$[H_\alpha^{(k)}] - [H_\beta^{(k)}] = [h_k(\beta\alpha)], \tag{8.39}$$

$$[G_\alpha^{(k)}] - [G_\beta^{(k)}] = 0. \tag{8.40}$$

For aqueous solutions of electrolytes the maintenance of a steady flow of an electrolyte component probably requires electrochemical adjuncts. (I discuss such systems more fully under the title *thermocells* in Chapter 9.) As long as the steady flow of a component between *terminal parts* can be maintained in *some* way, Eq. (8.40) must hold for the migrational equilibrium of that component.

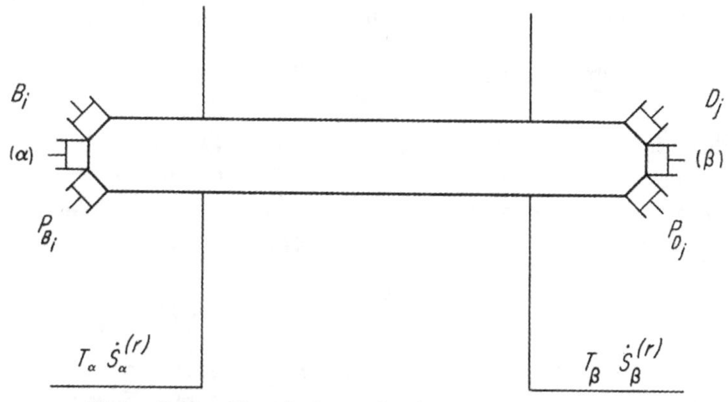

Figure 8–8. Chemical reaction in a polythermal field.

Exercise 8–7. Chemical Reaction in a Polythermal Field
Consider the chemical reaction (see Chapter 2)

$$\sum \nu_i B_i = \sum \nu_j D_j$$

and the situation indicated in Figure 8–8. Let the reactant reservoirs B_i be immersed in a thermostat of temperature T_α, and let the product reservoirs D_j be immersed in a thermostat of temperature T_β; let the reaction vessel function as a linkage between the thermostats. For the situation in which there is a steady flow of ξ reaction measures in unit time from reactants to products, show that the migrational equilibrium condition for the chemical reaction in the polythermal field is that

$$\sum \nu_j [G_\beta^{(j)}] - \sum \nu_i [G_\alpha^{(i)}] = 0. \tag{8.41}$$

From the definitions (for the case $P_\alpha = P_\beta = P$(atmospheric) = constant)

$$[S_\omega^{(j)}]^\partial \equiv -\left(\frac{\partial \mu_\omega^{(j)}}{\partial T_\omega}\right)_X \qquad j = 1, 2, \ldots, \nu, \tag{8.42}$$

$$\mu_\omega^{(k)} \equiv (\mu_\omega^{(k)})^0 + RT_\omega \ln (\gamma_\omega^{(k)} \zeta_\omega^{(k)}) \quad k = 2, 3, \ldots, \nu, \quad \zeta = m \text{ or } X, \tag{8.43}$$

and the Gibbs–Duhem relation for the solution phase (s) at ω,

$$S_\omega^{(s)}\, dT_\omega - V_\omega^{(s)}\, dP_\omega + \sum_{j=1}^{\nu} n_\omega^{(j)}\, d\mu_\omega^{(j)} = 0, \tag{8.44}$$

we see that

$$\sum_{j=1}^{\nu} \zeta_\omega^{(j)} [Q_\omega^{(j)}]^\partial = T_\omega \sum_{j=1}^{\nu} \zeta_\omega^{(j)} (\bar{S}_\omega^{(j)} - [S_\omega^{(j)}]^\partial) = 0, \tag{8.45}$$

$$[Q_\omega^{(j)}]^\partial = \sum_{k=2}^{\nu} \left(\frac{\partial \mu_\omega^{(j)}}{\partial \ln \zeta_\omega^{(k)}}\right)_{T_\omega, P, \zeta_\omega} \left(\frac{\partial \ln \zeta_\omega^{(k)}}{\partial \ln T_\omega}\right)_{X, P}. \tag{8.46}$$

For the special case of a 2-component solution

$$[Q_\omega^{(2)}]^\partial = RT_\omega^{\,2}\left\{1 + \left(\frac{\partial \ln \gamma_\omega^{(2)}}{\partial \ln \zeta_\omega^{(2)}}\right)_{T_\omega, P}\right\}\left(\frac{\partial \ln \zeta_\omega^{(2)}}{\partial T_\omega}\right)_{\chi, P} \qquad \zeta = m, X. \tag{8.47}$$

The subscript ζ_ω in Eq. (8.46) means that the other $\nu - 2$ concentration variables are to be kept constant during the differentiation; the $\gamma_\omega^{(j)}$ quantities are activity coefficients appropriate to the choice of concentration variables (m or X). The integration of Eq. (8.47) between the states α and β yields

$$\ln \frac{\zeta_\beta^{(2)}}{\zeta_\alpha^{(2)}} = -\left\langle \frac{[Q^{(2)}]^\partial}{R\{1 + (\partial \ln \gamma^{(2)}/\partial \ln \zeta^{(2)})_{T,P}\}}\right\rangle\left(\frac{1}{T_\beta} - \frac{1}{T_\alpha}\right). \tag{8.48}$$

It has become customary to perform Soret-type experiments in wide-diameter vertical tubes with small ($\sim 10°$) temperature differences between top and bottom and to report the results in terms of a *Soret coefficient* $\boldsymbol{\sigma} \equiv -\partial \ln m/\partial T$; from Eq. (8.47) we see that

$$\boldsymbol{\sigma} \approx -\frac{\ln (m_\beta^{(2)}/m_\alpha^{(2)})}{T_\beta - T_\alpha} = -\left\langle \frac{[Q^{(2)}]^\partial}{RT^2\{1 + (\partial \ln \gamma^{(2)}/\partial \ln m^{(2)})_{T,P}\}}\right\rangle. \tag{8.49}$$

In dealing with solutions of electrolytes, it is customary to express Eqs. (8.43) and (8.46) to (8.49) in terms of mean ionic activity coefficients $\gamma_\pm$ [9]. Figure 8–9 shows the results obtained by Chipman [10] for the system H_2O—HCl. The Soret coefficient defined in the preceding manner is taken to be a function of the average temperature and average concentration in the

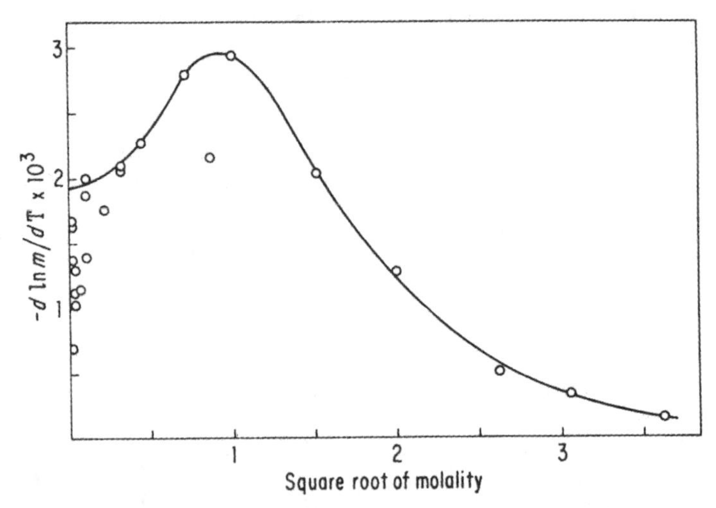

Figure 8–9. Soret effect for the system H_2O—HCl at 25°C (data of Chipman). (From the *J. Am. Chem. Soc.*, **48**, 2577 (1926), by permission.)

tube. A positive value of σ means that the substance is more concentrated at the colder end of the tube. Agar and Turner [11–13] have been especially active in this field; Table 8–4 is a compilation of some typical results obtained by Turner [13].

Table 8–4

Soret Effect for 1–1 Electrolytes at 0.01 Mole/Liter and a Mean Temperature of ~25°C
(Data of Snowdon and Turner [13])

Substance	$10^3\sigma(\text{deg}^{-1})$
LiF	2.35
LiCl	0.02
LiBr	0.08
LiI	−1.44
KF	3.81
KCl	1.41
KBr	1.44
KI	−0.08
KNO_3	0.64
KOH	13.34
$KClO_4$	0.79
KIO_4	2.47
$TlNO_3$	1.87
$AgNO_3$	3.40
HCl	9.03
$HClO_4$	8.29

Exercise 8–8. Equation (8.49) shows that variation in σ with concentration for a fixed temperature interval may reflect variation in one or both of the quantities $[Q^{(2)}]^{\partial}$ and $(\partial \ln \gamma^{(2)}/\partial \ln m^{(2)})_{T,P}$. For the system HCl—$H_2O$, Eq. (8.49) takes the form

$$\sigma \approx -\left\langle \frac{[Q^{(2)}]^{\partial}}{2RT^2\{1 + (\partial \ln \gamma_{\pm}/\partial \ln m)_{T,P}\}} \right\rangle,$$

where $\gamma_{\pm}$ is the mean ionic activity coefficient for HCl, and I have written simply m for $m^{(2)}$. Table 8–5 lists values of $\gamma_{\pm}$ for HCl at 25°C as reported by Harned and Owen [14]. Determine from the table approximate values of $(\partial \ln \gamma_{\pm}/\partial \ln m)_{T,P} \approx (m/\gamma_{\pm})(\Delta\gamma_{\pm}/\Delta m)$ and plot the quantity

$$\{1 + (\partial \ln \gamma_{\pm}/\partial \ln m)_{T,P}\}^{-1}$$

versus $m^{1/2}$. Compare the general shape of this plot with that of Chipman's plot (Figure 8–9).

Table 8–5

The Mean Ionic Activity Coefficient $\gamma_{\pm}$ for HCl(aq) at 25°C
(Compilation of Harned and Owen [14])

m	$\gamma_{\pm}$
0.001	0.9656
0.002	0.9521
0.005	0.9285
0.01	0.9048
0.02	0.8755
0.05	0.8304
0.1	0.7964
0.5	0.7571
1.0	0.8090
1.5	0.8962
2	1.0090
3	1.316

I should like to see the measurement and theoretical treatment of the Soret effect based firmly on the linkage concept developed in Chapter 5 in conjunction with the thermomolecular pressure effect. In wide-diameter tubes the possibility of convection currents is an ever present source of annoyance. Would it not be better to introduce a well-defined linkage between terminal parts, to calculate quantities appropriate to either end of the linkage (the quantities $[Q_{\omega}^{(i)}]^{\theta}$, say), to make plots such as $[Q_{\omega}^{(i)}]^{\theta}$ versus $[Q_{\chi}^{(i)}]^{\theta}$ for a given (T_{ω}, T_{χ}) bitherm, and to study a *series* of linkages of varying diameter so as to be able to extrapolate the results to an infinite-diameter limit? Such a procedure would produce a *complete* study: we would obtain apparatus-dependent results of importance for engineering applications, and we could *confidently* use the apparatus-independent, infinite-diameter limiting values in theoretical discussions.

Three-Component Soret Cell with Constraint, Full-Flux Linkage

Consider an aqueous solution containing both a sparingly soluble salt and a second quite soluble salt. Such a solution is a 3-component system; let components 1, 2, and 3 be water, sparingly soluble salt, and quite soluble salt, respectively. Figure 8–10 is a schematic rendering of the situation. The system of interest is composed of two terminal parts (α, β) at temperatures T_{α} and T_{β}; the terminal parts are connected by a gradient part that is the site of the temperature gradient in the system; the connecting link is of the full-flux

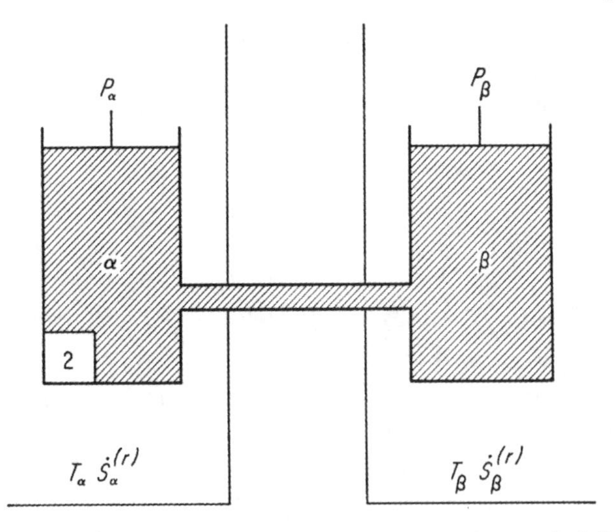

Figure 8–10. Schematic version of a 3-component Soret cell, full-flux linkage. Terminal part α contains some solid component 2.

type and $\dot{W}_0 = 0$. Fill the apparatus with the 3-component solution and saturate the terminal part at α with the sparingly soluble salt (let it contain some solid salt). The chemical potential of the second component in the α terminal part $\mu_\alpha^{(2)}$ is then fixed for a given T_α. Let $T_\alpha \neq T_\beta$; let $P_\alpha = P_\beta = P$ (atmospheric); and let the subscript $\chi(i)$ mean constant T_χ, P_χ, $\mu_\chi^{(i)}$. Thus we have prepared a polythermal field subject to a constraint; the chemical potential of component 2 in the α terminal part $\mu_\alpha^{(2)}$ is maintained constant; i.e., $\mu_\alpha^{(2)} = \mu_\alpha^{(2)}(\text{sat})$. I wonder if this constraint influences other parts of the polythermal field as well? Is $\mu_\beta^{(2)}$ equal to $\mu_\beta^{(2)}(\text{sat})$ under conditions of migrational equilibrium? Is $\mu_\beta^{(2)}$ independent of the molality of the third component at constant T_α, T_β? In order to explore the situation, we need to investigate operations such as $(\partial/\partial T_\beta)_{\alpha(2),m_\alpha^{(3)}}$ and $(\partial/\partial m_\alpha^{(3)})_{\alpha(2),T_\beta}$, where $m_\alpha^{(3)}$ is the molality of component 3 in the α terminal part. I lay down the following definitions:

$$\left(\frac{\partial \mu_\beta^{(2)}}{\partial T_\beta}\right)_{\alpha(2),m_\alpha^{(3)}} \equiv -[S_\beta^{(22)}]^\partial, \tag{8.50}$$

$$\left(\frac{\partial \mu_\beta^{(2)}}{\partial m_\alpha^{(3)}}\right)_{\alpha(2),T_\beta} \equiv RT_\beta[M_\beta^{(23)}]^\partial, \tag{8.51}$$

$$\mu_\beta^{(2)} \equiv \mu_\beta^{(2)}(\text{sat}) + RT_\beta \ln \theta_\beta^{(2)}, \tag{8.52}$$

$$\mu_\beta^{(2)}(m_\alpha^{(3)}) \equiv \mu_\beta^{(2)}(0) + RT_\beta m_\alpha^{(3)}[M_\beta^{(23)}] \quad \text{const. } \mu_\alpha^{(2)}, P, T_\alpha, T_\beta, \tag{8.53}$$

where $\mu_\beta^{(2)}(\text{sat})$ represents the value of the given chemical potential in a solution saturated with component 2 at temperature T_β and pressure P. We also have available a number of standard relationships:

$$\mu_\beta^{(2)} - \mu_\alpha^{(2)} = -\int_{T_\alpha}^{T_\beta} [S_b^{(22)}]^\partial \, dT_b, \tag{8.54}$$

$$\left(\frac{\partial \mu_\beta^{(2)}}{\partial T_\beta}\right)_{\alpha(2),m_\alpha^{(3)}} = -\left(\frac{\partial \mu_\beta^{(2)}}{\partial m_\alpha^{(3)}}\right)_{\alpha(2),T_\beta}\left(\frac{\partial m_\alpha^{(3)}}{\partial T_\beta}\right)_{\alpha(2),\mu_\beta^{(2)}}, \tag{8.55}$$

$$\mu_\beta^{(2)} = (\mu_\beta^{(2)})^0 + RT_\beta \ln a_\beta^{(2)} = (\mu_\beta^{(2)})^0 + RT_\beta \ln(\gamma_\beta^{(2)} m_\beta^{(2)}), \tag{8.56}$$

where $a_\beta^{(2)}$ is the activity of component 2 in terminal part β. We need two composition variables such as $(X^{(2)}, X^{(3)})$ or $(m^{(2)}, m^{(3)})$ to help fix the thermodynamic state of each terminal part.

Equations (8.50) to (8.56) lead to the following theorems:

$$[M_\beta^{(23)}]^\partial = \left(\frac{\partial \ln \theta_\beta^{(2)}}{\partial m_\alpha^{(3)}}\right)_{\alpha(2),T_\beta}, \tag{8.57}$$

$$\begin{aligned}\left(\frac{\partial \ln \theta_\beta^{(2)}}{\partial T_\beta}\right)_{\alpha(2),m_\alpha^{(3)}} &= \frac{\bar{H}_\beta^{(2)}(\text{sat}) - \bar{H}_\beta^{(2)} + \bar{H}_\beta^{(2)} - [H_\beta^{(22)}]^\partial}{RT_\beta^2} \\ &= \frac{\Delta \bar{H}_\beta^{(2)} + [Q_\beta^{(22)}]^\partial}{RT_\beta{}^2},\end{aligned} \tag{8.58}$$

$$[M_\beta^{(23)}] = \frac{1}{m_\alpha^{(3)}} \int_0^{m_\alpha^{(3)}} [M_\beta^{(23)}]^\partial \, dm_\alpha^{(3)}, \tag{8.59}$$

$$\left(\frac{\partial [M_\beta^{(23)}]}{\partial T_\beta}\right)_{\alpha(2),m_\alpha^{(3)}} = \frac{[H_\beta^{(22)}]^\partial - [H_\beta^{(22)}(0)]^\partial}{m_\alpha^{(3)} RT_\beta{}^2}, \tag{8.60}$$

$$-[M_\beta^{(23)}]^\partial = \frac{1}{RT_\beta} \int_{T_\alpha}^{T_\beta} \left(\frac{\partial [S_b^{(22)}]^\partial}{\partial m_\alpha^{(3)}}\right)_{\alpha(2),T_b} dT_b, \tag{8.61}$$

$$[S_\beta^{(22)}]^\partial = RT_\beta [M_\beta^{(23)}]^\partial \left(\frac{\partial m_\alpha^{(3)}}{\partial T_\beta}\right)_{\alpha(2),\mu_\beta^{(2)}}, \tag{8.62}$$

$$[M_\beta^{(23)}]^\partial = \left(\frac{\partial \ln a_\beta^{(2)}}{\partial m_\beta^{(3)}}\right)_{\alpha(2),T_\beta} \left(\frac{\partial m_\beta^{(3)}}{\partial m_\alpha^{(3)}}\right)_{\alpha(2),T_\beta}, \tag{8.63}$$

$$\begin{aligned}\left(\frac{\partial \ln m_\beta^{(2)}}{\partial T_\beta}\right)_{\alpha(2),m_\alpha^{(3)}} &= \frac{\bar{S}_\beta^{(2)} - [S_\beta^{(22)}]^\partial}{RT_\beta} + \left(\frac{\partial \ln \gamma_\beta^{(2)}}{\partial T_\beta}\right)_{m_\beta^{(2)},m_\beta^{(3)}} - \left(\frac{\partial \ln \gamma_\beta^{(2)}}{\partial T_\beta}\right)_{\alpha(2),m_\alpha^{(3)}} \\ &= \frac{[Q_\beta^{(22)}]^\partial}{RT_\beta{}^2} + \Delta\left(\frac{\partial \ln \gamma_\beta^{(2)}}{\partial T_\beta}\right),\end{aligned} \tag{8.64}$$

where $[H]^\partial \equiv \mu + T[S]^\partial$ and $[Q]^\partial \equiv \bar{H} - [H]^\partial$.

We can often evaluate the quantities in the preceding equations by inserting suitable electrodes into the solution in the β terminal part and making emf measurements. By way of example let the sparingly soluble salt be $PbCl_2$; let the quite soluble salt be KNO_3; and let Pb(s) and Ag(s), AgCl(s) electrodes be inserted into the solution at β. We then have at β the cell Pb(s)|solution|AgCl(s), Ag(s) with overall reaction

$$Pb(s) + 2AgCl(s) = 2Ag(s) + PbCl_2(solution)$$

for every 2 Faradays of electricity passing through the cell. For the given cell we have the relations

$$-2FE_\beta = \mu_\beta^{(PbCl_2)} + 2\mu_\beta^{(Ag)} - 2\mu_\beta^{(AgCl)} - \mu_\beta^{(Pb)}, \tag{8.65}$$

$$-2F\left(\frac{\partial E_\beta}{\partial m_\alpha^{(3)}}\right)_{\alpha(2),T_\beta} = RT_\beta[M_\beta^{(23)}]^\partial, \tag{8.66}$$

$$-2F\left(\frac{\partial E_\beta}{\partial T_\beta}\right)_{\alpha(2),m_\alpha^{(3)}} = -[S_\beta^{(22)}]^\partial - 2\bar{S}_\beta^{(Ag)} + 2\bar{S}_\beta^{(AgCl)} + \bar{S}_\beta^{(Pb)}, \tag{8.67}$$

$$-2F\{E_\beta - E_\beta(\text{sat})\} = RT_\beta \ln \theta_\beta^{(2)}, \tag{8.68}$$

$$-2F\{E_\beta - E_\beta(0)\} = RT_\beta m_\alpha^{(3)}[M_\beta^{(23)}], \tag{8.69}$$

where E_β is the emf of the cell at β and F is the Faraday.

The most direct experimental procedure is to keep T_α and T_β fixed, keep the terminal part at α saturated with $PbCl_2$, and measure the emf at β as $m_\alpha^{(3)}$ is varied. Repeat this procedure for the same T_α but a new T_β. From a family of E_β versus $m_\alpha^{(3)}$ plots, analyses of the solutions at α and β for the various $m_\omega^{(i)}$ quantities, and studies of the temperature dependence of E_β(sat), we can evaluate the various defined quantities appearing in the preceding equations. In terms of the various quantities introduced, we see that $[Q_\beta^{(22)}]^\partial \neq 0$ implies that there is a Soret effect; $\theta_\beta^{(2)} \neq 1$ implies that the Soret effect is not due merely to the temperature coefficient of solubility; and $[M_\beta^{(23)}]^\partial \neq 0$ implies that the distribution of the given component in the polythermal field depends on the chemical environment as well as the thermal environment.

We have here an example of another language function of thermodynamics: in addition to its efficient encoding and processing of experimental information, the language of thermodynamics (here and in the last section of Chapter 2, e.g.), sometimes suggests interesting new experiments.

References

1. K. Denbigh and G. Raumann, *Proc. Roy. Soc.* (London), **A210**, 518 (1952).
2. R. J. Bearman, *J. Phys. Chem.*, **61**, 708 (1957).
3. S. Chapman and T. Cowling, *The Mathematical Theory of Non-Uniform Gases*, second edition (Cambridge University Press, Cambridge, 1952).
4. K. E. Grew and T. L. Ibbs, *Thermal Diffusion in Gases* (Cambridge University Press, Cambridge, 1952), Chapters 3 and 4.

5. H. R. Heath, T. L. Ibbs, and N. E. Wild, *Proc. Roy. Soc.* (London), **A178**, 380 (1941).
6. B. E. Atkins, R. E. Bastick, and T. L. Ibbs, *Proc. Roy. Soc.* (London), **A172**, 142 (1939).
7. K. Clusius and G. Dickel, *Z. Phys. Chem.*, **B44**, 397, 451 (1939).
8. M. N. Saha and B. N. Srivastava, *A Treatise on Heat*, fourth edition (The Indian Press Private Ltd., Allahabad, 1958), p. 918.
9. H. J. V. Tyrrell, *Diffusion and Heat Flow in Liquids* (Butterworths, London, 1961), Chapter 4.
10. J. Chipman, *J. Am. Chem. Soc.*, **48**, 2577 (1926).
11. J. Agar, *Trans. Faraday Soc.*, **56**, 776 (1960).
12. J. Agar and J. Turner, *Proc. Roy. Soc.* (London), **A255**, 307 (1960).
13. P. Snowdon and J. Turner, *Trans. Faraday Soc.*, **56**, 1409 (1960).
14. H. S. Harned and B. B. Owen, *The Physical Chemistry of Electrolytic Solutions*, second edition (Reinhold Publishing Company, New York, 1950), p. 340

9

Thermocouples and Thermocells

One of the more interesting and practical of the polythermal field effects is the Seebeck effect: the state of migrational equilibrium with respect to the charge carriers in a thermocouple is accompanied by an electrical potential difference (the Seebeck potential difference) that depends only on the temperatures of the bi-material junctions. The situation to be analyzed is indicated schematically in Figure 9–1; the electric engine e at γ is a device for converting electric current to mechanical work (a d-c motor used to raise a weight, say).

Thermocouples

In Figure 9–1 let I be the negative electric current flowing through the system, let the electric engine e be such that the heat effects at γ are purely dissipative heat effects—i.e., $(\delta \dot{Q}_\gamma^{(r)}/\delta I)_T = 0$—and let a sign convention for the direction of the electric current flow be established. Now when a steady current I flows through the system, the entire system is stationary and

$$\dot{U}(\text{system}) = -\dot{Q}_\alpha^{(r)} - \dot{Q}_\beta^{(r)} - \dot{Q}_\gamma^{(r)} - \sum_{i=1}^{6} \dot{Q}_i^{(r)} - \sum \dot{q}(ij) + \dot{W}_0 = 0, \quad (9.1)$$

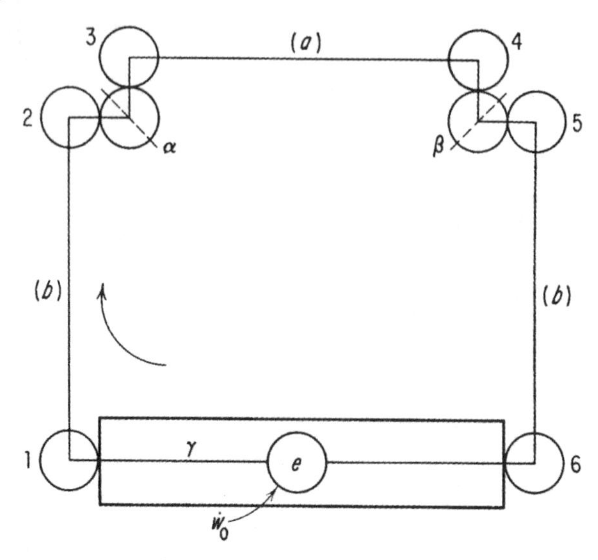

Figure 9–1. Schematic version of a thermocouple: e—electric engine at temperature T; 1, γ, 6—heat reservoirs at temperature T; 2, α, 3—heat reservoirs at temperature T_α; 4, β, 5—heat reservoirs at temperature T_β; wire segments e–1–2–α and β–5–6–e of material (b); wire segment α–3–4–β of material (a); junctions between materials (a) and (b) occur in reservoirs α and β; $\dot{W}_0$—rate at which work is supplied to the electric engine e in order to make negative current I flow in the direction indicated. For convenience let all current-carrying elements of the engine e be made of material (b).

$$\begin{aligned}\Theta &= \frac{\dot{Q}_\alpha^{(r)}}{T_\alpha} + \frac{\dot{Q}_\beta^{(r)}}{T_\beta} + \frac{\dot{Q}_\gamma^{(r)}}{T} + \sum_{i=1}^{6} \frac{\dot{Q}_i^{(r)}}{T_i} + \sum \frac{\dot{q}(ij)}{\langle T(ij)\rangle} \\ &= \frac{\dot{W}_0}{T} + \dot{Q}_\alpha^{(r)}\left(\frac{1}{T_\alpha} - \frac{1}{T}\right) + \dot{Q}_\beta^{(r)}\left(\frac{1}{T_\beta} - \frac{1}{T}\right) + \sum_{i=2}^{5} \dot{Q}_i^{(r)}\left(\frac{1}{T_i} - \frac{1}{T}\right) \\ &\qquad + \sum \dot{q}(ij)\left(\frac{1}{\langle T(ij)\rangle} - \frac{1}{T}\right) \\ &= I\left(\frac{\dot{W}_0}{IT}\right) + \dot{Q}_\alpha^{(r)}\left(\frac{1}{T_\alpha} - \frac{1}{T}\right) + \dot{Q}_\beta^{(r)}\left(\frac{1}{T_\beta} - \frac{1}{T}\right) + \sum_{i=2}^{5} \dot{Q}_i^{(r)}\left(\frac{1}{T_i} - \frac{1}{T}\right) \\ &\qquad + \sum \dot{q}(ij)\left(\frac{1}{\langle T(ij)\rangle} - \frac{1}{T}\right) \\ &= \sum Y_k \Omega_k, \end{aligned} \tag{9.2}$$

where the quantities $\dot{q}(ij)$ are the rates of lateral heat leakage from the links 12, 34, 56 to the surroundings. We can write Eq. (9.1) as

$$-\dot{W}_0 = -\dot{Q}_\alpha^{(r)} - \dot{Q}_\beta^{(r)} - \dot{Q}_\gamma^{(r)} - \sum \dot{Q}_i^{(r)} - \sum \dot{q}(ij), \tag{9.3}$$

and it follows that

$$\left(\frac{\delta(-\dot{W}_0)}{\bar{\delta}I}\right)_{T_\alpha,T_\beta,T,\langle T(ij)\rangle} = -\left(\frac{\delta\dot{Q}_\alpha^{(r)}}{\bar{\delta}I}\right)_{T_\alpha,\dots} - \left(\frac{\delta\dot{Q}_\beta^{(r)}}{\bar{\delta}I}\right)_{T_\alpha,\dots} - \sum\left(\frac{\delta\dot{Q}_i^{(r)}}{\bar{\delta}I}\right)_{T_\alpha,\dots} - \sum\left(\frac{\delta\dot{q}(ij)}{\bar{\delta}I}\right)_{T_\alpha,\dots}. \tag{9.4}$$

From Eqs. (6.38) and (9.2) we see that the operation $(\delta/\bar{\delta}I)_{T_\alpha,\dots}$ is of the form $(\delta/\bar{\delta}Y_{0x})_{\Omega'}$. We can also express Eq. (9.4) as

$$\begin{aligned}\Delta\psi &= \int_T^{T_\alpha}[c^{(b)}]\,dT + \pi_\alpha^{(ba)} + \int_{T_\alpha}^{T_\beta}[c^{(a)}]\,dT + \pi_\beta^{(ab)} + \int_{T_\beta}^{T}[c^{(b)}]\,dT \\ &= \int_{T_\alpha}^{T_\beta}\{[c^{(a)}] - [c^{(b)}]\}\,dT + \pi_\alpha^{(ba)} + \pi_\beta^{(ab)},\end{aligned} \tag{9.5}$$

where $\Delta\psi \equiv (\delta(-\dot{W}_0)/\bar{\delta}I)_{T_\alpha,\dots}$ is the Seebeck potential difference of the thermocouple,* $\pi_\omega^{(ba)} \equiv -(\delta\dot{Q}_\omega^{(r)}/\bar{\delta}I)_{T_\alpha,\dots}$ is the Peltier heat absorbed at the bimetallic junction as the current goes from material b to material a,† and I have made use of the assumed properties of the engine e $[(\delta\dot{Q}_\gamma^{(r)}/\bar{\delta}I)_T = 0]$ and of the definition, e.g.,‡

$$\left(\frac{\delta(\dot{Q}_1^{(r)} + \dot{Q}_2^{(r)} + \dot{q}(12))}{\bar{\delta}I}\right)_{T_1,T_2,\langle T(12)\rangle} \equiv -\int_{T_1}^{T_2}[c^{(b)}]\,dT \tag{9.6}$$

for the Thomson coefficient $[c]$ (defined in terms of the flow of negative electricity).

We see from Eq. (9.5) that

$$\begin{aligned}\left(\frac{\partial\Delta\psi}{\partial T_\beta}\right)_{T_\alpha} &= [c_\beta^{(a)}] - [c_\beta^{(b)}] + \left(\frac{\partial\pi_\beta^{(ab)}}{\partial T_\beta}\right)_{T_\alpha} \\ &= [c_\beta^{(a)}] - [c_\beta^{(b)}] + \frac{\pi_\beta^{(ab)}}{T_\beta} + T_\beta\frac{d(\pi_\beta^{(ab)}/T_\beta)}{dT_\beta},\end{aligned} \tag{9.7}$$

where for convenience I have written $\pi_\beta^{(ab)} = \pi_\beta^{(ab)}(T_\beta, a, b)$ as $T_\beta(\pi_\beta^{(ab)}/T_\beta)$. The application of assumption IV to the present situation leads to

$$-\left(\frac{\delta\Theta}{\bar{\delta}I}\right)_{T_\alpha,\dots} = \int_{T_\alpha}^{T_\beta}\frac{[c^{(a)}] - [c^{(b)}]}{T}\,dT + \frac{\pi_\alpha^{(ba)}}{T_\alpha} + \frac{\pi_\beta^{(ab)}}{T_\beta} = 0 \tag{9.8}$$

and, by the application of the operation $(\partial/\partial T_\beta)_{T_\alpha}$ to Eq. (9.8), to

$$\frac{[c_\beta^{(a)}] - [c_\beta^{(b)}]}{T_\beta} + \frac{d(\pi_\beta^{(ab)}/T_\beta)}{dT_\beta} = 0. \tag{9.9}$$

* Depending on the sign convention for the direction of electric current flow, we have $\Delta\psi = \pm(\psi' - \psi'')$ with ψ' and ψ'' being the electrical potentials of the binding posts of the electric engine e. See the note in Chapter 4.

† The Peltier heat has the property that $\pi_\omega^{(ba)} = -\pi_\omega^{(ab)}$.

‡ See Eq. (6.17).

Equations (9.7) and (9.9) lead to the Thomson relations [1] for the thermocouple:

$$\left(\frac{\partial \Delta\psi}{\partial T_\beta}\right)_{T_\alpha} = \frac{\pi_\beta^{(ab)}}{T_\beta}, \tag{9.10}$$

$$T_\beta\left(\frac{\partial^2 \Delta\psi}{\partial T_\beta{}^2}\right)_{T_\alpha} = [c_\beta^{(b)}] - [c_\beta^{(a)}], \tag{9.11}$$

$$\left(\frac{\partial \Delta\psi}{\partial T}\right)_{T_\alpha, T_\beta} = 0. \tag{9.12}$$

Thus the Thomson relations (Eqs. 9.10 and 9.11) depend only on the applicability of assumption IV; if IV is satisfied, the Thomson relations must hold whether we deal with metallic conductors or with semiconductors. Indeed, the wide applicability [2] of the Thomson relations is evidence for the general acceptance of assumption IV.* The treatment here is essentially that of Thomson [1] cast into a physically more reasonable (assumption IV) form.†

Inasmuch as the form of Eq. (9.8) is not immediately obvious, I here give a derivation. Consider a loop of a single, homogeneous piece of wire arranged as in Figure 9–2. Let the 12 link be of the partial-flux type, and let the 34 link be of the zero-flux type; i.e., an infinite number of auxiliary thermal reservoirs are stationed along the 34 link and exchange heat with the link in such a way as to maintain the temperature distribution in the 34 link in the $I = 0$ configuration even when $I \neq 0$. The total heat exchanged between the auxiliary reservoirs and the 34 link per unit current in the limit of vanishing current is just $(\delta\dot{q}(34)/\delta I)_{T_\gamma, T_\delta} = -\int_{T_\delta}^{T_\gamma} [c]\, dT$. The application of assumption IV to this situation leads to

$$\left(\frac{\delta \Theta}{\delta I}\right)_{T_1,\ldots} = \left(\frac{\delta(\dot{S}_1^{(r)} + \dot{S}_2^{(r)} + \dot{s}(12))}{\delta I}\right)_{T_1,\ldots} - \int_{T_\delta}^{T_\gamma} \frac{[c]}{T}\, dT = 0, \tag{9.13}$$

* Lancia and McGervey [3] have questioned the validity of the Thomson relations for inhomogeneous materials. They measured Peltier heats in a Bunsen-type ice calorimeter and made Seebeck measurements by using resistance heating of the thermocouple junctions; they used currents of several tenths of an ampere in these measurements. For thermocouples made from bismuth–tellurium alloys and copper wire, they found that the percentage difference, $\{(\partial\, \Delta\psi/\partial T - \pi/T)/\partial\, \Delta\psi/\partial T\} \times 100$, ranged from 5–17% for their Bi–Te alloys. The alloys showed a 5–10% variation in resistivity over a 30-cm length, and Lancia and McGervey attribute the apparent breakdown of the Thomson relations to the inhomogeneity of the samples.

I feel that the significance of the results of Lancia and McGervey would have been greater had they made measurements at various current levels and had they systematically extrapolated the results to zero current. As it is, the rather large currents used in the inhomogeneous medium somewhat cloud the issue: we cannot be sure that the deviations from the Thomson relations do not vanish in the limit of zero current.

† Remember that assumption IV is a necessary and sufficient condition for the existence of Onsager reciprocal relations in the linear current-affinity region; see Chapter 3.

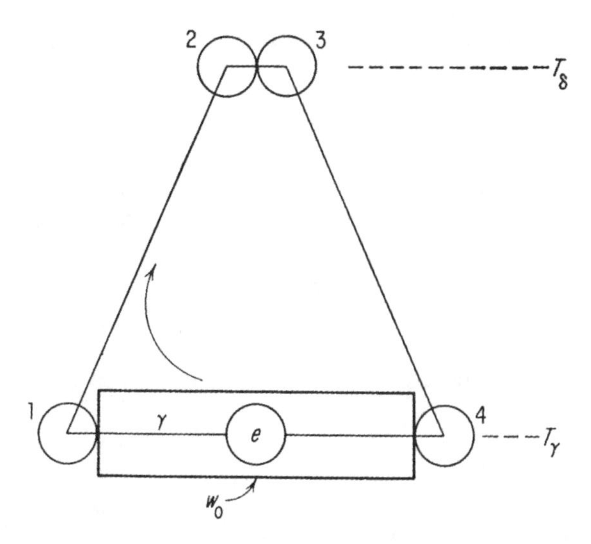

Figure 9–2. A single, homogeneous loop of wire attached to an electric engine e: 1, γ, 4—heat reservoirs at temperature T_γ; 2,3—heat reservoirs at temperature T_δ; $\dot{W}_0$—rate at which work is supplied to the electric engine so as to make negative current I flow in the direction indicated. Let all current-carrying elements of the engine e be made of the same wire as the loop.

where $\dot{s}(12)$ is the rate of gain of entropy in the surroundings of the 12 link associated with the heat quantity $\dot{q}(12)$, and the integral represents the entropy gain of the auxiliary (Thomson) reservoirs per unit current in the limit of vanishing current. Equation (9.13) shows that for any arbitrary ij link (current flowing from i to j) we have*

$$-\left(\frac{\delta(\dot{S}_i^{(r)} + \dot{S}_j^{(r)} + \dot{s}(ij))}{\delta I}\right)_{T_i,\ldots} = \int_{T_i}^{T_j} \frac{[c]}{T}\, dT \tag{9.14}$$

regardless of the nature (full, partial, or zero flux) of the linkage. For the case analyzed in Figure 9–1 we have

$$-\Theta = -\frac{\dot{Q}_\alpha^{(r)}}{T_\alpha} - \frac{\dot{Q}_\beta^{(r)}}{T_\beta} - \frac{\dot{Q}_\gamma^{(r)}}{T} - \sum \dot{S}_i^{(r)} - \sum \dot{s}(ij); \tag{9.15}$$

if we apply the operation $(\delta/\delta I)_{T_\alpha,\ldots}$ to Eq. (9.15), and if we make use of assumption IV, of the definition of $\pi_\omega^{(ab)}$, and of Eq. (9.14), we arrive at Eq. (9.8).

* Note the *order*, with respect to the direction of current flow, of the limits of integration in Eqs. (9.13) and (9.14).

Electrons in Homogeneous Isotropic Metallic Wires

If we turn to thermocouples composed of homogeneous isotropic metallic wires and if we assume that electrons are the only charge carriers, we can then carry out an analysis of the situation in terms of the electrochemical potential $\bar{\mu}$ of the electron and can arrive at some surprising results. Let $\bar{Z}$ represent a thermodynamic property associated with 1 mole of electrons in a given medium and let $\mathscr{Z} \equiv \bar{Z} - F\psi$, where F is the Faraday and ψ is the electric potential for electrons in the given medium; then

$$\mathscr{G} \equiv \bar{\mu} \equiv \mu - F\psi \; (\mu \equiv \bar{G} = \bar{U} + P\bar{V} - T\bar{S}).$$

In Figure 9–1 let materials (a) and (b) be homogeneous isotropic metallic wires and set

$$\Delta\psi = \psi_1 - \psi_6; \tag{9.16}$$

then

$$F\Delta\psi = \bar{\mu}_6 - \bar{\mu}_1 = \sum_{i=1}^{6} (-1)^i \bar{\mu}_i, \tag{9.17}$$

since [4] $\bar{\mu}_2 = \bar{\mu}_3$ and $\bar{\mu}_4 = \bar{\mu}_5$. We can now write Eq. (9.5) as

$$\sum_{i=1}^{6} (-1)^i \bar{\mu}_i = F\Delta\psi = F\left\{\int_{T}^{T_\alpha} [c^{(b)}]\, dT + \int_{T_\alpha}^{T_\beta} [c^{(a)}]\, dT + \int_{T_\beta}^{T} [c^{(b)}]\, dT \right.$$
$$\left. + T_\alpha\left(\frac{\pi_\alpha^{(ba)}}{T_\alpha}\right) + T_\beta\left(\frac{\pi_\beta^{(ab)}}{T_\beta}\right)\right\}. \tag{9.18}$$

At this point a digression is necessary in order to show that $F\pi_\omega^{(ab)}/T_\omega = \bar{S}_\omega^{(b)} - \bar{S}_\omega^{(a)}$. Consider the situation indicated in Figure 9–3. If the entropy change associated with the chemical reaction in the cell per Faraday of charge transferred around the loop is written as ΔS(reaction), then the total entropy change associated with the cell ΔS(cell) is [5]

$$\Delta S(\text{cell}) = \Delta S(\text{reaction}) + \bar{S}^{(-)} - \bar{S}^{(+)}. \tag{9.19}$$

For the reversible transfer of charge around the loop (limit of a sequence of steady currents), we have

$$\Delta S(\text{system}) + \Delta S(\text{surroundings}) = 0, \tag{9.20}$$

with

$$\Delta S(\text{system}) = \Delta S(\text{reaction}) \tag{9.21}$$

and

$$\Delta S(\text{surroundings}) = -\Delta S(\text{cell}) - \frac{F}{T}(\pi^{(-,b)} + \pi^{(b,a)} + \pi^{(a,+)}). \tag{9.22}$$

Therefore,

$$\bar{S}^{(+)} - \bar{S}^{(-)} = \frac{F}{T}(\pi^{(-,b)} + \pi^{(b,a)} + \pi^{(a,+)}). \tag{9.23}$$

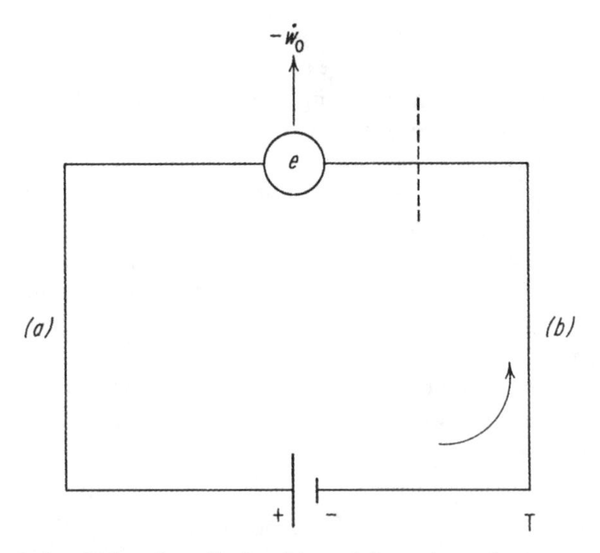

Figure 9–3. Galvanic cell plus bimetal loop in a thermostat. The anode (−) adjoins to metal (*b*); the cathode (+) adjoins to metal (*a*); the junction of metals (*a*) and (*b*) occurs at the point marked by the dotted line; there is an electric engine at *e*—for convenience let all current-carrying elements of the engine *e* be of material (*a*). When negative current flows in the direction indicated by the arrow, the engine at *e* delivers work at the rate $-\dot{W}_0$ to the surroundings.

A sufficient condition for the validity of Eq. (9.23) is that

$$\frac{F\pi^{(x,y)}}{T} = \bar{S}^{(y)} - \bar{S}^{(x)}. \tag{9.24}$$

Suppose instead that

$$\frac{F\pi^{(x,y)}}{T} = \bar{S}^{(y)} + f(y, T) - \{\bar{S}^{(x)} + f(x, T)\}; \tag{9.25}$$

then

$$\frac{F}{T}(\pi^{(-,b)} + \pi^{(b,a)} + \pi^{(a,+)}) = \bar{S}^{(+)} - \bar{S}^{(-)} + f(+, T) - f(-, T). \tag{9.26}$$

For Eq. (9.26) to be in accord with Eq. (9.23) it is necessary that

$$f(+, T) = f(-, T). \tag{9.27}$$

Inasmuch as Eq. (9.27) must hold regardless of the nature of the anode and cathode materials, it seems to be required that in general

$$f(z, T) = f(T). \tag{9.28}$$

But if $f(z, T) = f(T)$ only, then Eq. (9.25) reduces to Eq. (9.24).

Thus it seems that the normal situation for isotropic metals requires that $F\pi/T = \Delta\bar{S}$. Some doubt has been expressed concerning this result because the Peltier effect in experiments using anisotropic single metal crystals was found to depend on the orientation of the crystal relative to the second metal [6]. Bridgman [7] has shown that an internal heat effect is to be expected in anisotropic metal crystals whenever the electric current changes direction with respect to a principal axis of the crystal. These observations merely show that a very careful analysis is needed to establish just what is meant by a partial molar property $\bar{Z}$ in an anisotropic medium (operational definition) and just how current flow in a given direction is related to $\bar{Z}$; such an analysis has yet to be carried out.

The digression now being over, consider once again Eq. (9.18). Since $\bar{\mu} \equiv \mathscr{H} - T\bar{S}$, we can rewrite Eq. (9.18) (making use of Eq. 9.24 and of the fact that $\bar{S}_1 = \bar{S}_6$) as

$$\sum_{i=1}^{6} (-1)^i \mathscr{H}_i = F\left\{\int_T^{T_\alpha} [c^{(b)}]\, dT + \int_{T_\alpha}^{T_\beta} [c^{(a)}]\, dT + \int_{T_\beta}^{T} [c^{(b)}]\, dT\right\}. \quad (9.29)$$

Now for the ij link

$$\mathscr{H}_j - \mathscr{H}_i = \int_{T_i}^{T_j} \frac{d\mathscr{H}}{dT}\, dT; \quad (9.30)$$

so it follows that

$$\frac{d\mathscr{H}}{dT} = F[c]. \quad (9.31)$$

ZERO-FLUX LINKAGE

If we take each of the ij links in Figure 9–1 to be of the zero-flux type, i.e., $(\delta\dot{q}(ij)/\delta I)_{T_i,\ldots} = -\int_{T_i}^{T_j} [c]\, dT$, then our application of assumption IV to this case yields

$$-\left(\frac{\delta\Theta}{\delta\dot{n}}\right)_{T,\ldots} = F\left\{\frac{\pi_\alpha^{(ba)}}{T_\alpha} + \frac{\pi_\beta^{(ab)}}{T_\beta} + \int_T^{T_\alpha} [c^{(b)}]\, d\ln T + \int_{T_\alpha}^{T_\beta} [c^{(a)}]\, d\ln T + \int_{T_\beta}^{T} [c^{(b)}]\, d\ln T\right\} = 0, \quad (9.32)$$

where $\dot{n}$ is the number of Faradays traversing the thermocouple in unit time. Since $F\pi/T = \Delta\bar{S}$, $\bar{S}_1 = \bar{S}_6$, and, e.g., $\bar{S}_2 = \bar{S}_\alpha^{(b)}$, we can also write Eq. (9.32) as

$$\left\{\bar{S}_1 - \bar{S}_2 + F\int_T^{T_\alpha} [c^{(b)}]\, d\ln T\right\} + \left\{\bar{S}_3 - \bar{S}_4 + F\int_{T_\alpha}^{T_\beta} [c^{(a)}]\, d\ln T\right\} + \left\{\bar{S}_5 - \bar{S}_6 + F\int_{T_\beta}^{T} [c^{(b)}]\, d\ln T\right\} = 0, \quad (9.33)$$

each bracketed set of terms applying to a single link. The separate ij links being independent of one another, Eq. (9.33) requires that for each ij link

$$\bar{S}_i - \bar{S}_j + F\int_{T_i}^{T_j}[c]\,d\ln T = 0. \tag{9.34}$$

Now for electrons in wires it is customary to assume that the entropy is independent of the electrical potential; i.e., $\bar{S} \neq \bar{S}(\psi)$. Thus

$$\bar{S}_i - \bar{S}_j = \int_{T_j}^{T_i}\bar{C}_P\,d\ln T, \tag{9.35}$$

where $\bar{C}_P \equiv d\bar{H}/dT$ is the ordinary heat capacity associated with 1 mole of electrons in the given medium. Equations (9.31), (9.34), (9.35), and the definition $\mathscr{H} \equiv \bar{H} - F\psi$ lead to the relation

$$-\int_{T_i}^{T_j}\bar{C}_P\,d\ln T + \int_{T_i}^{T_j}\left(\bar{C}_P - F\frac{d\psi}{dT}\right)d\ln T = 0, \tag{9.36}$$

and hence to

$$\int_{T_i}^{T_j}\frac{d\psi}{dT}\,d\ln T = 0. \tag{9.37}$$

Thus assumption IV applied to zero-flux linkages leads to the conclusion that a *homogeneous* thermoelectric effect does not exist for electrons in wires; i.e., $d\psi/dT = 0$ and $\psi_j = \psi_i$ even though $T_j \neq T_i$. This conclusion immediately leads to the relations

$$F\Delta\psi = \mu_2 - \mu_3 + \mu_4 - \mu_5 = \mu_\alpha^{(b)} - \mu_\alpha^{(a)} + \mu_\beta^{(a)} - \mu_\beta^{(b)}, \tag{9.38}$$

$$F\left(\frac{\partial\,\Delta\psi}{\partial T_\beta}\right)_{T_\alpha} = \bar{S}_\beta^{(b)} - \bar{S}_\beta^{(a)} = \frac{F\pi_\beta^{(ab)}}{T_\beta}, \tag{9.39}$$

$$F[c] = \bar{C}_P, \tag{9.40}$$

$$FT_\beta\left(\frac{\partial^2\,\Delta\psi}{\partial T_\beta^{\,2}}\right)_{T_\alpha} = \bar{C}_P^{(b)}(\beta) - \bar{C}_P^{(a)}(\beta) = F\{[c_\beta^{(b)}] - [c_\beta^{(a)}]\}. \tag{9.41}$$

The very same relations—Eqs. (9.38) to (9.41)—appear if we take the linkages to be of the full-flux type.

FULL-FLUX LINKAGES

Equation (6.39) when applied to the situation indicated in Figure 9–1 with full-flux linkages yields the relation

$$-F\Delta\psi = -\sum_{i=1}^{6}(-1)^i\bar{\mu}_i = T\{[S_6] - [S_1]\} + T_\alpha\{[S_2] - [S_3]\} + T_\beta\{[S_4] - [S_5]\}. \tag{9.42}$$

We can rearrange Eq. (9.42) to give

$$\sum_{i=1}^{6}(-1)^i[\mathscr{H}_i] = 0, \tag{9.43}$$

with $[\mathscr{H}_i] \equiv \bar{\mu}_i + T_i[S_i]$. Equation (9.43) together with the independence of the links with respect to one another leads to

$$[\mathscr{H}_i] - [\mathscr{H}_j] = 0 \tag{9.44}$$

for each full-flux link in the system. Assumption IV implies that

$$[S_i] = [S_j] \tag{9.45}$$

for each full-flux link. Now the definition (for the ij full-flux link)

$$[S_i]^{\partial} \equiv -\left(\frac{\partial \bar{\mu}_i}{\partial T_i}\right)_j, \tag{9.46}$$

where the subscript j means that the thermodynamic state at j is to be kept constant (i.e., constant T_j, ψ_j, etc.), leads to

$$F\left(\frac{\partial \psi_i}{\partial T_i}\right)_j = [S_i]^{\partial} - \bar{S}_i \quad \text{(const. } P\text{)}. \tag{9.47}$$

Equation (9.47) shows clearly that if $[S_i]^{\partial} - \bar{S}_i = 0$ for a full-flux link then again there is no homogeneous thermoelectric effect; i.e., $(\partial\psi_i/\partial T_i)_j = 0$ and $\psi_i = \psi_j$. We see from Chapter 6 that Eq. (9.45) implies that

$$[S_i]^{\partial} = [S_i(i)]; \tag{9.48}$$

so if it turns out that $[S_i(i)] = \bar{S}_i$ for a full-flux link, then the nonexistence of the homogeneous thermoelectric effect will be established.

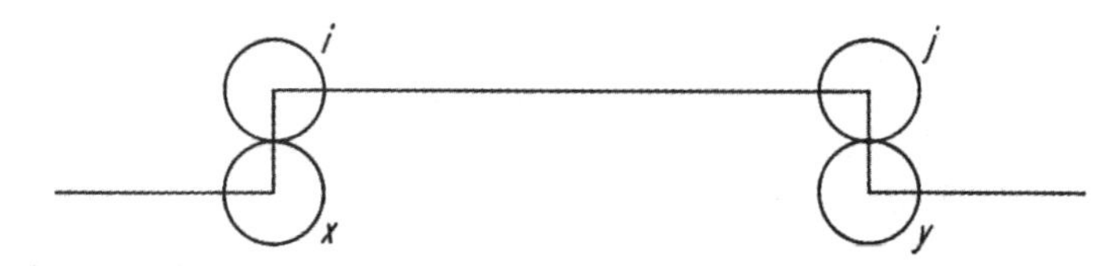

Figure 9–4. Full-flux linkage under equithermal conditions. The thermostats at i, j, x, y are all at the same temperature T; one single, homogeneous piece of metallic wire.

Consider the situation indicated in Figure 9–4. When a steady current of $\dot{n}$ Faradays per second flows through the ij link, the heat developed per second in the link is given by $R_{ij}(\dot{n}F)^2$, where R_{ij} is the resistance of the ij segment of wire (the part outside the thermostats). Since the endpoints of the ij segment are at the same temperature, the Thomson effects in the segment are self-canceling (the Thomson effect, however, introduces a certain asymmetry in the distribution of temperatures with respect to the midpoint of the segment; this asymmetry has been used by Keesom et al. [8] as a way of measuring Thomson coefficients). A certain fraction ϑ_i of the heat developed in the ij segment will find its way into the ith thermostat; thus

$$\dot{S}_i^{(r)} = \frac{1}{T}(R_i + \vartheta_i R_{ij})(\dot{n}F)^2, \tag{9.49}$$

where R_i is the resistance of the wire in the ith thermostat and $0 < \vartheta_i < 1$, and

$$\left(\frac{\delta \dot{S}_i^{(r)}}{\delta \dot{n}}\right)_T \equiv \{\bar{S}_i - [S_i(i)]\} = 0. \tag{9.50}$$

Equations (9.47), (9.48), and (9.50) show that a homogeneous thermoelectric effect does not exist across a full-flux linkage if assumption IV holds; Eqs. (9.38) to (9.41) follow in this case just as in the zero-flux case.

General Remarks

The relations derived for the thermocouple in this chapter hinge on the applicability of assumption IV; if IV is valid, then the Thomson relations (Eqs. 9.10 and 9.11) are generally true. For electrons in isotropic metallic wires there is no homogeneous thermoelectric effect, the Seebeck potential difference being due entirely to the temperature coefficient of the interfacial equilibrium condition $\psi_\omega^{(a)} - \psi_\omega^{(b)} = (\mu_\omega^{(a)} - \mu_\omega^{(b)})/F$; i.e., $\bar{\mu}_\omega^{(a)} = \bar{\mu}_\omega^{(b)}$—Eqs. (9.38) to (9.41) following at once.

Exercise 9–1. Liquid–Vapor Analog of the Thermocouple

Consider the situation indicated in Figure 9–5: two liquid-vapor interfaces situated in separate thermostats communicate with opposite sides of a manometer housed in yet another thermsotat. Show that the pressure difference ΔP registered by the manometer at γ is

$$\Delta P = P_0(\alpha) + \int_{T_\alpha}^{T} \frac{\bar{S}^{(g)} - [S^{(g)}]^\partial}{\bar{V}^{(g)}}\,dT - P_0(\beta) - \int_{T_\beta}^{T} \frac{\bar{S}^{(g)} - [S^{(g)}]^\partial}{\bar{V}^{(g)}}\,dT,$$

where $P_0(\omega)$ is the saturated vapor pressure of the liquid at temperature T_ω and the integrals are over the appropriate communicating tubes. Also show that the condition $(\partial\,\Delta P/\partial T)_{T_\alpha,T_\beta} = 0$ implies that $\bar{S}^{(g)} - [S^{(g)}]^\partial = 0$, and hence that

$$\Delta P = P_0(\alpha) - P_0(\beta),$$

$$\left(\frac{\partial\,\Delta P}{\partial T_\alpha}\right)_{T_\beta} = \frac{\Delta\bar{S}(\alpha)}{\Delta\bar{V}(\alpha)}.$$

Exercise 9–2. Simplify the situation shown in Figure 9–1 in the following ways: connect thermostats 2, α, 3 together to form a single thermostat

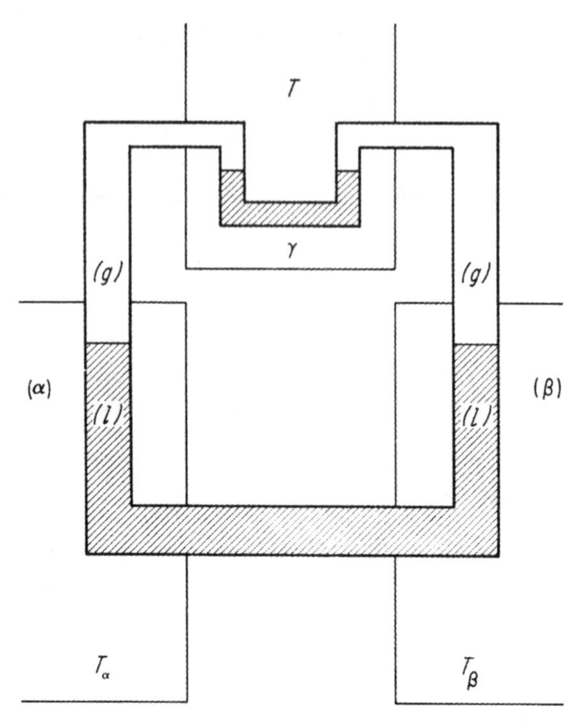

Figure 9–5. Liquid–vapor analog of the thermocouple; arbitrary linkage.

A ($\dot{Q}_A^{(r)} = \dot{Q}_2^{(r)} + \dot{Q}_\alpha^{(r)} + \dot{Q}_3^{(r)}$) and do a similar thing for thermostats 4, β, 5 ($\dot{Q}_B^{(r)} = \dot{Q}_4^{(r)} + \dot{Q}_\beta^{(r)} + \dot{Q}_5^{(r)}$); let $\langle T(ij)\rangle = T = T_R$ with T_R being room temperature, and take state R as a reference thermostatic state.

(i) Show that Eq. (9.2) then simplifies to

$$\Theta = I\left(\frac{\dot{W}_0}{IT_R}\right) + \dot{Q}_A^{(r)}\left(\frac{1}{T_A} - \frac{1}{T_R}\right) + \dot{Q}_B^{(r)}\left(\frac{1}{T_B} - \frac{1}{T_R}\right).$$

(ii) Show that the relations $(\partial \Omega_i/\partial Y_i)_{\Omega',R} > 0$ imply that

$$\left(\frac{\partial I}{\partial(\dot{W}_0/I)}\right)_{T_R,T_B,T_A} > 0, \qquad \left(\frac{\partial \dot{Q}_A^{(r)}}{\partial T_A}\right)_{T_R,T_B,\dot{W}_0/I} < 0, \qquad \text{etc.}$$

(iii) Show that the relations $(\partial Y_i/\partial \Omega_k)_{\Omega',R} = (\partial Y_k/\partial \Omega_i)_{\Omega',R}$ imply that

$$-T_A^2\left(\frac{\partial I}{\partial T_A}\right)_{T_R,T_B,\dot{W}_0/I} = T_R\left(\frac{\partial \dot{Q}_A^{(r)}}{\partial(\dot{W}_0/I)}\right)_{T_R,T_B,T_A},$$

$$T_B^2\left(\frac{\partial \dot{Q}_A^{(r)}}{\partial T_B}\right)_{T_R,T_A,\dot{W}_0/I} = T_A^2\left(\frac{\partial \dot{Q}_B^{(r)}}{\partial T_A}\right)_{T_R,T_B,\dot{W}_0/I}, \qquad \text{etc.}$$

(iv) In a similar fashion, work out the consequences of the assumption that

$$\left(\frac{\partial \Omega_i}{\partial Y_k}\right)_{Y',R} = \left(\frac{\partial \Omega_k}{\partial Y_i}\right)_{Y',R}.$$

Thermocells

I consider next thermocells consisting of molten metal halides or nitrates of the general form MX_ν or aqueous solutions of such halides or nitrates (yielding the ions $M^{\nu+}$ and X^-) in contact with electrodes of metal M; the formalism appropriate to the two cases is exactly the same. I find that thermocells can be described rigorously either with or without the use of current ratios (transference numbers).

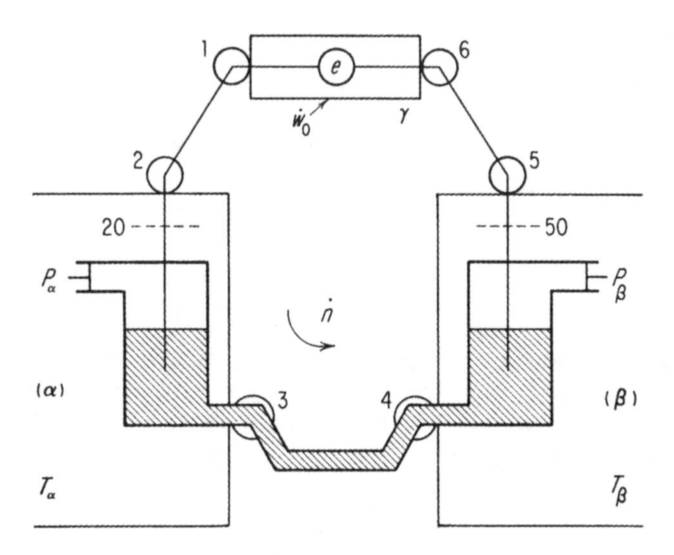

Figure 9–6. Schematic version of a thermocell. 1, γ, 6—heat reservoirs at temperature T; 2, α, 3—heat reservoirs at temperature T_α; 4, β, 5—heat reservoirs at temperature T_β. Wire of metal M dips into the molten salt or solution at α and β and is joined to copper wires at the points 20 and 50; the copper wires attach to an electric engine e at temperature T; all current-carrying elements of the engine e are to be of copper. When work is supplied to the engine e at the rate $\dot{W}_0$, a current of $\dot{n}$ Faradays/sec flows in the direction indicated. The pressures P_ω are solvent partial pressures for the aqueous case and atmospheric pressure for the molten salt case.

Consider the case shown schematically in Figure 9–6. The equation setting out the migrational equilibrium condition for this situation is Eq. (6.39); I rewrite it here for convenience:

$$\sum_j \sum_\rho \sum_i \mathscr{G}_\rho^{(i)} \left(\frac{\delta \dot{n}_\rho^{(i)}}{\delta \dot{n}}\right)_{\Omega'} + \sum_k T_k \left\{ \left(\frac{\delta \dot{S}_k(\text{in})}{\delta \dot{n}}\right)_{\Omega'} - \left(\frac{\delta \dot{S}_k(\text{out})}{\delta \dot{n}}\right)_{\Omega'} \right\}$$
$$+ \sum \left(\frac{\delta \dot{q}(kh)}{\delta \dot{n}}\right)_{\Omega'} = \left(\frac{\delta \dot{W}_0}{\delta \dot{n}}\right)_{\Omega'} = -F\Delta\psi, \quad (9.51)$$

where $\Delta\psi \equiv -(\delta \dot{W}_0/\delta \dot{n})_{\Omega'}/F$ is the potential difference of the thermocell. Let the superscripts *el*, $+$, and $-$ refer to electrons in the copper wires, to the cation $M^{\nu+}$, and to the anion X^-, respectively. Upon introducing the current ratios τ_+, and τ_-,

$$\tau_+ \equiv \left|\frac{\nu \dot{n}_\omega^{(+)}(\text{in})}{\dot{n}}\right|, \qquad \tau_- \equiv \left|\frac{\dot{n}_\omega^{(-)}(\text{in})}{\dot{n}}\right|, \qquad \tau_+ + \tau_- = 1, \qquad (9.52)$$

we can write Eq. (9.51) in the form

$$\begin{aligned} -F\Delta\psi = {} & \Delta G_\alpha + \Delta G_\beta + T\{[S_6^{(el)}] - [S_1^{(el)}]\} \\ & + T_\alpha \left\{ [S_2^{(el)}] + \left(\frac{\delta \dot{n}_\alpha^{(L)}}{\delta \dot{n}}\right)_{\Omega'} [S_\alpha^{(1)}] + \frac{\tau_+}{\nu}[S_\alpha^{(+)}] - \tau_-[S_\alpha^{(-)}] \right\} \\ & + T_\beta \left\{ \tau_-[S_\beta^{(-)}] - \frac{\tau_+}{\nu}[S_\beta^{(+)}] + \left(\frac{\delta \dot{n}_\beta^{(L)}}{\delta \dot{n}}\right)_{\Omega'} [S_\beta^{(1)}] - [S_5^{(el)}] \right\} \\ & + [h(12)] + [h(34)] + [h(56)], \end{aligned} \qquad (9.53)$$

where $\Delta G_\omega \equiv \{\delta(\dot{n}_\omega^{(g)}\mu_\omega^{(g)} + \dot{n}_\omega^{(M)}\mu_\omega^{(M)} + \dot{n}_\omega^{(1)}\mu_\omega^{(1)} + \dot{n}_\omega^{(MX_\nu)}\mu_\omega^{(MX_\nu)})/\delta \dot{n}\}_{\Omega'}$ — the superscripts referring to the gas phase, the metal phase, the solvent component in the 2-component case, and the metal halide or nitrate component, respectively*— and $\dot{n}_\omega^{(L)} \equiv \dot{n}_\omega^{(1)} + \dot{n}_\omega^{(g)}$. In the sequence of steady-flow states implicit in Eq. (9.51), it must be true that

$$\dot{n}_\alpha^{(MX_\nu)} + \dot{n}_\beta^{(MX_\nu)} = 0, \quad \dot{n}_\alpha^{(g)} + \dot{n}_\alpha^{(1)} + \dot{n}_\beta^{(g)} + \dot{n}_\beta^{(1)} \equiv \dot{n}_\alpha^{(L)} + \dot{n}_\beta^{(L)} = 0, \quad (9.54)$$

and at α, e.g., that

$$\dot{n}_\alpha^{(MX_\nu)} = \dot{n}_\alpha^{(+)}(\text{in}) - \frac{\dot{n}}{\nu} = -\frac{\dot{n}_\alpha^{(-)}(\text{out})}{\nu}. \qquad (9.55)$$

Our study of the thermocouple showed that we can let

$$T_i[S_i^{(el)}] - T_j[S_j^{(el)}] + [h(ij)] = \bar{\mu}_j^{(el)} - \bar{\mu}_i^{(el)}$$

* In the molten salt case there will be no $\dot{n}_\omega^{(g)}$ or $\dot{n}_\omega^{(1)}$.

for an ij link. Thus we can cast Eq. (9.53) into the form

$$
\begin{aligned}
-F\Delta\psi = {} & \left(\frac{\delta\dot{n}_\alpha^{(g)}}{\tilde{\delta}\dot{n}}\right)_{\Omega'}(\mu_\alpha^{(g)} - \mu_\alpha^{(1)}) + \left(\frac{\delta\dot{n}_\alpha^{(L)}}{\tilde{\delta}\dot{n}}\right)_{\Omega'}(\mu_\alpha^{(1)} + T_\alpha[S_\alpha^{(1)}]) \\
& + \frac{1}{\nu}\mu_\alpha^{(M)} - \frac{\tau_-}{\nu}\mu_\alpha^{(MX_\nu)} + T_\alpha[S_\alpha^{(\pm)}] \\
& + \left(\frac{\delta\dot{n}_\beta^{(g)}}{\tilde{\delta}\dot{n}}\right)_{\Omega'}(\mu_\beta^{(g)} - \mu_\beta^{(1)}) - \frac{1}{\nu}\mu_\beta^{(M)} \\
& + \left(\frac{\delta\dot{n}_\beta^{(L)}}{\tilde{\delta}\dot{n}}\right)_{\Omega'}(\mu_\beta^{(1)} + T_\beta[S_\beta^{(1)}]) + \frac{\tau_-}{\nu}\mu_\beta^{(MX_\nu)} - T_\beta[S_\beta^{(\pm)}] \\
& + \bar{\mu}_1^{(el)} - \bar{\mu}_2^{(el)} + \bar{\mu}_5^{(el)} - \bar{\mu}_6^{(el)} + [h(34)],
\end{aligned}
\tag{9.56}
$$

where $[S_\omega^{(\pm)}] \equiv (\tau_+/\nu)[S_\omega^{(+)}] - \tau_-[S_\omega^{(-)}] \equiv (1/\nu)[S_\omega^{(+)}] - (\tau_-/\nu)[S_\omega^{(MX_\nu)}]$. We can also write Eq. (9.56) as

$$
\begin{aligned}
-F\Delta\psi = {} & \left(\frac{\delta\dot{n}_\alpha^{(g)}}{\tilde{\delta}\dot{n}}\right)_{\Omega'}(\mu_\alpha^{(g)} - \mu_\alpha^{(1)}) + \left(\frac{\delta\dot{n}_\beta^{(g)}}{\tilde{\delta}\dot{n}}\right)_{\Omega'}(\mu_\beta^{(g)} - \mu_\beta^{(1)}) \\
& + \left(\frac{\delta\dot{n}_\alpha^{(L)}}{\tilde{\delta}\dot{n}}\right)_{\Omega'}\{[G_\alpha^{(1)}] - [G_\beta^{(1)}]\} - \frac{\tau_-}{\nu}\{[G_\alpha^{(MX_\nu)}] - [G_\beta^{(MX_\nu)}]\} \\
& + \frac{1}{\nu}\{\mu_\alpha^{(M)} + T_\alpha[S_\alpha^{(+)}] - \langle T(34)\rangle[S_\alpha^{(+)}]\} \\
& - \frac{1}{\nu}\{\mu_\beta^{(M)} + T_\beta[S_\beta^{(+)}] - \langle T(34)\rangle[S_\beta^{(+)}]\} + \bar{\mu}_1^{(el)} - \bar{\mu}_2^{(el)} \\
& + \bar{\mu}_5^{(el)} - \bar{\mu}_6^{(el)},
\end{aligned}
\tag{9.57}
$$

where $[G_\omega^{(i)}] \equiv \mu_\omega^{(i)} + T_\omega[S_\omega^{(i)}] - \langle T(34)\rangle[S_\omega^{(i)}]$, $[h(34)] = \langle T(34)\rangle[s(34)]$, and

$$
\begin{aligned}
[s(34)] + \left(\frac{\delta\dot{n}_\alpha^{(L)}}{\tilde{\delta}\dot{n}}\right)_{\Omega'}[S_\alpha^{(1)}] + \left(\frac{\tau_+}{\nu}\right)[S_\alpha^{(+)}] - \tau_-[S_\alpha^{(-)}] + \tau_-[S_\beta^{(-)}] - \left(\frac{\tau_+}{\nu}\right) \\
\times [S_\beta^{(+)}] + \left(\frac{\delta\dot{n}_\beta^{(L)}}{\tilde{\delta}\dot{n}}\right)_{\Omega'}[S_\beta^{(1)}] = 0.
\end{aligned}
$$

Now in the state of migrational equilibrium $\mu_\omega^{(g)} = \mu_\omega^{(1)}$ and $[G_\alpha^{(i)}] = [G_\beta^{(i)}]$, so Eq. (9.57) reduces to

$$
\begin{aligned}
-F\Delta\psi = {} & \bar{\mu}_1^{(el)} - \bar{\mu}_2^{(el)} + \bar{\mu}_5^{(el)} - \bar{\mu}_6^{(el)} \\
& + \frac{1}{\nu}\{(\mu_\alpha^{(M)} + T_\alpha[S_\alpha^{(+)}] - \langle T(34)\rangle[S_\alpha^{(+)}]) \\
& - (\mu_\beta^{(M)} + T_\beta[S_\beta^{(+)}] - \langle T(34)\rangle[S_\beta^{(+)}])\}.
\end{aligned}
\tag{9.58}
$$

If, however, we formally choose to leave in the current ratio τ_- [going back to Eq. (9.56)], then

$$
\begin{aligned}
-F\Delta\psi = {} & \bar{\mu}_1^{(el)} - \bar{\mu}_2^{(el)} + \bar{\mu}_5^{(el)} - \bar{\mu}_6^{(el)} + \\
& \frac{1}{\nu}\{(\mu_\alpha^{(M)} - \tau_-\mu_\alpha^{(MX_\nu)} + \nu T_\alpha[S_\alpha^{(\pm)}] - \nu\langle T(34)\rangle[S_\alpha^{(\pm)}]) \\
& - (\mu_\beta^{(M)} - \tau_-\mu_\beta^{(MX_\nu)} + \nu T_\beta[S_\beta^{(\pm)}] - \nu\langle T(34)\rangle[S_\beta^{(\pm)}])\}.
\end{aligned}
\tag{9.59}
$$

The current ratios τ are well-defined experimental quantities when measured *in situ*; they may or may not have any *direct* relationship to transference numbers determined in isothermal experiments. The fact that in the state of migrational equilibrium the potential difference of thermocells of the type being investigated does not depend on current ratios is well known [9, 10]. From Eq. (9.58) it follows that

$$\nu F\left(\frac{\partial \Delta\psi}{\partial T_\alpha}\right)_{T_\beta,T,\langle T\rangle} = \nu[S_\alpha^{(el)}]^\partial + \bar{S}_\alpha^{(M)} - [S_\alpha^{(+)}]^\partial, \tag{9.60}$$

$$\nu F\left(\frac{\partial \Delta\psi}{\partial T_\beta}\right)_{T_\alpha,T,\langle T\rangle} = [S_\beta^{(+)}]^\partial - \bar{S}_\beta^{(M)} - \nu[S_\beta^{(el)}]^\partial, \tag{9.61}$$

$$\begin{aligned}\nu F\left(\frac{\partial \Delta\psi}{\partial T}\right)_{T_\alpha,T_\beta,\langle T\rangle} &= \nu\{[S_6^{(el)}]^\partial - [S_1^{(el)}]^\partial\} \\ &= \nu(\bar{S}_6^{(el)} - \bar{S}_1^{(el)}) = 0,\end{aligned} \tag{9.62}$$

with, e.g., $[S_\alpha^{(+)}]^\partial \equiv \{\partial(T_\alpha[S_\alpha^{(+)}] - T_\beta[S_\beta^{(+)}] + \langle T\rangle([S_\beta^{(+)}] - [S_\alpha^{(+)}]))/\partial T_\alpha\}_{\beta,\langle T\rangle}$, $\langle T\rangle \equiv \langle T(34)\rangle$, and $[S_\alpha^{(el)}]^\partial \equiv -(\partial\bar{\mu}_2^{(el)}/\partial T_2)_1$. I have assumed that $(\partial\,\Delta\psi/\partial\langle T\rangle)_{T_\alpha,T_\beta} = 0$.*

I showed earlier in connection with the discussion of the thermocouple that for electrons in wires $[S^{(el)}]^\partial = \bar{S}^{(el)}$. As the entropy associated with the electron in copper wires is quite small [11], Eqs. (9.60) and (9.61) enable us to evaluate the $[S^{(+)}]^\partial$ quantity for the cation from thermocell potential difference measurements and third-law calculations of $\bar{S}^{(M)}$; thus, e.g.,

$$[S_\alpha^{(+)}]^\partial \approx \bar{S}_\alpha^{(M)} - \nu F\left(\frac{\partial \Delta\psi}{\partial T_\alpha}\right)_{T_\beta}. \tag{9.63}$$

Pitzer [10] has compiled a table of values of $[S_\omega^{(+)}]^\partial$ for the cation in a number of molten salt electrolytes; Table 9–1 shows a few typical entries from that table.

Table 9–1

Thermocells with Fused Salt Electrolyte (Units: cal/deg-mole)
(Compilation of Pitzer [10])

Electrolyte	T°K	$\bar{S}^{(M)}$	$-\nu Fd\,\Delta\psi/dT$	$[S^{(+)}]^\partial$
$AgNO_3$	500	13.37	7.6	21.0
AgCl	800	16.43	9.3	26
AgBr	750	16.00	11	27
AgI	850	16.84	10	27
$ZnCl_2$	~600	14.41	−6	8
$SnCl_2$	~600	20.59	+1	22

* This is the analog of assumption *Q*, Chapter 5.

For formal completeness I point out that Eq. (9.59) leads to

$$\nu F\left(\frac{\partial \Delta\psi}{\partial T_\alpha}\right)_{T_\beta,T,\langle T\rangle} = \nu[S_\alpha^{(el)}]^\partial + \bar{S}_\alpha^{(M)} - \tau_-[S_\alpha^{(MX_\nu)}]^\partial - \nu[S_\alpha^{(\pm)}]^\partial + \{\mu_\alpha^{(MX_\nu)} - \mu_\beta^{(MX_\nu)}\}\left(\frac{\partial \tau_-}{\partial T_\alpha}\right)_{T_\beta,T,\langle T\rangle}, \tag{9.64}$$

with, e.g., $[S_\alpha^{(\pm)}]^\partial \equiv \{\partial(T_\alpha[S_\alpha^{(\pm)}] - T_\beta[S_\beta^{(\pm)}] + \langle T\rangle([S_\beta^{(\pm)}] - [S_\alpha^{(\pm)}]))/\partial T_\alpha\}_{\beta,\langle T\rangle}$. The relation

$$[S_\alpha^{(+)}]^\partial = \tau_-[S_\alpha^{(MX_\nu)}]^\partial + \nu[S_\alpha^{(\pm)}]^\partial - (\Delta\mu^{(MX_\nu)})\left(\frac{\partial \tau_-}{\partial T_\alpha}\right)_{T_\beta,T,\langle T\rangle} \tag{9.65}$$

must then hold between the various quantities. For a molten salt, since there can be no Soret effect in a 1-component system, $[S_\omega^{(MX_\nu)}]^\partial = \bar{S}_\omega^{(MX_\nu)}$.

For 1-component molten salt thermocells there is nothing further to say; for 2-component aqueous solutions it is customary to measure an initial potential difference $(\Delta\psi)_0$ for the thermocell when there is an essentially steady flow of heat through the system and *before* the Soret effect has brought about any appreciable change in the compositions of the solutions at α and β. Such $(\Delta\psi)_0$ values are *not* steady-state values (the potential difference characteristic of the final state of complete migrational equilibrium is $\Delta\psi$); however, we can approximately reach the envisaged state of the system by allowing the solvent to flow at a steady rate from one terminal part to the other so as approximately to counteract by its dragging tendency the Soret motion of the solute. The situation to be analyzed, then, is one in which we place a solution of average composition $n^{(MX_\nu)}/n^{(1)} \equiv r$ into the apparatus indicated in Figure 9–6; we then set the pressures P_ω equal to the solvent partial pressures appropriate to solutions of composition r and temperature T_ω at thermostatic equilibrium. As in the case of the concentration cell (Chapter 13) it is necessary to agitate gently the solutions at α and β so as to keep them nearly uniform in composition. There will ultimately be a steady flow of solvent from one terminal part to the other; i.e., $\dot{n}_\alpha^{(g)} + \dot{n}_\beta^{(g)} = 0$. If now we superimpose a flow of electric current on the solvent flow and if we carry out the operation $\delta/\bar{\delta}\dot{n}$ at constant T_α, T_β, $\langle T(ij)\rangle$, P_α, P_β,* then the $\Delta\psi$ value of the limiting state should be approximately equal to $(\Delta\psi)_0$.

The appropriate equations are Eqs. (9.54) to (9.56); so

$$\begin{aligned} -F(\Delta\psi)_0 \approx \left(\frac{\delta\dot{n}_\alpha^{(L)}}{\bar{\delta}\dot{n}}\right)_{\Omega'} &\{[G_\alpha^{(1)}] - [G_\beta^{(1)}]\} - \frac{\tau_-}{\nu}\{[G_\alpha^{(MX_\nu)}] - [G_\beta^{(MX_\nu)}]\} \\ &+ \frac{1}{\nu}\{\mu_\alpha^{(M)} + T_\alpha[S_\alpha^{(+)}] - \langle T(34)\rangle[S_\alpha^{(+)}]\} \\ &- \frac{1}{\nu}\{\mu_\beta^{(M)} + T_\beta[S_\beta^{(+)}] - \langle T(34)\rangle[S_\beta^{(+)}]\} \\ &+ \bar{\mu}_1^{(el)} - \bar{\mu}_2^{(el)} + \bar{\mu}_5^{(el)} - \bar{\mu}_6^{(el)}, \end{aligned} \tag{9.66}$$

* If we neglect kinetic energy terms, the indicated operation is of the form $(\delta/\bar{\delta}\dot{n})_{\Omega'}$.

where I have made use of the "gentle agitation" condition $\mu_{\omega}^{(g)} = \mu_{\omega}^{(1)}$. The Hittorf assumption is that $(\delta \dot{n}_{\alpha}^{(L)}/\delta \dot{n})_{\Omega'} = 0$ (no coupling between the flow of the electric current and the flow of the solvent).* If we accept the Hittorf assumption, then the temperature coefficient of $(\Delta\psi)_0$ is

$$\nu F\left(\frac{\partial(\Delta\psi)_0}{\partial T_\alpha}\right)_{\beta,r,T,\langle T\rangle} \approx \bar{S}_\alpha^{(M)} - [S_\alpha^{(+)}]^\partial - \nu[S_\alpha^{(el)}]^\partial + \tau_-\left(\frac{\partial\Delta[G^{(MX_\nu)}]}{\partial T_\alpha}\right)_{\beta,r,T,\langle T\rangle} + \left(\frac{\partial\tau_-}{\partial T_\alpha}\right)_{\beta,r,T,\langle T\rangle}\{[G_\alpha^{(MX_\nu)}] - [G_\beta^{(MX_\nu)}]\}. \tag{9.67}$$

We can put Eq. (9.67) into the form of Eq. (9.64) (but not into the form of Eq. (9.60) as $[G_\alpha^{(i)}] - [G_\beta^{(i)}] \neq 0$). Further simplifications are possible if we assume that $(\partial\tau_-/\partial T_\alpha)_{\beta,r,T,\langle T\rangle} \approx 0$ and that $[S_\omega^{(el)}]^\partial \approx 0$:

$$\nu F\left(\frac{\partial(\Delta\psi)_0}{\partial T_\alpha}\right)_{\beta,r,T,\langle T\rangle} \approx \bar{S}_\alpha^{(M)} - [S_\alpha^{(+)}]^\partial + \tau_-\left(\frac{\partial\Delta[G^{(MX_\nu)}]}{\partial T_\alpha}\right)_{\beta,r,T,\langle T\rangle}. \tag{9.68}$$

References

1. W. Thomson, *Mathematical and Physical Papers*, Vol. I (Cambridge University Press, Cambridge, 1882), pp. 232–291.
2. D. G. Miller, *Chem. Revs.*, **60**, 15 (1960).
3. F. N. Lancia and J. D. McGervey, *Rev. Sci. Instruments*, **35**, 1302 (1964).
4. E. A. Guggenheim, *Thermodynamics* (North Holland Publishing Company, Amsterdam, 1950), Chapter 10.
5. C. Reid, *Principles of Chemical Thermodynamics* (Reinhold Publishing Corporation, New York, 1960), Chapter 9.
6. J. Agar, *Revs. Pure Appl. Chem.*, **8**, 19 (1958).
7. P. Bridgman, *The Thermodynamics of Electric Phenomena in Metals* (The Macmillan Company, New York, 1934), Chapter 6.
8. G. Borelius, W. Keesom, and C. Johanson, *Comm. Leiden*, 196a.
9. J. deBethune, *J. Electrochem. Soc.*, **107**, 829 (1960).
10. K. Pitzer, *J. Phys. Chem.*, **65**, 147 (1961).
11. M. Tempkin and A. Khoroshin, *Zhur. Fiz. Khim.*, **26**, 500, 773 (1952).

* See the discussion relative to concentration cells in Chapter 13.

10

Thermal Converters and Polythermal Processes

Thermal Converters

We can use some of the polythermal field effects considered in the previous chapters as the basis for devices of the heat-to-work type, devices that I hereafter call *thermal converters*. These devices are sites of heat currents due to maintained temperature differences; the heat currents give rise to currents of electricity, mass, or chemical reaction measures that are agents for the supply of work to the surroundings.

Figure 10–1 is a schematic rendering of one type of thermal converter, a *closed-loop* converter with full-flux linkages (heat is exchanged only with the heat reservoirs at α and β). The heat flows $\dot{Q}_\omega^{(r)}$ drive a current $\dot{n}$ around the loop with the result that work is delivered at the rate $-\dot{W}$ to the surroundings. Let T_α be greater than T_β, and let the direction of spontaneous current flow for a small temperature difference $T_\alpha - T_\beta$ be taken as the positive direction of current flow. For a closed-loop converter we have, then (in the steady state),

$$\dot{U}(\text{system}) = -\dot{Q}_\alpha^{(r)} - \dot{Q}_\beta^{(r)} + \dot{W} = 0, \tag{10.1}$$

$$\dot{U}(0) = -\dot{Q}_\alpha^{(r)}(0) - \dot{Q}_\beta^{(r)}(0) = 0, \tag{10.2}$$

or

$$-\dot{W} = -\dot{Q}_\alpha^{(r)} - \dot{Q}_\beta^{(r)} = -\Delta\dot{Q}_\alpha^{(r)} - \Delta\dot{Q}_\beta^{(r)}, \tag{10.3}$$

where $\dot{Z}(0)$ represents the rate of change of the Z quantity for the case $\dot{n} = 0$, $\Delta\dot{Z} \equiv \dot{Z}(\dot{n}) - \dot{Z}(0)$, and $\dot{Q}_\omega^{(r)}$ is the rate of influx of heat into the ω heat

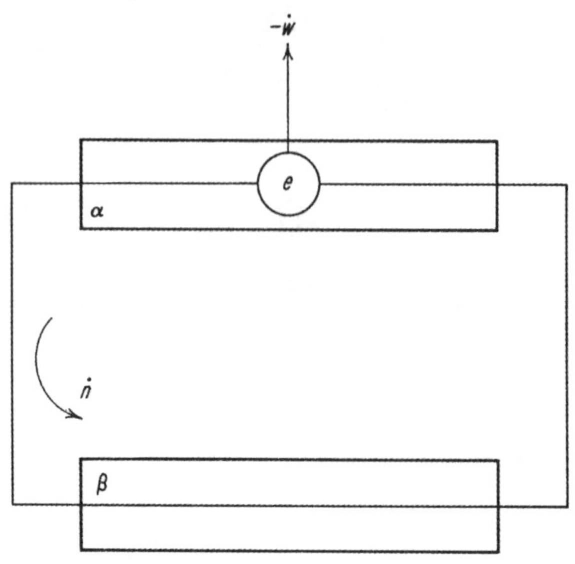

Figure 10–1. Closed-loop thermal converter with full-flux linkages; schematic. α—heat reservoir at temperature T_α; β—heat reservoir at temperature T_β; $\dot{n}$—appropriate current; e—appropriate engine, activated by the current $\dot{n}$, for delivering work at the rate $-\dot{W}$ to the surroundings; $T_\alpha > T_\beta$.

reservoir. We can express the efficiency of the thermal conversion process in two ways: in terms of the *total* heat flow $\dot{Q}_\omega^{(r)}$ or in terms of the *participating* heat flow $\Delta\dot{Q}_\omega^{(r)}$; i.e.,

$$\eta \equiv \frac{-\dot{W}}{-\dot{Q}_\alpha^{(r)}} \tag{10.4}$$

and

$$[\eta] \equiv \frac{-\dot{W}}{-\Delta\dot{Q}_\alpha^{(r)}}. \tag{10.5}$$

We can relate these efficiencies to the rate of entropy production associated with the process:

$$\eta = \frac{T_\alpha - T_\beta}{T_\alpha}\left(\frac{-\dot{W}}{-\dot{W} + T_\beta\Theta}\right), \tag{10.6}$$

$$[\eta] = \frac{T_\alpha - T_\beta}{T_\alpha}\left(\frac{-\dot{W}}{-\dot{W} + T_\beta\Delta\Theta}\right). \tag{10.7}$$

For the case of the thermocouple used as a source of work it has become customary to introduce a figure of merit ζ into discussions about efficiency [1]; the figure of merit is related to the maximum value of η for a given T_α, T_β by

$$\eta_{\max} = \frac{T_\alpha - T_\beta}{T_\alpha}\,\frac{\{1 + [\zeta(T_\alpha + T_\beta)/2]\}^{1/2} - 1}{\{1 + [\zeta(T_\alpha + T_\beta)/2]\}^{1/2} + (T_\beta/T_\alpha)}. \tag{10.8}$$

We see from Eqs. (10.6) and (10.8) that

$$\frac{\{1 + [\zeta(T_\alpha + T_\beta)/2]\}^{1/2} - 1}{\{1 + [\zeta(T_\alpha + T_\beta)/2]\}^{1/2} + (T_\beta/T_\alpha)} = \left(\frac{-\dot{W}}{-\dot{W} + T_\beta\Theta}\right)_{\max}. \tag{10.9}$$

Consider for a moment the expression for the rate of entropy production:

$$\begin{aligned} \Theta &= \frac{\dot{Q}_\alpha^{(r)}}{T_\alpha} + \frac{\dot{Q}_\beta^{(r)}}{T_\beta} \\ &= \frac{\dot{W}}{T_\beta} + \dot{Q}_\alpha^{(r)}\left(\frac{1}{T_\alpha} - \frac{1}{T_\beta}\right) \\ &= \dot{n}\left(\frac{\dot{W}}{\dot{n}T_\beta}\right) + \dot{Q}_\alpha^{(r)}\left(\frac{1}{T_\alpha} - \frac{1}{T_\beta}\right) \\ &= Y_1\Omega_1 + Y_2\Omega_2. \end{aligned} \tag{10.10}$$

If we invoke the condition of stability (Eq. 3.30) and the *haste-makes-waste* principle (Eq. 3.29), we can write

$$\left(\frac{\partial(\dot{W}/\dot{n})}{\partial\dot{n}}\right)_{T_\alpha,T_\beta} > 0 \quad \text{or} \quad \left(\frac{\partial(-\dot{W}/\dot{n})}{\partial\dot{n}}\right)_{T_\alpha,T_\beta} < 0 \tag{10.11}$$

and, since we are taking $\dot{n}$ to be a positive quantity,*

$$\left(\frac{\partial\Theta}{\partial\dot{n}}\right)_{T_\alpha,T_\beta} \geqslant 0. \tag{10.12}$$

Now the expected behavior of $-\dot{W}$ is that indicated in Figure 10–2 for a given converter and fixed values of T_α, T_β: we see that $-\dot{W}$ reaches a maximum for some value of $\dot{n}$ and then decreases, ultimately becoming zero at a value of $\dot{n}$ equal to $\dot{n}_{\max}$; in order to force a current of magnitude greater than $\dot{n}_{\max}$ around the converter loop it is necessary for $-\dot{W}$ to be negative—i.e., it is necessary to *supply* work to the system. The shape of the η versus $\dot{n}$ curve is similar to that of the $-\dot{W}$ versus $\dot{n}$ curve, η being zero at those points where $-\dot{W}$ is zero. However, the current-value $\dot{n}(\eta_{\max})$ that makes η a maximum is, in general, not the same as the value $\dot{n}(-\dot{W}_{\max})$ that makes $-\dot{W}$ a maximum: when $-\dot{W} = -\dot{W}_{\max}$, it follows that

$$\left(\frac{\partial\eta}{\partial\dot{n}}\right)_{T_\alpha,T_\beta} = \frac{\dot{W}T_\beta(T_\alpha - T_\beta)}{T_\alpha(-\dot{W} + T_\beta\Theta)^2}\left(\frac{\partial\Theta}{\partial\dot{n}}\right)_{T_\alpha,T_\beta} < 0.$$

The negative value of $(\partial\eta/\partial\dot{n})_{T_\alpha,T_\beta}$ at the point where $-\dot{W} = -\dot{W}_{\max}$ indicates that $\dot{n}(\eta_{\max}) < \dot{n}(-\dot{W}_{\max})$.

The shape of the curve in Figure 10–2 is consistent with the condition of stability (Eq. 10.11). Let $\{\partial(-\dot{W}/\dot{n})/\partial\dot{n}\}_{T_\alpha,T_\beta} \equiv -RT_\alpha[N]$, and for

* See Eq. (3.32).

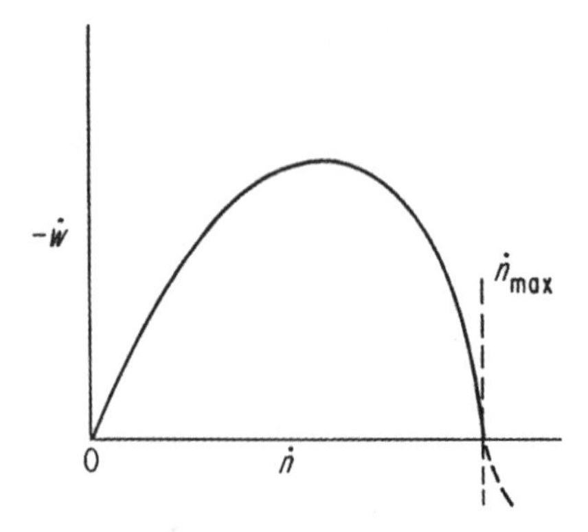

Figure 10–2. Schematic plot of $-\dot{W}$ versus $\dot{n}$ for a typical thermal converter; fixed T_α, T_β.

convenience take $[N]$ to be independent of $\dot{n}$ (this is equivalent to using a set of linear current-affinity relations); then

$$-\dot{W} = \dot{n}\{-\Delta\mathscr{G}_\alpha(0) - \dot{n}RT_\alpha[N]\}, \tag{10.13}$$

$$\dot{n}_{max} = \frac{-\Delta\mathscr{G}_\alpha(0)}{RT_\alpha[N]}, \tag{10.14}$$

$$-\dot{W}_{max} = \frac{(-\Delta\mathscr{G}_\alpha(0)/2)^2}{RT_\alpha[N]} = -\dot{W}(\dot{n}_{max}/2)$$

$$= \frac{\dot{n}_{max}}{4}\{-\Delta\mathscr{G}_\alpha(0)\}, \tag{10.15}$$

where $-\Delta\mathscr{G}_\alpha(0)$ is the Gibbs free energy difference (per unit of "charge") across the poles of the engine e in the limit of vanishing current; i.e., $-\Delta\mathscr{G}_\alpha(0) = \lim(-\dot{W}/\dot{n})$ as $\dot{n} \to 0$ (for the thermocouple: if the current is measured in Faradays, then $-\Delta\mathscr{G}_\alpha(0) = F\Delta\psi$ with $\Delta\psi$ being the no-current Seebeck potential difference). The quantity $\dot{n}_{max}$ is a *short-circuit* current: the converter loop is merely closed with no work being exchanged between system and surroundings. Within the assumptions made in deriving Eq. (10.15) we see that the maximum rate of delivery of work to the surroundings is just one-fourth of the product of the short-circuit current and the open-circuit "potential difference"—i.e., for the thermocouple

$$-\dot{W}_{max} = \tfrac{1}{4}I_{max}\,\Delta\psi(0) = \tfrac{1}{4}\dot{n}_{max}\,F\Delta\psi.$$

Exercise 10–1. From Eqs. (10.10), (3.17), and (3.18)—taking state β as the thermostatic reference state—find the relationship between the quantity $[N]$ and the quantities L_{ij} or the quantities K_{ij}.

Consider now the quantity $[\eta]$; this quantity reaches its maximum in the limit of vanishing current:

$$[\eta]_{\max} = \lim_{\dot{n}\to 0} [\eta]. \tag{10.16}$$

We see from Eq. (10.7) that

$$[\eta] = \frac{T_\alpha - T_\beta}{T_\alpha}\left\{\frac{(-\dot{W}/\dot{n})}{(-\dot{W}/\dot{n}) + T_\beta(\Delta\,\Theta/\dot{n})}\right\} \tag{10.17}$$

and that

$$[\eta]_{\max} = \frac{T_\alpha - T_\beta}{T_\alpha}\left\{\frac{(\delta(-\dot{W})/\bar{\delta}\dot{n})_{T_\alpha,T_\beta}}{(\delta(-\dot{W})/\bar{\delta}\dot{n})_{T_\alpha,T_\beta} + T_\beta(\delta\,\Theta/\bar{\delta}\dot{n})_{T_\alpha,T_\beta}}\right\}. \tag{10.18}$$

Now $\{\delta(-\dot{W})/\bar{\delta}\dot{n}\}_{T_\alpha,T_\beta} = -\Delta\mathcal{G}_\alpha(0)$, and by assumption IV $(\delta\,\Theta/\bar{\delta}\dot{n})_{T_\alpha,T_\beta} = 0$, so

$$[\eta]_{\max} = \frac{T_\alpha - T_\beta}{T_\alpha}. \tag{10.19}$$

Thus if we compute the efficiency on the basis of the participating heat flow, then the Carnot efficiency (Eq. 10.19) is the ideal limit. The efficiency computed on the basis of the total heat flow is much less than the Carnot efficiency since much of the heat merely passes by conduction from the higher temperature to the lower.

The engineering aspects of the thermocouple functioning as a thermal converter are receiving a good deal of attention;* in principle other sorts of thermal converters are potentially of engineering value. In Figure 10–1 if the engine e is a gas turbine and the apparatus contains a circulating gas, the links $\alpha\beta$ and $\beta\alpha$ being tubes of different diameter, then the difference in the thermomolecular pressure effect across the two links will maintain a pressure difference across the turbine and work can be delivered to the surroundings at a steady rate—a Brønsted thermal converter [3, 4].

In a similar fashion if two liquid–vapor interfaces at different temperatures (parts of a closed loop) communicate with a turbine, then the difference in vapor pressure for the two temperatures will establish the necessary pressure difference across the turbine. Consider the situation indicated schematically in Figure 10–3: two liquid–vapor interfaces maintained at different temperatures communicate with opposite sides of a turbine engine in a full-flux linkage. For a steady state with $\dot{n}$ moles of fluid circulating per second around the loop we have

$$-\dot{W} = -\dot{Q}_\alpha^{(r)} - \dot{Q}_\beta^{(r)} = -T_\alpha\dot{S}_\alpha^{(r)} - T_\beta\dot{S}_\beta^{(r)} \tag{10.20}$$

* See the many papers in the book cited in reference [1], and see also the book by Angrist [2].

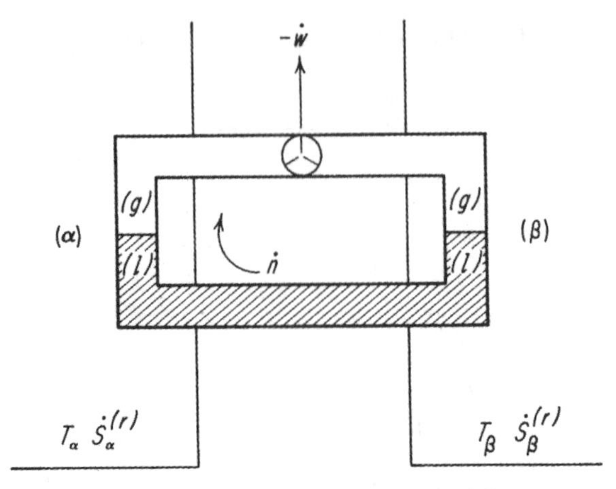

Figure 10–3. A closed-loop thermal converter of the vapor pressure–turbine type; linkages are of the full-flux type; $T_\alpha > T_\beta$.

and

$$\begin{aligned} \Theta &= \dot{S}_\alpha^{(r)} + \dot{S}_\beta^{(r)} = \dot{n}\left(\frac{\dot{W}}{\dot{n}T_\beta}\right) + \dot{Q}_\alpha^{(r)}\left(\frac{1}{T_\alpha} - \frac{1}{T_\beta}\right) \\ &= Y_1\Omega_1 + Y_2\Omega_2. \end{aligned} \tag{10.21}$$

In the limiting state ($\dot{n} \to 0$) we have

$$\left(\frac{\delta(-\dot{W})}{\delta\dot{n}}\right)_{T_\alpha, T_\beta} = -T_\alpha\left(\frac{\delta\dot{S}_\alpha^{(r)}}{\delta\dot{n}}\right)_{T_\alpha, T_\beta} - T_\beta\left(\frac{\delta\dot{S}_\beta^{(r)}}{\delta\dot{n}}\right)_{T_\alpha, T_\beta}, \tag{10.22}$$

$$\left(\frac{\delta\Theta}{\delta\dot{n}}\right)_{T_\alpha, T_\beta} = \left(\frac{\delta\dot{S}_\alpha^{(r)}}{\delta\dot{n}}\right)_{T_\alpha, T_\beta} + \left(\frac{\delta\dot{S}_\beta^{(r)}}{\delta\dot{n}}\right)_{T_\alpha, T_\beta}, \tag{10.23}$$

$$\mu_\omega^{(\text{liq})} = \mu_\omega^{(g)}. \tag{10.24}$$

In accordance with the discussion in Chapter 6, we can write $(\delta\dot{S}_\omega^{(r)}/\delta\dot{n})_{\Omega'} = [S_\omega(\text{in})] - [S_\omega(\text{out})]$; hence we have (assumption IV)

$$[S_\alpha^{(\text{liq})}] - [S_\alpha^{(g)}] + [S_\beta^{(g)}] - [S_\beta^{(\text{liq})}] = 0, \tag{10.25}$$

$$\begin{aligned} \left(\frac{\delta(-\dot{W})}{\delta\dot{n}}\right)_{T_\alpha, T_\beta} &= -T_\alpha\{[S_\alpha^{(\text{liq})}] - [S_\alpha^{(g)}]\} - T_\beta\{[S_\beta^{(g)}] - [S_\beta^{(\text{liq})}]\} \\ &= (T_\alpha - T_\beta)\{[S_\beta^{(g)}] - [S_\beta^{(\text{liq})}]\}. \end{aligned} \tag{10.26}$$

Alternatively, if we write for convenience $\mu_\alpha^{(g)} - \mu_\alpha^{(\text{liq})} + \mu_\beta^{(\text{liq})} - \mu_\beta^{(g)} = 0$ and add this expression to Eq. (10.26), we obtain

$$\left(\frac{\delta(-\dot{W})}{\delta\dot{n}}\right)_{T_\alpha, T_\beta} = [H_\alpha^{(g)}] - [H_\alpha^{(\text{liq})}] + [H_\beta^{(\text{liq})}] - [H_\beta^{(g)}]. \tag{10.27}$$

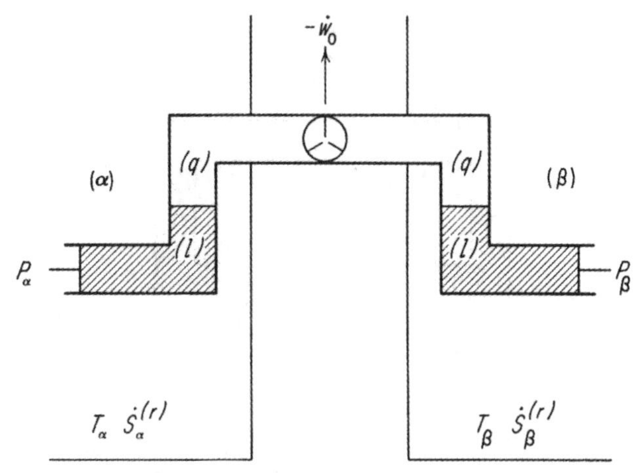

Figure 10–4. A source-sink thermal converter of the vapor pressure–turbine type; full-flux linkage; $T_\alpha > T_\beta$.

Now, there being no hindrance to the migration of liquid through the lower link, it should be true that $[H_\alpha^{(\text{liq})}] - [H_\beta^{(\text{liq})}] = 0$; hence Eq. (10.27) should reduce to

$$\left(\frac{\delta(-\dot{W})}{\delta\dot{n}}\right)_{T_\alpha, T_\beta} = [H_\alpha^{(g)}] - [H_\beta^{(g)}]. \tag{10.28}$$

Equation (10.28) shows that in a certain sense the maximum work (per unit "charge" transferred) obtainable out of a full-flux linkage is governed by the magnitude of the $\Delta[H]$ quantity across the linkage.

Exercise 10–2. Analyze the slightly different situation indicated in Figure 10–4 (make use of Eq. 6.39) and show that

$$\left(\frac{\delta(-\dot{W}_0)}{\delta\dot{n}}\right)_{T_\alpha, T_\beta} = [H_\alpha^{(g)}] - [H_\beta^{(g)}].$$

Steady Bithermal Mass Flow with Full-Flux Linkage

Consider the case indicated in Figure 10–5: the situation is the same as that analyzed in establishing the migrational equilibrium condition for the thermomolecular pressure effect; I now investigate the relations governing the simultaneous flow of heat and mass. Assume for a given gas and (full-flux) linkage that $\dot{n}_\alpha = \dot{n}_\alpha(T_\alpha, T_\beta, P_\alpha, P_\beta)$; it follows then

$$\left(\frac{\partial\dot{n}_\alpha}{\partial T_\alpha}\right)_{\beta, P_\alpha} = -\left(\frac{\partial\dot{n}_\alpha}{\partial P_\alpha}\right)_{\beta, T_\alpha}\left(\frac{\partial P_\alpha}{\partial T_\alpha}\right)_{\beta, \dot{n}_\alpha} \tag{10.29}$$

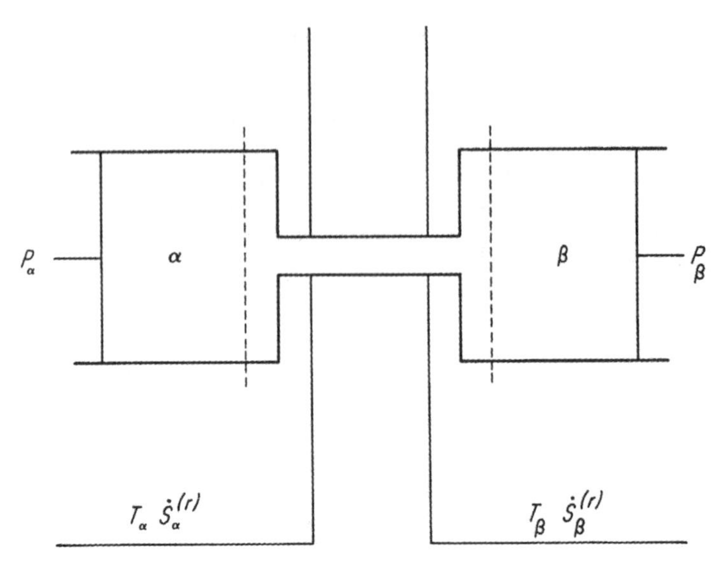

Figure 10–5. Steady bithermal flow of a gas across a capillary, full-flux linkage.

Equation (10.29) has an interesting consequence. Since $(\partial\dot{n}_\alpha/\partial P_\alpha)_{\beta,T_\alpha}$ is necessarily negative, it follows from Eq. (10.29) that $(\partial\dot{n}_\alpha/\partial T_\alpha)_{\beta,P_\alpha}$ and $(\partial P_\alpha/\partial T_\alpha)_{\beta,\dot{n}_\alpha}$ have the same sign. The thermomolecular pressure type of experiment at moderate values of $\dot{n}_\alpha$ yields positive values for $(\partial P_\alpha/\partial T_\alpha)_{\beta,\dot{n}_\alpha}$ [since $(\partial P_\alpha/\partial T_\alpha)_{\beta,\dot{n}_\alpha=0}$ is positive]; the implication, then, is that the change in state $\{T_\alpha(*),\ P_\alpha(*),\ T_\beta(*),\ P_\beta(*)\} \rightarrow \{T_\alpha(*) + \Delta T_\alpha,\ P_\alpha(*),\ T_\beta(*),\ P_\beta(*)\}$—the first state being one of migrational equilibrium—results in a mass flow *into* the α terminal part if ΔT_α is positive.

Exercise 10–3. The preceding considerations suggest that relations such as

$$-\left(\frac{\partial\dot{Q}_\alpha^{(r)}}{\partial\dot{n}_\alpha}\right)_{\beta,T_\alpha} = T_\alpha\bar{V}_\alpha\left(\frac{\partial P_\alpha}{\partial T_\alpha}\right)_{\beta,\dot{n}_\alpha} \tag{10.30}$$

may hold for the case considered with $\dot{Q}_\alpha^{(r)} = \dot{Q}_\alpha^{(r)}(T_\alpha, P_\alpha, T_\beta, P_\beta)$ for a given gas and linkage. Investigate the applicability of Eq. (10.30) in the limit $\dot{n}_\alpha \rightarrow 0$, showing the consequences of its assumed truth. On the basis of the findings for the $\dot{n}_\alpha \rightarrow 0$ case, decide on the reasonableness of the equation for the case $\dot{n}_\alpha \neq 0$.

References

1. J. Kaye and J. Walsh (editors), *Direct Conversion of Heat to Electricity* (John Wiley and Sons, New York, 1960), pp. 16-4 and 17-3.
2. S. W. Angrist, *Direct Energy Conversion* (Allyn and Bacon, Boston, 1965).
3. J. N. Brønsted, *Principles and Problems in Energetics* (Interscience Publishers, New York, 1955), p. 110.
4. M. Tribus, *Thermostatics and Thermodynamics* (D. Van Nostrand Company, Princeton, New Jersey, 1961), Chapter 16.

11

Heat Conduction

One-Dimensional Heat Conduction

Consider the situation indicated schematically in Figure 11–1. The system of interest, taken to be in the shape of a bar for convenience, forms a full-flux linkage between two thermostats α and β; the cross-sectional area of the bar *at the point where it enters the* ω *thermostat* is B_ω; and the length of the bar outside the thermostat is Λ. We determine the coefficient of thermal conductivity κ for the system from the relations

$$-\left(\frac{\partial \dot{Q}_\alpha^{(r)}}{\partial T_\alpha}\right)_\beta \equiv \frac{\kappa_\alpha B_\alpha}{\Lambda}, \tag{11.1}$$

$$\begin{aligned} -\dot{Q}_\alpha^{(r)} &= \int_{T_\beta}^{T_\alpha} \left(\frac{\kappa_\alpha B_\alpha}{\Lambda}\right)_\beta dT_\alpha \\ &= \left\langle \frac{\kappa B}{\Lambda} \right\rangle (T_\alpha - T_\beta), \end{aligned} \tag{11.2}$$

where $\dot{Q}_\omega^{(r)} = T_\omega \dot{S}_\omega^{(r)}$ is the rate of influx of heat into the ω heat reservoir via the link Λ. Since we are taking the link to be a full-flux one, it follows that $\dot{Q}_\alpha^{(r)} + \dot{Q}_\beta^{(r)} = 0$, and, in general, that

$$\kappa_\omega = -\frac{\Lambda}{B_\omega}\left(\frac{\partial \dot{Q}_\omega^{(r)}}{\partial T_\omega}\right)_\chi = \frac{\Lambda}{B_\omega}\left(\frac{\partial \dot{Q}_\chi^{(r)}}{\partial T_\omega}\right)_\chi. \tag{11.3}$$

Now in the state of steady heat conduction

$$\dot{U}(\text{system}) = -\dot{Q}_\alpha^{(r)} - \dot{Q}_\beta^{(r)} = 0, \tag{11.4}$$

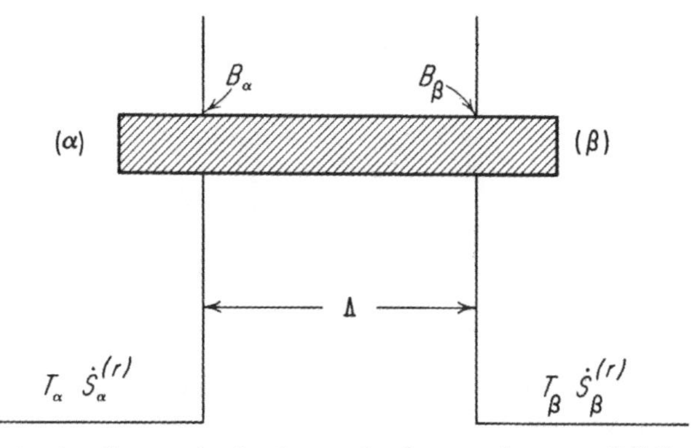

Figure 11–1. System in the form of a bar serving as a full-flux link between two thermostats α and β.

$$\begin{aligned}\Theta &= \frac{\dot{Q}_\alpha^{(r)}}{T_\alpha} + \frac{\dot{Q}_\beta^{(r)}}{T_\beta} = \dot{Q}_\alpha^{(r)}\left(\frac{1}{T_\alpha} - \frac{1}{T_\beta}\right) \\ &= Y_1 \Omega_1. \end{aligned} \tag{11.5}$$

If we take state β as our reference thermostatic state,* we can write

$$Y_1 = L_{11}(\beta)\Omega_1 \tag{11.6}$$

or

$$\dot{Q}_\alpha^{(r)} = L_{11}(\beta)\left(\frac{1}{T_\alpha} - \frac{1}{T_\beta}\right) = -\frac{L_{11}(T_\alpha - T_\beta)}{T_\alpha T_\beta}. \tag{11.7}$$

It follows from Eqs. (11.2) and (11.7) that

$$\langle \kappa \rangle = \left(\frac{L_{11}(\beta)}{T_\alpha T_\beta}\right)\frac{\Lambda}{B}; \tag{11.8}$$

hence for fixed T_β, if L_{11} does not depend on T_α, then $\langle \kappa \rangle$ does, and vice versa. In addition, from the relation $L_{11} = (\partial Y_1/\partial \Omega_1)_\beta = \{\partial \dot{Q}_\alpha^{(r)}/\partial(1/T_\alpha)\}_{T_\beta}$ and Eq. (11.1), we see that

$$\frac{L_{11}}{T_\alpha^2} = \frac{\kappa_\alpha B_\alpha}{\Lambda}. \tag{11.9}$$

Although the thermal conductivity κ for a solid material does depend on temperature [1], we have no grounds for believing that κT^2 should be a more slowly varying function of temperature than κ itself.

Multiple Heat Currents

Consider now the case outlined in Figure 11–2: a branched, full-flux linkage (a solid, heat-conducting material) connecting four separate

* See Chapter 3.

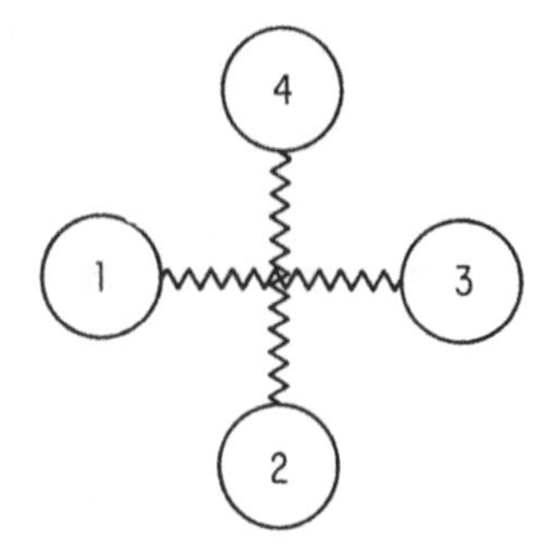

Figure 11–2. A branched, full-flux linkage (a solid, heat-conducting material) connecting four separate heat reservoirs (thermostats) 1, 2, 3, 4.

thermostats. Let terminal part 1 be in state I, part 2 in state II, and so on. The state of steady heat exchange is characterized by

$$\dot{U}(\text{system}) = -\sum_{i=1}^{4} \dot{Q}_i^{(r)} = 0 \tag{11.10}$$

and

$$\begin{aligned}\Theta &= \sum_{i=1}^{4} \frac{\dot{Q}_i^{(r)}}{T_i} \\ &= \dot{Q}_1^{(r)}\left(\frac{1}{T_1} - \frac{1}{T_4}\right) + \dot{Q}_2^{(r)}\left(\frac{1}{T_2} - \frac{1}{T_4}\right) + \dot{Q}_3^{(r)}\left(\frac{1}{T_3} - \frac{1}{T_4}\right) \\ &= Y_1\Omega_1 + Y_2\Omega_2 + Y_3\Omega_3. \end{aligned} \tag{11.11}$$

Apparently, then, the situation is a 3-current situation. Now, can some constraint be applied to the system such that Eq. (11.10) can be resolved into the separate equations $\dot{Q}_1^{(r)} + \dot{Q}_3^{(r)} = 0$ and $\dot{Q}_2^{(r)} + \dot{Q}_4^{(r)} = 0$? That is, can we set up two independent *heat currents* in the linkage?

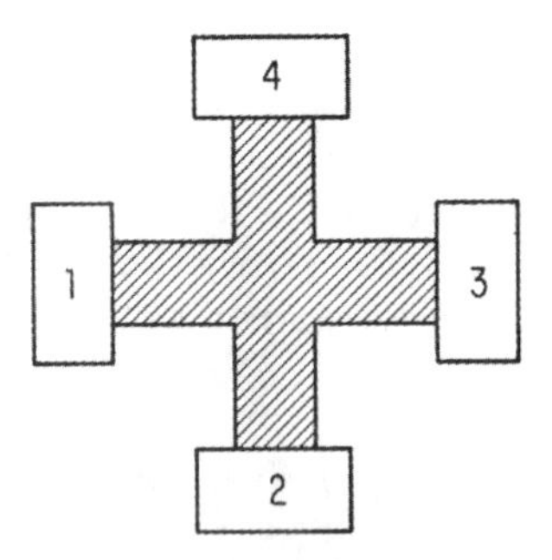

Figure 11–3. A symmetric cross-shaped piece of heat-conducting material functioning as a branched, full-flux linkage between the four heat reservoirs (thermostats) 1, 2, 3, 4.

In order to see the matter clearly, consider the situation in Figure 11–3: a homogeneous, cross-shaped piece is cut from a metal plate; the ends of the arms of the cross butt up against thermostats and exchange heat with those thermostats; i.e., the cross-shaped piece functions as a branched, full-flux linkage of an especially symmetric type. In the state of steady heat exchange, Eqs. (11.10) and (11.11) hold. Let $T_4 = T_{IV}$ and take state IV as reference thermostatic state. In the region of linear current-affinity relations we have, then,

$$\dot{Q}_i^{(r)} = \sum_{j=1}^{3} L_{ij}(\mathrm{IV})\left(\frac{1}{T_j} - \frac{1}{T_4}\right) \qquad i = 1, 2, 3, 4, \tag{11.12}$$

$$L_{4j} = -\sum_{i=1}^{3} L_{ij} \qquad j = 1, 2, 3. \tag{11.13}$$

Exercise 11–1. Due to the symmetry of the cross-shaped linkage, of the nine basic coefficients in the **L** matrix, only two are independent. Consider an "old" state $\{T_1 = T_{I}, T_2 = T_{II}, T_3 = T_{III}, T_4 = T_{IV}\}$ and a "new$_1$" state $\{T_1 = T_{I}, T_2 = T_{IV}, T_3 = T_{III}, T_4 = T_{II}\}$; then the symmetry of the linkage requires that $\dot{Q}_1^{(r)}(\text{old}) = \dot{Q}_1^{(r)}(\text{new}_1)$, and $\dot{Q}_3^{(r)}(\text{old}) = \dot{Q}_3^{(r)}(\text{new}_1)$. Similarly, for the "new$_2$" state $\{T_1 = T_{III}, T_2 = T_{II}, T_3 = T_{I}, T_4 = T_{IV}\}$ the relations $\dot{Q}_2^{(r)}(\text{old}) = \dot{Q}_2^{(r)}(\text{new}_2)$ and $\dot{Q}_4^{(r)}(\text{old}) = \dot{Q}_4^{(r)}(\text{new}_2)$ hold. Also, for the "new$_3$" state $\{T_1 = T_{II}, T_2 = T_{III}, T_3 = T_{IV}, T_4 = T_{I}\}$, the relations $\dot{Q}_i^{(r)}(\text{new}_3) = \dot{Q}_{i+1}^{(r)}(\text{old})$ $i = 1, 2, 3, 4$ are valid (with the convention that $5 \equiv 1$). Show that the preceding relations constrain the matrix

$$\begin{bmatrix} L_{11} & L_{12} & L_{13} \\ L_{21} & L_{22} & L_{23} \\ L_{31} & L_{32} & L_{33} \end{bmatrix}$$

to have the form

$$\begin{bmatrix} L_{11} & L_{12} & -(L_{11} + 2L_{12}) \\ L_{12} & L_{11} & L_{12} \\ -(L_{11} + 2L_{12}) & L_{12} & L_{11} \end{bmatrix} \tag{11.14}$$

(Note that the symmetry guarantees the validity of the Onsager relations.)

Consider now the following questions. (i) What do the conditions of stability (Eq. 3.30) imply for the present situation? The conditions of stability take the form, e.g., $\{\partial \dot{Q}_1^{(r)}/\partial(1/T_1)\}_{T_2,T_3,T_4} > 0$, or, alternatively,

$$(\partial \dot{Q}_1^{(r)}/\partial T_1)_{T_2,T_3,T_4} < 0$$

—a relation that is certainly true regardless of the nature (linear or nonlinear) of the current-affinity relations. (ii) For the case $T_1 = T_{\text{I}}$, $T_2 = T_{\text{II}}$, $T_4 = T_{\text{IV}}$, what must be the temperature T_3 so that $\dot{Q}_3^{(r)} = 0$? The question refers to an ordinary state of "migrational equilibrium":

$$\Omega_1\left(\frac{\delta Y_1}{\delta Y_3}\right)_{\Omega'} + \Omega_2\left(\frac{\delta Y_2}{\delta Y_3}\right)_{\Omega'} + \Omega_3 = 0. \tag{11.15}$$

In the region of linear current-affinity relations $(\delta Y_i/\delta Y_k)_{\Omega'} = L_{ik}/L_{kk}$, so

$$L_{13}\Omega_1 + L_{23}\Omega_2 + L_{33}\Omega_3 = 0, \tag{11.16}$$

or, if we make use of the matrix elements in relation (11.14),

$$-(L_{11} + 2L_{12})\left(\frac{1}{T_{\text{I}}} - \frac{1}{T_{\text{IV}}}\right) + L_{12}\left(\frac{1}{T_{\text{II}}} - \frac{1}{T_{\text{IV}}}\right) + L_{11}\left(\frac{1}{T_3} - \frac{1}{T_{\text{IV}}}\right) = 0; \tag{11.17}$$

and the problem is solved if we know the coefficients L_{11} and L_{12}. (iii) Under what conditions is it possible to have $\dot{Q}_1^{(r)} + \dot{Q}_3^{(r)} = 0$? The relation $\dot{Q}_1^{(r)} + \dot{Q}_3^{(r)} = 0$ implies (in the linear current-affinity region) that

$$(L_{11} + L_{31})\Omega_1 + (L_{12} + L_{32})\Omega_2 + (L_{13} + L_{33})\Omega_3 = 0, \tag{11.18}$$

or, if we use relation (11.14), that

$$-\frac{1}{T_1} + \frac{1}{T_2} - \frac{1}{T_3} + \frac{1}{T_4} = 0. \tag{11.19}$$

Exercise 11–2. Consider a special case of Eq. (11.19): $T_1 = T_2 = T_{\text{II}}$ (say) and $T_3 = T_4 = T_{\text{IV}}$ (say). Equation (11.11) then reduces to

$$\Theta = \dot{Q}_1^{(r)}\left(\frac{1}{T_1} - \frac{1}{T_4}\right) + \dot{Q}_2^{(r)}\left(\frac{1}{T_2} - \frac{1}{T_4}\right). \tag{11.20}$$

Although Eq. (11.20) seems to indicate a 2-current situation, show that the situation is in fact only a 1-current situation.

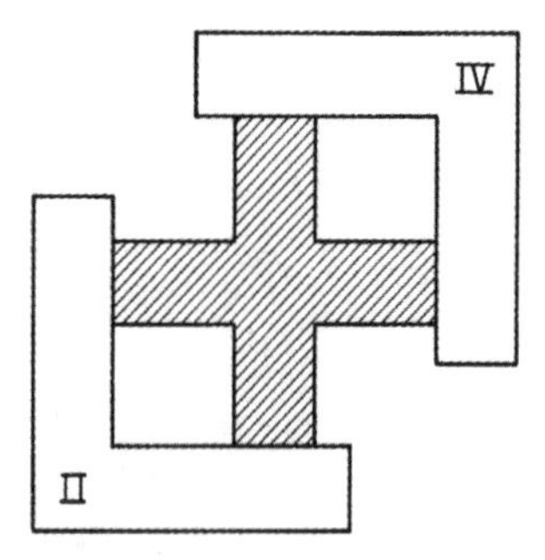

Figure 11–4. Reduction of the apparent two-current case to an equivalent one-current case.

Hint: Show that $\dot{Q}_1^{(r)}$ and $\dot{Q}_2^{(r)}$ are not independent for the special case considered—see Figure 11–4.

Exercise 11–3. Consider Figure 11–3 once again. Let the center point of the cross have the temperature T_0, and let the distance from the midpoint of the cross to each of the thermostats be Λ. Now let

$$-\dot{Q}_i^{(r)} = \frac{\kappa B}{\Lambda}(T_i - T_0) \qquad i = 1, 2, 3, 4, \tag{11.21}$$

where κ is the coefficient of thermal conductivity and B is the cross-sectional area of each arm of the cross. Show that for the condition $\dot{Q}_1^{(r)} + \dot{Q}_3^{(r)} = 0$ to hold Eq. (11.21) requires that

$$T_1 - T_2 + T_3 - T_4 = 0. \tag{11.22}$$

Show that if $|(T_i - T_4)/T_4| \ll 1$ for $i = 1, 2, 3, 4$, then Eqs. (11.19) and (11.22) are approximately equivalent.

An interesting problem in the heat-conduction field is the mapping of the three-dimensional flow of heat in an anisotropic crystal. For a discussion of the bearing of the Onsager reciprocal relations on this problem, see the treatments of Miller [2] and de Groot and Mazur [3].

References

1. R. C. L. Bosworth, *Heat Transfer Phenomena* (John Wiley and Sons, New York, 1952), Chapter 4.
2. D. G. Miller, *Chem. Revs.*, **60**, 15 (1960).
3. S. de Groot and P. Mazur, *Non-Equilibrium Thermodynamics* (North Holland Publishing Company, Amsterdam, 1962), Chapter 11.

12

Heat Radiation

General Remarks

The (equilibrium) thermodynamic treatment of thermal radiation fields (heat radiation) is given in full in the classic text of Planck [1]; in this chapter I treat a few topics pertaining to heat radiation by the steady-state methods developed in Part I. First I find it necessary to lay down a few conventions. The source of the thermal radiation being studied is always to be a heat reservoir (thermostat) of specified temperature; the reservoir is to be insulated against (other) radiative heat losses and is to be in communication with a cylinder-and-piston arrangement. The inside walls and piston face of the cylinder-and-piston arrangement are to be perfectly reflecting so that the radiation field can be considered to be localized inside the apparatus. A schematic version of the basic apparatus showing the communicating wall of the thermostat and the cylinder-and-piston arrangement appears in Figure 12–1. The system to be studied is the radiation trapped inside the cylinder-and-piston arrangement, the radiation interacting thermodynamically with the heat reservoir. It is convenient to use thermodynamic density functions in the course of the analysis; I indicate these density functions by lower case letters; thus let $z \equiv Z/V$, e.g.

Consider the situation indicated schematically in Figure 12–1, and let the volume of the radiation field be increased at a steady rate $\dot{V}$; then it follows that

$$\dot{U}(\text{system}) = -T\dot{S}^{(r)} - P\dot{V}, \tag{12.1}$$

and

$$\dot{H}(\text{system}) + T\dot{S}^{(r)} = 0, \tag{12.2}$$

where $\dot{S}^{(r)}$ represents the rate of accumulation of entropy in the heat reservoir. If we can associate well-defined thermodynamic density functions with the radiation field, so that $\dot{H}(\text{system}) = h\dot{V}$, e.g., we then arrive at the relations

$$h\dot{V} + T\dot{S}^{(r)} = 0 \tag{12.3}$$

and

$$\begin{aligned} \Theta &= \dot{S}^{(r)} + s\dot{V} = \dot{V}\left(s - \frac{h}{T}\right) = \dot{V}\left(\frac{-g}{T}\right) \\ &= Y_1\Omega_1, \end{aligned} \tag{12.4}$$

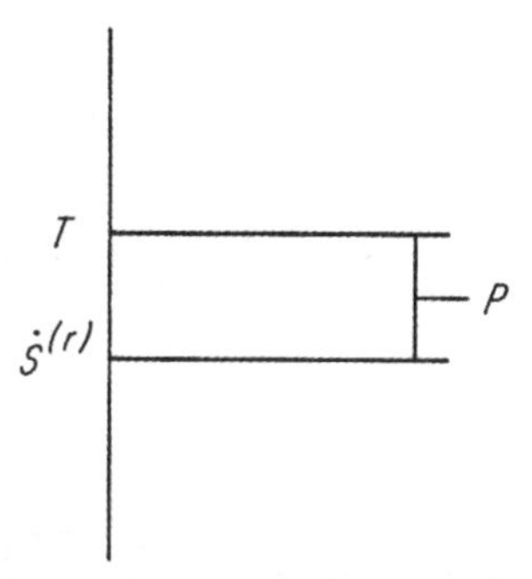

Figure 12–1. Thermostat plus cylinder-and-piston arrangement for studying heat radiation.

where $g \equiv h - Ts$. We find the properties of the equilibrium radiation field (complete radiation, blackbody radiation) by applying the operation $(\delta/\delta\dot{V})_T$ to Eq. (12.4) and by making use of assumption IV:

$$g \equiv h - Ts = 0. \tag{12.5}$$

Thus for equilibrium radiation the Gibbs free energy density of the radiation field has the value zero. Since in general $g = g(T, P)$, the relation $g(T, P) = 0$ implies that the equilibrium radiation field is a univariant system (in the phase rule sense [2]), and the properties of the radiation field are completely determined by specifying any *one* intensive property of the (equilibrium) field. Thus we have relations such as $P = P(T)$, $T = T(P)$, and $z = z(T)$; it is usually most convenient to take the temperature T as the independent variable. (Note that for equilibrium radiation $T = h/s$.)

We can spectrally decompose the equilibrium radiation field into monochromatic radiation "intervals" [1] having monochromatic density functions z_λ associated with them, and we can show that the Gibbs free energy density g_λ associated with *any* wavelength interval has the value zero for equilibrium radiation.

Consider the situation indicated schematically in Figure 12–2. Partition an equilibrium radiation field by a movable filter (permeable only to radiation of wavelength between λ and $\lambda + d\lambda$) into a region containing radiation of wavelength λ (to the right of the filter) and another region (to the left of the filter) containing complete radiation. Let the filter be moved by external forces. In the following analysis I indicate the properties of the complete radiation by unsubscripted symbols and those of the monochromatic radiation by symbols with subscript λ.

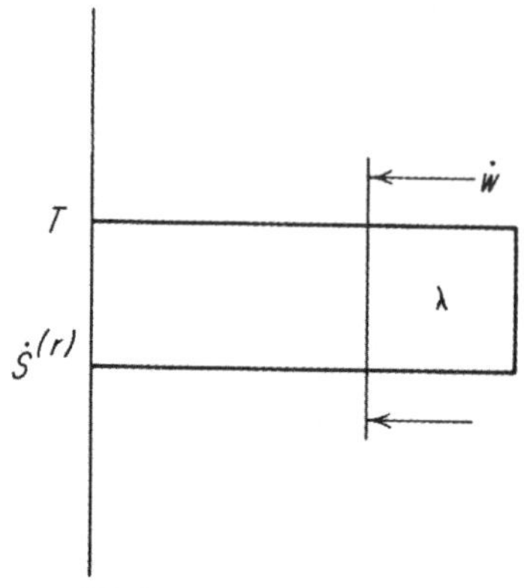

Figure 12–2. An equilibrium radiation field partitioned by a movable filter permeable only to radiation of wavelength between λ and $\lambda + d\lambda$; the filter is moved by external forces.

Let us move the filter at a steady rate so that $\dot{V} + \dot{V}_\lambda = 0$; the standard steady-state analysis then yields the relations

$$u\dot{V} + u_\lambda \dot{V}_\lambda = -T\dot{S}^{(r)} + \dot{W} \tag{12.6}$$

and

$$\Theta = \dot{S}^{(r)} + s\dot{V} + s_\lambda \dot{V}_\lambda = \dot{V}_\lambda\left(\frac{a}{T} - \frac{a_\lambda}{T} + \frac{\dot{W}}{T\dot{V}_\lambda}\right), \tag{12.7}$$

where $a \equiv u - Ts$. The application of the operation $(\delta/\delta\dot{V}_\lambda)_T$ to Eq. (12.7) and the use of assumption IV lead to the relation

$$a - a_\lambda + \left(\frac{\delta\dot{W}}{\delta\dot{V}_\lambda}\right)_T = 0, \tag{12.8}$$

where I have made use of the definition lim $(\dot{W}/\dot{V}_\lambda)$ as $\dot{V}_\lambda \to 0 \equiv (\delta\dot{W}/\delta\dot{V}_\lambda)$. Now, if the pressure of the radiation to the left of the filter is P and that to the right of the filter is P_λ, then

$$\left(\frac{\delta\dot{W}}{\delta\dot{V}_\lambda}\right)_T = P - P_\lambda \tag{12.9}$$

and

$$g - g_\lambda = 0. \tag{12.10}$$

Exercise 12–1. For equilibrium radiation it can be shown [1, 3] that $P = u/3$ and that $(\partial U/\partial V)_T = u$; hence show that the thermodynamic relation $(\partial U/\partial V)_T = T(\partial P/\partial T)_V - P$ leads directly to the Stefan–Boltzmann relation

$$u = \Gamma T^4,$$

where Γ is a constant.

Show the explicit dependence on T of the density functions h, a, s; and verify the relation $s = -da/dT$.

The classical theory of heat radiation [1, 3] concerns itself primarily with two relations derived from a detailed analysis of the electromagnetic field and with the thermodynamic consequences of those two relations: the relations so singled out for attention are the radiation–pressure relation $P = u/3$ and the Planck law for the spectral decomposition of the radiant energy density. For describing nonequilibrium radiation, we usually introduce a number of coefficients (emissivity, absorptivity, etc.) so as to relate the properties of the observed radiation to those of complete (blackbody) radiation. In determining the temperature of a radiating body we match some property of the radiating body with the corresponding property of a blackbody and then determine a *blackbody matching temperature* (of which there are several: color temperature, brightness temperature, effective temperature, and so on [4]). In order to see some of the problems that arise in treating nonequilibrium radiation consider the case of steady forced radiation. Refer again to Figure 12–1, and consider the case where the radiation field is expanding at a steady rate $\dot{V}$. We can put Eq. (12.4) into the form

$$-(h - Ts) = \frac{T\Theta}{\dot{V}} > 0. \tag{12.11}$$

Relation (12.11) shows that for the nonequilibrium radiation field $(h/s) < T$. Now, what is the "temperature" of the radiation undergoing the forced radiation process? Is it the thermostat temperature T? Is it a blackbody temperature T_1* associated with the pressure P—i.e., $T_1* = T_1*(P)$? Is it a temperature T_2* defined by $T_2* \equiv h/s$? Or is it something else? Is it meaningful to ascribe a temperature to the radiation itself? Does $T_1* = T_2*$? We can easily ask many questions for which we have no ready answers. It seems fairly evident that any blackbody matching temperature such as T_i* will be smaller than the thermostat temperature T (in order that the forced radiation process proceed, it is necessary that the pressure P be less than the equilibrium pressure $P*(T)$, and so on). Perhaps we could consider the radiation field to

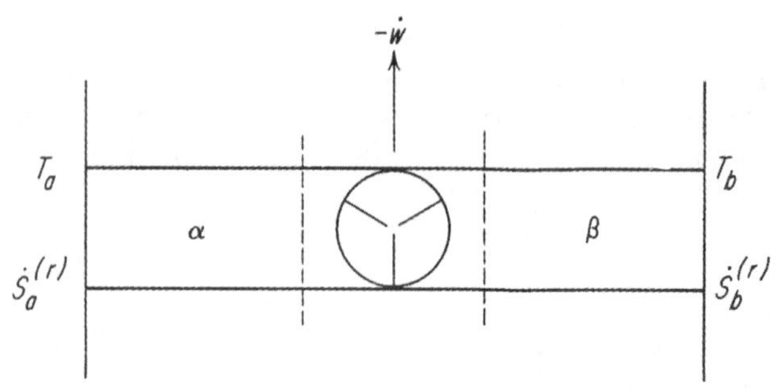

Figure 12–3. A turbine with perfectly reflecting blades placed in a radiation field; the difference in radiation pressure across the turbine results in the delivery of work at the rate $-\dot{W}$ to the surroundings; $T_a > T_b$.

be interacting thermodynamically with the surface (instead of with the bulk) of the thermostat, the radiating surface being at a lower temperature than the interior—just as in the forced vaporization process [5, 6] the region in the vicinity of the liquid–vapor interface is at a lower temperature than that of the encompassing thermostat. The various blackbody matching temperatures might then be reasonable estimates of the temperature of the radiating *surface.*

Radiation Turbine

Consider now the situation shown in Figure 12–3. A radiation turbine with perfectly reflecting blades is placed in communication with two sources of radiation ($T_a > T_b$); the difference in radiation pressure drives the turbine so as to deliver work at the rate $-\dot{W}$ to the surroundings. When the system is in a steady state, we have

$$-\dot{W} = -T_a\dot{S}_a^{(r)} - T_b\dot{S}_b^{(r)} \tag{12.12}$$

and

$$\begin{aligned}\Theta = \dot{S}_a^{(r)} + \dot{S}_b^{(r)} &= \frac{\dot{Q}_a^{(r)}}{T_a} + \frac{\dot{Q}_b^{(r)}}{T_b} \\ &= \dot{V}\left(\frac{\dot{W}}{\dot{V}T_b}\right) + \dot{Q}_a^{(r)}\left(\frac{1}{T_a} - \frac{1}{T_b}\right) \\ &= Y_1\Omega_1 + Y_2\Omega_2,\end{aligned} \tag{12.13}$$

where $\dot{V}$ is the volume swept out per unit time by the turbine blades. Now, e.g., let $z_a(T_a)$ represent a property of the equilibrium radiation appropriate to temperature T_a, and introduce the definition, e.g.,

$$\left(\frac{\delta\dot{S}_a^{(r)}}{\delta\dot{V}}\right)_{T_a,T_b} \equiv [s_a] - s_a. \tag{12.14}$$

Then, on applying the operation $(\delta/\tilde{\delta}\dot{V})_{T_a,T_b}$ to Eqs. (12.12) and (12.13), we obtain*

$$\left(\frac{\delta(-\dot{W})}{\delta\dot{V}}\right)_{T_a,T_b} = (T_a - T_b)(s_b - [\mathbf{s}_\beta]) \tag{12.15}$$

and

$$\left(\frac{\delta(-\dot{W})}{\tilde{\delta}\dot{V}}\right)_{T_a,T_b} = h_a - T_a[\mathbf{s}_\alpha] - (h_b - T_b[\mathbf{s}_\beta])$$
$$= [\mathbf{g}_\alpha] - [\mathbf{g}_\beta], \tag{12.16}$$

where $[\mathbf{g}] \equiv h - T[\mathbf{s}]$. The role of the nonequilibrium functions $[\mathbf{z}]$ in determining maximum-work or migrational equilibrium conditions is strikingly similar to the role of the analogous equilibrium functions z. Assumption IV implies that $[\mathbf{s}_\alpha] - s_a + s_b - [\mathbf{s}_\beta] = 0$.

The radiation turbine is another example of a thermal converter (see the general discussion of thermal converters in Chapter 10).

Inhomogeneous Radiation Field

As a final example of the application of steady-state methods to the study of radiation fields, I investigate the internal structure of an inhomogeneous radiation field. Consider the case indicated schematically in Figure 12–4. For the case $T_a \neq T_b$ there is a net flow of radiation through the reflecting cylinder from one thermostat to the other, and the properties of the radiation field vary in the direction marked x in the figure. In order to sample the radiation at an arbitrary point x in the field, I adjoin a probe cylinder-and-piston arrangement to the field at the point of interest. I indicate the properties of the radiation in the test probe by the subscript x, and I indicate reference properties of equilibrium radiation at temperature T_a or T_b by a subscript a or b, respectively.

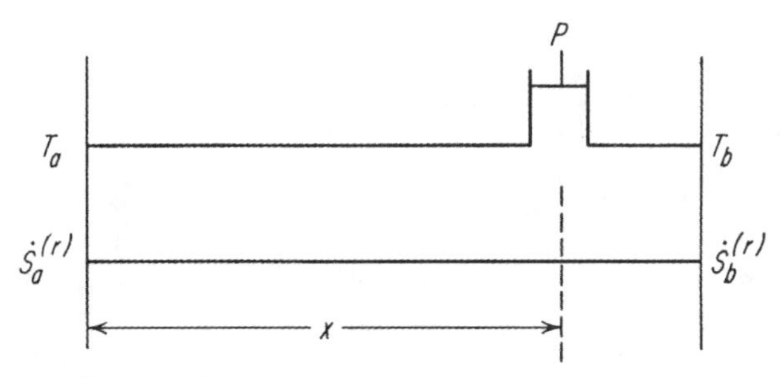

Figure 12–4. An inhomogeneous radiation field. A probe cylinder-and-piston arrangement is located at position x in the field; $T_a \neq T_b$.

* Since, e.g., $T_a s_a = h_a$.

The following relations hold for the steady-state situation:

$$h_x \dot{V}_x + T_a \dot{S}_a^{(r)} + T_b \dot{S}_b^{(r)} = 0, \tag{12.17}$$

$$\begin{aligned} \Theta &= \dot{S}_a^{(r)} + \dot{S}_b^{(r)} + s_x \dot{V}_x \\ &= \dot{V}_x\left(s_x - \frac{h_x}{T_b}\right) + \dot{Q}_a^{(r)}\left(\frac{1}{T_a} - \frac{1}{T_b}\right) \\ &= Y_1 \Omega_1 + Y_2 \Omega_2. \end{aligned} \tag{12.18}$$

To determine the properties of the *stationary* field between the two radiation sources, we apply the operation $(\delta/\tilde{\delta}\dot{V}_x)_{T_a,T_b}$ to Eqs. (12.17) and (12.18):

$$h_x + T_a\left(\frac{\delta \dot{S}_a^{(r)}}{\tilde{\delta}\dot{V}_x}\right)_{T_a,T_b} + T_b\left(\frac{\delta \dot{S}_b^{(r)}}{\tilde{\delta}\dot{V}_x}\right)_{T_a,T_b} = 0. \tag{12.19}$$

$$\left(\frac{\delta \Theta}{\tilde{\delta}\dot{V}_x}\right)_{T_a,T_b} = \left(\frac{\delta \dot{S}_a^{(r)}}{\tilde{\delta}\dot{V}_x}\right)_{T_a,T_b} + \left(\frac{\delta \dot{S}_b^{(r)}}{\tilde{\delta}\dot{V}_x}\right)_{T_a,T_b} + s_x = 0. \tag{12.20}$$

Now let

$$\text{grad} \equiv \left(\frac{\partial}{\partial x}\right)_{T_a,T_b} \tag{12.21}$$

and, e.g.,

$$\left(\frac{\delta \dot{S}_a^{(r)}}{\tilde{\delta}\dot{V}_x}\right)_{T_a,T_b} \equiv -\vartheta_x^{(a)} s_a, \tag{12.22}$$

where s_a is the blackbody value for temperature T_a; then, since, e.g., $h_a = T_a s_a$, we have (at any point x in the stationary field)

$$h_x = \vartheta_x^{(a)} h_a + \vartheta_x^{(b)} h_b, \tag{12.23}$$

$$s_x = \vartheta_x^{(a)} s_a + \vartheta_x^{(b)} s_b, \tag{12.24}$$

$$\lim_{T_a \to T_b} (\vartheta_x^{(a)} + \vartheta_x^{(b)}) = 1, \tag{12.25}$$

$$\text{grad}\, h_x - T_b\, \text{grad}\, s_x + s_a(T_b - T_a)\, \text{grad}\, \vartheta_x^{(a)} = 0, \tag{12.26}$$

$$\begin{aligned} \left(\frac{\partial h_x}{\partial T_b}\right)_{T_a} &= s_x + T_b\left(\frac{\partial s_x}{\partial T_b}\right)_{T_a} + (T_b - T_a)s_a\left(\frac{\partial \vartheta_x^{(a)}}{\partial T_b}\right)_{T_a} - \vartheta_x^{(a)} s_a \\ &= \vartheta_x^{(b)} s_b + T_b\left(\frac{\partial s_x}{\partial T_b}\right)_{T_a} + (T_b - T_a)s_a\left(\frac{\partial \vartheta_x^{(a)}}{\partial T_b}\right)_{T_a}. \end{aligned} \tag{12.27}$$

Once again we are in a position to ask a number of intriguing questions to which there are no obvious answers. Can we unambiguously define a temperature T_x for the radiation in the test probe (i.e., at any point x of the stationary radiation field) and if so, what is the relation of T_x to the quantities $\vartheta_x^{(a)} T_a + \vartheta_x^{(b)} T_b$, grad h_x/grad s_x, h_x/s_x, grad(h_x/s_x)? Is $g_x = 0$? Is the relation $\vartheta_x^{(a)} + \vartheta_x^{(b)} = 1$ true other than in the limit $T_a \to T_b$?

Conclusion

We have obtained by steady-state thermodynamic methods the following results for the radiation field: (i) A simple, direct derivation of the condition on the Gibbs free energy density of complete radiation—without making use of photon statistics or the detailed laws of electromagnetic radiation. (ii) The derivation of the maximum-work condition for a radiation turbine. (iii) A purely *thermodynamic* treatment of a stationary inhomogeneous radiation field. In addition to the results just mentioned, the steady-state formalism has shown that many interesting questions can be couched in its terms; we see exemplified here again one of the language functions (that of giving a lead to experiment) of a good thermodynamic formalism. We see then that the thermodynamics of heat radiation can still be a lively topic of conversation and a fruitful field for research.

References

1. M. Planck, *The Theory of Heat Radiation* (Dover Publications, New York, 1959).
2. J. Zernike, *Chemical Phase Theory* (Kluwer's Publishing Co., Ltd., Deventer, The Netherlands, 1958), p. 18.
3. M. Saha and B. Srivastava, *A Treatise on Heat*, fourth edition (The Indian Press Private, Ltd., Allahabad, 1958), Chapter 15.
4. C. Payne-Gaposchkin, *Temperature, Its Measurement and Control in Science and Industry*, Vol. 2 (edited by H. Wolfe) (Reinhold Publishing Corporation, New York, 1955), Chapter 4.
5. R. J. Tykodi and T. A. Erikson, *J. Chem. Phys.*, **31**, 1521 (1959); **33**, 46 (1960).
6. T. A. Erikson, *J. Phys. Chem.*, **64**, 820 (1960).

13

Migrational Equilibrium in Monothermal Fields: Applications

ν-Component Concentration Field

I considered 2-component concentration fields in Chapter 4; consider now a monothermal concentration field of ν ($\nu \geqslant 2$) components. Let the concentration field be established by the flow of component k through the ν-component MIX region (see Figure 4–1). Let the terminal parts containing the other components be arranged around the MIX region in such a way that for any stationary component (the ith component, say) $[r_{\alpha\beta}^{(ki)}] \approx [r_{\lambda}^{(ki)}]$, where $[r_{\lambda}^{(ki)}] \equiv \text{grad}\, \mu_{\lambda}^{(i)}/(\Delta\mathcal{G}_k/\Lambda_k)$, and Λ_k is the distance (measured through the MIX region) between the terminal parts containing the kth (flowing) component. By considering each component in turn and by establishing a steady flow of the chosen component (as well as of component k) between terminal parts, we can find the conditions for the migrational equilibrium of the chosen component in the concentration field in the usual way. The analysis provides us with a set of coupling coefficients from which I single out only the $[r_{\lambda}^{(ki)}]$ coefficients for discussion. For notational convenience I define one additional coefficient $[r_{\lambda}^{(kk)}]$ so that

$$[r_{\lambda}^{(kk)}] \equiv \frac{\text{grad}\, \mu_{\lambda}^{(k)}}{\Delta\mathcal{G}_k/\Lambda_k}. \tag{13.1}$$

Since the monothermal concentration field is completely described by the parameters T, $\Delta\mathscr{G}_k/\Lambda_k$, and $\nu - 1$ of the average mole fractions $\langle X^{(i)}\rangle$, the functional dependence of the $[r_\lambda^{(ki)}]$ coefficient is given (mainly) by

$$[r_\lambda^{(ki)}] = [r_\lambda^{(ki)}](T, \ldots, \langle X^{(i)}\rangle, \ldots, \Delta\mathscr{G}_k/\Lambda_k), \tag{13.2}$$

where I normally take as composition variables the average mole fractions of the stationary components. I prefer average composition variables to local ones because the average composition variables can be readily determined experimentally and are properties of the whole field.

We can determine the average composition of the field quite simply by stopping the flow of the field-inducing component and analyzing a sample drawn from the MIX region after that region has come to equilibrium. In addition, we often know from the way in which we prepared the solution in the MIX region exactly what its average composition is. Should we find, however, that the values of the $[r_\lambda^{(ki)}]$ coefficients vary from point to point in the concentration field, then it will become necessary in Eq. (13.2) to show explicit dependence on the quantities grad $X_\lambda^{(i)}$.

Consider now the case where the roles of the jth and the kth components are reversed: the jth component is now to be the flowing component, flowing over the same path that the kth component flowed over in the previous case; and the kth component is to be one of the stationary components. In the concentration field induced by the flow of the jth component, the coupling coefficients are of the form $[r_\lambda^{(ji)}]$. Now, what is the relation of the new coefficient $[r_\lambda^{(jk)}]$ to the old coefficient $[r_\lambda^{(kj)}]$? I conjecture that

$$[r_\lambda^{(jk)}][r_\lambda^{(kj)}] \approx 1 \tag{13.3}$$

for the same conditions of temperature, $\Delta\mathscr{G}/\Lambda$, and average composition. I base this conjecture on the following physical argument. The $[r_\lambda]$ coefficients seem to have the physical significance of coefficients measuring the drag effect of the flowing component on the stationary components. Such a drag effect depends on the ratio of the effective cross-sectional area of the stationary species to the similar quantity for the flowing species. According to such an interpretation, the exchange of roles between the two species leads to the type of relation suggested in Eq. (13.3).

The Gibbs–Duhem relation applied to the ν-component MIX region yields

$$\bar{S}_\lambda \text{ grad } T_\lambda - \bar{V}_\lambda \text{ grad } P_\lambda + \sum_{i=1}^{\nu} X_\lambda^{(i)} \text{ grad } \mu_\lambda^{(i)} = 0. \tag{13.4}$$

If we divide Eq. (13.4) through by $\Delta\mathscr{G}_k/\Lambda_k$, we get

$$\bar{S}_\lambda\left(\frac{\text{grad } T_\lambda}{\Delta\mathscr{G}_k/\Lambda_k}\right) - \bar{V}_\lambda\left(\frac{\text{grad } P_\lambda}{\Delta\mathscr{G}_k/\Lambda_k}\right) + \sum_{i=1}^{\nu} X_\lambda^{(i)} [r_\lambda^{(ki)}] = 0, \tag{13.5}$$

$$[r_\lambda^{(kk)}]\left\{\bar{S}_\lambda\left(\frac{\text{grad } T_\lambda}{\text{grad } \mu_\lambda^{(k)}}\right) - \bar{V}_\lambda\left(\frac{\text{grad } P_\lambda}{\text{grad } \mu_\lambda^{(k)}}\right) + X_\lambda^{(k)}\right\} + \sum_{i \neq k} X_\lambda^{(i)} [r_\lambda^{(ki)}] = 0; \tag{13.6}$$

and I conjecture that the relation

$$\sum_{i=1}^{\nu} \langle X^{(i)} \rangle [r_{\lambda}^{(ki)}] = 0 \tag{13.7}$$

comes pretty close to the truth.

If we limit ourselves to one-dimensional concentration fields, there is not much more to say. We can generate complicated three-dimensional concentration fields by having several components flow simultaneously through the MIX region at steady rates; the analysis of such situations, however, requires the use of vector and tensor quantities.

Chemical Reaction in a Concentration Field

Consider the reversible chemical reaction

$$\sum \nu_i B_i = \sum \nu_j D_j \tag{13.8}$$

(the reactants B_i being transformed to products D_j with stoichiometric coefficients ν_i and ν_j, respectively) taking place in a reactor of constant volume in a thermostat of temperature T. Let the reactants be individually supplied to and the products individually removed from the reaction vessel

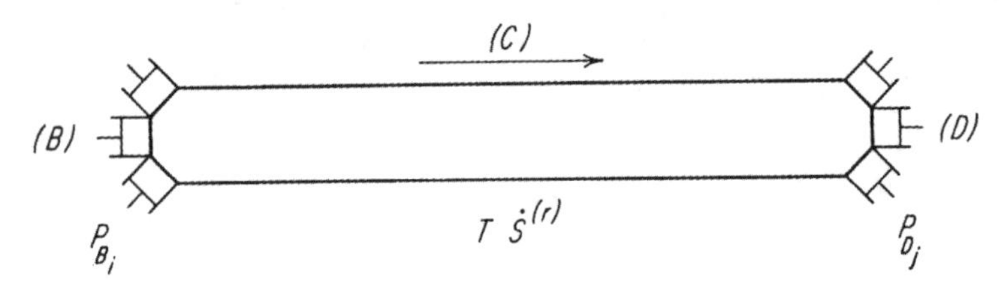

Figure 13–1. Chemical reaction in a monothermal concentration field.

via semipermeable membranes communicating with terminal parts at temperature T containing the appropriate substances. The situation just described is indicated schematically in Figure 13–1. If we supply the reactants to and remove the products from the reaction vessel at suitable steady rates, then the number of measures reacting in unit time $\dot{\xi}$ is just

$$\dot{\xi} = -\frac{\dot{n}_i}{\nu_i} = \cdots = \frac{\dot{n}_j}{\nu_j} = \cdots, \tag{13.9}$$

where $\dot{n}_k$, e.g., is the rate of influx of matter into the kth terminal part containing the product species D_k. The condition for no net reaction (equilibrium) in the reactor is, of course, that

$$\Delta G \equiv \sum \nu_j \mu_j - \sum \nu_i \mu_i = 0, \tag{13.10}$$

where μ_k, e.g., is the chemical potential of the D_k species computed for the state of affairs existing in the kth terminal part of the system.

Consider now the condition for no net reaction when the reactor is traversed by the steady flow of a nonreactive or reaction-neutral component C. In Figure 13–1 let there be included among the terminal parts at i a part C_i containing the nonreactive component C at temperature T and pressure P_{C_i}; similarly, let there be at j a terminal part C_j containing C at temperature T and pressure P_{C_j}; finally let there be a steady flow of component C through the reactor such that $\dot{n}_{C_i} + \dot{n}_{C_j} = 0$. When there is a steady flow of component C through the reactor and a steady chemical reaction ($\dot{\xi} \neq 0$) inside the reactor, we have

$$\Theta = -\frac{\dot{G}(\text{system})}{T} = -\frac{1}{T}\{\dot{n}_{C+}(\mathscr{G}_{C+} - \mathscr{G}_{C-}) + \dot{\xi}\,\Delta G\}$$
$$= Y_1\Omega_1 + Y_2\Omega_2, \tag{13.11}$$

where $\mathscr{G}_C = \mu_C + \mathscr{K}_C$. I have used subscripts + and − to identify the flow source and flow sink with respect to component C, and I have neglected kinetic energy terms associated with the reactive components.

If we apply the operation $(\delta/\delta\dot{\xi})_{T,\Delta\mathscr{G}_C}$ to Eq. (13.11) and if we make use of assumption IV, we obtain

$$\Delta G = -[R_{C+}](\mathscr{G}_{C+} - \mathscr{G}_{C-}) = -[R_{C+}]\,\Delta\mathscr{G}_C \tag{13.12}$$

as the no-reaction condition in the concentration field; as usual we have

$$[R_{C+}] \equiv (\delta\dot{n}_{C+}/\delta\dot{\xi})_{T,\Delta\mathscr{G}_C}.$$

Since we compute the chemical potentials in Eq. (13.12) *in the terminal parts of the system*—the terminal parts containing reactants being at the opposite end of the concentration field from those containing products—it is not surprising that $\Delta G \neq 0$. We saw in the preceding section of this chapter that a flowing component tends to induce a gradient in the chemical potential of a stationary component in a mixture.

Exercise 13–1. Show that if we assume local equilibrium with respect to the chemical reaction in the concentration field, i.e., if we assume that at each point λ of the concentration field $\Delta G_\lambda = 0$, then we must have

$$\sum \nu_j[r_\lambda^{(Cj)}] - \sum \nu_i[r_\lambda^{(Ci)}] = 0,$$

where $[r_\lambda^{(Cj)}] \equiv \text{grad}\,\mu_\lambda^{(j)}/(\Delta\mathscr{G}_C/\Lambda_C)$, e.g.

Concentration Cells ("Gentle Agitation")

We can also use the methods evolved for treating concentration fields to treat electrochemical cells with transference. We can handle electrochemical cells that do not have concentration-dependent terms in the cell reaction in a straightforward fashion. Consider now a cell such as $Pb(s)|PbCl_2(m)|AgCl(s), Ag(s)$, the potential difference of which is concentration-dependent. Although we cannot make the reaction

$$\tfrac{1}{2}Pb(s) + AgCl(s) = \tfrac{1}{2}PbCl_2(m) + Ag(s) \tag{13.13}$$

take place at a steady rate between terminal parts (all at temperature T) of a system (the concentration of the solute changes continuously during the course of the reaction), we can make the reaction

$$\tfrac{1}{2}Pb(s) + AgCl(s) + \frac{55}{2m} H_2O(g) = Ag(s) + \text{solution}(\tfrac{1}{2}PbCl_2 + \frac{55}{2m} H_2O) \tag{13.14}$$

take place in such a manner. In Figure 13–2 let P be the vapor pressure of water over a solution of molality m with respect to $PbCl_2$. By suitable operation of the electric engine e, we can send a steady current of $\dot{n}$ Faradays per second through the lead chloride solution causing the reaction described in Eq. (13.14) to take place in a steady manner; thus

$$\begin{aligned}\Theta &= -\frac{\dot{G}(\text{system})}{T} + \frac{\dot{W}_0}{T} \\ &= -\frac{1}{T}\{\dot{n}^{(1)}\mu^{(1)} + \tfrac{1}{2}\dot{n}\mu^{(2)} + \dot{n}\mu^{(\text{Ag})} - \tfrac{1}{2}\dot{n}\mu^{(\text{Pb})} - \dot{n}\mu^{(\text{AgCl})} + \dot{n}^{(g)}\mu^{(g)}\} \\ &\qquad + \frac{\dot{W}_e}{T} + \frac{\dot{\theta}N}{T} \\ &= \frac{\dot{n}}{T}\left\{\frac{\dot{W}_e}{\dot{n}} - \frac{55}{2m}(\mu^{(1)} - \mu^{(g)}) - [\tfrac{1}{2}\mu^{(2)} + \mu^{(\text{Ag})} - \tfrac{1}{2}\mu^{(\text{Pb})} - \mu^{(\text{AgCl})}]\right\} + \frac{\dot{\theta}N}{T} \\ &= Y_1\Omega_1 + Y_2\Omega_2,\end{aligned} \tag{13.15}$$

where $\dot{W}_e$ is the rate of supply of work to the electric engine in order to maintain the current $\dot{n}$ through the system, $\dot{\theta}N$ is the rate at which work is being dissipated by the torque-driven stirring motor, the superscript g refers to the gas phase, the superscripts 1 and 2 refer to the components (water and lead chloride, respectively) of the solution phase, $\dot{n}^{(g)} + \dot{n}^{(1)} = 0$, and $\dot{n}^{(1)} = (55/2m)\dot{n}$; the series of terms inside the square brackets in the third form of Eq. (13.15) is just the change in the Gibbs function $\Delta G(13)$ associated with the reaction given in Eq. (13.13). To find the condition for no flow of current in the circuit, we must perform the operation $(\delta/\tilde{\delta}\dot{n})_{T,\ldots}$ on Eq. (13.15). There is some uncertainty as to whether we should hold $\dot{\theta}$ or N constant in the $\delta/\tilde{\delta}\dot{n}$

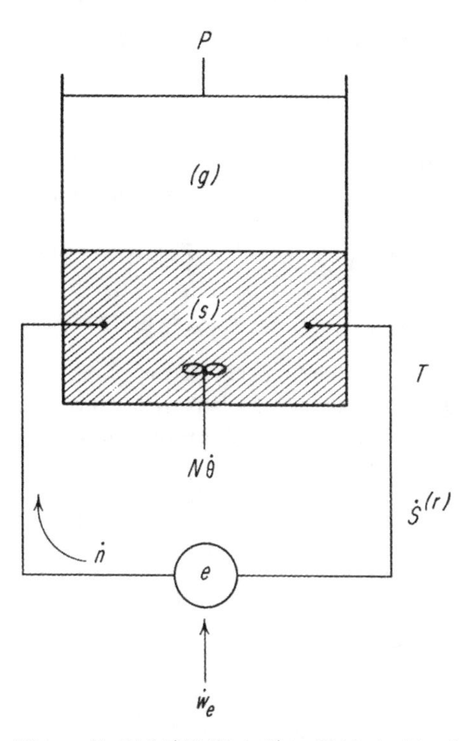

Figure 13–2. The cell $Pb(s)|PbCl_2(m)|AgCl(s),Ag(s)$. An aqueous solution of molality m in $PbCl_2$ is in contact with water vapor at pressure P. Work is supplied to (or received from) an electric engine e at the rate $\dot{W}_e$ so as to maintain a current of $\dot{n}$ Faradays/sec through the system. The solution is stirred gently by a torque-driven motor, the work done by the stirring motor being given by $N\dot{\theta}$, with N being the applied torque and $\dot{\theta}$ being the velocity of rotation (angular) of the stirring propeller.

operation;* in this chapter the uncertainty is of no consequence since I assume in each case that $(\delta(\dot{\theta}N)/\tilde{\delta}\dot{n})_{T,\ldots} \approx 0$.

If we use assumption IV and the assumption just mentioned, we get

$$-\Delta G(13) = F\Delta\psi, \tag{13.16}$$

since at equilibrium $\mu^{(1)} = \mu^{(g)}$ and $(\delta\dot{W}_e/\tilde{\delta}\dot{n})_{T,\ldots} = -F\Delta\psi$, with F being the Faraday and $\Delta\psi$ being the potential difference of the cell in the limit of vanishing current. Equation (13.16) is, of course, a standard result; I produce it here merely to show how concentration-dependent cells are handled by steady-flow methods.

* See the *special fields* section of the Appendix.

Concentration Cells with Transference ("Gentle Agitation")

Consider now a concentration cell with transference (Figure 13-3). If we cause a steady current of $\dot{n}$ Faradays per second to pass through the cell by supplying work at the rate $\dot{W}_e$ to the electric engine e, then we have

$$\Theta = -\frac{1}{T}\{\dot{n}_\alpha^{(1)}\mu_\alpha^{(1)} + \dot{n}_\alpha^{(g)}\mu_\alpha^{(g)} + \dot{n}_\alpha^{(2)}\mu_\alpha^{(2)} + \dot{n}_\beta^{(1)}\mu_\beta^{(1)} + \dot{n}_\beta^{(g)}\mu_\beta^{(g)} + \dot{n}_\beta^{(2)}\mu_\beta^{(2)}\} + \frac{\dot{W}_e}{T} + \frac{\dot{\theta}_\alpha N_\alpha}{T} + \frac{\dot{\theta}_\beta N_\beta}{T}, \quad (13.17)$$

where I have neglected kinetic energy terms. $\dot{\theta}_\omega N_\omega$ is the work of "gentle agitation" supplied by the torque-driven stirring motor at ω so as to keep the solution part at ω nearly uniform in composition; the partial pressures of

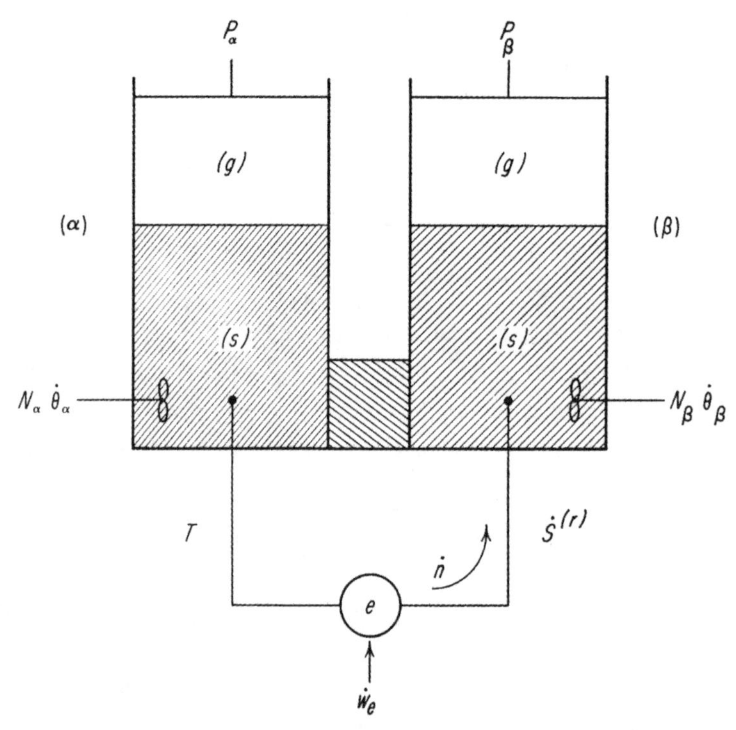

Figure 13-3. The cell Cu|$CuSO_4(m_\alpha)$:$CuSO_4(m_\beta)$|Cu. The solutions at α and β are connected by a porous membrane; each solution is subjected to "gentle agitation." The pressures P_α and P_β are saturated vapor pressures of water as appropriate to the molalities (m) of the solutions. An electric engine e is used to maintain a steady current through the system.

component 1 (water) over the solutions are P_α and P_β. I assume that because of the agitation $\mu_\omega^{(1)} = \mu_\omega^{(g)}$. Let $\dot{n}_\omega^{(L)} \equiv \dot{n}_\omega^{(1)} + \dot{n}_\omega^{(g)}$ represent the rate of influx of component 1 into the solution part at ω due to flow across the porous membrane (link), then the steady state is characterized by

$$\dot{n}_\alpha^{(2)} + \dot{n}_\beta^{(2)} = 0 \quad \text{and} \quad \dot{n}_\alpha^{(L)} + \dot{n}_\beta^{(L)} = 0. \tag{13.18}$$

We can now write Eq. (13.17) as

$$\begin{aligned}\Theta &= -\left\{\dot{n}_\alpha^{(L)}\left(\frac{\mu_\alpha^{(1)} - \mu_\beta^{(1)}}{T}\right) + \dot{n}_\alpha^{(2)}\left(\frac{\mu_\alpha^{(2)} - \mu_\beta^{(2)}}{T}\right)\right\} + \dot{n}\left(\frac{\dot{W}_e}{\dot{n}T}\right) + \dot{\theta}_\alpha \frac{N_\alpha}{T} + \dot{\theta}_\beta \frac{N_\beta}{T} \\ &= Y_1\Omega_1 + Y_2\Omega_2 + Y_3\Omega_3 + Y_4\Omega_4 + Y_5\Omega_5.\end{aligned} \tag{13.19}$$

The application of the operation $(\delta/\bar{\delta}\dot{n})_{T,\Delta\mu^{(1)},\Delta\mu^{(2)},\ldots}$ to Eq. (13.19) [together with assumption IV, the assumption that $(\delta\dot{\theta}_\omega N_\omega/\bar{\delta}\dot{n})_{T,\ldots} \approx 0$, and the relation $(\delta\dot{W}_e/\bar{\delta}\dot{n})_{T,\ldots} = -F\Delta\psi$] leads to the relation

$$-F\Delta\psi = \left(\frac{\delta\dot{n}_\alpha^{(2)}}{\bar{\delta}\dot{n}}\right)_{T,\ldots}(\mu_\alpha^{(2)} - \mu_\beta^{(2)}) + \left(\frac{\delta\dot{n}_\alpha^{(L)}}{\bar{\delta}\dot{n}}\right)_{T,\ldots}(\mu_\alpha^{(1)} - \mu_\beta^{(1)}). \tag{13.20}$$

Let $(\delta\dot{n}_\alpha^{(2)}/\bar{\delta}\dot{n})_{T,\ldots} \equiv \tau_-/2$, then

$$-F\Delta\psi = \frac{\tau_-}{2}(\mu_\alpha^{(2)} - \mu_\beta^{(2)}) + \left(\frac{\delta\dot{n}_\alpha^{(L)}}{\bar{\delta}\dot{n}}\right)_{T,\ldots}(\mu_\alpha^{(1)} - \mu_\beta^{(1)}). \tag{13.21}$$

The usual (Hittorf) assumption made in this case is that $(\delta\dot{n}_\alpha^{(L)}/\bar{\delta}\dot{n})_{T,\ldots} = 0$; i.e., that there is no coupling between the flow of electric current and the flow of solvent across the membrane.

I showed earlier in the chapter that the steady flow of one *component* tends to produce a dragging effect on the stationary components. For the flow of electric current through a solution the positive and negative ions move in opposite directions, thus exerting more of a *shearing* effect than a dragging effect on the solvent molecules. It could well be the case that component flow gives rise to coupling effects ($[R] \neq 0$), whereas counteropposed ionic flow does not $[(\delta\dot{n}_\alpha^{(L)}/\bar{\delta}\dot{n})_{T,\ldots} = 0]$. If we use the Hittorf assumption in conjunction with Eq. (13.21), we get, finally,

$$-F\Delta\psi = \frac{\tau_-}{2}(\mu_\alpha^{(2)} - \mu_\beta^{(2)}). \tag{13.22}$$

Exercise 13–2. Show that for moderately dilute solutions $|\mu_\alpha^{(1)} - \mu_\beta^{(1)}| \ll |\mu_\alpha^{(2)} - \mu_\beta^{(2)}|$, and hence that Eq. (13.22) is approximately true even if the Hittorf assumption does not hold (provided that $(\delta\dot{n}_\alpha^{(L)}/\bar{\delta}\dot{n})_{T,\ldots}$ is not inordinately large).

Hint: Suppose that $m_\alpha^{(2)} = 0.1$ and $m_\beta^{(2)} = 1$, then the ratio of the molalities $m_\omega^{(2)}$ is 1:10 but the ratio of the mole fractions $X_\omega^{(1)}$ is only 56.55:55.65 or 1.016:1.

For the case we are considering (copper electrodes in a copper sulfate solution) the classical thermodynamic treatment [1] yields

$$-F\Delta\psi = \int_{\beta}^{\alpha} \frac{t_-}{2}\, d\mu^{(2)}, \tag{13.23}$$

where t_- is the Hittorf transference number of the anion. The classical result and our result are approximately equivalent if we identify τ_- with the average Hittorf transference number of the anion over the concentration range m_β to m_α.

An Example

Perhaps an example will help clarify some of the conceptions thus far introduced. Consider the situation indicated schematically in Figure 13–4. An aqueous solution of copper sulfate contains copper electrodes a fixed distance Λ_{Cu} apart and has two liquid–vapor interfaces a distance Λ_{vap} apart. Let us apply a potential difference $\Delta\psi$ to the copper electrodes via an electric engine e so as to set up a steady current of electricity through the solution. For such a steady state it follows that

$$0 = -\dot{G}(\text{system}) = T\Theta - \dot{W}_0, \tag{13.24}$$

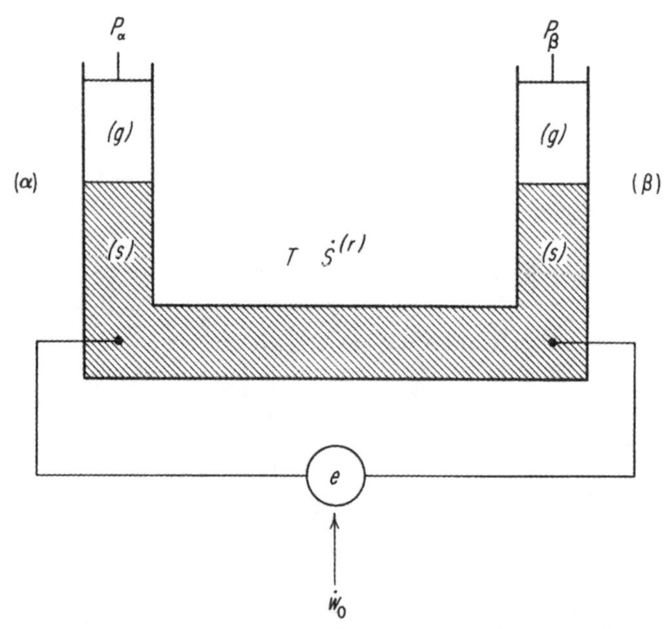

Figure 13–4. Copper electrodes in a copper sulfate solution. The electric engine e receives work at the rate $\dot{W}_0$ and maintains a steady electric current through the solution.

where the number of Faradays of electricity passing through the solution per second is $\dot{n}$, and $\dot{W}_0$ is the rate at which work is supplied to the electric engine; i.e., $\dot{W}_0 = \dot{n}F\Delta\psi$.* If, now, we keep $\Delta\psi$ constant and if we cause a steady current of solvent to pass through the solution by varying the pressures P_α, P_β, we have (neglecting kinetic energy terms)

$$\begin{aligned}\Theta &= -\dot{n}_\alpha^{(g)}\left(\frac{\mu_\alpha^{(g)} - \mu_\beta^{(g)}}{T}\right) + \dot{n}\,\frac{F\Delta\psi}{T} \\ &= Y_1\Omega_1 + Y_2\Omega_2.\end{aligned} \tag{13.25}$$

Our application of the operation $(\delta/\bar{\delta}\dot{n}_\alpha^{(g)})_{T,\Delta\psi}$ to Eq. (13.25) and our use of assumption IV lead to the relation

$$\mu_\alpha^{(g)} - \mu_\beta^{(g)} = \left(\frac{\delta\dot{n}}{\bar{\delta}\dot{n}_\alpha^{(g)}}\right)_{T,\Delta\psi} F\Delta\psi \equiv [R^{(21)}]F\Delta\psi, \tag{13.26}$$

where I have labeled the solvent component 1, have labeled the copper sulfate component 2, and have indicated by the superscript on the $[R]$ coefficient that the concentration field has been set up by a flow related to component 2. In the present case the concentration field is set up between the copper electrodes so the appropriate gradients are $\Delta\mu^{(g)}/\Lambda_{Cu}$ and $\Delta\psi/\Lambda_{Cu}$. Thus, in this case, $[r^{(21)}] \equiv [R^{(21)}](\Lambda_{Cu}/\Lambda_{Cu}) = [R^{(21)}]$ and

$$[r^{(21)}] \approx \frac{RT\ln(P_\alpha/P_\beta)}{F\Delta\psi}. \tag{13.27}$$

In principle it should be possible for us to reverse the roles of the two components by letting the concentration field be set up by the flow of component 1, and we can carry through the analysis so as to get the migrational equilibrium condition for component 2 (by making use of the operation $(\delta/\bar{\delta}\dot{n})_{T,\Delta\mu^{(1)}}$). The analog of Eq. (13.26) is then

$$F\Delta\psi = \left(\frac{\delta\dot{n}_\alpha^{(g)}}{\bar{\delta}\dot{n}}\right)_{T,\Delta\mu^{(1)}} (\mu_\alpha^{(g)} - \mu_\beta^{(g)}) \equiv [R^{(12)}](\mu_\alpha^{(g)} - \mu_\beta^{(g)}). \tag{13.28}$$

In this case the concentration field extends from one liquid–vapor interface to the other, a distance of Λ_{vap}, and the copper electrodes are inside the concentration field. The appropriate average gradients are $\Delta\mu^{(g)}/\Lambda_{vap}$ and $\Delta\psi/\Lambda_{Cu}$. Equation (13.28) thus yields

$$\frac{F\Delta\psi}{\Lambda_{Cu}} = [R^{(12)}]\left(\frac{\Lambda_{vap}}{\Lambda_{Cu}}\right)\left(\frac{\mu_\alpha^{(g)} - \mu_\beta^{(g)}}{\Lambda_{vap}}\right) \tag{13.29}$$

or

$$[r^{(12)}] = \frac{F\Delta\psi/\Lambda_{Cu}}{\Delta\mu^{(g)}/\Lambda_{vap}} = \left(\frac{F\Delta\psi}{RT\ln(P_\alpha/P_\beta)}\right)\left(\frac{\Lambda_{vap}}{\Lambda_{Cu}}\right). \tag{13.30}$$

* We must establish a convention about the direction of electric current flow. The sign of $\Delta\psi = \psi' - \psi''$ depends on our sign convention for the direction of electric current flow.

The work of a previous section implies that

$$[r^{(12)}][r^{(21)}] \approx 1 \tag{13.31}$$

if the coefficients are computed for two states such that $\Delta\mu^{(g)}/\Lambda_{\text{vap}}$ for Eq. (13.30) is the same as $F\Delta\psi/\Lambda_{\text{Cu}}$ for Eq. (13.27)* and the temperature and the average composition of the solutions are the same.

Although in principle we can test Eq. (13.31) experimentally, it is unlikely that we can measure the coefficient $[r^{(12)}]$ with any ease due to the extreme slowness of the diffusion process in solutions. It is quite likely that in order to test relations of the types (13.3) and (13.31) it will be necessary to go to 3-component systems so that two sets of electrodes can be inserted into the solution and two different potential differences applied. Then if we call the species participating in the two electrode reactions components 2 and 3, it should be possible for us to determine experimentally whether the relations

$$[r_\lambda^{(23)}][r_\lambda^{(32)}] \approx 1 \tag{13.32}$$

and

$$\sum_{i=1}^{3} \langle X^{(i)}\rangle [r_\lambda^{(3i)}] = 0 \quad \text{(say)} \tag{13.33}$$

hold.

Reference

1. G. N. Lewis and M. Randall, *Thermodynamics*, second edition (edited by K. S. Pitzer and G. Brewer) (McGraw-Hill Book Company, New York, 1961), Chapter 24.

* In Eq. (13.27) divide the numerator and the denominator of the fraction by Λ_{Cu}.

14

Migrational Equilibrium in Polycurrent Fields

General Remarks

The problems of migrational equilibrium that I have dealt with thus far have been problems pertaining to monocurrent fields; i.e., the system* has been the site of a single steady current of heat $\dot{Q}$, of a chemical component $\dot{n}_k$, of an electric current I, or the site of a steady chemical reaction consuming $\dot{\xi}$ reaction measures per unit time. The steady current in each case had associated with it a gradient in an intensive property, and the problem was to determine the migrational equilibrium condition in the particular gradient field. For the set of current variables $\dot{Q}$, $\dot{n}_k$, $\dot{n}_j$, I, $\dot{\xi}$ the various monocurrent effects are related to coefficients of the type $\delta(\text{current}_1)/\bar{\delta}(\text{current}_2)$: $\delta\dot{Q}/\bar{\delta}\dot{n}_k$, thermomolecular pressure effect; $\delta\dot{Q}/\bar{\delta}I$, thermocouple; $\delta\dot{Q}/\bar{\delta}\dot{\xi}$, reaction equilibrium in a polythermal field; $\delta\dot{n}_k/\bar{\delta}\dot{n}_j$, concentration field; $\delta\dot{n}_k/\bar{\delta}I$, concentration field with transference; $\delta\dot{n}_k/\bar{\delta}\dot{\xi}$, reaction equilibrium in a concentration field; $\delta I/\bar{\delta}\dot{\xi}$, relation between the Gibbs free energy change and the potential difference for a cell.

Consider the case now where there are two or more independent currents traversing the system* and we seek a migrational equilibrium condition in such a polycurrent field. Let us represent a heat current by $\dot{Q}$ and let J_i stand generally for any of the currents $\dot{n}_k$, $\dot{n}_j$, I, or $\dot{\xi}$; also let us use the notation (*independent current*$_1$, *independent current*$_2$)/*current with respect to which*

* In the state of migrational equilibrium.

migrational equilibrium is sought to describe bicurrent fields. Bicurrent effects, then, are either of the form $(\dot{Q}, J_1)/J_2$ or $(J_1, J_2)/J_3$. For the first case we describe the phenomena in terms of the quantities $[S]$ and $[R^{(12)}] \equiv \delta J_1/\delta J_2$, where we adapt the $[S]$ quantity to the new context; for the second case the appropriate quantities are $[R^{(13)}] \equiv \delta J_1/\delta J_3$ and $[R^{(23)}] \equiv \delta J_2/\delta J_3$.

To illustrate the general procedure, I investigate two cases of the type $(\dot{Q}, J_1)/J_2$; I analyzed the type $(\dot{Q}, \dot{n}_k)/I$ in a previous chapter in discussing the initial potential difference of thermocells.

Polythermal Concentration Field

The situation that I propose to investigate is indicated schematically in Figure 14–1. A 2-component MIX region communicates with terminal parts containing pure components 1 and 2 in each of two heat reservoirs kept at temperatures T_α and T_β; the MIX region functions as a link between the two heat reservoirs; and we maintain a steady flow of component 1 from *a* to *b*. Our problem is to find the migrational equilibium condition for component 2 in the polythermal concentration field.

In the usual manner I superimpose a steady flow of component 2 on the flows of heat and of component 1 and then pass to the limit of vanishing flow

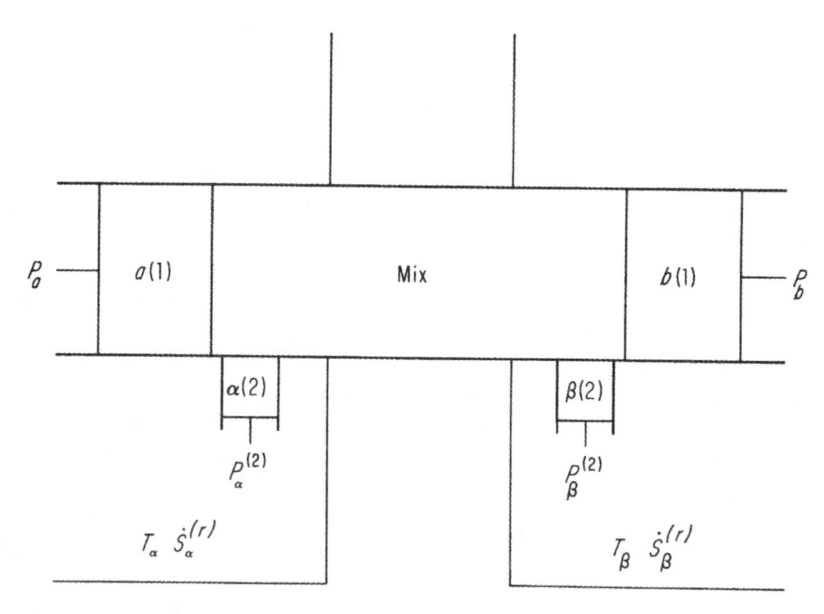

Figure 14–1. Steady flow of heat and component 1 through a 2-component MIX region; arbitrary linkage.

of component 2 via a sequence of steady-flow states, keeping $\mathscr{G}_a$, $\mathscr{G}_b$, T_α, T_β, $\langle T(\alpha\beta)\rangle$ constant. Equation (6.39) yields in this case*

$$\left(\frac{\delta \dot{G}_a}{\tilde{\delta}\dot{n}_\alpha^{(2)}}\right)_{\Omega'} + \left(\frac{\delta \dot{G}_b}{\tilde{\delta}\dot{n}_\alpha^{(2)}}\right)_{\Omega'} + \mu_\alpha^{(2)} - \mu_\beta^{(2)} + \left(\frac{\delta \dot{q}(\alpha\beta)}{\tilde{\delta}\dot{n}_\alpha^{(2)}}\right)_{\Omega'} + T_\alpha\left\{\left(\frac{\delta \dot{S}_\alpha(\text{in})}{\tilde{\delta}\dot{n}_\alpha^{(2)}}\right)_{\Omega'} - \left(\frac{\delta \dot{S}_\alpha(\text{out})}{\tilde{\delta}\dot{n}_\alpha^{(2)}}\right)_{\Omega'}\right\} + T_\beta\left\{\left(\frac{\delta \dot{S}_\beta(\text{in})}{\tilde{\delta}\dot{n}_\alpha^{(2)}}\right)_{\Omega'} - \left(\frac{\delta \dot{S}_\beta(\text{out})}{\tilde{\delta}\dot{n}_\alpha^{(2)}}\right)_{\Omega'}\right\} = 0. \quad (14.1)$$

As the component-flows either initiate or terminate in each heat reservoir, we may formally write the last two terms on the left-hand side of Eq. (14.1) in terms of $\dot{S}_\omega(\text{in})$ with the sign of the term being taken care of by the signs of the quantities $\dot{n}_z$ and $\dot{n}_\omega^{(2)}$; thus let

$$\left(\frac{\delta \dot{S}_\omega(\text{in})}{\tilde{\delta}\dot{n}_\alpha^{(2)}}\right)_{\Omega'} \equiv [S_z]\left(\frac{\delta \dot{n}_z}{\tilde{\delta}\dot{n}_\alpha^{(2)}}\right)_{\Omega'} + [S_\omega^{(2)}]\left(\frac{\delta \dot{n}_\omega^{(2)}}{\tilde{\delta}\dot{n}_\alpha^{(2)}}\right)_{\Omega'}; \quad (14.2)$$

it follows then that

$$[R_\alpha^{(12)}](\mathscr{G}_a - \mathscr{G}_b) + \mu_\alpha^{(2)} - \mu_\beta^{(2)} + T_\alpha\{[R_\alpha^{(12)}][S_a] + [S_\alpha^{(2)}]\} - T_\beta\{[R_\alpha^{(12)}][S_b] + [S_\beta^{(2)}]\} - [h(\beta\alpha)] = 0 \quad (14.3)$$

and that

$$[R_\alpha^{(12)}]\{[\mathscr{G}_a] - [\mathscr{G}_b]\} + [G_\alpha^{(2)}] - [G_\beta^{(2)}] = 0, \quad (14.4)$$

where $\mathscr{G} \equiv \mu + \mathscr{K}$ with $\mathscr{K}$ being the molar macroscopic kinetic energy and $[R_\alpha^{(12)}] \equiv (\delta\dot{n}_a/\tilde{\delta}\dot{n}_\alpha^{(2)})_{\Omega'}$. Compare Eq. (14.4) to Eq. (4.10).

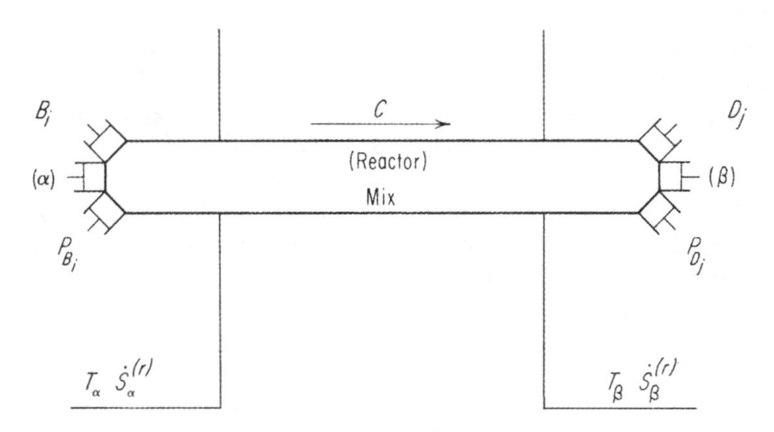

Figure 14–2. Chemical reaction in a polythermal concentration field.

* There is no work other than pressure–volume work exchanged between system and surroundings; hence $\dot{W}_0 = 0$.

Assumption IV implies that

$$[R_\alpha^{(12)}][S_a] + [S_\alpha^{(2)}] - [R_\alpha^{(12)}][S_b] - [S_\beta^{(2)}] = [s(\beta\alpha)]. \qquad (14.5)$$

Consider now a chemical reaction in a polythermal concentration field. In Figure 14–2 let a reaction-neutral component (C) flow between terminal parts at α and β, and let reactants B_i communicate with the MIX region at α and let products D_j communicate similarly at β. Let the chemical reaction be represented by

$$\sum \nu_i B_i = \sum \nu_j D_j, \qquad (14.6)$$

and let $\dot{\xi}$ reaction measures per second flow from reactants to products ($\dot{\xi} = -\dot{n}_i/\nu_i = \dot{n}_j/\nu_j$); then a set of equations analogous to Eqs. (14.1) to (14.4) holds:

$$\left(\frac{\delta \dot{G}_\alpha^{(C)}}{\delta \dot{\xi}}\right)_{\Omega'} + \left(\frac{\delta \dot{G}_\beta^{(C)}}{\delta \dot{\xi}}\right)_{\Omega'} - \sum \nu_i \mu_\alpha^{(B_i)} + \sum \nu_j \mu_\beta^{(D_j)} + T_\alpha \left(\frac{\delta \dot{S}_\alpha(\text{in})}{\delta \dot{\xi}}\right)_{\Omega'} + T_\beta \left(\frac{\delta \dot{S}_\beta(\text{in})}{\delta \dot{\xi}}\right)_{\Omega'} + \left(\frac{\delta \dot{q}}{\delta \dot{\xi}}\right)_{\Omega'} = 0, \qquad (14.7)$$

$$\left(\frac{\delta \dot{S}_\alpha(\text{in})}{\delta \dot{\xi}}\right)_{\Omega'} \equiv [S_\alpha^{(C)}]\left(\frac{\delta \dot{n}_\alpha^{(C)}}{\delta \dot{\xi}}\right)_{\Omega'} + \sum [S_\alpha^{(B_i)}]\left(\frac{\delta \dot{n}_\alpha^{(B_i)}}{\delta \dot{\xi}}\right)_{\Omega'} \quad \text{(e.g.)}, \qquad (14.8)$$

$$[R_\alpha^{(C)}]\{[\mathscr{G}_\alpha^{(C)}] - [\mathscr{G}_\beta^{(C)}]\} + \sum \nu_j [G_\beta^{(D_j)}] - \sum \nu_i [G_\alpha^{(B_i)}] = 0, \qquad (14.9)$$

with $[R_\alpha^{(C)}] \equiv (\delta \dot{n}_\alpha^{(C)}/\delta \dot{\xi})_{\Omega'}$.

Exercise 14–1. In Figure 14–3 let there be a 2-component metal amalgam phase in contact with mercury vapor at two temperatures T_α, T_β and let the amalgam be traversed by a steady electric current I. Analyze the situation according to the principles we have established and display the migrational equilibrium condition in terms of the coefficient $[R_\alpha^{(I)}] \equiv (\delta I/\delta \dot{n}_\alpha^{(g)})_{\Delta\psi, T_\alpha, T_\beta, \langle T\rangle}$, employing "gentle agitation" at α and β if necessary. Comment on the physical interpretation of the $[R_\alpha^{(I)}]$ coefficient and on the likelihood that the coefficient is significantly different from zero.

Catalog of Binary Effects

Instead of concentrating on the currents traversing a system, we can consider the gradients in intensive properties induced by the various flows through the system; we can consider one gradient to be the "cause" of some of

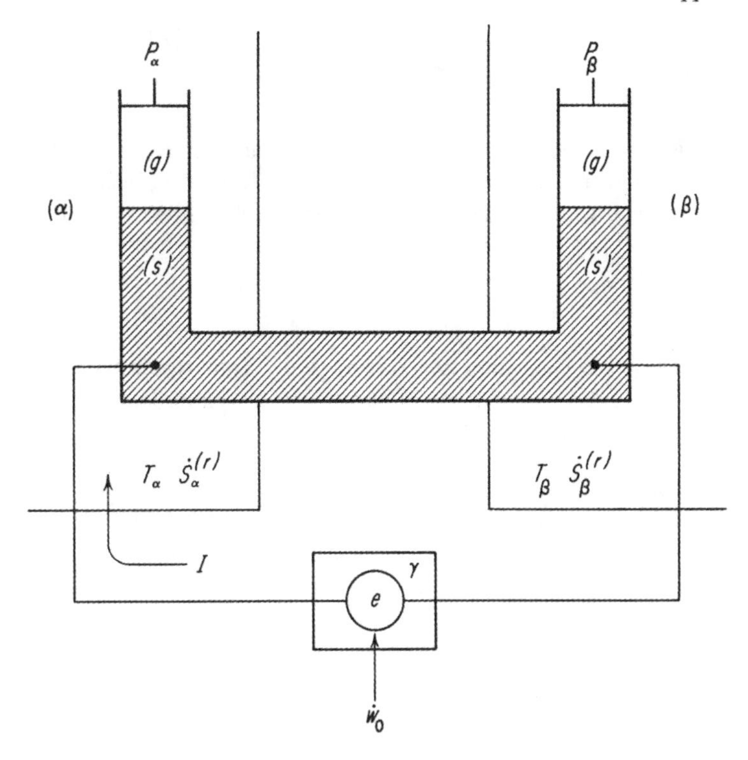

Figure 14–3. Metal amalgam in contact with mercury vapor and traversed by a steady electric current I in a polythermal field; e—an electric engine receiving work at the rate $\dot{W}_0$ and maintaining a potential difference $\Delta\psi$; "gentle agitation" at α and β if necessary.

the other gradients and we can call the ratio of two such gradients an "effect." If we take the set of variables T, P, φ, ψ, μ_2, μ_3, ..., μ_ν (X_2, X_3, ..., X_ν) —temperature, pressure, gravitational potential, electric potential, $\nu - 1$ concentration variables, either chemical potentials μ_i or mole fractions X_i—then the various binary steady-state effects (if we omit ordinary thermostatic effects such as osmotic pressure and the piezoelectric effect) together with the appropriate migrational equilibrium conditions are as follows:

grad P/grad T—thermopiestic effect, $[G]$ = const.
grad φ/grad T—thermogravitational effect, $[G] + M\varphi$ = const.
grad ψ/grad T—thermoelectric effect, $[G] + zF\psi$ = const.
grad X_i/grad T—thermoconcentration effect (Soret effect), $[G_i]$ = const.
grad μ_i/grad T—thermocomposition effect, $[G_i] + [R][\mathcal{G}_k]$ = const.
grad μ_i/grad P—piezocomposition effect, $\mu_i + [R]\mathcal{G}_k$ = const.

grad μ_i/grad φ—gravicomposition effect, $\mu_i + M\varphi + [R]\mathscr{G}_k = \text{const.}$
grad μ_i/grad ψ—electrocomposition effect, $\mu_i + zF\psi + [R]\mathscr{G}_k = \text{const.}$
grad μ_i/grad μ_k—chemicomposition effect, $\mu_i + [R]\mathscr{G}_k = \text{const.}$

where M is the molecular weight, z is the "valence charge," and for those effects involving grad μ_i there is interaction between a concentration field involving a flowing component and another field (thermal, piestic, gravitational etc.).

PART III. COMMENTS

15

Conclusion

General Remarks

I have shown that nonequilibrium situations involving stationary states and steady-rate processes can be handled in an entirely thermodynamic fashion; I needed only the concepts *heat*, *work*, *energy*, and concepts derived therefrom. The various bracketed quantities that I introduced are perfectly good *thermodynamic* quantities that give us the means whereby to describe pertinent features of irreversible processes; these quantities, however, are characteristic of the complete system (contents plus container) being analyzed. Here, then, is a dilemma of nonequilibrium thermodynamics: we can easily and rigorously describe nonequilibrium situations in terms of general (bracketed) thermodynamic quantities; the quantities that we so use, however, are generally apparatus-dependent. We may accept this dilemma as a challenge. Can we in each case so devise a series of experiments as to be able to extrapolate the phenomena to an apparatus-independent limit? Thus far experimenters have given this matter but little attention—the thermomolecular pressure case and the forced vaporization case being the only two cases where an apparatus-independent limit has been established *experimentally*.* Of course, for engineering purposes it is no great tragedy if quantities prove to be apparatus-dependent, provided that such quantities depend only on general parameters of the apparatus (tube diameters, thermal conductivity of the wall material, etc.).

* In the thermomolecular pressure case, the Knudsen equation represents an apparatus-independent low-pressure limiting equation with a straightforward kinetic theory interpretation.

Although we tend to look upon the apparatus-dependence of the steady-state thermodynamic quantities as a nuisance, to be circumvented by suitable extrapolation procedures, from another point of view this apparatus-dependence is a characteristic and distinguishing feature of nonequilibrium thermodynamics. In (equilibrium) thermostatics we talk about walls and constraints when discussing a given system or process; our discussion, however, usually deals with but the functional characteristics, in the context of the thermostatic language, of the walls and constraints; and we avoid, as much as possible, saying anything about their physical structure. In non-equilibrium thermodynamics we continue to concern ourselves with walls and constraints, but now we must go into more detail concerning these items and their influence on the process being studied. Although we avoid giving gratuitous detail to our walls and constraints, we find, nevertheless, that their properties enter more explicitly into the discussions of nonequilibrium thermodynamics than they do into the discussions of thermostatics (see, e.g., the discussion of linkage-classes at the end of Chapter 5). It is, after all, matters of this sort that make steady states different from static ones.

The Assumptions I–IV

Consider again the basic assumptions I–IV in the light of the many analyses that we have carried through. Assumptions I and II are quite clear and of general validity. We may phrase assumptions III and IV thus:

(III) We find the condition of migrational equilibrium with respect to a given quantity in a field with ν independent steady currents by inducing a steady flow of the given quantity between terminal parts, keeping constant an appropriate set of variables, and observing the behavior of the system in the $(\nu + 1)$ current field as it approaches the original state (with ν currents) via a sequence of steady states.

(IV) If a system that is the site of $\nu + 1$ steady (independent) currents converges via a sequence of steady states on a state involving only ν steady currents in such a way that the affinities conjugate to the ν nonvanishing currents are maintained constant, then the state with ν currents is a state of minimum rate of entropy production relative to the (steady) states with $\nu + 1$ currents.

There is, of course, a dual form of the preceding statement with the roles of currents and affinities reversed. Although assumption III seems to be of general validity, the form in which I have heretofore stated assumption IV is one of restricted range: I have made no allowance for the presence of magnetic or centrifugal fields. I include a brief discussion of these *special fields* in the Appendix.

I now wish to make some additional remarks about the rate of entropy production and about assumption IV. We get an explicit resolution of the rate of entropy production into currents and affinities by eliminating a current between the First Law energy balance equation and the equation $\Theta = \dot{S}(\text{system}) + \dot{S}(\text{surroundings})$. In the case of a monothermal field we always eliminate the quantity $\dot{Q}^{(r)}$ to obtain the equation

$$\Theta = -\dot{G}(\text{system})T^{-1} + \dot{W}_0 T^{-1} = -\sum \dot{n}_i \mathscr{G}_i T^{-1} + \sum Y_{0x}\mathscr{F}_{0x}T^{-1}.$$

In a polythermal field we usually have several currents that can be eliminated between the basic pair of equations; thus we wind up with a set of equivalent resolutions of the rate of entropy production into currents and affinities. As an example, consider the thermomolecular pressure effect (Figure 5–1). The two basic equations are Eqs. (5.6) and (5.7):

$$\dot{n}_\alpha \Delta\bar{H} + \dot{Q}_\alpha^{(r)} + \dot{Q}_\beta^{(r)} + \dot{q} = 0, \tag{5.6}$$

$$\Theta = \dot{n}_\alpha \Delta\bar{S} + \frac{\dot{Q}_\alpha^{(r)}}{T_\alpha} + \frac{\dot{Q}_\beta^{(r)}}{T_\beta} + \frac{\dot{q}}{\langle T\rangle}, \tag{5.7}$$

where $\Delta\bar{Z} \equiv \bar{Z}_\alpha - \bar{Z}_\beta$. Upon eliminating one current at a time between Eqs. (5.6) and (5.7), we get the following set of resolutions of Θ into currents and affinities:

$$\Theta = \dot{n}_\alpha\left(\Delta\bar{S} - \frac{\Delta\bar{H}}{T_\beta}\right) + \dot{Q}_\alpha^{(r)}\left(\frac{1}{T_\alpha} - \frac{1}{T_\beta}\right) + \dot{q}\left(\frac{1}{\langle T\rangle} - \frac{1}{T_\beta}\right) \tag{15.1}$$

$$= \dot{n}_\alpha\left(\Delta\bar{S} - \frac{\Delta\bar{H}}{T_\alpha}\right) + \dot{Q}_\beta^{(r)}\left(\frac{1}{T_\beta} - \frac{1}{T_\alpha}\right) + \dot{q}\left(\frac{1}{\langle T\rangle} - \frac{1}{T_\alpha}\right) \tag{15.2}$$

$$= \dot{n}_\alpha\left(\Delta\bar{S} - \frac{\Delta\bar{H}}{\langle T\rangle}\right) + \dot{Q}_\alpha^{(r)}\left(\frac{1}{T_\alpha} - \frac{1}{\langle T\rangle}\right) + \dot{Q}_\beta^{(r)}\left(\frac{1}{T_\beta} - \frac{1}{\langle T\rangle}\right) \tag{15.3}$$

$$= \dot{Q}_\alpha^{(r)}\left(\frac{1}{T_\alpha} - \frac{\Delta\bar{S}}{\Delta\bar{H}}\right) + \dot{Q}_\beta^{(r)}\left(\frac{1}{T_\beta} - \frac{\Delta\bar{S}}{\Delta\bar{H}}\right) + \dot{q}\left(\frac{1}{\langle T\rangle} - \frac{\Delta\bar{S}}{\Delta\bar{H}}\right). \tag{15.4}$$

Note the following things about the set of relations (15.1) to (15.4). If we are interested in the condition of migrational equilibrium with respect to mass flow ($\dot{n}_\alpha = 0$), we will not use Eq. (15.4). In addition, we will tend not to use relation (15.3) because of its inconvenience in the case of full-flux linkage. We are left then with the two remaining resolutions (15.1) and (15.2).

We *could* combine these two expressions to get a resultant expression that is more symmetric in the indices α and β:

$$\Theta = \dot{n}_\alpha\left(\Delta\bar{S} - \frac{\Delta\bar{H}}{T_m}\right) + \tfrac{1}{2}(\dot{Q}_\alpha^{(r)} - \dot{Q}_\beta^{(r)})\left(\frac{1}{T_\alpha} - \frac{1}{T_\beta}\right) + \dot{q}\left(\frac{1}{\langle T\rangle} - \frac{1}{T_m}\right), \quad (15.5)$$

where $T_m^{-1} \equiv \tfrac{1}{2}(T_\alpha^{-1} + T_\beta^{-1})$, but at the cost of having to consider the combination $\tfrac{1}{2}(\dot{Q}_\alpha^{(r)} - \dot{Q}_\beta^{(r)})$ as a *single current*.

I prefer *not* to deal with combinations such as Eq. (15.5), and I feel that the separate resolutions of Θ (Eqs. 15.1 to 15.4) are each equally valid. I also feel that the range of validity of the *petit* principles of entropy production (Chapter 3) is the same for each member of this set of equations. In this regard consider the relation $(\delta\Omega_1/\bar{\delta} Y_1)_{\Omega',\beta} > 0$ for Eqs. (15.1) and (15.2). For Eq. (15.1) we have, as we saw in Chapter 5,

$$(\delta\Omega_1/\bar{\delta} Y_1)_{\Omega',\beta} = -T_\beta^{-1}\{\bar{V}_\alpha - T_\alpha(\partial\bar{V}_\alpha/\partial T_\alpha)_{P_\alpha} + T_\beta(\partial\bar{V}_\alpha/\partial T_\alpha)_{P_\alpha}\}(\delta P_\alpha/\bar{\delta}\dot{n}_\alpha)_{\beta,T_\alpha,\langle T\rangle},$$

whereas for Eq. (15.2) we have $(\delta\Omega_1/\bar{\delta} Y_1)_{\Omega',\beta} = -T_\alpha^{-1}\bar{V}_\alpha(\delta P_\alpha/\bar{\delta}\dot{n}_\alpha)_{\beta,T_\alpha,\langle T\rangle}$. Thus the condition $(\delta\Omega_1/\bar{\delta} Y_1)_{\Omega',\beta} > 0$ leads directly to the condition of stability $(\delta P_\alpha/\bar{\delta}\dot{n}_\alpha)_{\beta,T_\alpha,\langle T\rangle} < 0$ in the case of Eq. (15.2); in Eq. (15.1), however, we obtain the condition of stability $(\delta P_\alpha/\bar{\delta}\dot{n}_\alpha)_{\beta,T_\alpha,\langle T\rangle} < 0$ only if the quantity in the braces { } is positive. We saw in Chapter 5 that it is difficult to decide *a priori* about the sign of { } for all physically compatible pairs of states α,β subject to the migrational equilibrium constraint. Now it is conceivable that for some pair or pairs of states α,β (migrational equilibrium constraint) the quantity { } might be negative; in such circumstances we should have the range of validity of the relation $(\delta\Omega_1/\bar{\delta} Y_1)_{\Omega',\beta} > 0$ depending on the choice of current-affinity representation for Θ: a lucky choice of current-affinity representation would give us an unrestricted range of validity for the relation $(\delta\Omega_1/\bar{\delta} Y_1)_{\Omega',\beta} > 0$, whereas an unlucky choice would give us a restricted range of validity for the equivalent relation. Now I maintain that all of the representations (15.1) to (15.4) are equally valid with respect to the *petit* principles of entropy production; consequently I maintain that the quantity { } *must* be positive for all physically compatible pairs of states α,β subject to the migrational equilibrium constraint.

The preceding considerations are pertinent to the *grand* principles of entropy production as well as to the *petit* principles. I mentioned in Chapter 5 that the *grand* principle $(\partial\Omega_i/\partial Y_i)_{\Omega',\sigma} > 0$ might be of limited validity. I now wish to discuss the dependence of the rate of entropy production on the currents or the affinities and the implications of this dependence for the range of validity of the *grand* principle just mentioned. Consider the thermomolecular pressure case once more (full-flux linkage). The rate of entropy production Θ is fully determined when the two states α and β are given; i.e., $\Theta = \Theta(T_\alpha, P_\alpha, T_\beta, P_\beta) = \Theta(\alpha, \beta)$. Now take state β as the reference thermostatic state and write $\dot{n}_\alpha = \dot{n}_\alpha(T_\alpha, P_\alpha, \beta)$, $\dot{Q}_\alpha^{(r)} = \dot{Q}_\alpha^{(r)}(T_\alpha, P_\alpha, \beta)$. In order to invert these relations so as to get $T_\alpha = T_\alpha(\dot{n}_\alpha, \dot{Q}_\alpha^{(r)}, \beta)$, $P_\alpha = P_\alpha(\dot{n}_\alpha, \dot{Q}_\alpha^{(r)}, \beta)$ and $\Theta = \Theta(\dot{n}_\alpha, \dot{Q}_\alpha^{(r)}, \beta)$, we must have a nonvanishing value for the Jacobian of the transformation; i.e., $\partial(\dot{n}_\alpha, \dot{Q}_\alpha^{(r)})/\partial(T_\alpha, P_\alpha) \neq 0$. The Jacobian of the transformation has the value

$$\frac{\partial(\dot{n}_\alpha, \dot{Q}_\alpha^{(r)})}{\partial(T_\alpha, P_\alpha)} = \left(\frac{\partial \dot{n}_\alpha}{\partial T_\alpha}\right)_{\beta, P_\alpha} \left(\frac{\partial \dot{Q}_\alpha^{(r)}}{\partial P_\alpha}\right)_{\beta, T_\alpha} - \left(\frac{\partial \dot{n}_\alpha}{\partial P_\alpha}\right)_{\beta, T_\alpha} \left(\frac{\partial \dot{Q}_\alpha^{(r)}}{\partial T_\alpha}\right)_{\beta, P_\alpha}$$
$$= -\left(\frac{\partial \dot{n}_\alpha}{\partial P_\alpha}\right)_{\beta, T_\alpha} \left(\frac{\partial \dot{Q}_\alpha^{(r)}}{\partial T_\alpha}\right)_{\beta, P_\alpha} \left[1 - \left(\frac{\partial P_\alpha}{\partial T_\alpha}\right)_{\beta, \dot{n}_\alpha} \left(\frac{\partial P_\alpha}{\partial T_\alpha}\right)^{-1}_{\beta, \dot{Q}_\alpha^{(r)}}\right]. \quad (15.6)$$

Although the factor $(\partial \dot{n}_\alpha/\partial P_\alpha)_{\beta, T_\alpha}$ is certainly nonvanishing and the factor $(\partial \dot{Q}_\alpha^{(r)}/\partial T_\alpha)_{\beta, P_\alpha}$ is probably nonvanishing, it is difficult to say anything definite about the remaining factor [ ]. Thus, although the current form of the rate of entropy production $\Theta = \Theta(\dot{n}_\alpha, \dot{Q}_\alpha^{(r)}, \beta)$ is intuitively very reasonable, we cannot be sure that the shift from $\Theta = \Theta(\alpha, \beta)$ to $\Theta = \Theta(\dot{n}_\alpha, \dot{Q}_\alpha^{(r)}, \beta)$ will always proceed smoothly and unambiguously.

The case is clearer if we deal with the affinity form of the rate of entropy production $\Theta = \Theta(\Omega_1, \Omega_2, \beta)$. The Jacobian of the transformation now has the value

$$\frac{\partial(\Omega_1, \Omega_2)}{\partial(T_\alpha, P_\alpha)} = -\frac{1}{T_\alpha^2 T_\beta}\left\{\bar{V}_\alpha - T_\alpha\left(\frac{\partial \bar{V}_\alpha}{\partial T_\alpha}\right)_{P_\alpha} + T_\beta\left(\frac{\partial \bar{V}_\alpha}{\partial T_\alpha}\right)_{P_\alpha}\right\}, \quad (15.7)$$

and the Jacobian vanishes when the quantity in the braces { } has the value zero. Thus, when the quantity { } has the value zero, not only does the *grand* principle $(\partial\Omega_1/\partial\dot{n}_\alpha)_{\beta, T_\alpha} > 0$ fail for Eq. (15.1), but even the simple shift from $\Theta = \Theta(\alpha, \beta)$ to $\Theta = \Theta(\Omega_1, \Omega_2, \beta)$ becomes ambiguous. I argued earlier that the *petit* principles of entropy production for ordinary steady state situations were valid for any of the equivalent current-affinity representations of Θ; I now argue that to secure a maximum range of validity for the *grand* principles of entropy production we should (if possible) choose a current-affinity representation of Θ and a thermostatic reference state such that neither of the two Jacobians of transformation* can vanish. If we can make such a choice, it may be that then the *grand* principles of entropy production, as well as the *petit* principles, for ordinary steady-state situations will be of unlimited validity.

The Stability Problem for Steady States

Consider the following statement:

(A) A steady state has minimum rate of entropy production *relative to transient states* with the same boundary conditions.

* The transformation of Θ to the current form or to the affinity form.

Statement A and the verbal statement of assumption IV seemingly refer to two different physical situations; statement A is Prigogine's well-known theorem [1–4] about the stability of a steady state relative to a class of transient states.

Prigogine's Theorem

Prigogine and his coworkers [3, 5, 6] have shown that for a certain class of transient states* (i) close to (thermostatic) equilibrium, statement A holds, (ii) far away from equilibrium, statement A is not generally true. The procedure followed in their analysis is to consider a system in a steady state and, by instantaneous step-function forcing, to change the boundary conditions of the system, thereafter holding the new boundary conditions constant. The properties of the system are studied as it adjusts to the new set of boundary conditions. The system thus starts out in a steady state and then drifts through a series of *transient states* toward some final state compatible with the new boundary conditions. The rate of entropy production in the final state is compared to the rate of entropy production in the *transient states.* Prigogine and coworkers have shown that far away from equilibrium for systems with time-independent boundary conditions (i) the rate of entropy production in the final state need not have extremum properties relative to the rate of entropy production in the transient states, and (ii) the final state may not be a steady state at all—the system may oscillate or circulate about some average configuration. Denbigh [7] and Klein [8] have also pointed out the limitations of Prigogine's theorem.

This sort of stability problem (steady state versus related transient states) has been of concern to a number of people; thus Bak [9] and Ono [10] have published excellent overviews of the present status of the stability problem, and Li [4] has sought to approach the stability question by defining a thermokinetic potential and examining its extremum properties. What is being sought is a steady-state analog of the thermostatic criterion of equilibrium, so let us take a moment to review that thermostatic criterion.

The condition of equilibrium for an isolated system in thermostatics is that the entropy of the system be a maximum. In making use of the entropy-maximum principle, we must find a class of states with which to compare the equilibrium state, and that poses something of a problem since the entropy function is only defined for equilibrium states. The way out of the dilemma is

* If we wish to talk about rates of change of thermodynamic quantities in *transient states*, we must make extra, far-reaching assumptions about the local thermodynamic state at an arbitrary point in the system and about the description of that state in terms of thermostatic functional forms (see Chapter 11 of the book cited in reference [1] and page 93 of the book cited in reference [3]).

to consider the virtual states to which the equilibrium state is to be compared as equilibrium states also, but *more constrained* equilibrium states than the one that we are investigating [11, 12]. What we do in effect is to associate with a nonequilibrium situation a constrained equilibrium state that approximates the actual gradients in temperature, composition, etc., through the use of a sufficient number of partitioning adiabatic or diathermal walls. The entropy-maximum principle then compares the class of constrained equilibrium states associated with the actual nonequilibrium states to the (unconstrained) final equilibrium state.

Consider, for example, the following case. Divide an isolated system into two equal parts (α, β) by a fixed, rigid adiabatic partition and consider all states such that $U_\alpha + U_\beta =$ constant. The state of maximum entropy for the given situation is then the equilibrium state that subsists in the absence of the adiabatic partition [11, 12]. Mathematically the situation is as follows.

Let ΔS represent the difference in entropy between the equilibrium state and a nearby, more constrained state; i.e.,

$$\Delta S = \Delta S_\alpha + \Delta S_\beta \tag{15.8}$$

subject to the constraint that

$$\Delta U_\alpha + \Delta U_\beta = 0. \tag{15.9}$$

The two subsystems (α, β) being of constant volume, we may develop each of the quantities ΔS_α and ΔS_β into a Taylor's series in terms of the appropriate quantity ΔU_α or ΔU_β; thus we have

$$\begin{aligned}\Delta S &= \frac{\partial S_\alpha}{\partial U_\alpha}\Delta U_\alpha + \tfrac{1}{2}\frac{\partial^2 S_\alpha}{\partial U_\alpha^{\,2}}(\Delta U_\alpha)^2 + \cdots + \frac{\partial S_\beta}{\partial U_\beta}\Delta U_\beta + \tfrac{1}{2}\frac{\partial^2 S_\beta}{\partial U_\beta^{\,2}}(\Delta U_\beta)^2 + \cdots \\ &= \left(\frac{1}{T_\alpha} - \frac{1}{T_\beta}\right)\Delta U_\alpha - \tfrac{1}{2}\left(\frac{1}{T_\alpha^{\,2}C_V(\alpha)} + \frac{1}{T_\beta^{\,2}C_V(\beta)}\right)(\Delta U_\alpha)^2 + \cdots. \end{aligned} \tag{15.10}$$

Since the equilibrium state is to be the state of maximum entropy, we require that the coefficient of ΔU_α vanish and that the coefficient of $(\Delta U_\alpha)^2$ be negative. We thus get the results $T_\alpha = T_\beta$, and $C_V > 0$: at equilibrium the temperature is uniform throughout the system and the heat capacity at constant volume is positive.* Upon generalizing this procedure, we obtain the conditions of equilibrium—$T, P, \mu^{(i)}, \ldots$, each uniform throughout the system—and the conditions of intrinsic stability—$C_V > 0$, $(\partial V/\partial P)_T < 0, \ldots$.

In seeking a steady-state analog of the thermostatic entropy-maximum principle, Prigogine [3] and others [1, 2, 4, 9, 10] have reckoned the rate of entropy production for a given state, steady or nonsteady, by applying the Gibbs relation $T\,dS = dU + P\,dV - \sum \mu^{(i)}\,dn^{(i)}$ in a local form to each and every point (gradient-bearing or nongradient-bearing) of the system. This

* At equilibrium $C_V(\alpha) = C_V(\beta) = \frac{1}{2}C_V(\text{total system})$.

local form application of the Gibbs relation to gradient-bearing systems constitutes a new definition of entropy for nonequilibrium systems.*

In terms of this new entropy definition the Prigogine theorem (statement A) holds near thermostatic equilibrium, but further away from equilibrium Prigogine's theorem fails and has to be replaced by a more elaborate evolution principle [3, 5, 6, 13].

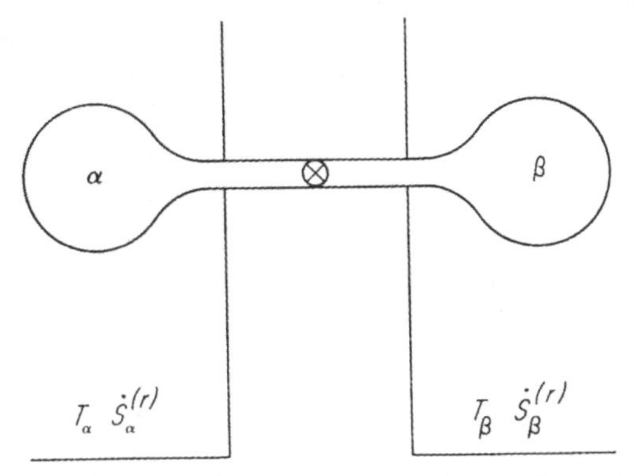

Figure 15–1. Thermomolecular pressure effect, schematic; full-flux linkage.

I now wish to show that statement A reduces to a special case of assumption IV for those cases for which Prigogine's theorem is valid and that for those cases for which Prigogine's theorem is not valid the failure is due largely to the new entropy definition, there being alternative ways of developing a steady-state stability principle. Consider yet again the thermomolecular pressure case (full-flux linkage) as indicated in Figure 15–1. Let two separately thermostatted bulbs (α, β) be connected by an efficiently insulated capillary tube and let the tube have a stopcock halfway between the bulbs. With the stopcock closed, fill the bulbs with the gas in question; wait a few minutes, and then open the stopcock. In general, upon the stopcock being opened, the gas migrates from one bulb to the other until eventually a state of migrational equilibrium is reached. Prigogine's theorem (statement A) says that from the time the stopcock is opened to the final establishment of migrational equilibrium the rate of entropy production for the system steadily decreases, reaching its minimum in the state of migrational equilibrium. Now if the capillary tube offers a high resistance to the flow of the gas, then the bulbs at α and β function nearly as terminal parts of the system, and the system effectively drifts through a sequence of steady states; i.e., over a small time interval the changes in thermodynamic quantities in the bulbs are much larger than the analogous changes in the capillary, and the changes in the

* See the earlier footnote devoted to this point on page 365.

bulbs are approximately proportional to the rates of mass flow into or out of the bulbs:

$$\dot{n}_\alpha(t) + \dot{n}_\beta(t) \approx 0, \tag{15.11}$$

$$\Theta(t) \approx \dot{n}_\alpha(t)\bar{S}_\alpha(t) + \dot{n}_\beta(t)\bar{S}_\beta(t) + \frac{\dot{Q}_\alpha^{(r)}(t)}{T_\alpha} + \frac{\dot{Q}_\beta^{(r)}(t)}{T_\beta}, \tag{15.12}$$

at a given instant of time t. If we assume that the behavior in time of the system is approximately equivalent to motion along a sequence of steady states toward a state of migrational equilibrium such that $\dot{n}_\omega = 0$—i.e., if we assume that upon replacing the walls of the bulbs by cylinder-and-piston arrangements such that $P_\alpha = P_\alpha(t)$ and $P_\beta = P_\beta(t)$, we should have $\dot{n}_\omega \approx \dot{n}_\omega(t)$, $\Theta \approx \Theta(t)$, etc.—then clearly statement A becomes just a special case of assumption IV.

Consider now the following point. The rate of entropy production is well defined in a steady state, the entropy changes in reservoirs or in terminal parts of the system being calculable according to classically approved recipes. In nonsteady states (transient states) the rate of entropy production is undefined in the classical sense. The considerations of Prigogine [3, 5, 6, 13] for nonsteady states are based on the acceptance of the local-form Gibbs relation for all points of a nonequilibrium system; i.e., Prigogine defines a rate of entropy production in a nonsteady state via a certain thermostatic functional form. Now there are several ways of defining a rate of entropy production for nonsteady states. By analogy to what we do in thermostatics—where we associate a constrained equilibrium state with each nonequilibrium state in order to be able to speak of an entropy-maximum principle—we can associate a "more constrained" steady state with each transient state, the "more constrained" steady state approximating the gradient distributions and current fluctuations of the transient state. We can then compare the rate of entropy production in the associated "more constrained" steady states to that of the final state (less constrained steady state) upon which the transient states converge for given boundary constraints.

Consider by way of example the simple heat conduction case (Figure 11–1). Let the system start out in the state $T_a = T_b = T_\beta$ and then instantaneously make the change $T_a \rightarrow T_\alpha$. The system will drift through a series of transient states ($\dot{Q}_\alpha^{(r)\prime} + \dot{Q}_\beta^{(r)\prime} \neq 0$) toward a final steady state ($\dot{Q}_\alpha^{(r)} + \dot{Q}_\beta^{(r)} = 0$), and the temperature distribution along the bar $T'(\lambda)$ will gradually change over to the distribution $T(\lambda)$ characteristic of the steady state.* Now let us associate with each transient state a "more constrained" steady state as indicated schematically in Figure 15–2. For this "more constrained" steady state we have

$$\dot{Q}_\alpha^{(r)} + \dot{Q}_\beta^{(r)} + \dot{Q}_x^{(r)} = 0 \tag{15.13}$$

* I am using primes to distinguish transient-state quantities from steady-state quantities.

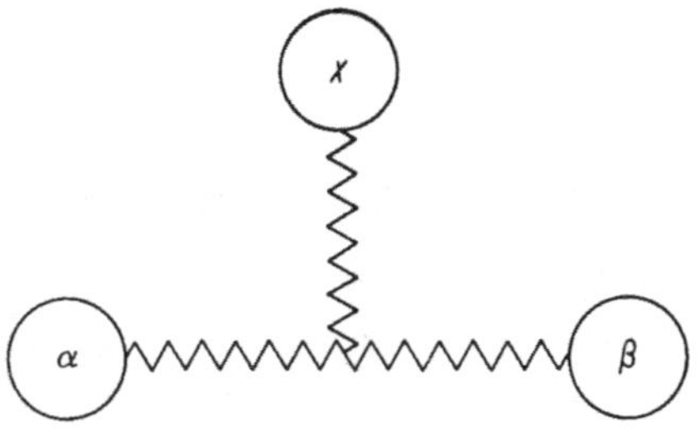

Figure 15–2. The "more constrained" steady state associated with a given transient state in the linkage.

and

$$\Theta = \frac{\dot{Q}_\alpha^{(r)}}{T_\alpha} + \frac{\dot{Q}_\beta^{(r)}}{T_\beta} + \frac{\dot{Q}_x^{(r)}}{T_x}$$
$$= \dot{Q}_\alpha^{(r)}\left(\frac{1}{T_\alpha} - \frac{1}{T_\beta}\right) + \dot{Q}_x^{(r)}\left(\frac{1}{T_x} - \frac{1}{T_\beta}\right). \tag{15.14}$$

Now let us choose the temperature T_x and the point of attachment of the x-reservoir such that $-\dot{Q}_x^{(r)} = \dot{Q}_\alpha^{(r)} + \dot{Q}_\beta^{(r)} \approx \dot{Q}_\alpha^{(r)\prime} + \dot{Q}_\beta^{(r)\prime}$ and $T(\lambda) \approx T'(\lambda)$. With respect to these "more constrained" steady states that approximate (in some respects) the properties of the actual transient states, we can say that the final state ($\dot{Q}_\alpha^{(r)} + \dot{Q}_\beta^{(r)} = 0$) is the state of minimum rate of entropy production since, according to assumption IV,

$$\left(\frac{\delta\Theta}{\bar{\delta}\dot{Q}_x^{(r)}}\right)_{T_\alpha, T_\beta} = 0, \quad \left(\frac{\delta^2\Theta}{\bar{\delta}(\dot{Q}_x^{(r)})^2}\right)_{T_\alpha, T_\beta} > 0 \tag{15.15}$$

and

$$\Theta(\dot{Q}_\alpha^{(r)}, \dot{Q}_x^{(r)} = 0, \beta) < \Theta(\dot{Q}_\alpha^{(r)}, \dot{Q}_x^{(r)} \neq 0, \beta) \tag{15.16}$$

for fixed T_α, T_β.

Thus statement A can be given an unlimited range of validity if we interpret it to mean that the rate of entropy production in the (final) steady state is to be compared to that of the "more constrained" steady states associated with the transient states in question. By coupling together assumption IV and the concept of the "more constrained" steady state associated with a fluctuation state or with a transient state, we arrive at a generally valid entropy production-minimum principle that plays a role for ordinary steady states analogous to that played by the entropy-maximum principle for static states.

Oscillating or Circulating States

Now what of the possibility that a given system, for a given set of boundary constraints, may not settle down into a steady state but may perhaps circulate or oscillate about some average configuration [3]? To illustrate the point, consider again the simple heat conduction case (Figure 11–1). Suppose that the $\alpha\beta$ link does not merely transmit energy from the α

reservoir to the β reservoir (assume $T_\alpha > T_\beta$), but that it periodically stores up and releases energy as well; i.e., suppose that no matter how long we wait it will still be true that $\dot{Q}_\alpha^{(r)} + \dot{Q}_\beta^{(r)} = f(t)$, where $f(t)$ is some periodically varying function of the time. What effect would such a situation have on our basic thermodynamic considerations?

We are familiar with just this sort of thing in equilibrium thermostatics. We know that our macroscopic thermostatic quantities show fluctuations in time and that what we actually measure are time averages of the given quantities: the period of a fluctuation is usually very short compared to the time taken by an experimental measurement so we obtain a value that is an average over many fluctuation cycles. It is to the time-averaged values of our macroscopic quantities that we apply the laws of thermodynamics.

Of course, we do the same thing in steady-state thermodynamics: our steady currents show fluctuations in time and we merely arrange to average our current measurements over many fluctuation cycles. If the fluctuations are small in magnitude, of the order of experimental error, the averaging procedure is usually taken for granted. If the fluctuations are of a spectacular order of magnitude [3], we need to remind ourselves that thermodynamics concerns itself with the average, long-term trends in the macroscopic properties of systems. Thus for the hypothetical heat conduction case, it is still true that, over the long-term, energy flows on the average from reservoir α to reservoir β. Let me express this observation somewhat more formally. I define the time average $\langle Z \rangle_t$ of a macroscopic quantity Z via the relation

$$\langle Z \rangle_t \equiv \frac{1}{t_2 - t_1} \int_{t_1}^{t_2} Z(t)\, dt, \tag{15.17}$$

where the interval of time $t_2 - t_1 \equiv \Delta t$ is to be taken large compared to the period of fluctuation in question.* Now, regardless of the scale of the fluctuations, oscillations, or circulations, I maintain that the fundamental relations†

$$\langle \Theta \rangle_t = \sum \langle Y_i \rangle_t \Omega_i, \tag{15.18}$$

$$\left(\frac{\delta \langle \Theta \rangle_t}{\bar{\delta} \langle Y_k \rangle_t} \right)_{\Omega'} = 0, \quad \left(\frac{\delta^2 \langle \Theta \rangle_t}{\bar{\delta} \langle Y_k \rangle_t^{\,2}} \right)_{\Omega'} > 0, \tag{15.19}$$

* If in doubt, let $|\Delta t| \to \infty$.

† Note the following points. The entropy associated with material in the terminal parts of the system or in reservoirs adjoined to the system is well defined in the classical sense, whereas the entropy associated with material in the gradient parts of the system is not. The ambiguity in the entropy concept for the gradient parts of the system, however, is immaterial in the following two cases: (i) in a steady state the rate of change of the entropy of a gradient part, *no matter how that entropy is defined*, is just exactly zero; (ii) in an oscillating or circulating state, we may make the time-average rate of change of the entropy of a gradient part, *no matter how that entropy is defined*, approach zero as closely as we please by averaging over a sufficiently long time interval.

are valid in terms of the time-averaged quantities. For the hypothetical heat conduction case we have, then, the relations

$$\langle \dot{Q}_\alpha^{(r)} \rangle_t + \langle \dot{Q}_\beta^{(r)} \rangle_t = 0, \tag{15.20}$$

$$\begin{aligned} \langle \Theta \rangle_t &= \frac{\langle \dot{Q}_\alpha^{(r)} \rangle_t}{T_\alpha} + \frac{\langle \dot{Q}_\beta^{(r)} \rangle_t}{T_\beta} \\ &= \langle \dot{Q}_\alpha^{(r)} \rangle_t \left(\frac{1}{T_\alpha} - \frac{1}{T_\beta} \right), \end{aligned} \tag{15.21}$$

$$0 = \left(\frac{\delta \langle \Theta \rangle_t}{\delta \langle \dot{Q}_\alpha^{(r)} \rangle_t} \right)_\beta = \left(\frac{1}{T_\alpha} - \frac{1}{T_\beta} \right). \tag{15.22}$$

On the average, over a long period of time, there will be a net transfer of energy between reservoirs α and β unless $T_\alpha = T_\beta$.

Now what of the entropy production-minimum principle of the previous section? I showed in that section that a steady state is stable relative to an "instantaneous" fluctuation of the system by computing the rate of entropy production via the "more constrained" steady-state technique. We must remember, however, that "instantaneous" in macroscopic thermodynamics means a time interval short on the human scale but long enough on the microscopic scale so that the system shall have undergone many microscopic fluctuation cycles. The time scale in macroscopic thermodynamics is based on the minimum time needed to measure to a given degree of accuracy* the macroscopic thermodynamic properties of the system. If the period of the microscopic fluctuations is short, the macroscopic time scale will also be short and we may speak of the "instantaneous" properties of the system. If we are dealing with oscillating or circulating systems, it will take a certain minimum time of observation to establish average properties such as $\langle \dot{Q}_\omega^{(r)} \rangle_t$ to a desired degree of accuracy; time intervals much shorter than this minimum time of observation will not have any *macroscopic thermodynamic* significance. Hence for oscillating or circulating systems we may say that the final *average* state for given boundary conditions is stable *relative to fluctuations averaged over the thermodynamic time scale appropriate to these systems*: if over the appropriate time interval we find that $\langle \dot{Q}_\alpha^{(r)} \rangle_t' + \langle \dot{Q}_\beta^{(r)} \rangle_t' \neq 0$ for the hypothetical heat conduction case, we then associate with this average fluctuation state an averaged version of the situation shown in Figure 15–2 such that $-\langle \dot{Q}_x^{(r)} \rangle_t = \langle \dot{Q}_\alpha^{(r)} \rangle_t + \langle \dot{Q}_\beta^{(r)} \rangle_t \approx \langle \dot{Q}_\alpha^{(r)} \rangle_t' + \langle \dot{Q}_\beta^{(r)} \rangle_t'$ and $\langle \Theta \rangle_t = \langle \dot{Q}_\alpha^{(r)} \rangle_t (T_\alpha^{-1} - T_\beta^{-1}) + \langle \dot{Q}_x^{(r)} \rangle_t (T_x^{-1} - T_\beta^{-1})$; and we find that

$$\langle \Theta \rangle_t (\langle \dot{Q}_x^{(r)} \rangle_t = 0) < \langle \Theta \rangle_t (\langle \dot{Q}_x^{(r)} \rangle_t \neq 0)$$

for fixed T_α, T_β.†

* To determine $\langle \dot{Q}_\omega^{(r)} \rangle_t$, e.g., to an order of accuracy of 0.1% takes a longer observation time (more fluctuation cycles must be averaged over) than it does to determine $\langle \dot{Q}_\omega^{(r)} \rangle_t$ to 1.0%, and so on.

† For some interesting comments on time scales in thermodynamics, see page 107 of the book cited in reference [3], pp. 146–148 of the book cited in reference [4] of Chapter 3, and the last chapter of the book by Van Rysselberghe [14].

The general validity of the entropy production-minimum principle for ordinary situations then hinges on our recognition of an appropriate time scale for a given thermodynamic system and on our using, in the formulation of the principle, quantities that are averaged over the characteristic time scale of the system. Fluctuating or oscillating systems, then, present us with no fundamental *thermodynamic* problems; the larger the scale of the fluctuation, the longer the time interval over which we average. Although the really spectacular fluctuations or oscillations pose no *thermodynamic* problems, they do challenge our statistical mechanical abilities: we shall, of course, be greatly interested in the detailed structure of the fluctuations or oscillations and in elucidating the underlying microscopic causes of the phenomenon [3].

Quasi-Steady States

In establishing the conditions of migrational equilibrium, I have thus far always used states that are rigorously steady; now, of course, the same general techniques will work for quasi-steady states as well. If we can treat a system as temporally drifting through a sequence of (essentially) steady states, the situation is quasi-steady and we can handle it by our standard methods.

In a given steady state the currents Y_i are interconnected via the First Law energy balance equation and via a set of mass balance (stoichiometric) relations. We may say that the currents Y_i are subject to a set of *constraints*, and we may express the equations of constraint in the form $f_k(Y_i) = 0$, $k = 1, 2, \ldots$. In the case of the thermomolecular pressure effect, for example, the equations of constraint are $\dot{n}_\alpha \overline{H}_\alpha + \dot{n}_\beta \overline{H}_\beta + \dot{Q}_\alpha^{(r)} + \dot{Q}_\beta^{(r)} + \dot{q} = 0$ and $\dot{n}_\alpha + \dot{n}_\beta = 0$. Now it is the steady-state equations of constraint plus the relation $\dot{Z}(\text{system}) = \sum \overline{Z}_i \dot{n}_i$ for the rate of change of a thermodynamic property Z of the system, the sum being over the terminal parts of the system, that enable us to transform the expression for the rate of entropy production into the form $\Theta = \sum_i Y_i \Omega_i$.

If a state is to be quasi-steady, then the equations of constraint (as they would hold in the steady state) must be nearly satisfied, i.e., $f_k(Y_i) \approx 0$; and the rate of change of a thermodynamic property Z of the system must be nearly expressible as $\sum \overline{Z}_i \dot{n}_i$, i.e., $\dot{Z}(\text{system}) - \sum \overline{Z}_i \dot{n}_i \approx 0$. Formally, then, the conditions of quasi-steadiness are

$$\frac{|f_k(Y_i)|}{\sum_i |\lambda_k^{(i)} Y_i|} \ll 1 \qquad k = 1, 2, \ldots, \tag{15.23}$$

and

$$\frac{|\dot{Z}(\text{system}) - \sum \overline{Z}_i \dot{n}_i|}{\sum |\overline{Z}_i \dot{n}_i|} \ll 1, \tag{15.24}$$

where $f_k(Y_i) \equiv \sum_i \lambda_k^{(i)} Y_i$.

What I am saying in Eqs. (15.23) and (15.24) is this: characteristic sums that add up rigorously to zero in the steady state must add up to values that are small compared to the individual terms of the sums if the state in question is to be considered quasi-steady.

"Continuous" versus "Discontinuous" Systems

The analyses that we have carried out have all been of the global type; i.e., we have analyzed finite *systems* rather than local effects at single points in space. This global versus local dichotomy has also been referred to in the literature as a *continuous* versus *discontinuous* one.* Now the global form of analysis is the more appropriate one for describing experiments: we rarely *measure* a flux or a gradient; instead we measure a total current and divide by an area or we measure a finite difference between two spatial points and divide by the distance between the points. At those points where we measure a property of the system we assume that the property is well defined; i.e., we assume that we make contact with the system at a terminal part. We can accommodate some spatial operations in the global formalism: see the treatment of concentration fields (Chapters 4 and 13) and the treatment of the radiation field (Chapter 12). The global formulation has the advantage that properties of the total system are known: in the concentration field, e.g., we perform the operation grad while keeping the average concentration of certain components constant; in the radiation field we perform the operation grad while keeping certain bounding temperatures constant. Whereas the global formulation readily yields conservation relations of the form $[Z] =$ constant, the corresponding local form relations are not at all apparent. For the experimenter, then, the global formulation is the more natural, the more operational one.

Statistical Mechanics of the Steady State

Of course, we cannot remain satisfied with a purely macroscopic treatment of steady states; ultimately we must develop a satisfactory statistical mechanics of the steady state to form a bridge between the properties of atoms and molecules and the measurable (time-averaged) properties of steady-state systems. Now the statistical mechanical structure that we strive to erect will depend on our view of the thermodynamics of the steady state. Thus, e.g., the statistical considerations of Boltzmann, culminating in the *H*-theorem, have a distinctly different flavor from the statistical considerations of Gibbs,

* See Chapters 4 and 5 of the book cited in reference [1] and Chapter 5 of the book cited in reference [3].

which culminated in the "thermodynamic analogies." Now much of the statistical work to date has aimed at justifying and generalizing the Onsager reciprocal relations and, hence, has been (in the thermodynamic sense) rather narrowly conceived. I hope that I have shown in this book that a wider, richer conception of the thermodynamics of steady states is called for and that the statistical explication of the language of steady-state thermodynamics is an exciting and challenging task.

For the interested reader I list a few references spanning a spectrum of approaches to the problem of applying statistical mechanics to nonequilibrium situations [15–17]; I find the work of Cox [15] especially interesting.

Final Comments

Now *any* thermodynamic formalism (devised for equilibrium states or for irreversible processes) is largely a language for describing experiments and for correlating the results of those experiments; a good formalism, therefore, should be *operational* and *natural*—it should be couched in terms of (idealized) laboratory operations. I feel that we have seen in the preceding pages the development of just such a formalism—a formalism entirely classical in spirit, easy of application, and laboratory oriented.

The task of bringing to fruition the potentialities inherent in the application of thermodynamics to nonequilibrium situations lies with the experimentalist; the interesting work is still largely to be done. We need renewed, informed experimentation—experimentation conducted in such a way as to yield well-defined results expressible in terms of a simple, operational language. While we remain in the macroscopic domain, let us preserve the full generality and rigor of the thermodynamic way of doing things. Let us not limit ourselves to near-equilibrium situations, and let us make no appeals (at this stage) to the principle of microscopic reversibility.

After we have learned what to expect at the macroscopic level, *then*, by all means, let us try to explain such behavior statistically in terms of atoms and molecules.

References

1. S. de Groot, *Thermodynamics of Irreversible Processes* (North-Holland Publishing Company, Amsterdam, 1951), Chapter 10.
2. S. de Groot and P. Mazur, *Non-Equilibrium Thermodynamics* (North-Holland Publishing Company, Amsterdam, 1962), Chapter 5.
3. I. Prigogine, *Introduction to the Thermodynamics of Irreversible Processes*, second edition (John Wiley and Sons, New York, 1961), Chapters 6 and 7.
4. J. C. M. Li, *J. Appl. Phys.*, **33**, 616 (1962).

5. P. Glansdorff, *Molecular Phys.*, **3**, 277 (1960).
6. P. Glansdorff and I. Prigogine, *Physica*, **30**, 351 (1964).
7. K. Denbigh, *Trans. Faraday Soc.*, **48**, 389 (1952).
8. M. J. Klein, *Proceedings of the International Symposium on Transport Processes in Statistical Mechanics—Brussels 1956* (I. Prigogine, editor) (Interscience Publishers, New York, 1958), p. 311.
9. T. A. Bak, *Advances in Chemical Physics* (I. Prigogine, editor), Vol. III (Interscience Publishers, New York, 1961), p. 33.
10. S. Ono, *ibid.*, p. 267.
11. L. Tisza, *Annals Phys.*, **13**, 1 (1961).
12. H. B. Callen, *Thermodynamics* (John Wiley and Sons, New York, 1960).
13. I. Prigogine and P. Glansdorff, *Physica*, **31**, 1242 (1965).
14. P. Van Rysselberghe, *Thermodynamics of Irreversible Processes* (Hermann, Paris, and Blaisdell Publishing Co., New York, 1963), Chapter 15.
15. R. T. Cox, *Statistical Mechanics of Irreversible Change* (The Johns Hopkins Press, Baltimore, 1955).
16. H. N. V. Temperley, *Proceedings of the International Symposium on Transport Processes in Statistical Mechanics—Brussels 1956* (I. Prigogine, editor) (Interscience Publishers, New York, 1958), p. 45.
17. I. Prigogine, *Statistical Mechanics of Irreversible Processes* (John Wiley and Sons, New York, 1962).

APPENDIXES

Appendix A

Supplementary Discussions

Thermotics: The Science of Heat

In recent years there has been evidence of increasing dissatisfaction with the use of the word "thermodynamics" to describe (practically) the entire field of the science of heat. It has often been remarked that the classical material collected under the title *thermodynamics* would be more aptly described by the title *thermostatics* [1, 2], and there seems to be a growing sentiment for restricting the word "thermodynamics" to its literal signification, i.e., to the motional aspects of the science of heat or to those aspects for which time variation is important—aspects which are prominently displayed in this book and in the discipline entitled *thermodynamics of irreversible processes.**

I wish to point out the existence of a word that can readily do the work that the word "thermodynamics" formerly did, thus freeing "thermodynamics" for its new, restricted usage: the word that I wish to call attention to is "thermotics." (Brønsted's word† "energetics" would be equally good or perhaps even better; however, as used at present, the word "energetics" carries with it the connotation of the Brønsted formalism and thus has become a highly specialized word.)

"Thermotics" is a word of good Greek origin meaning *the science of heat* and having obvious affinities to a large fraction of the words in a scientist's vocabulary. The *Oxford English Dictionary* indicates that the use of "thermotics" in written English antedates the use of "thermodynamic"

* See the preface of this book and references [5–9] in that preface.

† See reference [1] of Chapter 1.

and "thermodynamics" by some ten to twenty years (1837 versus 1849 and 1854, respectively). However, "thermotics" seems not to have taken hold, and since at least the time of Gibbs, "thermodynamics" has been the word customarily used by writers on the subject. In light of the discussion in the first paragraph it seems worthwhile to attempt to revive and to popularize the use of the word "thermotics," since through its agency we can establish a logical and satisfactory nomenclature.

In line with the foregoing I recommend that the general science of heat be called *thermotics* and that the well-known laws be referred to as the First, Second, and Third Laws of Thermotics. We can conveniently subdivide thermotics into *thermo-statics*, *thermo-staedics*, and *thermo-dynamics*. Thermo-statics pertains to the ordinary, classical equilibrium aspects of thermotics; thermo-dynamics pertains to those aspects for which time variation is important; and thermo-staedics pertains to aspects that are temporally steady or stationary. I write "thermo-staedics" with *ae* instead of *ea* to indicate that the pronunciation should be "-stē′dĭks" so as to keep it phonetically distinct from "-statics." We may speak of the *state* of a thermotic system as static, staedic, or dynamic. All thermotic *processes* are *dynamic*, but under special circumstances we may call processes *quasi-static* or *quasi-staedic* if they involve a change from a well-defined initial state to a well-defined final state along a path that can be considered (approximately) as a locus of proper static or staedic states, respectively.

I have written the subdivisions of thermotics (thermo-statics, etc.) with a hyphen so as to give visual warning of special usage. If we rigidly adhere to this convention, we can make allowance for the use (if such usage is still desired) of the unhyphenated form of the word "thermodynamics" in the old sense to cover all those aspects of the science of heat amenable to analysis via the concept of *equilibrium state function*.

We may indicate subfields of the general science of heat (the overlap or intersection of the science of heat with another science) by placing the appropriate adjective in front of the word "thermotics"; thus, e.g., we may speak of chemical or electrical thermotics, and it would seem appropriate to label the statistical approach to the science of heat as *statistical thermotics* [3]. Lastly, that part of the kinetic theory of gases that ordinarily finds its way into textbooks on heat we could aptly call *molecular thermotics*.

Nonlinear Current-Affinity Relations

Within the limitations mentioned earlier (no magnetic fields, etc.; see the next section on Special Fields), we can apply assumption IV and its dual to current-affinity relations of any complexity to get relations among the phenomenological coefficients. Consider the general current-affinity relation (relative to a fixed thermostatic reference state σ)

$$Y_i = \sum_j L_{ij}\Omega_j + \frac{1}{2!}\sum_j\sum_k L_{ijk}\Omega_j\Omega_k + \frac{1}{3!}\sum_j\sum_k\sum_m L_{ijkm}\Omega_j\Omega_k\Omega_m + \cdots; \quad \text{(A.1)}$$

in actual practice it is customary to lump together all the terms of a given type, $\Omega_j^a\Omega_k^b\Omega_m^c\ldots$ say, with a single coefficient $L_{ijkm\ldots}$. For the case of two currents and two affinities Eq. (A.1) takes the form, if we stop with terms of the second degree,

$$Y_1 = L_{11}\Omega_1 + L_{12}\Omega_2 + \tfrac{1}{2}L_{111}\Omega_1^2 + L_{112}\Omega_1\Omega_2 + \tfrac{1}{2}L_{122}\Omega_2^2, \quad \text{(A.2)}$$

$$Y_2 = L_{21}\Omega_1 + L_{22}\Omega_2 + \tfrac{1}{2}L_{211}\Omega_1^2 + L_{212}\Omega_1\Omega_2 + \tfrac{1}{2}L_{222}\Omega_2^2. \quad \text{(A.3)}$$

The application of assumption IV and its dual to Eqs. (A.2) and (A.3) leads, after much tedious algebra [4], to the following relations:

$$L_{12} = L_{21}, \quad \text{(A.4)}$$

$$L_{122} = L_{212}, \quad \text{(A.5)}$$

$$L_{112} = L_{211}, \quad \text{(A.6)}$$

$$L_{12}L_{222} = L_{22}L_{212}, \quad \text{(A.7)}$$

$$L_{21}L_{111} = L_{11}L_{112}, \quad \text{(A.8)}$$

$$L_{11}L_{122} = L_{12}L_{112}, \quad \text{(A.9)}$$

$$L_{22}L_{211} = L_{21}L_{212}, \quad \text{(A.10)}$$

where each of the L coefficients has been treated as a function of the thermostatic reference state (σ) only. Now the set of relations (A.4) to (A.10) is highly restrictive: there are seven equations of constraint connecting ten coefficients, and for the case $L_{12} \neq 0$ the relation

$$L_{222} = \left(\frac{L_{22}^2}{L_{11}L_{12}}\right)L_{111} \quad \text{(A.11)}$$

follows from the equations of constraint. But it is physically implausible to expect a relation between L_{111} and L_{222}; hence actual physical situations satisfying all seven of Eqs. (A.4) to (A.10) are either very rare or nonexistent.

Furthermore, for currents and affinities as defined in this text it should be the case that reversal of all the affinities in a given ordinary situation results in the exact reversal of all the currents; i.e., the transformation $\Omega_i \to -\Omega_i$ for all i should result in the relations $Y_i \to -Y_i$ for all i. The foregoing implies, then, that in the general expansion (Eq. A.1), only terms of odd degree should appear (the coefficients of terms of even degree are to be set identically equal to zero). The simplest nonlinear current-affinity relation should then be one involving first-order and third-order terms only. If we write equations analogous to Eqs. (A.2) and (A.3) with terms of first degree and third degree only and if we apply assumption IV, we get ten equations of constraint connecting twelve coefficients; and a number of physically implausible relations follow. Thus the restriction to terms of odd degree in Eq. (A.1) does not appreciably change the situation.

It seems, then, that outside of the linear current-affinity region we must either give up the entropy-production principle (assumption IV) or show that the expansion in Eq. (A.1), with constant coefficients, is generally invalid. Consider for a moment the linear current-affinity region; it is possible to write simple current-affinity relations in this region because the thermodynamic affinities are proportional to the true physical forces: for small values of the gradient, the (isothermal) gradient of a chemical potential is proportional to the gradient of a pressure or to the gradient of a concentration; the gradient of reciprocal temperature is proportional to the gradient of temperature, and so on. The proportionality between the thermodynamic affinities and the physical forces, however, normally holds for only a small part of the region where the physical laws are linear; outside of this *overlap region* the physical laws are still linear, but the current-affinity relations are not expressible in the form (A.1) with constant coefficients.

As an example consider the heat conduction case discussed in Chapter 11 (see Figure 11–1). Let the average temperature $\langle T\rangle \equiv \frac{1}{2}(T_\alpha + T_\beta)$ characterize the thermostatic reference state, and let the state of steady heat flow be characterized by $\langle T\rangle$ and $\Delta T \equiv T_\alpha - T_\beta$. The appropriate linear physical law is, then,

$$\dot{Q}_\alpha^{(r)} = -M_{11}(\langle T\rangle)\,\Delta T, \tag{A.12}$$

where $M_{11} = \kappa(\langle T\rangle)B/\Lambda$; and the current-affinity relation is

$$\begin{aligned}\dot{Q}_\alpha^{(r)} &= L_{11}\left(\frac{1}{T_\alpha} - \frac{1}{T_\beta}\right)\\ &= -\frac{L_{11}\,\Delta T}{\langle T\rangle^2\{1 - \frac{1}{4}(\Delta T/\langle T\rangle)^2\}}\end{aligned} \tag{A.13}$$

From Eqs. (A.12) and (A.13) it follows that

$$L_{11} = M_{11}\langle T\rangle^2\left\{1 - \frac{1}{4}\left(\frac{\Delta T}{\langle T\rangle}\right)^2\right\} \tag{A.14}$$

and that the overlap region is defined by the condition $(\Delta T/\langle T\rangle)^2 \ll 1$. In the overlap region we can treat L_{11} as effectively a function of $\langle T\rangle$ only; outside the overlap region, however, we must consider L_{11} to be a function of both $\langle T\rangle$ and ΔT. Now suppose we try to improve things by letting L_{11} equal

$M_{11}\langle T\rangle^2$ and by extending Eq. (A.13) to include a term quadratic in $(T_\alpha^{-1} - T_\beta^{-1})$. We should then find that

$$L_{111} = \frac{1}{2} M_{11}\, \Delta T \langle T\rangle^2 \left\{1 - \frac{1}{4}\left(\frac{\Delta T}{\langle T\rangle}\right)^2\right\}, \tag{A.15}$$

and the situation would be worse instead of better. On the other hand, if we were to set L_{11} equal to $M_{11}\langle T\rangle^2$, L_{111} equal to zero, and were to extend Eq. (A.13) to include a cubic term, we should find that

$$L_{1111} = -\frac{3}{2} M_{11}\langle T\rangle^4 \left\{1 - \frac{1}{4}\left(\frac{\Delta T}{\langle T\rangle}\right)^2\right\}^2; \tag{A.16}$$

and, although the situation would not be worse than in the linear case, neither would it be better: for L_{1111} to be effectively a function of $\langle T\rangle$ only, it would be necessary for the condition $(\Delta T/\langle T\rangle)^2 \ll 1$ to be satisfied. However, if the given condition were satisfied, then the linear current-affinity relation would be quite adequate and there would be no need for the extra complication of the cubic term.

Perhaps I can make the matter clearer by considering the relation between ΔT and $\Omega_1 \equiv (T_\alpha^{-1} - T_\beta^{-1})$:

$$\Omega_1 = -\frac{\Delta T}{\langle T\rangle^2 - (\frac{1}{2}\Delta T)^2}. \tag{A.17}$$

Hence

$$\Delta T = \frac{2}{\Omega_1}\{1 - [1 + (\Omega_1\langle T\rangle)^2]^{1/2}\} \tag{A.18}$$

and

$$\Delta T = -\Omega_1\langle T\rangle^2 + \frac{1}{4}\Omega_1^3\langle T\rangle^4 - \frac{1}{8}\Omega_1^5\langle T\rangle^6 + \cdots$$
$$0 \leqslant (\Omega_1\langle T\rangle)^2 < 1. \tag{A.19}$$

The condition $|\Omega_1\langle T\rangle| < 1$ is equivalent to the condition $|\Delta T/\langle T\rangle| < 0.83$. Inside the range $0 \leqslant |\Delta T/\langle T\rangle| < 0.83$, then, we can write Eq. (A.12) as

$$\dot{Q}_\alpha^{(T)} = M_{11}\langle T\rangle^2\Omega_1 - \left(\frac{1}{3!}\right)\frac{3}{2} M_{11}\langle T\rangle^4\Omega_1^3 + \left(\frac{1}{5!}\right) 15 M_{11}\langle T\rangle^6\Omega_1^5 - \cdots. \tag{A.12a}$$

There are two things to note about Eq. (A.12a): (i) Equation (A.12a) is equivalent to Eq. (A.12) only if we use the full infinite series expression; if we cut off the series in Eq. (A.12a) after a finite number of terms, then the coefficient of the last term will depend on ΔT and, in fact, will contain $\{1 - \frac{1}{4}(\Delta T/\langle T\rangle)^2\}$ as a factor. (ii) If $|\Delta T/\langle T\rangle| > 0.83$, it is impossible to express Eq. (A.12) as a power series in Ω_1. By way of example consider the case of stainless steel in the temperature range 5–50°K, a range over which the thermal conductivity of the steel is very nearly a linear function of the temperature [5]. If we take 27.5°K as our midpoint and take temperature intervals ΔT about this point, then for the range $0 \leqslant |\Delta T/\langle T\rangle| \leqslant 1.63$ Eq. (A.12) is a fairly good representation of the facts; yet for the part of the range $0.83 < |\Delta T/\langle T\rangle| \leqslant 1.63$ it is not possible to express $\dot{Q}_{\alpha}^{(r)}$ as a power series in Ω_1 (with constant coefficients).

The current-affinity relations for other physical situations behave in a manner analogous to that of the heat conduction case.

The conclusions that I draw from the preceding discussion are the following:

(i) The entropy-production principle (assumption IV) holds for *all* ordinary steady-state situations.

(ii) For the thermodynamic currents and affinities as defined in Chapter 3 it is correct to write linear current-affinity relations with (effectively) constant coefficients (coefficients that are functions of the thermostatic reference state variables only) in the overlap region.

(iii) Outside the overlap region it is in general not possible to express the currents as powers of the affinities with constant coefficients.*

Special Fields

The considerations that I have thus far advanced hold for ordinary situations—i.e., situations not involving magnetic or centrifugal fields. In the presence of the two special fields some modification of the heretofore standard procedure is required. The reversal of the direction of a magnetic field $\mathbf{B}$ or of a centrifugal field $\dot{\boldsymbol{\theta}}$ sometimes leads to a distinguishably different state of the system (due to the presence of Lorentz or Coriolis forces [7, 8]). If we let $\mathbf{f}$ stand indifferently for either of the special field vectors $\mathbf{B}$ or $\dot{\boldsymbol{\theta}}$, then we must in general distinguish between properties of the system in state $\mathbf{f}$ and properties of the system in state $-\mathbf{f}$. In the presence of a magnetic or centrifugal field,

* J. C. M. Li (see reference [4] of Chapter 15) has proposed a situation that seems to allow the application of Eqs. (A.2) and (A.3) with constant coefficients; his L coefficients (there are a few arithmetical errors in Li's derivation) satisfy Eqs. (A.4) to (A.7) and (A.10) but not Eqs. (A.8) and (A.9). Li's analysis, however, seems to be overidealized: the steady diffusion of hydrogen and nitrogen through an iron membrane at a rate sufficiently large to require higher-order terms in the current-affinity relations will certainly generate a temperature difference between the interfaces where the gases dissolve and evaporate; hence the simple isothermal Fick's law expression will not be adequate to describe the process. Also, the assumption of gas-solution equilibrium at each gas–membrane interface becomes increasingly invalid as the rate of diffusion increases. Thus Li's analysis of his example is incomplete, and his present results have no bearing on points (i) and (iii) above. See also the discussion of Rastogi, Srivastava, and Singh [6].

then, the assumption analogous to assumption IV for ordinary situations is that

$$\left(\frac{\delta \Theta(\mathbf{f})}{\delta Y_k}\right)_{\Omega',\mathbf{f}} + \left(\frac{\delta \Theta(-\mathbf{f})}{\delta Y_k}\right)_{\Omega',-\mathbf{f}} = 0,$$

$$\left(\frac{\delta^2 \Theta(\mathbf{f})}{\delta Y_k^{\,2}}\right)_{\Omega',\mathbf{f}} + \left(\frac{\delta^2 \Theta(-\mathbf{f})}{\delta Y_k^{\,2}}\right)_{\Omega',-\mathbf{f}} > 0. \tag{A.20}$$

I refer to Eq. (A.20) as assumption IV.S. If we multiply Eq. (A.20) through by the factor $\frac{1}{2}$, then we may state assumption IV.S thus:

(IV.S) If in the presence of a magnetic or centrifugal field we average the rates of entropy production (steady-state situations) for two opposed directions of the field vector **f**, then the *average rate of entropy production* in an affine sequence of steady states with $\nu + 1$ currents tends to a minimum for the state with ν currents.

The quantities that are averaged in assumption IV.S are the quantities $\Theta(Y_k, \Omega', \sigma, \mathbf{f})$ and $\Theta(Y_k, \Omega', \sigma, -\mathbf{f})$; i.e., we compare the rate of entropy production in two states that have the same value of Y_k, of Ω_i $(i \neq k)$, the same reference state σ, and reversed directions of the field vector **f**.

In the region of linear current-affinity relations Eq. (A.20) implies that

$$\frac{L_{ik}(\mathbf{f})}{L_{kk}(\mathbf{f})} + \frac{L_{ik}(-\mathbf{f})}{L_{kk}(-\mathbf{f})} = \frac{L_{ki}(\mathbf{f})}{L_{kk}(\mathbf{f})} + \frac{L_{ki}(-\mathbf{f})}{L_{kk}(-\mathbf{f})}. \tag{A.21}$$

Equation (A.21) is weaker than, but compatible with, the Onsager relations for special fields [7–9]:

$$L_{ik}(\mathbf{f}) = L_{ki}(-\mathbf{f}). \tag{A.21a}$$

Exercise A–1. Show that in the linear current-affinity region the Onsager relations (Eq. A.21a) imply that

$$\left(\frac{\delta \Theta}{\delta Y_k}\right)_{Y',\mathbf{f}} = \left(\frac{\delta \Theta}{\delta Y_k}\right)_{Y',-\mathbf{f}}.$$

There is a class of special field situations governed by a more stringent entropy-production principle than that of Eq. (A.20). For effects that vanish with the field **f**, I introduce the following extra assumption:

(IV.SX) The relation $0 = \delta\Theta/\delta Y_k$ holds for special field effects that vanish with the field **f** provided that (i) field-inducing currents (electric currents in solenoids and currents associated with angular motion) are kept constant, and (ii) currents that couple with the field **f** are kept constant.

I rationalize assumption IV.SX somewhat in the following manner. Even in equilibrium thermodynamics, magnetic fields have their puzzling aspects; the differential element of work in magnetic systems has an unusual form [10, 11]: $\int \mathbf{H}\cdot d\mathbf{B}\,dV$ instead of $\int \mathbf{B}\cdot d\mathbf{H}\,dV$—the roles of the intensive and extensive factors being just the reverse of usual. Magnetic energy is in some ways analogous to kinetic energy [10]; hence it is not too surprising to find magnetic fields and centrifugal fields lumped together into a common class of "odd" variables [7]. Since for magnetic systems the intensive and extensive variables play reversed roles (relative to the usual case), we would expect that for steady-state situations involving magnetic or centrifugal fields there should be some reversal of the roles of currents and affinities; assumption IV.SX shows that such is indeed the case for those effects that vanish with the field $\mathbf{f}$.

In order to make clear the physical significance of assumption IV.SX, I work out a number of examples; the cases that I consider are the equilibrium centrifuge, the Hall effect, the Nernst effect, the Righi–Leduc effect, and the Ettingshausen effect.

The Equilibrium Centrifuge

In the equilibrium centrifuge the tangential velocity of the fluid in the rotating container varies with the distance from the axis of rotation, so we are really dealing with a problem of migrational equilibrium in a velocity field. It is well known in the study of mechanics that a system rotating with a constant angular velocity $\dot{\boldsymbol{\theta}}$ is equivalent to a static system with an imposed centrifugal force, the force having as its potential the expression $-\frac{1}{2}mr^2\dot{\boldsymbol{\theta}}^2$, where m is the mass of the particle acted upon and r is the distance from the axis of rotation. In Chapter 4 I considered a general potential field in which the potential energy per gram of material was φ, and we reached the conclusion that for migrational equilibrium in such a field it is necessary that the condition

$$\mu + M\varphi = \text{constant} \tag{A.22}$$

be satisfied at each point in the field, M being the molecular weight and μ being the chemical potential of the substance in question. We expect, then, that the condition for migrational equilibrium in the centrifuge is that [12]

$$\mu - \tfrac{1}{2}Mr^2\dot{\boldsymbol{\theta}}^2 = \text{constant}. \tag{A.23}$$

The problem at hand is to derive Eq. (A.23) by considering a sequence of steady-flow states in the centrifugal field (1-component fluid).

Consider now the schematic version of a centrifuge shown in Figure A–1; the axis of rotation is perpendicular to the plane of the diagram. Suppose that there are two terminal parts (α, β) at distances r_α and r_β from the axis of rotation, and suppose that the pressures P_α and P_β are maintained by devices

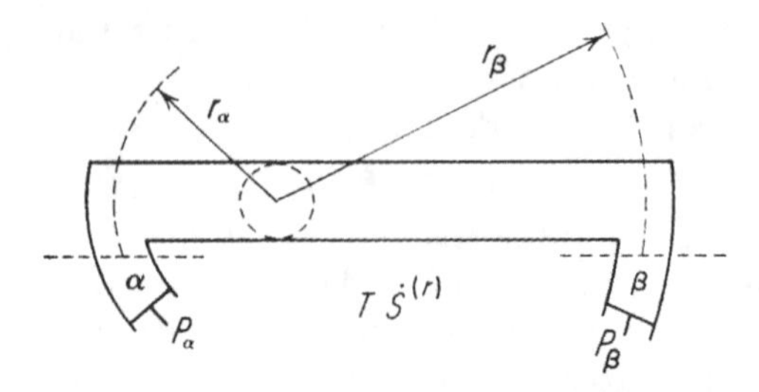

Figure A–1. Schematic version of a centrifuge.

coupled to the walls of the containing vessel. The apparatus is rigidly fixed to a shaft that is subjected to an external torque N, thus bringing about the necessary rotation. Let us assume that all the walls, shafts, etc., are of metal and are good thermal conductors; let us also assume that some stationary metallic part of the apparatus is in contact with a heat bath of temperature T. All exchanges of heat between the system and the surroundings are to take place by conduction through the metallic parts of the apparatus and are to be registered eventually in the heat reservoir of temperature T.

For the case of constant angular velocity $\dot{\boldsymbol{\theta}}$ and steady exchange of mass between the terminal parts ($\dot{n}_\alpha + \dot{n}_\beta = 0$), we have (see Eq. 4.1)

$$\Theta = -\frac{1}{T}\sum \mathscr{G}_i\dot{n}_i + \frac{\dot{W}_0}{T} = -\dot{n}_\alpha\frac{\mathscr{G}_\alpha - \mathscr{G}_\beta}{T} + \dot{\boldsymbol{\theta}}\frac{N}{T}$$

$$= Y_1\Omega_1 + Y_2\Omega_2. \tag{A.24}$$

For small rates of transfer (so as to minimize the frictional dissipation in transferring the mass from part α to part β) the relation

$$N\dot{\boldsymbol{\theta}} = \dot{\boldsymbol{\theta}}\frac{d(\mathscr{I}\dot{\boldsymbol{\theta}})}{dt} + T\Theta, \tag{A.25}$$

where $\mathscr{I}$ is the moment of inertia of the system about the axis of rotation, is valid. We can combine Eqs. (A.24) and (A.25) to get

$$\dot{n}_\alpha(\mathscr{G}_\alpha - \mathscr{G}_\beta) = \dot{\boldsymbol{\theta}}\frac{d(\mathscr{I}\dot{\boldsymbol{\theta}})}{dt}, \tag{A.26}$$

where $\mathscr{G} = \mu + \mathscr{K} = \mu + \frac{1}{2}Mr^2\dot{\boldsymbol{\theta}}^2$, the translational kinetic energy generated in transferring mass from one terminal part to the other being neglected.

The application of the operation $(\delta/\delta\dot{n}_\alpha)_{T,\dot{\boldsymbol{\theta}}}$ (assumption IV.SX.i) to Eq. (A.26) yields

$$\mu_\alpha + \mathscr{K}_\alpha - (\mu_\beta + \mathscr{K}_\beta) = \dot{\boldsymbol{\theta}}^2\left(\frac{\delta(d\mathscr{I}/dt)}{\delta\dot{n}_\alpha}\right)_{T,\dot{\boldsymbol{\theta}}}$$

$$= \dot{\boldsymbol{\theta}}^2\left(\frac{\delta(\dot{n}_\alpha Mr_\alpha{}^2 + \dot{n}_\beta Mr_\beta{}^2)}{\delta\dot{n}_\alpha}\right)_{T,\dot{\boldsymbol{\theta}}}$$

$$= Mr_\alpha{}^2\dot{\boldsymbol{\theta}}^2 - Mr_\beta{}^2\dot{\boldsymbol{\theta}}^2; \tag{A.27}$$

but $\mathscr{K} = \frac{1}{2}Mr^2\dot{\theta}^2$, so

$$\mu_\alpha - \tfrac{1}{2}Mr_\alpha^2\dot{\theta}^2 = \mu_\beta - \tfrac{1}{2}Mr_\beta^2\dot{\theta}^2. \tag{A.28}$$

Thus we have produced the equivalent of Eq. (A.23) by steady-flow methods, and assumption IV.SX.i was a necessary element of our derivation.

Exercise A–2. Show that for the case

$$\Theta = Y_1\Omega_1 + Y_2\Omega_2$$

the following implication holds for linear current-affinity relations:

$$0 = \left(\frac{\delta\Theta}{\delta Y_1}\right)_{Y_2} \rightarrow K_{12} = -K_{21} \quad (\text{and } L_{12} = -L_{21}).$$

Transverse Thermomagnetic and Galvanomagnetic Effects

The transverse thermomagnetic and galvanomagnetic effects refer to the coupling of crossed currents of heat or electricity in the presence of a magnetic field. The coupling effects vanish with the magnetic field, hence assumption IV.SX.ii applies to each of them. In the form of my general discussion and in the definitions of the various coefficients, I follow the treatment given by Bridgman [13]. The system of interest is a homogeneous isotropic metallic plate of rectangular section with length Λ, breadth b, and thickness d, together with some auxiliary loops and engines. Figure A–2 shows the orientation of the plate relative to a set of Cartesian axes. A magnetic field of magnitude B_z, pointing in the z-direction, spans a part of the metallic plate; and currents of heat or electricity flow in the x-direction and in the

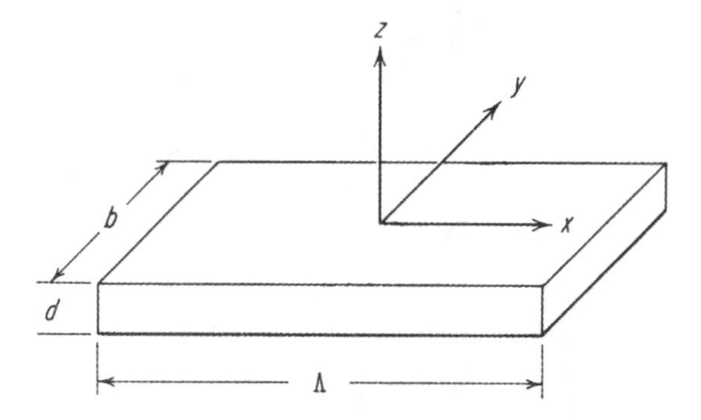

Figure A–2. Homogeneous isotropic metallic plate of rectangular section—length Λ, breadth b, thickness d.

y-direction. The appropriate coefficient in each case is proportional to the product of $-(\delta\Omega_x/\delta Y_y)_{Y_x}$ and d/B_z. The assumption that $0 = (\delta\Theta/\delta Y_y)_{Y_x}$ implies that $K_{xy} = -K_{yx}$ (and $L_{xy} = -L_{yx}$) in the linear current-affinity region.

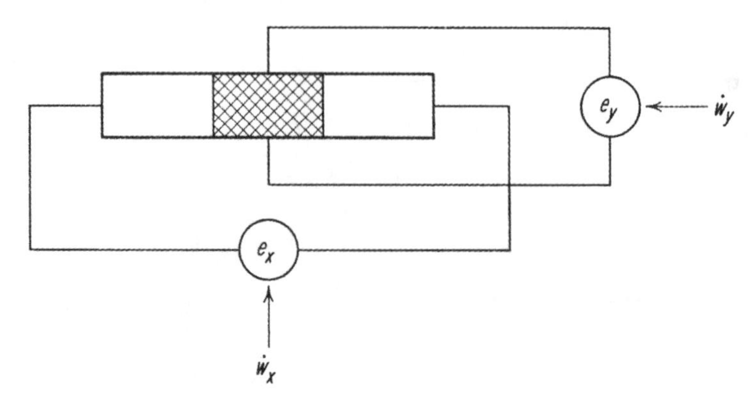

Figure A–3. The Hall effect. The cross-hatched part of the metal plate represents the region spanned by the magnetic field B_z. Electric engines at e_x and e_y maintain electric currents I_x and I_y through the metal plate.

THE HALL EFFECT

Consider the situation shown in Figure A–3: the electric engines e_x and e_y cause the electric currents I_x and I_y to flow through the metal plate; the cross-hatched part of the figure shows the region spanned by the magnetic field B_z; and the entire system is in a thermostat of temperature T. For a fixed magnetic field B_z and steady currents I_x and I_y we have

$$\dot{U}(\text{system}) = -\dot{Q}^{(r)} + \dot{W}_x + \dot{W}_y = 0 \tag{A.29}$$

and

$$\Theta = \frac{\dot{Q}^{(r)}}{T} = I_x \frac{\Delta\psi_x}{T} + I_y \frac{\Delta\psi_y}{T} = Y_1\Omega_1 + Y_2\Omega_2, \tag{A.30}$$

where, e.g., $\dot{W}_x \equiv I_x \Delta\psi_x$ for a given convention regarding the direction of current flow. The application of the operation $(\delta/\delta I_y)_{T,I_x}$ (assumption IV.SX.ii) to Eq. (A.30) leads to the relation

$$\frac{\Delta\psi_y}{I_x} = -\left(\frac{\delta\,\Delta\psi_x}{\delta I_y}\right)_{T,I_x}. \tag{A.31}$$

The Hall coefficient Γ_H is defined by [13]

$$\Gamma_H \equiv \frac{d}{B_z}\left(\frac{\Delta\psi_y}{I_x}\right)_{I_y=0}; \tag{A.32}$$

hence

$$\Gamma_H = -\frac{d}{B_z}\left(\frac{\delta\,\Delta\psi_x}{\bar{\delta} I_y}\right)_{T,I_x}, \tag{A.33}$$

or, in the linear current-affinity region,

$$\Gamma_H = -\frac{TK_{12}d}{B_z}. \tag{A.34}$$

THE NERNST EFFECT

Consider the situation shown schematically in Figure A–4. Heat flows by conduction from the reservoir at α to the reservoir at β, and the engine e_y maintains a steady electric current I_y in the y-direction. Let us choose the temperature T_γ such that $\dot{Q}_\gamma^{(r)} = 0$ when $I_y = 0$. In the steady state we have

$$\dot{U}(\text{system}) = -\dot{Q}_\alpha^{(r)} - \dot{Q}_\beta^{(r)} - \dot{Q}_\gamma^{(r)} + \dot{W}_y = 0 \tag{A.35}$$

and

$$\begin{aligned}\Theta &= \frac{\dot{Q}_\alpha^{(r)}}{T_\alpha} + \frac{\dot{Q}_\beta^{(r)}}{T_\beta} + \frac{\dot{Q}_\gamma^{(r)}}{T_\gamma} \\ &= \dot{Q}_\alpha^{(r)}\left(\frac{1}{T_\alpha} - \frac{1}{T_\beta}\right) + I_y\frac{\Delta\psi_y}{T_\beta} + \dot{Q}_\gamma^{(r)}\left(\frac{1}{T_\gamma} - \frac{1}{T_\beta}\right) \\ &= Y_1\Omega_1 + Y_2\Omega_2 + Y_3\Omega_3.\end{aligned} \tag{A.36}$$

The operation $(\delta/\bar{\delta} I_y)_{\dot{Q}_\alpha^{(r)},T_\beta,T_\gamma}$ is of the form $(\delta/\bar{\delta} Y_2)_{Y_1,\Omega_3}$; hence the application of this operation to Eq. (A.36) (assumption IV.SX.ii) leads to the relation

$$\frac{\Delta\psi_y}{\dot{Q}_\alpha^{(r)}} = -T_\beta\left(\frac{\delta\Omega_1}{\bar{\delta} Y_2}\right)_{Y_1,\Omega_3}, \tag{A.37}$$

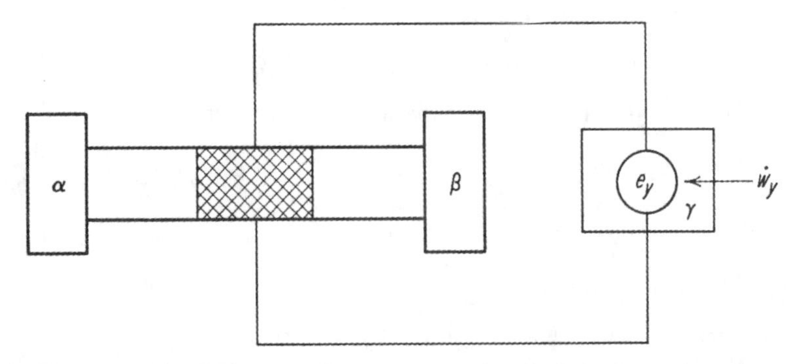

Figure A–4. The Nernst effect. The cross-hatched part represents the region spanned by the magnetic field B_z. α, β, γ—heat reservoirs at temperatures T_α, T_β, T_γ, respectively. The electric engine e_y maintains a current I_y in the y-direction. The temperature T_γ is chosen such that $\dot{Q}_\gamma^{(r)} = 0$ when $I_y = 0$; $T_\alpha > T_\beta$.

where I have assumed that $(\delta \dot{Q}_{\gamma}^{(r)}/\delta I_y)_{\dot{Q}_{\alpha}^{(r)},T_\beta,T_\gamma} \approx 0$. In the limiting path used in obtaining Eq. (A.37), I have treated the currents $\dot{Q}_{\alpha}^{(r)}$ and $\dot{Q}_{\gamma}^{(r)}$ unsymmetrically. The current $\dot{Q}_{\gamma}^{(r)}$ mainly represents work dissipation in the form of joule heat and frictional effects in the engine e_y; the dissipation occurs at the place γ, and hence $\dot{Q}_{\gamma}^{(r)}$ is a current that is *not* expected to couple with the magnetic field. On the other hand the current $\dot{Q}_{\alpha}^{(r)}$ represents (mainly) a heat flow from α to β through the magnetic field; hence coupling of $\dot{Q}_{\alpha}^{(r)}$ with the magnetic field *is* expected. Assumption IV.SX.ii, then—$\dot{Q}_{\alpha}^{(r)}$ coupling, $\dot{Q}_{\gamma}^{(r)}$ noncoupling—dictates that the form of the limiting path be $(\delta/\bar{\delta} Y_2)_{Y_1,\Omega_3}$.

The Nernst coefficient Γ_N is defined [13] by

$$\Gamma_N \equiv \frac{\kappa d}{B_z}\left(\frac{\Delta\psi_y}{\dot{Q}_{\alpha}^{(r)}}\right)_{I_y=0}, \tag{A.38}$$

where κ is the coefficient of thermal conductivity; hence

$$\Gamma_N = -T_\beta\left(\frac{\kappa d}{B_z}\right)\left(\frac{\delta\Omega_1}{\bar{\delta} Y_2}\right)_{Y_1,\Omega_3}, \tag{A.39}$$

or, in the linear current-affinity region,

$$\Gamma_N = -T_\beta\left(\frac{|K|_{21}}{K_{33}}\right)\left(\frac{\kappa d}{B_z}\right), \tag{A.40}$$

where $|K|_{21}$ is the cofactor of element K_{21} in the determinant of the matrix of coefficients K_{ij}.

It has been suggested that a thermal converter (Chapter 10) based on the Nernst effect may ultimately prove to be practical [14].

Exercise A–3. Show that for the case

$$\Theta = Y_1\Omega_1 + Y_2\Omega_2 + Y_3\Omega_3$$

the following implication holds in the linear current-affinity region:

$$0 = \left(\frac{\delta\Theta}{\bar{\delta} Y_2}\right)_{Y_1,\Omega_3} \leftarrow K_{12} = -K_{21},\, K_{13} = -K_{31},\, K_{23} = K_{32}.$$

THE RIGHI–LEDUC EFFECT

Consider the situation shown in Figure A–5. Heat flows from the reservoir at α to the reservoir at β. There is to be a second heat flow in the y-direction. In order to manipulate easily the two independent heat currents, let that in the y-direction be driven by a radiation turbine (Chapter 12). Let the two areas of dimensions Ad be insulated against heat losses except for two diametrically opposed spots that connect to hollow, flexible tubes with

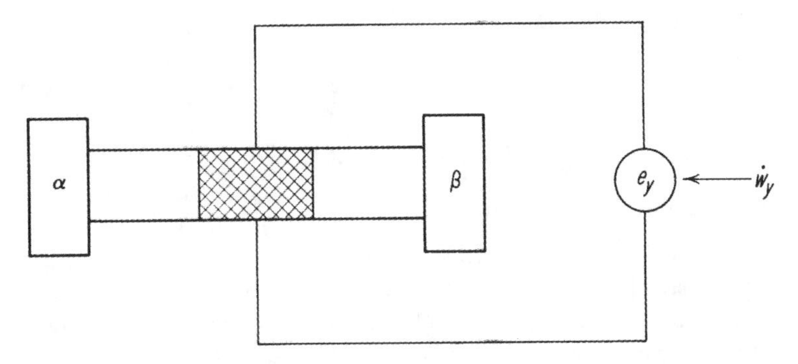

Figure A–5. The Righi–Leduc effect. The cross-hatched part represents the region spanned by the magnetic field. α, β—heat reservoirs at temperatures T_α, T_β, with $T_\alpha > T_\beta$. The engine e_y is a radiation turbine, its blades sweeping out a volume $\dot{V}_y$ per second.

perfectly reflecting walls. The flexible, perfectly reflecting tubes connect to opposite sides of a radiation turbine e_y; the perfectly reflecting blades of the turbine sweep out a volume $\dot{V}_y$ per second. In the steady state, then,

$$\dot{U}(\text{system}) = -\dot{Q}_\alpha^{(r)} - \dot{Q}_\beta^{(r)} + \dot{W}_y = 0 \tag{A.41}$$

and

$$\Theta = \frac{\dot{Q}_\alpha^{(r)}}{T_\alpha} + \frac{\dot{Q}_\beta^{(r)}}{T_\beta} = \dot{Q}_\alpha^{(r)}\left(\frac{1}{T_\alpha} - \frac{1}{T_\beta}\right) + \dot{V}_y\left(\frac{\dot{W}_y}{\dot{V}_y T_\beta}\right)$$

$$= Y_1\Omega_1 + Y_2\Omega_2. \tag{A.42}$$

The application of the operation $(\delta/\tilde{\delta}\dot{V}_y)_{\dot{Q}_\alpha^{(r)},T_\beta}$ to Eq. (A.42) (assumption IV.SX.ii) leads to

$$\frac{1}{\dot{Q}_\alpha^{(r)}T_\beta}\left(\frac{\delta\dot{W}_y}{\tilde{\delta}\dot{V}_y}\right)_{\dot{Q}_\alpha^{(r)},T_\beta} = -\left(\frac{\delta\Omega_1}{\tilde{\delta}Y_2}\right)_{Y_1,\beta} \tag{A.43}$$

since lim $(\dot{W}_y/\dot{V}_y)$ as $\dot{V}_y \to 0 \equiv \delta\dot{W}_y/\tilde{\delta}\dot{V}_y$. By analogy to Eq. (12.15) we should be able to factorize the quantity $(\delta\dot{W}_y/\tilde{\delta}\dot{V}_y)_{Y_1,\beta}$ into the form $(\delta\dot{W}_y/\tilde{\delta}\dot{V}_y)_{Y_1,\beta} = \Delta s\,\Delta T_y$; we can then rewrite Eq. (A.43) as

$$\frac{\Delta T_y}{\dot{Q}_\alpha^{(r)}} = -\frac{T_\beta}{\Delta s}\left(\frac{\delta\Omega_1}{\tilde{\delta}Y_2}\right)_{Y_1,\beta}. \tag{A.44}$$

The Righi–Leduc coefficient Γ_{R-L} is defined [13] by

$$\Gamma_{R-L} \equiv \frac{\kappa d}{B_z}\left(\frac{\Delta T_y}{\dot{Q}_\alpha^{(r)}}\right)_{\dot{V}_y=0}; \tag{A.45}$$

hence

$$\Gamma_{R-L} = -\frac{T_\beta}{\Delta s}\left(\frac{\kappa d}{B_z}\right)\left(\frac{\delta\Omega_1}{\tilde{\delta}Y_2}\right)_{Y_1,\beta}, \tag{A.46}$$

or, in the linear current-affinity region,

$$\Gamma_{R-L} = -\frac{K_{12}T_{\beta}\kappa d}{B_z\,\Delta s}, \tag{A.47}$$

where κ is the coefficient of thermal conductivity.

THE ETTINGSHAUSEN EFFECT

Consider the case shown schematically in Figure A–6. The engine e_x maintains a steady electric current I_x in the x-direction, and the radiation turbine at e_y drives a heat current in the y-direction (see the discussion of the radiation turbine in the treatment of the Righi–Leduc effect). A thermostat keeps the ends of the metal plate and the engine e_x at temperature T. In the steady state we have

$$\dot{U}(\text{system}) = -\dot{Q}^{(r)} + \dot{W}_x + \dot{W}_y = 0 \tag{A.48}$$

and

$$\begin{aligned}\Theta = \frac{\dot{Q}^{(r)}}{T} &= \frac{\dot{W}_x}{T} + \frac{\dot{W}_y}{T}\\ &= \frac{1}{T}\left\{I_x\,\Delta\psi_x + \dot{V}_y\left(\frac{\dot{W}_y}{\dot{V}_y}\right)\right\}\\ &= Y_1\Omega_1 + Y_2\Omega_2.\end{aligned} \tag{A.49}$$

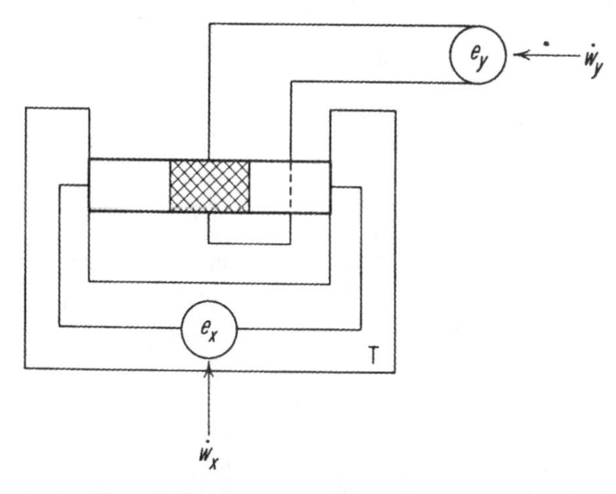

Figure A–6. The Ettingshausen effect. The cross-hatched part represents the region spanned by the magnetic field. The electric engine e_x maintains a current I_x in the x-direction. The engine e_y is a radiation turbine, its blades sweeping out a volume $\dot{V}_y$ per second. The two areas of dimensions bd and the engine e_x are maintained at temperature T.

The application of the operation $(\delta/\tilde{\delta}\dot{V}_y)_{T,I_x}$ to Eq. (A.49) (assumption IV.SX.ii) yields (see the discussion of the Righi–Leduc effect)

$$\frac{1}{I_x}\left(\frac{\delta \dot{W}_y}{\tilde{\delta}\dot{V}_y}\right)_{T,I_x} = \frac{\Delta s\,\Delta T_y}{I_x} = -\left(\frac{\delta\,\Delta\psi_x}{\tilde{\delta}\dot{V}_y}\right)_{T,I_x}. \tag{A.50}$$

The Ettingshausen coefficient Γ_E is defined [13] by

$$\Gamma_E \equiv \frac{d}{B_z}\left(\frac{\Delta T_y}{I_x}\right)_{\dot{V}_y=0}. \tag{A.51}$$

Hence,

$$\Gamma_E = -\left(\frac{d}{B_z\,\Delta s}\right)\left(\frac{\delta\,\Delta\psi_x}{\tilde{\delta}\dot{V}_y}\right)_{T,I_x}, \tag{A.52}$$

or, in the region of linear current-affinity relations,

$$\Gamma_E = -\frac{TK_{12}\,d}{B_z\,\Delta s}. \tag{A.53}$$

REMARK

In the linear current-affinity region each of the thermomagnetic and galvanomagnetic effects satisfies the condition $K_{12} = -K_{21}$ *in the given magnetic field*; i.e.,

$$K_{12}(\mathbf{B}) = -K_{21}(\mathbf{B}). \tag{A.54}$$

It is also true for each of the effects that* $K_{12}(\mathbf{B}) = -K_{12}(-\mathbf{B})$; hence for the Hall, Righi–Leduc, and Ettingshausen effects the coefficients satisfy the relations $K_{ij}(\mathbf{B}) = K_{ji}(-\mathbf{B})$ and thus also satisfy Eq. (A.20). For the case of the Nernst effect it is evident from the physics of the situation that $K_{13} = 0$ and that K_{23} does not depend on the magnetic field; hence the relations $K_{13}(\mathbf{B}) = K_{31}(-\mathbf{B})$ and $K_{23}(\mathbf{B}) = K_{32}(-\mathbf{B})$ are formally satisfied; the coefficients for the Nernst effect thus also satisfy Eq. (A.20).

The Thermocouple in a Magnetic Field

Consider as a final example the case of the thermocouple in a magnetic field. In Figure A–7 the part of the circuit *outside* the dotted circle is in a homogeneous magnetic field $\mathbf{B}$; the field–no-field junctions occur in reservoirs 10 and 60; and material (*b*) is to have isotropic properties. The effects at the field–no-field junctions are such that $\{\delta(\dot{Q}_{10}^{(r)} + \dot{Q}_{60}^{(r)})/\tilde{\delta}I\}_{T_\alpha,\ldots} = 0$. Since the thermoelectric effect does not vanish with the magnetic field, we cannot use assumption IV.SX and we must fall back on Eq. (A.20). In the linear current-affinity region it can be shown† that Eq. (A.21a) implies that

* See Chapter 13 of the book cited in reference [8].

† See page 145 of the book cited in reference [2] and Chapter 13 of the book cited in reference [8].

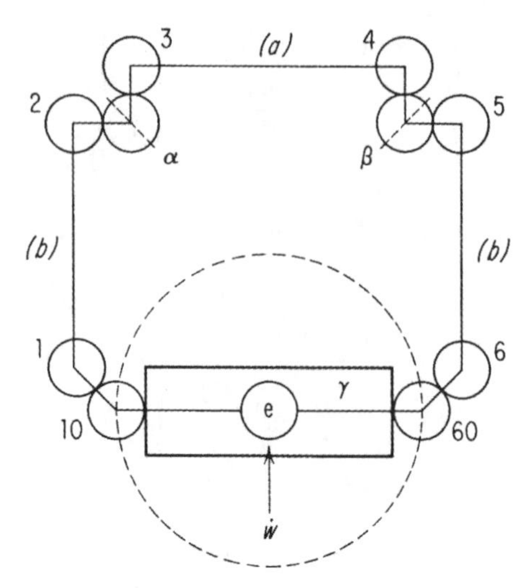

Figure A–7. Thermocouple in a magnetic field. The part of the circuit *outside* the dotted circle is in a homogeneous magnetic field **B**. The field–no-field junctions occur in thermostats 10 and 60; the rest of the figure is the same as Figure 9–1.

$d\,\Delta\psi(\mathbf{B})/dT = \pi(-\mathbf{B})/T$. The working assumption for arbitrary steady-state situations, then, is that the relations

$$\left(\frac{\partial\,\Delta\psi(\mathbf{B})}{\partial T_\beta}\right)_{T_\alpha,\mathbf{B}} = \frac{\pi_\beta^{(ab)}(-\mathbf{B})}{T_\beta}, \tag{A.55}$$

$$0 = [c_\beta^{(a)}(\mathbf{B})] - [c_\beta^{(b)}(\mathbf{B})] + T_\beta\left(\frac{\partial(\pi_\beta^{(ab)}(-\mathbf{B})/T_\beta)}{\partial T_\beta}\right)_{T_\alpha,-\mathbf{B}}, \tag{A.56}$$

$$T_\beta\left(\frac{\partial^2\,\Delta\psi(\mathbf{B})}{\partial T_\beta^{\,2}}\right)_{T_\alpha,\mathbf{B}} = [c_\beta^{(b)}(\mathbf{B})] - [c_\beta^{(a)}(\mathbf{B})], \tag{A.57}$$

are rigorously true for the thermocouple in a magnetic field.

Exercise A–4. Show that Eqs. (A.55) to (A.57) and Eq. (9.7) lead to the relations

$$\left(\frac{\partial\pi_\beta^{(ab)}(\mathbf{B})}{\partial T_\beta}\right)_{T_\alpha,\mathbf{B}} = \left(\frac{\partial\pi_\beta^{(ab)}(-\mathbf{B})}{\partial T_\beta}\right)_{T_\alpha,-\mathbf{B}}$$

and

$$-\Delta(\pi_\beta^{(ab)}/T_\beta) = \Delta\{[c_\beta^{(a)}] - [c_\beta^{(b)}]\},$$

where $\Delta Z \equiv Z(\mathbf{B}) - Z(-\mathbf{B})$.

If material (*a*) as well as material (*b*) has isotropic properties, then $\pi_{\omega}^{(ab)}(\mathbf{B}) = \pi_{\omega}^{(ab)}(-\mathbf{B})$ and Eqs. (9.10) to (9.12) hold in the given field with the proviso that the Seebeck potential difference, the Peltier heats, and the Thomson coefficients are each functions of the magnetic field **B**. If material (*a*) has anisotropic properties, bismuth being a prime example, then we have to use Eqs. (A.55) and (A.56) since it may be that $\pi_{\omega}^{(ab)}(\mathbf{B}) \neq \pi_{\omega}^{(ab)}(-\mathbf{B})$ in such cases [15].

References

1. E. A. Guggenheim, *Thermodynamics* (North-Holland Publishing Company, Amsterdam, 1950), p. 1.
2. S. de Groot, *Thermodynamics of Irreversible Processes* (North-Holland Publishing Company, Amsterdam, 1951), p. 1.
3. D. ter Haar, *Elements of Statistical Mechanics* (Rinehart and Company, New York, 1954), p. xvii.
4. M. L. Lieberman, M.S. Thesis, Illinois Institute of Technology (1963).
5. *American Institute of Physics Handbook* (McGraw-Hill Book Company, New York, 1957), p. **4**–77.
6. R. P. Rastogi, R. C. Srivastava, and K. Singh, *Trans. Faraday Soc.*, **61**, 854 (1965).
7. D. Fitts, *Nonequilibrium Thermodynamics* (McGraw-Hill Book Company, New York, 1962), p. 154.
8. S. de Groot and P. Mazur, *Non-Equilibrium Thermodynamics* (North-Holland Publishing Company, Amsterdam, 1962), Chapter 4.
9. L. Onsager, *Phys. Rev.*, **38**, 2265 (1931).
10. A. Sommerfeld, *Electrodynamics* (Academic Press, Publishers, New York, 1952), pp. 27 and 28.
11. E. A. Guggenheim, *Thermodynamics* (North-Holland Publishing Company, Amsterdam, 1950), pp. 374 and 375.
12. L. Landau and E. Lifshitz, *Statistical Physics* (Addison-Wesley Publishing Company, Reading, Mass., 1958), p. 74.
13. P. W. Bridgman, *The Thermodynamics of Electrical Phenomena in Metals and a Condensed Collection of Thermodynamic Formulas* (Dover Publications, New York, 1961), Chapter 7.
14. S. W. Angrist, *Direct Energy Conversion* (Allyn and Bacon, Boston, 1965), Chapter 9.
15. R. Wolfe and G. Smith, *Phys. Rev.*, **129**, 1086 (1963).

Appendix B

Glossary of Terms

[The page numbers cited below are holdovers from the 1967 edition. To get the correct page number for this edition, add 195 to the cited page number.]

(*Number following definition indicates page where term is first used.*)

affine sequence—a sequence of steady states all of which have the same values for a set of affinities Ω', 7, 28.

bitherm or **bithermal relation**—a form of the equation of correlation for the thermomolecular pressure effect, 55.

canonical set of steady currents—currents that are time derivatives of extensive thermodynamic quantities, 26.

energetics—Brønsted's reformulation of the principles of thermodynamics, 3.

equation of correlation—the "equation of state" for the thermomolecular pressure effect, 54.

forced vaporization—steady vaporization of a liquid *without boiling*, 13.

full-flux linkage—a linkage functioning as a lossless heat conductor, 51.

global property—a property of the entire system, to be distinguished from a local property; thus, e.g., the resistance of a length of wire is a global property whereas the resistivity at a point along the wire is a local property, 32.

gradient part—a part of the system that contains gradients in intensive variables, 6.

***grand* principles of entropy production**—principles asserted to hold for any value of a certain current; to be distinguished from *petit* principles of entropy production, 33.

link or **linkage**—a part of the system connecting two terminal parts in separate thermostats, 47.

local property—property at a point of the system; to be distinguished from a global property, 32.

migrational equilibrium—a chemical species is said to be in migrational equilibrium in a spatial field if there is no macroscopic tendency for the given species to migrate from one place in the field to another, 10.

monothermal process—a process in which the entire system is in heat communication with a single external heat reservoir during the process, 8.

ordinary steady-state situation—one not involving magnetic or centrifugal fields, 26.

overlap region—a range of conditions for which the pertinent physical laws are linear and for which the thermodynamic affinities are proportional to the physical forces, 186.

partial-flux linkage—a linkage functioning as a heat conductor with some lateral heat losses, 51.

***petit* principles of entropy production**—principles asserted to hold at a point where a certain current vanishes; to be distinguished from *grand* principles of entropy production, 33.

polythermal process—a process in which the terminal parts of the system are separately in heat communication with heat reservoirs of temperatures T_i during the process, 8.

quasi-static process—sequence of equilibrium states, 5.

quasi-steady process—sequence of steady states, 5.

special steady-state situation—one involving magnetic or centrifugal fields, 26.

static state—equilibrium state, 5.

stationary part—a part of the system the state of which remains pointwise invariant in time, 6.

terminal part—a part of the system that is spatially uniform with respect to the intensive variables, 6.

thermal converter—a device of the heat-to-work type, 122.

thermo-dynamics—aspects of thermotics for which time variation is important, 184.

thermo-staedics—aspects of thermotics that are temporally steady or stationary, 184.

thermostatics—ordinary equilibrium thermodynamics.

thermo-statics—ordinary, classical equilibrium aspects of thermotics, 184.

thermotics—the science of heat, 183.

zero-flux linkage—a theoretical construct used to maintain a fixed temperature gradient in a linkage during steady-flow operations, 51.

Answers to Exercises

2–1 From the equation in the exercise we see that $RT^2(\partial[N]/\partial T)_{\dot{\xi}} = RT^2\,d[N]/dT = -61.0R$; hence we have $-\dot{\xi}RT^2(\partial[N]/\partial T)_{\dot{\xi}} = \dot{\xi}61.0R = 610R = 1220$ cal. At the specified temperature $\Delta\bar{H}_{\text{vap}} = 10{,}500$ cal; the computed term is thus about 12% of $\Delta\bar{H}_{\text{vap}}$ at a flow rate of 10 μmoles sec^{-1}.

2–2 By hypothesis we have

$$\ln\frac{P}{P_0} = -\dot{\xi}^{(i)}[N^{(i)}] = -\dot{\xi}^{(k)}[N^{(k)}] = \cdots$$

for a given setting of the pressure. Averaging the terms on each side of this equation for the m pressure settings, we get

$$\left\langle \ln\frac{P}{P_0}\right\rangle = -\langle\dot{\xi}^{(i)}[N^{(i)}]\rangle = -\langle\dot{\xi}^{(k)}[N^{(k)}]\rangle = \cdots.$$

Since we are taking the $[N]$ coefficients as constant for each tube, we have, e.g.,

$$\langle\dot{\xi}^{(i)}[N^{(i)}]\rangle = \langle\dot{\xi}^{(i)}\rangle[N^{(i)}],$$

and consequently

$$\langle\dot{\xi}^{(i)}\rangle[N^{(i)}] = \langle\dot{\xi}^{(k)}\rangle[N^{(k)}] = \cdots.$$

If we now write the basic equation $\ln P = \ln P_0 - \dot{\xi}^{(i)}[N^{(i)}]$ in the form

$$\ln P = \ln P_0 - \theta^{(i)}\langle\dot{\xi}^{(i)}\rangle[N^{(i)}],$$

the m data points for the ith tube, when plotted as $\ln P$ versus $\theta^{(i)}$, will lie on a straight line of slope $-\langle\dot{\xi}^{(i)}\rangle[N^{(i)}]$. The other tubes give straight-line plots of exactly the same slope. Since the data points from all the tubes lie on straight lines of the same slope and since each straight line passes through the point $(0, \ln P_0)$, all the data fall on a single straight line when plotted in the form $\ln P$ versus $\theta^{(i)}$.

2–3 If we neglect kinetic energy terms, then under steady-flow conditions Eq. (2.16) takes the form

$$-(\mu_2^{(g)} - \mu_1^{(g)}) = [N]RT\dot{n}_2^{(g)},$$

where $\mu_i^{(g)}$ and $\dot{n}_i^{(g)}$ represent the chemical potential of the gas in the ith terminal part and the rate of influx of mass into the ith terminal part, respectively. Let $\mu_i^{(M)}$ stand for the chemical potential of the gas dissolved in the diaphragm just at the i interface, and make the following approximations:

$$\mu_i^{(M)} \approx \mu^\circ(T) + RT \ln C_i,$$

$$\mu_2^{(g)} - \mu_1^{(g)} \approx \mu_2^{(M)} - \mu_1^{(M)},$$

$$\mu_2^{(M)} - \mu_1^{(M)} \approx \frac{\partial \mu^{(M)}}{\partial x} \Delta x \approx \frac{RT}{\langle C \rangle} \left(\frac{\partial C}{\partial x}\right) \Delta x,$$

where C_i represents the concentration of dissolved gas in the diaphragm at the i interface and $\langle C \rangle$ represents the average concentration of the dissolved gas. Combining all these equations and writing J for $\dot{n}_2^{(g)}/B$, we get

$$J \approx -\frac{\Delta x}{B[N]\langle C \rangle} \left(\frac{\partial C}{\partial x}\right).$$

Upon comparing this equation to the Fick's law expression, we see that $D \approx \Delta x / B[N]\langle C \rangle$.

2–4 At the melting temperature $\mu(*)_{\text{sol}} - \mu(*)_{\text{liq}} = 0$; hence

(i) $$\mu_{\text{sol}} - \mu_{\text{liq}} = \mu_{\text{sol}} - \mu(*)_{\text{sol}} - (\mu_{\text{liq}} - \mu(*)_{\text{liq}}) = \int_{T*}^{T} (\bar{S}_{\text{liq}} - \bar{S}_{\text{sol}})\, dT > 0$$

as $\bar{S}_{\text{liq}} - \bar{S}_{\text{sol}} > 0$ at the melting point and in a neighborhood of the melting point. Similarly,

(ii) $$\mu_{\text{sol}} - \mu_{\text{liq}} = \mu(*)_{\text{sol}} - \mu_{\text{liq}} = \mu(*)_{\text{liq}} - \mu_{\text{liq}} = \int_{T*}^{T} \bar{S}_{\text{liq}}\, dT > 0.$$

From Eq. (2.16) it follows that

(i) $$RT[N]_{\text{mel}}\, \dot{n}_{\text{liq}} = \mu_{\text{sol}} - \mu_{\text{liq}} = \int_{T*}^{T} (\bar{S}_{\text{liq}} - \bar{S}_{\text{sol}})\, dT \approx \Delta\bar{H}(*)_{\text{mel}} \int_{T*}^{T} d \ln T = T*\, \Delta\bar{S}(*)_{\text{mel}} \ln \frac{T}{T*};$$

hence

$$\frac{T*}{T}\ln\frac{T}{T*} \approx \dot{n}_{\text{liq}}\left(\frac{R[N]_{\text{mel}}}{\Delta\bar{S}(*)_{\text{mel}}}\right).$$

From Eq. (2.15) it follows that

(ii) $$RT[N]_{\text{mel}}\,\dot{n}_{\text{liq}} - (T* - T)\bar{S}_{\text{sol}} = \mu_{\text{sol}} - \mu_{\text{liq}} = \mu(*)_{\text{liq}} - \mu_{\text{liq}}$$
$$= \int_{T*}^{T} \bar{S}_{\text{liq}}\,dT = \langle\bar{S}_{\text{liq}}\rangle(T - T*);$$

hence

$$\frac{\Delta T}{T} = \dot{n}_{\text{liq}}\left(\frac{R[N]_{\text{mel}}}{\langle\bar{S}_{\text{liq}}\rangle - \bar{S}_{\text{sol}}}\right) \approx \dot{n}_{\text{liq}}\left(\frac{R[N]_{\text{mel}}}{\Delta\bar{S}(*)_{\text{mel}}}\right).$$

Finally, observe that

$$\frac{T*}{T}\ln\frac{T}{T*} = \frac{T*}{T}\ln\left(1 + \frac{\Delta T}{T*}\right)$$
$$= \frac{T*}{T}\left(\frac{\Delta T}{T*} - \frac{1}{2}\left(\frac{\Delta T}{T*}\right)^2 + \cdots\right)$$
$$\approx \frac{\Delta T}{T}.$$

2–5 For the case considered, Eq. (2.28) takes the form

$$\ln\mathcal{Q} = \ln K - [N_V]\dot{C}_D.$$

Let* $\mathcal{Q} \equiv K + \Delta$, then

$$\ln\left(1 + \frac{\Delta}{K}\right) = -[N_V](k_f C_B - k_b C_D) = -[N_V]k_b C_D\left(\frac{K}{\mathcal{Q}} - 1\right)$$
$$= [N_V]k_b C_D\left\{\frac{\Delta/K}{1 + (\Delta/K)}\right\}.$$

Close to equilibrium we have $(\Delta/K) \ll 1$, $\ln\{1 + (\Delta/K)\} \approx \Delta/K$, and consequently

$$[N_V] \approx \frac{1}{k_b C_D^{(\text{eq})}}.$$

3–1 Since in the linear current-affinity region

$$\left(\frac{\partial Y_k}{\partial \Omega_j}\right)_{\Omega',\sigma} = L_{kj},$$

Eq. (3.31) is simply a statement of the Onsager reciprocal relations $L_{kj} = L_{jk}$.

From Eq. (3.33) we see that, in the linear current-affinity region,

$$\left(\frac{\partial^2\Theta}{\partial Y_k^2}\right)_{\Omega',\sigma} = 2\left(\frac{\partial\Omega_k}{\partial Y_k}\right)_{\Omega',\sigma} = \frac{2}{L_{kk}} > 0,$$

since $L_{kk} > 0$ due to the positive definite character of Θ.

* Remember that $\mathcal{Q} = C_D/C_B$, $K = C_D^{(\text{eq})}/C_B^{(\text{eq})} = k_f/k_b$, and that $\dot{C}_D = k_f C_B - k_b C_D$.

From the relations (for the linear current-affinity case)

$$\left(\frac{\partial \Omega_k}{\partial Y_k}\right)_{Y',J} = K_{kk} > 0 \quad \text{and} \quad \left(\frac{\partial \Omega_k}{\partial Y_k}\right)_{\Omega',\sigma} = \frac{1}{L_{kk}} > 0$$

and the relation $L_{ij} = |K|_{ji}/|K|$ with $|K|$ being the determinant of the matrix of coefficients K_{ij} and $|K|_{ji}$ being the cofactor of element K_{ji} in the determinant $|K|$, we see that

$$\left(\frac{\partial \Omega_k}{\partial Y_k}\right)_{Y',\sigma} - \left(\frac{\partial \Omega_k}{\partial Y_k}\right)_{\Omega',\sigma} = K_{kk} - \frac{|K|}{|K|_{kk}}$$

$$= K_{kk}\left(1 - \frac{|K|}{K_{kk}|K|_{kk}}\right).$$

The matrix of coefficients K_{ij} is symmetric for ordinary steady-state situations; hence (see reference [6]) $K_{kk}|K|_{kk} > |K| > 0$ and

$$\left(\frac{\partial \Omega_k}{\partial Y_k}\right)_{Y',\sigma} - \left(\frac{\partial \Omega_k}{\partial Y_k}\right)_{\Omega',\sigma} > 0.$$

See also the discussion in the *Explorations* section.

3–2 For fixed values of the affinities Ω_i $(i \neq k)$ the graph of Θ versus Y_k will be concave upward in the vicinity of the point $Y_k = 0$ and will have a minimum at that point. It follows then that on either side of the point $Y_k = 0$ the quantities Y_k and $(\partial \Theta/\partial Y_k)_{\Omega',\sigma}$ have the same sign; hence

$$Y_k(\partial \Theta/\partial Y_k)_{\Omega',\sigma} \geqslant 0.$$

3–3 From the relations (for the linear current-affinity region, with free use of the symmetry conditions $L_{ij} = L_{ji}$ and $K_{ij} = K_{ji}$)

$$\Theta = \sum_i Y_i \Omega_i = \sum_i \Omega_i \sum_k L_{ik}\Omega_k,$$

$$\left(\frac{\partial \Theta}{\partial \Omega_j}\right)_{\Omega',\sigma} = \sum_i \Omega_i L_{ij} + \sum_k L_{jk}\Omega_k = 2\sum_i L_{ji}\Omega_i = 2Y_j,$$

$$\left(\frac{\partial \Omega_j}{\partial Y_i}\right)_{Y',\sigma} = K_{ji},$$

it follows that

$$\sum_j \left\{\left(\frac{\partial \Theta}{\partial \Omega_j}\right)_{\Omega',\sigma} - Y_j\right\}\left(\frac{\partial \Omega_j}{\partial Y_i}\right)_{Y',\sigma} = \sum_j Y_j K_{ji} = \Omega_i.$$

Equation (3.47) can be treated similarly.

4–1 Direct computation shows that $(\partial \dot{V}/\partial\, \Delta\psi)_{T,P_\beta,\Delta P} = L_{12}/T$,

$$(\partial I/\partial\, \Delta\psi)_{T,P_\beta,\Delta P} = L_{22}/T,$$

and hence that $(\delta \dot{V}/\delta I)_{T,P_\beta,\Delta P} = L_{12}/L_{22}$. Also by direct computation we see that $(\dot{V}/I)_{T,P_\beta,\Delta P=0} = L_{12}/L_{22}$.

4-2 The values of $(\Delta\psi/\Delta P)_{T,I=0}$ appear to be fairly constant for each set of data, the average values being 2.62 and 144 for the cellulose plug and the capillary, respectively; the linear relations (4.27) and (4.28) thus appear to describe the data adequately. From Eq. (4.22) we see that the expressions $\dot{V}\,\Delta P$ and $I\,\Delta\psi$ must be expressed in the same energy units. If we select ergs for our energy unit, then when ΔP is expressed in centimeters of mercury, $\Delta\psi$ in millivolts, I in milliamperes, and $\dot{V}$ in $cm^3\ sec^{-1}$ Eq. (4.22) takes the form

$$\Theta = 13.6 \times 981\dot{V}\left(\frac{\Delta P}{T}\right) + 10I\left(\frac{\Delta\psi}{T}\right),$$

and Eq. (4.29) reduces to

$$\left|\left(\frac{\Delta\psi}{\Delta P}\right)_{T,I=0}\right| = \frac{13.6 \times 981}{10}\left|\left(\frac{\dot{V}}{I}\right)_{T,\Delta P=0}\right|.$$

For the cellulose plug, then (for the case $\Delta P = 0$),

$$\dot{V} = 10(13.6 \times 981)^{-1}\,2.62I = 0.00196I.$$

(i) For a current of 1 ma, $\dot{V} = 0.00196\ cm^3\ sec^{-1}$.

(ii) For a volume flow of 1 $cm^3\ sec^{-1}$, $I = 510$ ma.
For the glass capillary tube ($\Delta P = 0$)

$$\dot{V} = 10(13.6 \times 981)^{-1}\,144I = 0.108I.$$

(i) For a current of 1 ma, $\dot{V} = 0.108\ cm^3\ sec^{-1}$.
(ii) For a volume flow of 1 $cm^3\ sec^{-1}$, $I = 9.3$ ma.

5-1 Given the relation $P_\omega/P_\chi = (T_\omega/T_\chi)^{1/2}$, it follows that
(i) $\bar{S}_\omega - [S_\omega]^\partial = \bar{V}_\omega(\partial P_\omega/\partial T_\omega)_\chi = R(\partial \ln P_\omega/\partial \ln T_\omega)_\chi = R/2$ and
(ii) $[Q_\omega]^\partial = T_\omega\{\bar{S}_\omega - [S_\omega]^\partial\} = RT_\omega/2$. From the relation (i), i.e., $[S_\omega]^\partial = \bar{S}_\omega - R/2$, it follows that

(iii) $$[C_{\omega\omega}]^\partial = T_\omega(\partial\bar{S}_\omega/\partial T_\omega)_\chi = \bar{C}_P - T_\omega(\partial\bar{V}_\omega/\partial T_\omega)_{P_\omega}(\partial P_\omega/\partial T_\omega)_\chi = \bar{C}_P - \bar{V}_\omega(\partial P_\omega/\partial T_\omega)_\chi = \bar{C}_P - R/2,\ \text{and}$$

(iv) $$[C_{\omega\chi}]^\partial = T_\omega(\partial\bar{S}_\omega/\partial T_\chi)_\omega = 0.$$

5-2 (i) From the definition of the quantity $\langle[S]^\partial\rangle$ it follows that

$$-\langle[S]^\partial\rangle = \frac{\mu_\chi - \mu_\omega}{T_\chi - T_\omega} = \frac{1}{T_\chi - T_\omega}\left\{\int_{T_\omega}^{T_\chi} - \left(\bar{S}_\omega + \bar{C}_P \ln\frac{T}{T_\omega}\right) dT + RT_\chi \int_{P_\omega}^{P_\chi} d\ln P\right\},$$

the first integral being taken at constant pressure and the second being taken at constant temperature (see reference [1] of Chapter 1, pp. 112 and 113).

(ii) Upon subtracting the expression for $\bar{S}_\chi - \langle[S]^\theta\rangle$ from that for $\bar{S}_\omega - \langle[S]^\theta\rangle$, we get

$$\bar{S}_\omega - \bar{S}_\chi = \bar{C}_P \ln \frac{T_\omega}{T_\chi} - R \ln \frac{P_\omega}{P_\chi}$$

which is the correct expression for an ideal gas.

5–3 Let A and B represent the numerator and the denominator, respectively, of the right-hand side of Eq. (5.39). We then get

$$\left(\frac{\partial P_1}{\partial T_1}\right)_2 = \frac{P_2}{2B(T_1T_2)^{1/2}} \quad \text{and} \quad \left(\frac{\partial \ln P_1}{\partial \ln T_1}\right)_2 = \frac{(T_1/T_2)^{1/2}}{2A}.$$

It appears impossible to express derivatives such as $(\partial P_1/\partial T_1)_2$ and $(\partial \ln P_1/\partial \ln T_1)_2$ as functions of the variables T_1, P_1 only; hence Eq. (5.39) is an illegitimate equation of correlation. It is also clear by inspection that Eq. (5.39) cannot be rearranged into the symmetric form $f(1) = f(2)$.

5–4 Note that $P_\alpha = \frac{1}{2}(P_\alpha + P_\beta) + \frac{1}{2}(P_\alpha - P_\beta)$ and that

$$P_\beta = \tfrac{1}{2}(P_\alpha + P_\beta) - \tfrac{1}{2}(P_\alpha - P_\beta).$$

(i) Use Eq. (5.38).
(ii) The proper equation of the form (5.32) is just

$$\left\langle \frac{\bar{S}_\alpha - [S_\alpha]^\theta}{R} \right\rangle_\beta = \frac{\ln (P_\alpha/P_\beta)}{\ln (T_\alpha/T_\beta)}.$$

(iii)
$$\left\{ \frac{\bar{S}_\alpha - \langle[S]^\theta\rangle}{R} \right\} - \left\langle \frac{\bar{S}_\alpha - [S_\alpha]^\theta}{R} \right\rangle_\beta = \frac{\bar{C}_P}{R} - \frac{T_\beta}{T_\alpha - T_\beta}\left(\frac{\bar{C}_P}{R}\ln\frac{T_\alpha}{T_\beta} - \ln\frac{P_\alpha}{P_\beta}\right) - \frac{\ln (P_\alpha/P_\beta)}{\ln (T_\alpha/T_\beta)}.$$

For $P_\alpha/P_\beta = (T_\alpha/T_\beta)^{1/2}$ we get

$$\{\} - \langle\rangle = \frac{\bar{C}_P}{R} - \frac{1}{2} - \frac{T_\beta}{T_\alpha - T_\beta}\left\{\left(\frac{\bar{C}_P}{R} - \frac{1}{2}\right)\ln\frac{T_\alpha}{T_\beta}\right\}$$

and for $P_\alpha = P_\beta$ we get

$$\{\} - \langle\rangle = \frac{\bar{C}_P}{R} - \frac{T_\beta}{T_\alpha - T_\beta}\left\{\frac{\bar{C}_P}{R}\ln\frac{T_\alpha}{T_\beta}\right\}.$$

6–1 The derivations are all straightforward.

6–2 (i) If $(\partial[H_\omega]/\partial T_\omega)_\chi = [C_{\omega\omega}]$, then by Eqs. (6.9) and (6.10)

$$[C_{\omega\omega}] - [C_{\chi\omega}] - (\partial[h(\chi\omega)]/\partial T_\omega)_\chi = 0 \text{ and } [S_\omega]^\theta = [S_\omega].$$

But $[S_\omega]$ is in general a function of several of the intensive variables T_α, P_α, T_β, P_β, $\langle T\rangle$, whereas $[S_\omega]^\partial$ is a function of the ω state variables only; hence the relation $(\partial[H_\omega]/\partial T_\omega)_x = [C_{\omega\omega}]$ cannot have any general validity.

(ii) Set $(\partial[G_\alpha]/\partial T_\alpha)_{\beta,\langle T\rangle} = [C_{\beta\alpha}](1 - \langle T\rangle T_\beta^{-1})$ equal to $-[S_\alpha]$; then

$$[S_\alpha] = [C_{\beta\alpha}](\langle T\rangle T_\beta^{-1} - 1).$$

Now let

$$T_\alpha \to T_\beta\colon [S_a(\beta)] = [S_\beta]^\partial = [C_{ba}(\beta)](\langle T\rangle T_\beta^{-1} - 1).$$

The right-hand side of this last equality is a function of $\langle T\rangle$, whereas the left-hand side is not; hence the relation $(\partial[G_\alpha]/\partial T_\alpha)_{\beta,\langle T\rangle} = -[S_\alpha]$ cannot be generally valid.

(iii) $(\partial[G_\alpha]^\partial/\partial T_\alpha)_{\beta,\langle T\rangle} = [C_{\alpha\alpha}]^\partial(1 - \langle T\rangle T_\alpha^{-1})$ is clearly a function of $\langle T\rangle$, whereas $[S_\alpha]^\partial$ is not; hence the relation $(\partial[G_\alpha]^\partial/\partial T_\alpha)_{\beta,\langle T\rangle} = -[S_\alpha]^\partial$ cannot have any general validity.

The proofs in parts (i) to (iii) all depend on the acceptance of assumption Q.

6–3 In the limit $T_\alpha \to T_\beta$, Eq. (5.18) reduces to Eq. (6.24) and Eq. (5.22) reduces to Eq. (6.25). Equations (6.24) and (6.25) together with the relation $[h] = \langle T\rangle[s]$ yield the relation (6.26).

6–4 For the case considered, the relation $(\delta\Theta/\delta\tilde{\Omega}_2)_{Y_1,\beta} = 0$ implies that $(\partial\Theta/\partial T_\alpha)_{\dot{n}_\alpha,\beta}|_{T_\alpha=T_\beta} = 0$.

The application of this result to the relation

$$\Theta = \dot{n}_\alpha\left(\frac{\bar{H}_\beta - \bar{H}_\alpha}{T_\beta} + \bar{S}_\alpha - \bar{S}_\beta\right) + \dot{Q}_\alpha^{(r)}\left(\frac{1}{T_\alpha} - \frac{1}{T_\beta}\right)$$

yields

$$\bar{V}_\alpha\left(\frac{\partial P_\alpha}{\partial T_\alpha}\right)_{\dot{n}_\alpha,\beta}\Bigg|_{T_\alpha=T_\beta} = -\frac{\dot{Q}_\alpha^{(r)}}{T_\alpha\dot{n}_\alpha}\Bigg|_{T_\alpha=T_\beta} = -\frac{\dot{S}_\alpha^{(r)}}{\dot{n}_\alpha}\Bigg|_{T_\alpha=T_\beta}.$$

Upon passing to the limit $\dot{n}_\alpha \to 0$, the preceding relation implies that

$$\bar{S}_a(\beta) - [S_a(\beta)]^\partial = \bar{S}_a(\beta) - [S_a(\beta)]$$

and, hence, that $[S_a(\beta)]^\partial = [S_a(\beta)]$.

Since $[S_a(\beta)]^\partial = [S_b(\beta)]^\partial$ and $[S_a(\beta)] = [S_b(\beta)]$, it follows that

$$[S_b(\beta)]^\partial = [S_b(\beta)].$$

6–5 In Eq. (6.39) let $\dot{n} = \dot{n}_\alpha$, $\dot{W}_0 = 0$, $\mathscr{G}_\rho = \mu_\rho$ (there being no additional—kinetic, gravitational, etc.—forms of energy involved); then, if the mass flow is thought of as taking place in the α direction, it follows that

$$\mu_\alpha - \mu_\beta + T_\alpha\left(\frac{\delta\dot{S}_\alpha(\text{in})}{\delta\dot{n}_\alpha}\right)_{\Omega'} - T_\beta\left(\frac{\delta\dot{S}_\beta(\text{out})}{\delta\dot{n}_\alpha}\right)_{\Omega'} + \left(\frac{\delta\dot{q}(\alpha\beta)}{\delta\dot{n}_\alpha}\right)_{\Omega'} = 0$$

and

$$\mu_\alpha - \mu_\beta + T_\alpha[S_\alpha] - T_\beta[S_\beta] - [h(\beta\alpha)] = 0.$$

6–6 (i) Use Eq. (6.39) with the same conventions as in Ex. 6–5:

$$\mu_\alpha - \mu_\gamma + T_\alpha[S_\alpha] + T_\beta\{[S_\beta(\text{in})] - [S_\beta(\text{out})]\} - T_\gamma[S_\gamma] = 0.$$

Since the linkages are full-flux ones, it follows from assumption IV that

$$[S_\beta(\text{in})] = [S_\gamma] \quad \text{and} \quad [S_\beta(\text{out})] = [S_\alpha];$$

hence

$$[H_\alpha] - [H_\gamma] + T_\beta\{[S_\gamma] - [S_\alpha]\} = 0.$$

(ii) In Eq. (6.39) let $\dot{n} = \dot{n}_\chi$, $\dot{W}_0 = 0$, $\mathscr{G}_\rho = \mu_\rho$, etc.:

$$\mu_\alpha[R_{\alpha\chi}] + \mu_\beta[R_{\beta\chi}] + \mu_\gamma[R_{\gamma\chi}] + T_\alpha[S_\alpha][R_{\alpha\chi}] + T_\beta[S_\beta][R_{\beta\chi}] + T_\gamma[S_\gamma][R_{\gamma\chi}] + \left(\frac{\delta\dot{q}(\alpha\beta\gamma)}{\delta\dot{n}_\chi}\right)_{\Omega'} = 0$$

or

$$[R_{\alpha\chi}][H_\alpha] + [R_{\beta\chi}][H_\beta] + [R_{\gamma\chi}][H_\gamma] = [h(\psi\omega\chi)],$$

where $(\delta\dot{q}(\alpha\beta\gamma)/\delta\dot{n}_\chi)_{\Omega'} \equiv -[h(\psi\omega\chi)]$.

Equation (6.47) follows directly from Eq. (6.40). Equation (6.48) is a direct consequence of the relation $\dot{n}_\alpha + \dot{n}_\beta + \dot{n}_\gamma = 0$. Now set

$$[G_\omega] \equiv [H_\omega] - \langle T(\alpha\beta\gamma)\rangle[S_\omega];$$

then Eq. (6.50) follows from Eqs. (6.46) to (6.49).

6–7 In Ex. 6–6(i) the relations

$$\dot{n}_\alpha(\bar{H}_\alpha - \bar{H}_\gamma) + \dot{Q}_\alpha^{(r)} + \dot{Q}_\beta^{(r)} + \dot{Q}_\gamma^{(r)} = 0$$

and

$$\Theta = \frac{\dot{Q}_\alpha^{(r)}}{T_\alpha} + \frac{\dot{Q}_\beta^{(r)}}{T_\beta} + \frac{\dot{Q}_\gamma^{(r)}}{T_\gamma} + \dot{n}_\alpha\bar{S}_\alpha + \dot{n}_\gamma\bar{S}_\gamma$$
$$= \dot{n}_\alpha\left(\frac{\bar{H}_\gamma - \bar{H}_\alpha}{T_\gamma} + \bar{S}_\alpha - \bar{S}_\gamma\right) + \dot{Q}_\alpha^{(r)}\left(\frac{1}{T_\alpha} - \frac{1}{T_\gamma}\right) + \dot{Q}_\beta^{(r)}\left(\frac{1}{T_\beta} - \frac{1}{T_\gamma}\right)$$

show that there are three independent currents. There are 2 degrees of freedom for the reference thermostatic state γ (T_γ, P_γ). It follows then that

$$F_\gamma + \nu = 2 + 3 = 5.$$

In Ex. 6–6(ii) the relations

$$\dot{n}_\alpha\bar{H}_\alpha + \dot{n}_\beta\bar{H}_\beta + \dot{n}_\gamma\bar{H}_\gamma + \dot{Q}_\alpha^{(r)} + \dot{Q}_\beta^{(r)} + \dot{Q}_\gamma^{(r)} + \dot{q}(\alpha\beta\gamma) = 0$$

and

$$\begin{aligned}\Theta &= \frac{\dot{Q}_\alpha^{(r)}}{T_\alpha} + \frac{\dot{Q}_\beta^{(r)}}{T_\beta} + \frac{\dot{Q}_\gamma^{(r)}}{T_\gamma} + \frac{\dot{q}(\alpha\beta\gamma)}{\langle T(\alpha\beta\gamma)\rangle} + \dot{n}_\alpha \bar{S}_\alpha + \dot{n}_\beta \bar{S}_\beta + \dot{n}_\gamma \bar{S}_\gamma \\ &= \dot{n}_\alpha\left(\frac{\bar{H}_\gamma - \bar{H}_\alpha}{T_\gamma} + \bar{S}_\alpha - \bar{S}_\gamma\right) + \dot{n}_\beta\left(\frac{\bar{H}_\gamma - \bar{H}_\beta}{T_\gamma} + \bar{S}_\beta - \bar{S}_\gamma\right) \\ &\quad + \dot{Q}_\alpha^{(r)}\left(\frac{1}{T_\alpha} - \frac{1}{T_\gamma}\right) + \dot{Q}_\beta^{(r)}\left(\frac{1}{T_\beta} - \frac{1}{T_\gamma}\right) \\ &\quad + \dot{q}(\alpha\beta\gamma)\left(\frac{1}{\langle T(\alpha\beta\gamma)\rangle} - \frac{1}{T_\gamma}\right)\end{aligned}$$

indicate five independent currents. Hence $F_\gamma + \nu = 2 + 5 = 7$.

8–1 Equation (5.38) is merely based on the definition $-\langle [S]^\partial \rangle \equiv \Delta\mu/\Delta T$; the definition is applicable to the present case also; hence, with some rearrangement, we get

$$\begin{aligned}\ln \frac{P_\chi}{P_\omega} &= \frac{\bar{C}_P^{(g)}}{R} \ln \frac{T_\chi}{T_\omega} + \frac{T_\chi - T_\omega}{T_\chi}\left\{\frac{\bar{S}_\omega^{(g)} - \langle [S]^\partial \rangle - \bar{C}_P^{(g)}}{R}\right\} \\ &= -\frac{(T_\chi^{-1} - T_\omega^{-1})}{R}\left\{\frac{T_\chi T_\omega}{T_\chi - T_\omega} \bar{C}_P^{(g)} \ln \frac{T_\chi}{T_\omega} \right. \\ &\quad \left. + T_\omega(\bar{S}_\omega^{(g)} - \langle [S]^\partial \rangle - \bar{C}_P^{(g)})\right\}.\end{aligned}$$

It follows then that

$$\langle [Q]^\partial \rangle = T_\omega\{\bar{S}_\omega^{(g)} - \langle [S]^\partial \rangle - \bar{C}_P^{(g)}\} + \frac{T_\chi T_\omega}{T_\chi - T_\omega} \bar{C}_P^{(g)} \ln \frac{T_\chi}{T_\omega}.$$

8–2 We see by direct inspection of Eq. (8.7) that (i) if $\langle [Q]^\partial \rangle$ is to be independent of pressure (for given T_ω, T_χ) then $\bar{S}_\omega^{(g)} - \langle [S]^\partial \rangle$ must be independent of pressure and that (ii) if Eq. (8.7) is to be invariant with respect to interchange of the subscripts ω, χ, then $T_\omega\{\bar{S}_\omega^{(g)} - \langle [S]^\partial \rangle - \bar{C}_P^{(g)}\}$ must equal $T_\chi\{\bar{S}_\chi^{(g)} - \langle [S]^\partial \rangle - \bar{C}_P^{(g)}\}$.

8–3

gas	$[Q]^\partial$	$\Delta\bar{H}$	$[Q^{(M)}]^\partial$
CO_2	−1800	−2800	+1000
N_2	−260	+100	−360
H_2	+100	+800	−700

All values are expressed in calories.

8–4 From Eq. (8.12) with $S_\alpha^{(m)} = n_\alpha^{(1)}\bar{S}_\alpha^{(1)} + n_\alpha^{(2)}\bar{S}_\alpha^{(2)}$ and $(\partial P_\alpha/\partial T_\alpha)_\beta \approx 0$ we have

$$n_\alpha^{(1)}(\bar{S}_\alpha^{(1)} - [S_\alpha^{(1)}]^\partial) + n_\alpha^{(2)}(\bar{S}_\alpha^{(2)} - [S_\alpha^{(2)}]^\partial) = 0.$$

Upon multiplying this relation through by $T_\alpha/(n_\alpha^{(1)} + n_\alpha^{(2)})$ we get Eq. (8.16).

For an ideal 2-component gaseous mixture we have for each component the relation

$\mu^{(i)} = (\mu^{(i)})^0 + RT \ln P^{(i)} = (\mu^{(i)})^0 + RT \ln (X^{(i)}P)$; hence it follows that

$$\left(\frac{\partial \mu^{(i)}}{\partial X^{(i)}}\right)_{T,P} = \frac{RT}{X^{(i)}}.$$

Equation (8.13) thus takes the form

$$\frac{RT_\alpha}{X_\alpha^{(1)}}\left(\frac{\partial X_\alpha^{(1)}}{\partial T_\alpha}\right)_\beta = \bar{S}_\alpha^{(1)} - [S_\alpha^{(1)}]^\partial = \frac{[Q_\alpha^{(1)}]^\partial}{T_\alpha}$$

for the case $(\partial P_\alpha/\partial T_\alpha)_\beta \approx 0$; Eq. (8.15) then follows at once.

8–5 From the relations $X_\beta^{(1)} - X_\alpha^{(1)} = \mathbf{S}$ and $X_\beta^{(1)} + X_\alpha^{(1)} = 2\langle X^{(1)}\rangle$ we have $X_\alpha^{(1)} = \langle X^{(1)}\rangle - (\mathbf{S}/2)$ and $X_\beta^{(1)} = \langle X^{(1)}\rangle + (\mathbf{S}/2)$. It follows then that

$$\ln\frac{X_\beta^{(1)}}{X_\alpha^{(1)}} = \ln\left\{\frac{\langle X^{(1)}\rangle + (\mathbf{S}/2)}{\langle X^{(1)}\rangle - (\mathbf{S}/2)}\right\} = \ln\left\{\frac{1 + (\mathbf{S}/2\langle X^{(1)}\rangle)}{1 - (\mathbf{S}/2\langle X^{(1)}\rangle)}\right\} \approx 2\left(\frac{\mathbf{S}}{2\langle X^{(1)}\rangle}\right) = \frac{\mathbf{S}}{\langle X^{(1)}\rangle}.$$

Similarly, from the relations $X_\beta^{(2)} - X_\alpha^{(2)} = -\mathbf{S}$ and $X_\beta^{(2)} + X_\alpha^{(2)} = 2\langle X^{(2)}\rangle$ it follows that

$$\ln\frac{X_\beta^{(2)}}{X_\alpha^{(2)}} \approx -\frac{\mathbf{S}}{\langle X^{(2)}\rangle}$$

and that

$$\ln \mathbf{q} = \ln\frac{(X_\beta^{(1)}/X_\alpha^{(1)})}{(X_\beta^{(2)}/X_\alpha^{(2)})} \approx \left(\frac{\mathbf{S}}{\langle X^{(1)}\rangle}\right) + \left(\frac{\mathbf{S}}{\langle X^{(2)}\rangle}\right) = \frac{\mathbf{S}(\langle X^{(1)}\rangle + \langle X^{(2)}\rangle)}{\langle X^{(1)}\rangle\langle X^{(2)}\rangle}$$
$$= \frac{\mathbf{S}}{\langle X^{(1)}\rangle\langle X^{(2)}\rangle}.$$

8–6 From Eq. (8.15) we have

$$\left(\frac{\partial \ln X_\alpha^{(1)}}{\partial T_\alpha}\right)_\beta = \frac{[Q_\alpha^{(1)}]^\partial}{RT_\alpha^{\,2}}$$

and

$$\int_{X_\alpha^{(1)}=X_\alpha^{(1)}}^{X_\alpha^{(1)}=X_\beta^{(1)}} d\ln X_\alpha^{(1)} = \int_{T=T_\alpha}^{T=T_\beta} \frac{[Q_\alpha^{(1)}]^\partial}{RT^2}\,dT,$$

$$\ln\frac{X_\beta^{(1)}}{X_\alpha^{(1)}} = -\frac{\langle [Q^{(1)}]^\partial\rangle}{R}\left(\frac{1}{T_\beta} - \frac{1}{T_\alpha}\right).$$

We see then that

$$\begin{aligned}\langle [Q^{(1)}]^\partial\rangle &= T_\alpha T_\beta (T_\beta - T_\alpha)^{-1} R \ln\frac{X_\beta^{(1)}}{X_\alpha^{(1)}}\\ &= -RT_\alpha T_\beta (T_\alpha - T_\beta)^{-1} \ln\frac{T_\alpha}{T_\beta} \ln\frac{X_\beta^{(1)}}{X_\alpha^{(1)}}\left(\ln\frac{T_\alpha}{T_\beta}\right)^{-1}\\ &= RT_r\left\langle\frac{[Q^{(1)}]^\partial}{RT}\right\rangle\\ &= \langle [Q^{(1)}]^\partial\rangle_r,\end{aligned}$$

where I have made use of Eqs. (8.24) to (8.26).

8–7 The fundamental relations

$$\dot{\xi}\{\sum \nu_j \overline{H}_j - \sum \nu_i \overline{H}_i\} + T_\alpha \dot{S}_\alpha^{(r)} + T_\beta \dot{S}_\beta^{(r)} + \dot{q}(\alpha\beta) = 0$$

and

$$\Theta = \dot{S}_\alpha^{(r)} + \dot{S}_\beta^{(r)} + \dot{s}(\alpha\beta) + \dot{\xi}\{\sum \nu_j \overline{S}_j - \sum \nu_i \overline{S}_i\},$$

where, e.g., $\overline{H}_i$ is short for $\overline{H}_{B_i}$, and the relation, e.g.,

$$\left(\frac{\delta \dot{S}_\alpha^{(r)}}{\delta \dot{\xi}}\right)_{\Omega'} \equiv \sum \nu_i(\overline{S}_i - [S_i])$$

allow us to write an equation of the form of Eq. (8.41).

8–8 The plot shows that in the dilute region the variation in the Soret coefficient is due mainly to variation in $[Q^{(2)}]^\partial$ whereas in the more concentrated region the variation is due mainly to variation in $\partial \ln \gamma_\pm / \partial \ln m$.

9–1 Let $\Delta P_\gamma \equiv P_{\gamma(\alpha)} - P_{\gamma(\beta)}$; then

$$P_{\gamma(\alpha)} - P_0(\alpha) = \int_{T_\alpha}^{T} \left(\frac{\partial P}{\partial T}\right)_\alpha dT = \int_{T_\alpha}^{T} \left\{\frac{\overline{S}^{(g)} - [S^{(g)}]^\partial}{\overline{V}^{(g)}}\right\}_\alpha dT,$$

and similarly for $P_{\gamma(\beta)} - P_0(\beta)$; hence

$$\Delta P_\gamma = P_0(\alpha) + \int_{T_\alpha}^{T} \{\cdots\}_\alpha \, dT - P_0(\beta) - \int_{T_\beta}^{T} \{\cdots\}_\beta \, dT.$$

It follows that if

$$\left(\frac{\partial \Delta P_\gamma}{\partial T}\right)_{T_\alpha, T_\beta} = \left\{\frac{\overline{S}^{(g)} - [S^{(g)}]^\partial}{\overline{V}^{(g)}}\right\}_\alpha - \left\{\frac{\overline{S}^{(g)} - [S^{(g)}]^\partial}{\overline{V}^{(g)}}\right\}_\beta = 0,$$

then, since the two linkages are independent of onc another, the quantity inside each pair of braces must vanish: $\overline{S}^{(g)} - [S^{(g)}]^\partial = 0$ in each connecting tube. The remaining relations of the exercise follow at once from this result.

9–2 The relations are all straightforward.

10–1 If we express the current $\dot{n}$ as a linear function of the affinities, we get

$$\dot{n} = L_{11}\left(\frac{\dot{W}}{\dot{n}T_\beta}\right) + L_{12}\left(\frac{1}{T_\alpha} - \frac{1}{T_\beta}\right).$$

Upon rearrangement we get

$$\frac{\dot{W}}{\dot{n}} = \frac{\dot{n}T_\beta}{L_{11}} - \frac{L_{12}(T_\alpha - T_\beta)}{L_{11}T_\alpha}.$$

Comparing this equation to Eq. (10.13) we see that

$$\Delta\mathscr{G}_\alpha(0) = -\frac{L_{12}(T_\alpha - T_\beta)}{L_{11}T_\alpha} \quad \text{and} \quad RT_\alpha[N] = \frac{T_\beta}{L_{11}}.$$

10–2 From Eq. (6.39) it follows that

$$-\mu_\alpha^{(\text{liq})} + \mu_\beta^{(\text{liq})} - T_\alpha[S_\alpha^{(g)}] + T_\beta[S_\beta^{(g)}] = \left(\frac{\delta \dot{W}_0}{\delta \dot{n}}\right)_{\Omega'},$$

but $\mu_\omega^{(\text{liq})} = \mu_\omega^{(g)}$; hence

$$\left(\frac{\delta(-\dot{W}_0)}{\delta \dot{n}}\right)_{\Omega'} = [H_\alpha^{(g)}] - [H_\beta^{(g)}].$$

10–3 In the limit $\dot{n}_\alpha \to 0$, Eq. (10.30) implies that

$$\bar{V}_\alpha\left(\frac{\partial P_\alpha}{\partial T_\alpha}\right)_{\beta,\dot{n}_\alpha=0} = -\left(\frac{\delta \dot{S}_\alpha^{(r)}}{\delta \dot{n}_\alpha}\right)_{\beta,T_\alpha} = \bar{S}_\alpha - [S_\alpha],$$

but

$$\bar{V}_\alpha\left(\frac{\partial P_\alpha}{\partial T_\alpha}\right)_{\beta,\dot{n}_\alpha=0} = \bar{S}_\alpha - [S_\alpha]^\partial.$$

The implication then is that $[S_\alpha] = [S_\alpha]^\partial$, but this is in general not true, since $[S_\alpha]^\partial = [S_\alpha(\dot{\alpha})] \neq [S_\alpha]$—see Eqs. (6.22) and (6.23). Equation (10.30) is thus unlikely to have any validity whatsoever.

11–1 For the first symmetry relation $\dot{Q}_1^{(r)}(\text{old}) = \dot{Q}_1^{(r)}(\text{new}_1)$, write

$$\dot{Q}_1^{(r)}(\text{old}) = L_{11}\Omega_1 + L_{12}\Omega_2 + L_{13}\Omega_3,$$
$$\dot{Q}_1^{(r)}(\text{new}_1) = L_{11}\Omega_1' + L_{12}\Omega_2' + L_{13}\Omega_3',$$

where $\Omega_i = T_i^{-1} - T_4^{-1}$, and I have used primes to distinguish "new_1" values of the affinities. It follows from the conditions of the problem that $\Omega_1' = \Omega_1 - \Omega_2$, $\Omega_2' = -\Omega_2$, $\Omega_3' = \Omega_3 - \Omega_2$; hence the relation $\dot{Q}_1^{(r)}(\text{old}) = \dot{Q}_1^{(r)}(\text{new}_1)$ implies that

$$-(L_{11} + L_{12} + L_{13}) = L_{12}$$

or

$$L_{13} = -(L_{11} + 2L_{12}).$$

The other symmetry relations can be treated in a similar manner.

11–2 From the symmetry of the case it follows that $\dot{Q}_1^{(r)} = \dot{Q}_2^{(r)}$ and hence that the two currents are not independent. Thus

$$\Theta = \dot{Q}_1^{(r)}\left(\frac{1}{T_1} - \frac{1}{T_4}\right) + \dot{Q}_2^{(r)}\left(\frac{1}{T_2} - \frac{1}{T_4}\right)$$
$$= (\dot{Q}_1^{(r)} + \dot{Q}_2^{(r)})\left(\frac{1}{T_{\text{I}}} - \frac{1}{T_{\text{IV}}}\right) = 2\dot{Q}_1^{(r)}\left(\frac{1}{T_{\text{I}}} - \frac{1}{T_{\text{IV}}}\right).$$

11–3 From the relation $\dot{Q}_1^{(r)} + \dot{Q}_3^{(r)} = 0$ it follows that

$$\frac{\kappa B}{\Lambda}(T_0 - T_1 + T_0 - T_3) = 0$$

or

$$2T_0 - T_1 - T_3 = 0.$$

Similarly, the relation $\dot{Q}_2^{(r)} + \dot{Q}_4^{(r)} = 0$ implies that

$$2T_0 - T_2 - T_4 = 0.$$

Upon eliminating T_0 from these equations we get

$$T_4 - T_3 + T_2 - T_1 = 0.$$

If $|(T_i - T_4)/T_4| \ll 1$, then $T_i^{-1} - T_4^{-1} \approx (T_4 - T_i)/T_4^2$ and Eq. (11.19) reduces, approximately, to Eq. (11.22).

12–1 The derivations are straightforward:

$$\left(\frac{\partial U}{\partial V}\right)_T = T\left(\frac{\partial P}{\partial T}\right)_V - P, \qquad u = T\left(\frac{1}{3}\frac{du}{dT}\right) - \frac{u}{3},$$

$$4u = T\frac{du}{dT}, \qquad 4\frac{dT}{T} = \frac{du}{u}, \qquad u = \Gamma T^4.$$

$$h = u + P = \frac{4u}{3} = \frac{4\Gamma T^4}{3}, \qquad s = \frac{h}{T} = \frac{4\Gamma T^3}{3},$$

$$a = u - Ts = \Gamma T^4 - \left(\frac{4\Gamma T^4}{3}\right) = -\frac{\Gamma T^4}{3} = -P,$$

$$-\frac{da}{dT} = -\left(-\frac{4\Gamma T^3}{3}\right) = \frac{4\Gamma T^3}{3} = s.$$

13–1 Write the condition of local equilibrium in the form

$$\sum \nu_j \mu_\lambda^{(j)} - \sum \nu_i \mu_\lambda^{(i)} = 0.$$

Perform the operation grad upon this equation, and divide the resultant equation through by $\Delta\mathcal{G}_C/\Lambda_C$.

13–2 The numerical example in the hint adequately illustrates the point:

$$|\mu_\alpha^{(1)} - \mu_\beta^{(1)}| \approx 0.016RT,$$
$$|\mu_\alpha^{(2)} - \mu_\beta^{(2)}| \approx 2.3RT.$$

14–1 From the relations

$$\bar{H}_\alpha^{(g)}\dot{n}_\alpha^{(g)} + \bar{H}_\beta^{(g)}\dot{n}_\beta^{(g)} + \dot{Q}_\alpha^{(r)} + \dot{Q}_\beta^{(r)} + \dot{Q}_\gamma^{(r)} + \dot{q}(\alpha\beta) + \dot{q}(\beta\gamma) + \dot{q}(\alpha\gamma) = \dot{W}_0$$

and

$$\Theta = \bar{S}_\alpha^{(g)}\dot{n}_\alpha^{(g)} + \bar{S}_\beta^{(g)}\dot{n}_\beta^{(g)} + \dot{S}_\alpha^{(r)} + \dot{S}_\beta^{(r)} + \dot{S}_\gamma^{(r)} + \dot{\mathrm{s}}(\alpha\gamma) + \dot{\mathrm{s}}(\alpha\beta) + \dot{\mathrm{s}}(\beta\gamma),$$

it follows that [if we assume that (i) $\langle T(\alpha\beta)\rangle = \langle T(\beta\gamma)\rangle = \langle T(\alpha\gamma)\rangle$ (the linkages are exchanging heat with a common environment), that (ii)

$$(\delta \dot{Q}_{\gamma}^{(r)}/\bar{\delta}\dot{n}_{\alpha}^{(g)})_{\Omega'} \approx 0,$$

and that (iii)

$$(\delta \dot{W}_0/\bar{\delta}\dot{n}_{\alpha}^{(g)})_{\Omega'} = \Delta\psi(\delta I/\bar{\delta}\dot{n}_{\alpha}^{(g)})_{\Omega'}]$$

$$[H_\alpha] - [H_\beta] - [h(\beta\alpha)] = [R_\alpha^{(I)}]\left\{\Delta\psi - \frac{\partial \dot{q}(\alpha\gamma)}{\partial I} - \frac{\partial \dot{q}(\beta\gamma)}{\partial I}\right\}$$

and

$$[S_\alpha] - [S_\beta] - [s(\beta\alpha)] + [R_\alpha^{(I)}]\left\{\frac{\partial \dot{s}(\alpha\gamma)}{\partial I} + \frac{\partial \dot{s}(\beta\gamma)}{\partial I}\right\} = 0,$$

where

$$\frac{\partial}{\partial I} \equiv \left(\frac{\partial}{\partial I}\right)_{\Omega'}\Bigg|_{\dot{n}_{\alpha}^{(g)}=0}.$$

Since

$$\frac{\partial \dot{q}(\chi\omega)}{\partial I} = \langle T(\chi\omega)\rangle \frac{\partial \dot{s}(\chi\omega)}{\partial I},$$

it follows that

$$[G_\alpha] - [G_\beta] = [R_\alpha^{(I)}]\,\Delta\psi.$$

If an amalgam is placed in a thermal field, the Soret effect builds up a concentration gradient along the temperature gradient. If, after Soret equilibrium is established, we pass a steady electric current through the amalgam, will the electric current cause any appreciable change in the Soret concentration gradient? Will the flow of electrons through the medium exert more of a dragging effect on one component than on the other? In view of the disparity in size and mass of the electron relative to the atoms of the amalgam, it seems unlikely that any selective dragging tendency will be observed. It seems plausible then that $[R_\alpha^{(I)}] \approx 0$; however, we can only find out for certain by performing the experiment.

A–1 In the linear current-affinity region it follows, if we make use of Eq. (A.21a), that

$$\begin{aligned}\left(\frac{\delta\Theta}{\bar{\delta} Y_k}\right)_{Y',\mathbf{f}} &= \Omega_k + \sum_{i,i\neq k} Y_i\left(\frac{\delta\Omega_i}{\bar{\delta} Y_k}\right)_{Y',\mathbf{f}}\\ &= \sum_{i,i\neq k} Y_i\{K_{ki}(\mathbf{f}) + K_{ik}(\mathbf{f})\}\\ &= \sum_{i,i\neq k} Y_i\{K_{ik}(-\mathbf{f}) + K_{ki}(-\mathbf{f})\}\\ &= \left(\frac{\delta\Theta}{\bar{\delta} Y_k}\right)_{Y',-\mathbf{f}}.\end{aligned}$$

A–2 Given that

$$\Theta = Y_1\Omega_1 + Y_2\Omega_2$$

and that

$$\left(\frac{\delta\Theta}{\delta Y_1}\right)_{Y_2} = \Omega_1 + Y_2\left(\frac{\delta\Omega_2}{\delta Y_1}\right)_{Y_2} = 0$$

it follows, when $Y_1 = 0$, that

$$\Omega_1 = K_{12}Y_2$$

and

$$\left(\frac{\delta\Omega_2}{\delta Y_1}\right)_{Y_2} = K_{21}.$$

Hence

$$0 = (K_{12} + K_{21})Y_2.$$

But Y_2 is arbitrary, therefore,

$$K_{12} + K_{21} = 0.$$

A–3 From the relations

$$\Theta = Y_1\Omega_1 + Y_2\Omega_2 + Y_3\Omega_3$$

and

$$\left(\frac{\delta\Theta}{\delta Y_2}\right)_{Y_1,\Omega_3} = Y_1\left(\frac{\delta\Omega_1}{\delta Y_2}\right)_{Y_1,\Omega_3} + \Omega_2 + \Omega_3\left(\frac{\delta Y_3}{\delta Y_2}\right)_{Y_1,\Omega_3},$$

it follows, when $Y_2 = 0$, that

$$\begin{aligned}\left(\frac{\delta\Theta}{\delta Y_2}\right)_{Y_1,\Omega_3} &= Y_1\left\{K_{12} + K_{21} + \left(\frac{\delta Y_3}{\delta Y_2}\right)_{Y_1,\Omega_3}(K_{13} + K_{31})\right\} + Y_3(K_{23} - K_{32})\\ &= K_{33}^{-1}\{Y_1[K_{33}(K_{12} + K_{21}) - (K_{32}K_{13} + K_{31}K_{23})]\\ &\qquad + \Omega_3(K_{23} - K_{32})\}.\end{aligned}$$

Thus sufficient conditions for the vanishing of $(\delta\Theta/\delta Y_2)_{Y_1,\Omega_3}$ are $K_{12} = -K_{21}$, $K_{13} = -K_{31}$, $K_{23} = K_{32}$; but the necessary conditions are only $K_{23} = K_{32}$, $K_{33}(K_{12} + K_{21}) - K_{32}(K_{13} + K_{31}) = 0$.

A–4 It follows from Eqs. (9.7) and (A.55) to (A.57) that

$$\begin{aligned}\frac{\pi_\beta^{(ab)}(-\mathbf{B})}{T} &= [c_\beta^{(a)}(\mathbf{B})] - [c_\beta^{(b)}(\mathbf{B})] + \frac{\pi_\beta^{(ab)}(\mathbf{B})}{T_\beta} + T_\beta\left\{\frac{\partial(\pi_\beta^{(ab)}(\mathbf{B})/T_\beta)}{\partial T_\beta}\right\}_{T_\alpha,\mathbf{B}}\\ &= -T_\beta\left\{\frac{\partial(\pi_\beta^{(ab)}(-\mathbf{B})/T_\beta)}{\partial T_\beta}\right\}_{T_\alpha,-\mathbf{B}} + \frac{\pi_\beta^{(ab)}(\mathbf{B})}{T_\beta} + T_\beta\left\{\frac{\partial(\pi_\beta^{(ab)}(\mathbf{B})/T_\beta)}{\partial T_\beta}\right\}_{T_\alpha,\mathbf{B}}\end{aligned}$$

and hence, that

$$\left(\frac{\partial\pi_\beta^{(ab)}(-\mathbf{B})}{\partial T_\beta}\right)_{T_\alpha,-\mathbf{B}} = \left(\frac{\partial\pi_\beta^{(ab)}(\mathbf{B})}{\partial T_\beta}\right)_{T_\alpha,\mathbf{B}}.$$

The other relation follows similarly from Eq. (A.56) and the result just established.

BOOK III. IDEAS ABOUT THE OLD IDEA

Reprinted Papers

Reprinted here are six papers that developed further the ideas of Book II. In the last paper, there are a few errors in subscripting practice: in Eqs. (20), interchange the subscripts i, k on the L and K coefficients; in Eqs. (21), interchange the subscripts i, k throughout.

On the Thermocouple in a Magnetic Field

Nonequilibrium Thermodynamics: Current-Affinity Relations and Thermokinetic Potentials

Thermodynamics of Steady States: A Weak Entropy-Production Principle

Thermodynamics of Steady States: "Resistance Change" Transitions in Steady-state Systems

Thermodynamics of Steady States: Is the Entropy-Production Surface Convex in the Thermodynamic Space of Steady Currents?

Thermodynamics of Steady States: Definitions, Relations and Conjectures

Reprinted from THE JOURNAL OF CHEMICAL PHYSICS, Vol. 47, No. 5, 1879, 1 September 1967
Printed in U. S. A.

On the Thermocouple in a Magnetic Field

R. J. TYKODI
Department of Chemistry, Southeastern Massachusetts Technological Institute, North Dartmouth, Massachusetts

(Received 24 February 1967)

Consider a thermocouple formed of materials a and b with the temperatures of the bimaterial junctions being T_α and T_β. Let the Seebeck potential difference be $\Delta\psi$; let $\pi_\beta^{(ab)}$ be the Peltier heat (per unit current) absorbed at the ab junction when negative current flows from material a to material b and the junction is at temperature T_β; and let $[c_\beta^{(a)}]$, e.g., be the Thomson coefficient (defined in terms of the flow of negative electricity) for material a at a point where the temperature is T_β. In my recently published book[1] I derived the following basic relation for the thermocouple:

$$\left(\frac{\partial\Delta\psi}{\partial T_\beta}\right)_{T_\alpha}=[c_\beta^{(a)}]-[c_\beta^{(b)}]+\frac{\pi_\beta^{(ab)}}{T_\beta}+T_\beta\frac{d(\pi_\beta^{(ab)}/T_\beta)}{dT_\beta}. \quad (1)$$

With the help of an entropy-production principle, it was shown that Eq. (1) could be resolved into the Thomson relations for the thermocouple:

$$(\partial\Delta\psi/\partial T_\beta)_{T_\alpha}=\pi_\beta^{(ab)}/T_\beta, \quad (2)$$

$$\frac{[c_\beta^{(a)}]-[c_\beta^{(b)}]}{T_\beta}+\frac{d(\pi_\beta^{(ab)}/T_\beta)}{dT_\beta}=0, \quad (3)$$

$$T_\beta(\partial^2\Delta\psi/\partial T_\beta^2)_{T_\alpha}=[c_\beta^{(b)}]-[c_\beta^{(a)}]. \quad (4)$$

For the case of the thermocouple in a magnetic field $\mathbf{B}$ (or $-\mathbf{B}$), Eq. (1) still holds, but Eqs. (2)–(4) have to be recast. The analog of Eq. (2) is the equation[2,3]

$$[\partial\Delta\psi(\mathbf{B})/\partial T_\beta]_{T_\alpha,\mathbf{B}}=\pi_\beta^{(ab)}(-\mathbf{B})/T_\beta, \quad (5)$$

and in my book I took as the analogs of Eqs. (3) and (4) the relations

$$\frac{[c_\beta^{(a)}(\mathbf{B})]-[c_\beta^{(b)}(\mathbf{B})]}{T_\beta}+\left\{\frac{\partial[\pi_\beta^{(ab)}(-\mathbf{B})/T_\beta]}{\partial T_\beta}\right\}_{T_\alpha,-\mathbf{B}}=0 \quad (6)$$

and

$$T_\beta[\partial^2\Delta\psi(\mathbf{B})/\partial T_\beta^2]_{T_\alpha,\mathbf{B}}=[c_\beta^{(b)}(\mathbf{B})]-[c_\beta^{(a)}(\mathbf{B})]. \quad (7)$$

I now wish to show that in the light of experiments such as those of Wolfe and Smith[4-6] the relations (6) and (7) are untenable and that the proper analogs of Eqs. (3) and (4) are the relations

$$\left\langle\frac{[c_\beta^{(a)}]-[c_\beta^{(b)}]}{T_\beta}+\left[\frac{\partial(\pi_\beta^{(ab)}/T_\beta)}{\partial T_\beta}\right]_{T_\alpha,\mathbf{B}}\right\rangle_{\mathrm{Av}}=0 \quad (8)$$

and

$$T_\beta\langle(\partial^2\Delta\psi/\partial T_\beta^2)_{T_\alpha,\mathbf{B}}\rangle_{\mathrm{Av}}=\langle[c_\beta^{(b)}]-[c_\beta^{(a)}]\rangle_{\mathrm{Av}}, \quad (9)$$

where $\langle Z\rangle_{\mathrm{Av}}\equiv\frac{1}{2}[Z(\mathbf{B})+Z(-\mathbf{B})]$ and the subscript B means constant $\mathbf{B}$ or constant $-\mathbf{B}$ as appropriate to the term being operated on.

If we write Eq. (1) twice, once in field $\mathbf{B}$ and once in field $-\mathbf{B}$, if we add these two forms of Eq. (1), if we then make use of the two forms of Eq. (5), and if we multiply the resulting equation through by $\frac{1}{2}$, we get Eq. (8). Upon combining Eqs. (5) and (8), we get Eq. (9). Now if we try to resolve Eq. (8) into two simpler equations of the form (3) with *each* of the quantities in one of the equations depending on the *same* direction of the magnetic field ($\mathbf{B}$, say), then, because of Eq. (1), we have to write a "same-direction" version of Eq. (2). Equation (5) and the same-direction version of Eq. (2) will only be compatible in the case that $\pi_\beta^{(ab)}(\mathbf{B})$ always equals $\pi_\beta^{(ab)}(-\mathbf{B})$. The results of Wolfe and Smith show that such need not be the case; hence we cannot accept the same-direction resolutions of Eq. (8) as being generally valid.

Turning now to Eqs. (6) and (7), we find that Eqs. (1) and (5)–(7) imply[1] that

$$[\partial\pi_\beta^{(ab)}(\mathbf{B})/\partial T_\beta]_{T_\alpha,\mathbf{B}}=[\partial\pi_\beta^{(ab)}(-\mathbf{B})/\partial T_\beta]_{T_\alpha,-\mathbf{B}}. \quad (10)$$

Equation (10) shows that for a given field strength the difference $\pi_\beta^{(ab)}(\mathbf{B})-\pi_\beta^{(ab)}(-\mathbf{B})$ is independent of temperature. We know, however, that in the vicinity of 0°K the individual Peltier heats tend to vanish; hence the difference between the two Peltier heats also vanishes at 0°K. This observation and Eq. (10) together imply that $\pi_\beta^{(ab)}(\mathbf{B})$ equals $\pi_\beta^{(ab)}(-\mathbf{B})$ at all temperatures and for all values of the field strength; such a result, however, is in conflict with the findings of Wolfe and Smith.

In the face of experimental results such as those of Wolfe and Smith, I conclude, then, that for the thermocouple in a magnetic field the analogs of Eqs. (2)–(4) are Eqs. (5), (8), and (9). It is in general not possible to further decompose Eqs. (8) and (9) into simpler equations of the forms (6) and (7), etc.; the full forms (8) and (9) must be used as such if we seek the perfectly general analogs of Eqs. (3) and (4).

[1] R. J. Tykodi, *Thermodynamics of Steady States* (The Macmillan Co., New York, 1967).
[2] S. de Groot, *Thermodynamics of Irreversible Processes* (North-Holland Publ. Co., Amsterdam, 1951) p. 145.
[3] S. de Groot and P. Mazur, *Non-Equilibrium Thermodynamics* (North-Holland Publ. Co., Amsterdam, 1962) Chap. 13.
[4] R. Wolfe and G. Smith, Phys. Rev. **129**, 1086 (1963).
[5] G. Smith, R. Wolfe, and S. Haszko, Proc. Int. Conf. Semiconductor Physics, 7th, Paris (1964) p. 399.
[6] G. Smith and R. Wolfe, J. Phys. Soc. (Japan), **21**, S 651 (1966).

Reprinted from:

THE JOURNAL OF CHEMICAL PHYSICS VOLUME 57, NUMBER 1 1 JULY 1972

Nonequilibrium Thermodynamics: Current–Affinity Relations and Thermokinetic Potentials

R. J. TYKODI

Department of Chemistry, Southeastern Massachusetts University, North Dartmouth, Massachusetts 02747

(Received 30 December 1970)

Two examples, heat conduction in wires and the Poiseuille flow of a gas through a capillary tube, are used to show that the presence or absence of higher order symmetry relations among the phenomenological coefficients can depend on the choice of boundary conditions. A family of thermokinetic potentials is developed for each example, and the over-all utility of the thermokinetic potential concept is discussed.

The application of thermodynamic concepts to nonequilibrium situations has a long history, going back as far as the work of William Thomson in the nineteenth century; the modern period of active development of the subject dates from the work of Onsager in the early 1930's; full references to the literature can be found in recent monographs on the subject.[1–3] The basic approach, which was initiated by Onsager, is to consider the currents (fluxes) characterizing a nonequilibrium situation to be functions of a set of thermodynamic affinities (forces), to expand the currents in a Taylor's series in terms of the affinities about a reference thermodynamic equilibrium state, and to use thermodynamic arguments to derive relationships among the coefficients in the Taylor's series expansion.

A rich literature has grown up devoted to the linear current-affinity region (the Taylor's series limited to the first order terms), and the prime result has been the establishment of first order symmetries (Onsager reciprocal relations) among the phenomenological coefficients.[1–3] The successful treatment of the linear current-affinity region has, however, not been followed by equally successful excursions into the nonlinear region. Efforts to extend plausible thermodynamic ideas shown to hold in the linear region into the nonlinear region have repeatedly met with failure. In fact the history of work on the nonlinear region is largely the history of negative results: One seemingly good idea after another has been exploded by the discovery of a suitable counterexample. The development of the subject seems to have entered a stage of therapeutic criticism: we seem to be learning about the limitations of various thermodynamic constraints thought to hold for nonequilibrium situations.[4–6]

This paper also is an exercise in therapeutic criticism. I point out two simple cases where the nonlinear terms in the current-affinity relations can be calculated explicitly, and I show by direct example the need to be more concerned about the boundary conditions pertaining to a given situation and, correspondingly, about the choice of current-affinity representation. I also show the inadequacy of some thermodynamic ideas involving thermokinetic potentials.

One popular way of attacking nonlinear relations has been to search for extremum (maximum or minimum) principles—entropy production,[1–3] thermokinetic potentials,[7,8] local potentials,[6] etc.—which single out a steady state from a class of transient states with given boundary conditions. I have shown elsewhere[3] that the comparison of a steady state with a class of transient states can, for thermodynamic purposes, be replaced by the comparison of the given steady state with a class of "more constrained" steady states. My discussion in this paper, therefore, will be couched in an *all steady state* language.

In this paper I wish to discuss the relationships between the rate of entropy production, thermokinetic potentials, higher order symmetries among the phenomenological coefficients in current-affinity relations, and the boundary conditions for two special cases—

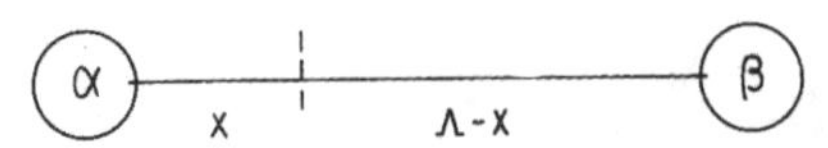

FIG. 1. Steady flow of heat along a wire of cross sectional area B and length Λ. The ends of the wire are maintained at temperatures T_α, T_β; a coordinate x has been laid off along the wire.

heat conduction in wires and the Poiseuille flow of a gas through a capillary; these two examples will serve to illustrate the (critical) points I have in mind.

HEAT CONDUCTION EXAMPLE

Stretch a homogeneous metallic wire of cross sectional area B and of length Λ between two thermostats α, β; insulate the wire in such a way that there are no lateral heat losses. Now, how can we determine the steady state temperature distribution along the wire? Proceed as follows: consider a point on the wire at a distance x from the α thermostat—Fig. 1; cut the wire at point x and let the two exposed faces of the wire communicate with a heat reservoir (thermostat) of temperature T_x—Fig. 2; consider only states of steady heat flow in Figs. 1 and 2.

When the inflow of heat into the reservoir at x across one face of the wire just balances the outflow of heat across the other face, the temperature T_x of Fig. 2 corresponds to the (steady state) temperature at point x in Fig. 1, and the two situations are entirely equivalent: the situation described in Fig. 1 is thus imbedded in the more general ("more constrained") situation of Fig. 2. Let $\dot{Q}_\alpha^{(r)}$, $\dot{Q}_\beta^{(r)}$, $\dot{Q}_x^{(r)}$ represent the rate of influx of heat into the appropriate reservoir; then by finding the conditions that guarantee that $\dot{Q}_x^{(r)}=0$ in Fig. 2 (for any point x), we determine the steady state temperature distribution along the wire for Fig. 1. In the steady state, then, we have the following relations:

$$\dot{Q}_\alpha^{(r)}+\dot{Q}_\beta^{(r)}+\dot{Q}_x^{(r)}=0, \tag{1}$$

$$\Theta=\dot{Q}_\alpha^{(r)}T_\alpha^{-1}+\dot{Q}_\beta^{(r)}T_\beta^{-1}+\dot{Q}_x^{(r)}T_x^{-1} \tag{2}$$

$$=\dot{Q}_\alpha^{(r)}(T_\alpha^{-1}-T_x^{-1})+\dot{Q}_\beta^{(r)}(T_\beta^{-1}-T_x^{-1})\equiv Y_1\Omega_1+Y_2\Omega_2 \tag{3}$$

$$=\dot{Q}_\alpha^{(r)}(T_\alpha^{-1}-T_\beta^{-1})+\dot{Q}_x^{(r)}(T_x^{-1}-T_\beta^{-1})\equiv y_1\omega_1+y_2\omega_2, \tag{4}$$

where $\Theta\equiv\dot{S}$ (system) $+\dot{S}$ (surroundings) is the rate of entropy production, Y_i (or y_i) is a *current*, and Ω_i (or ω_i) is an *affinity*. (Note that in general there are several, entirely equivalent, ways of representing the rate of entropy production in terms of currents and affinities.[3])

According to Fourier's law for heat conduction, the heat flux vector is proportional to the gradient of the temperature, the proportionality constant being the thermal conductivity κ. If we multiply the heat flux vector by the cross sectional area of the wire and integrate from a heat source to a heat sink, we get, for the quantities $\dot{Q}^{(r)}$, the relations

$$\dot{Q}_\alpha^{(r)}=\frac{B}{x}\int_{T_\alpha}^{T_x}\kappa\,dT, \tag{5}$$

$$\dot{Q}_\beta^{(r)}=\frac{B}{\Lambda-x}\int_{T_\beta}^{T_x}\kappa\,dT,$$

$$\dot{Q}_x^{(r)}=-\dot{Q}_\alpha^{(r)}-\dot{Q}_\beta^{(r)}, \tag{6}$$

where κ is the thermal conductivity of the wire. For the case of constant thermal conductivity, $\kappa\neq f(T)$, Fourier's law tells us that the steady state value of T_x in Fig. 1 is

$$T_x=[(\Lambda-x)/\Lambda]T_\alpha+(x/\Lambda)T_\beta=T_\alpha+(x/\Lambda)(T_\beta-T_\alpha) \tag{7}$$

(i.e., the steady state distribution of temperature along the wire is linear)—just the value T_x that makes $\dot{Q}_x^{(r)}$ vanish in Fig. 2.

Inasmuch as I am concerned, in this paper, with developing counterexamples showing that some thermodynamically plausible statements are not universally valid or are inadequate for purposes of practical calculation, I find it sufficient for my purposes to consider the special case $x=\Lambda/2$, $\kappa\neq f(T)$, and a wire of geometry such that $2B\,\kappa/\Lambda=1$; for this special case we have

$$\dot{Q}_\alpha^{(r)}=T_x-T_\alpha,\qquad \dot{Q}_\beta^{(r)}=T_x-T_\beta, \tag{8}$$

$$\dot{Q}_x^{(r)}=T_\alpha+T_\beta-2T_x, \tag{9}$$

and the value of T_x that makes $\dot{Q}_x^{(r)}$ vanish in Fig. 2 is, of course,

$$T_x=\tfrac{1}{2}(T_\alpha+T_\beta)\equiv\langle T\rangle. \tag{10}$$

By limiting myself to the special case just outlined, I save myself the labor of writing all those coefficients involving area, length, and thermal conductivity that would normally appear in the dozens of equations to follow.

The strategy, then, for the remainder of this section of the paper, is to test various thermodynamically inspired hypotheses against the Fourier's law solutions [Eqs. (8)–(10)] of the special case heat conduction problem.

Current-Affinity Relations

In bringing thermodynamic arguments to bear on nonequilibrium situations, it is customary[1–3] to express

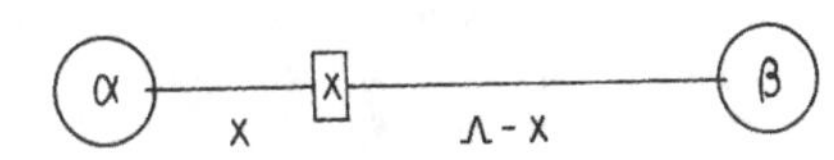

FIG. 2. A heat reservoir (thermostat) of temperature T_x has been spliced into the wire at the point x defined in Fig. 1.

the rate of entropy production in terms of currents and affinities,

$$\Theta = \sum_i Y_i \Omega_i \geq 0 \text{ (e.g.)}, \qquad (11)$$

and to express the currents as Taylor's series in the affinities relative to a fixed thermodynamic equilibrium state (σ, say),

$$Y_i = \sum_j L_{ij}(\sigma)\Omega_j + \tfrac{1}{2}\sum_{j,k} L_{ijk}(\sigma)\Omega_j\Omega_k + \cdots. \qquad (12)$$

In this regard it is essential to note that any legitimate[3] current-affinity representation is as good as any other and also that any choice of reference thermodynamic equilibrium state is as good as any other.

For the linear current-affinity region (the Taylor's series restricted to first order terms only) we find[1-3] that, for a given equilibrium reference state σ, the reversal of the affinities Ω_i exactly reverses the currents Y_i and that, for situations not involving magnetic or centrifugal fields, several thermodynamically inspired extremum principles imply that

$$L_{ij}(\sigma) = L_{ji}(\sigma), \qquad (13)$$

i.e., imply that the Onsager reciprocal relations hold. Now, does the reversal property persist outside the linear current-affinity region; and are there any *thermodynamically necessary* symmetry relations among the higher order phenomenological coefficients $L_{ijk\cdots}(\sigma)$? I intend to answer these and related questions by producing simple counterexamples. Let us, therefore, proceed with the study.

Consider Eqs. (3) and (8). For fixed T_z, we can solve for T_α in terms of T_z and Ω_1 and for T_β in terms of T_z and Ω_2:

$$T_\alpha = T_z/(1+T_z\Omega_1), \qquad T_\beta = T_z/(1+T_z\Omega_2). \qquad (14)$$

Upon expanding the quantities $(1+T_z\Omega_i)^{-1}$ in series, we see that the currents $Y_i = Y_i(T_z, \Omega_1, \Omega_2)$ can be expressed as power series in the affinities Ω_j, with constant coefficients $L_{ijk\cdots}(T_z)$:

$$Y_i = L_{i1}\Omega_1 + L_{i2}\Omega_2 + \tfrac{1}{2}L_{i11}\Omega_1^2 + L_{i12}\Omega_1\Omega_2 + \tfrac{1}{2}L_{i22}\Omega_2^2 + \cdots, \qquad (15)$$

$$Y_1 \equiv \dot{Q}_\alpha^{(r)} = T_z - T_\alpha = T_z - T_z(1+T_z\Omega_1)^{-1}$$
$$= T_z^2\Omega_1 - T_z^3\Omega_1^2 + T_z^4\Omega_1^3 - \cdots, \qquad (16)$$

$$Y_2 \equiv \dot{Q}_\beta^{(r)} = T_z - T_\beta = T_z - T_z(1+T_z\Omega_2)^{-1}$$
$$= T_z^2\Omega_2 - T_z^3\Omega_2^2 + T_z^4\Omega_2^3 - \cdots. \qquad (17)$$

Note that in this case

$$L_{12} = L_{21} = 0, \qquad L_{112} = L_{211} = 0, \qquad L_{122} = L_{212} = 0, \qquad (18)$$

and that

$$Y_i(T_z, \Omega_1, \Omega_2) \neq -Y_i(T_z, -\Omega_1, -\Omega_2). \qquad (19)$$

From similar considerations we see that (for fixed T_z) the currents $y_i = y_i(T_z, \omega_1, \omega_2)$ can be expressed as power series in the affinities ω_j, with constant coefficients $l_{ijk\cdots}(T_z)$:

$$y_i = l_{i1}\omega_1 + l_{i2}\omega_2 + \tfrac{1}{2}l_{i11}\omega_1^2 + l_{i12}\omega_1\omega_2 + \tfrac{1}{2}l_{i22}\omega_2^2 + \cdots, \qquad (20)$$

$$y_1 \equiv \dot{Q}_\alpha^{(r)} = T_z - T_\alpha = T_z - T_z(1+T_z\omega_1 - T_z\omega_2)^{-1}$$
$$= T_z^2\omega_1 - T_z^2\omega_2 - T_z^3\omega_1^2 + 2T_z^3\omega_1\omega_2 - T_z^3\omega_2^2 + \cdots, \qquad (21)$$

$$y_2 \equiv \dot{Q}_z^{(r)} = T_\alpha + T_\beta - 2T_z = T_z(1+T_z\omega_1 - T_z\omega_2)^{-1}$$
$$+ T_z(1-T_z\omega_2)^{-1} - 2T_z$$
$$= -T_z^2\omega_1 + T_z^2\omega_2 + T_z^3\omega_1^2 - 2T_z^3\omega_1\omega_2 + 2T_z^3\omega_2^2 + \cdots. \qquad (22)$$

In this case we have

$$l_{12} = l_{21} \neq 0, \qquad l_{112} = l_{211} \neq 0, \qquad l_{122} = l_{212} \neq 0, \qquad (23)$$

and

$$y_i(T_z, \omega_1, \omega_2) \neq -y_i(T_z, -\omega_1, -\omega_2). \qquad (24)$$

Now let T_β be fixed and vary $\dot{Q}_\alpha^{(r)}$, $\dot{Q}_\beta^{(r)}$, $\dot{Q}_z^{(r)}$ by varying T_α and T_z. Solve for T_α and T_z in terms of T_β, Ω_1, Ω_2 or T_β, ω_1, ω_2. The currents Y_i and y_i can again be developed into power series in terms of the affinities Ω_j and ω_j, with new coefficients $L_{ijk\cdots}(T_\beta)$ and $l_{ijk\cdots}(T_\beta)$. For fixed T_β we find that

$$L_{12} = L_{21} = 0, \qquad L_{112} \neq L_{211}, \qquad L_{122} = L_{212} = 0, \qquad (25)$$

$$l_{12} = l_{21} \neq 0, \qquad l_{112} \neq l_{211}, \qquad l_{122} \neq l_{212}, \qquad (26)$$

$$Y_i(T_\beta, \Omega_1, \Omega_2) \neq -Y_i(T_\beta, -\Omega_1, -\Omega_2), \qquad (27)$$

$$y_i(T_\beta, \omega_1, \omega_2), \neq -y_i(T_\beta, -\omega_1, -\omega_2). \qquad (28)$$

Li[7,8] has considered the problem of the existence of higher order symmetries among the phenomenological coefficients; he has proposed a nomenclature whereby if symmetries exist for all orders up to and including the kth, the currents are said to show kth order independence. All nonequilibrium situations not involving magnetic or centrifugal fields show first order independence (Onsager reciprocal relations). In his early work[7,8] Li indicated that he expected higher order independence to hold for most systems. We see from the above considerations relative to the heat conduction case that the currents Y_i and y_i show second order independence at constant T_z but only first order independence at constant T_β.

Thermokinetic Potentials

Much of the effort expended in applying thermodynamically inspired ideas to the treatment of nonequilibrium situations has been aimed at producing a satisfactory extremum principle which would do for steady state situations what the entropy-maximum principle does for equilibrium situations. The basic problem that we seek to solve is, What are the thermodynamically necessary conditions guaranteeing that, for given affinities and given boundary conditions, a given flow (current) must vanish, i.e., what are the conditions of migrational equilibrium?[3] Li has shown[7,8] that under very general conditions a fundamental Pfaffian form such as $Y_1 d\Omega_1 + Y_2 d\Omega_2$ or $y_1 d\omega_1 + y_2 d\omega_2$ can be integrated to yield a thermokinetic potential F. The potential F has the property of never being negative and of assuming a minimum in a given steady state relative to a class of transient states with the same boundary conditions. Upon transcribing Li's ideas into the "more constrained" *all steady state* language, we find that (e.g.) the Pfaffian $dF_\sigma = Y_1 d\Omega_1 + Y_2 d\Omega_2$ integrates, for a fixed set of boundary conditions σ, to the function $F = F(\Omega_1, \Omega_2, \sigma)$ with the properties

$$0 = (\partial F/\partial \Omega_i)_{\Omega_j,\sigma}|_{Y_i=0} \qquad i,j=1,2, i\neq j, \quad (29)$$

$$0 < (\partial^2 F/\partial \Omega_i^2)_{\Omega_j,\sigma}|_{Y_i=0}. \quad (30)$$

The thermokinetic potential F thus has a local minimum (at constant conditions σ) at each of the points where one of the currents Y_i vanishes; it has an absolute minimum at the point where all the currents Y_i vanish simultaneously. (The function F would seem to solve all our problems: to determine the conditions for $Y_i = 0$, merely differentiate F with respect to Ω_i, keeping σ and Ω_j constant, and set the resulting expression equal to zero. I intend to show, however, that, from a practical point of view, the preceding recipe involving the F function is worthless.)

Let us now explore the (special case) heat conduction example of Fig. 2 in terms of the thermokinetic potential. At constant T_x we have

$$dF_x = Y_1 d\Omega_1 + Y_2 d\Omega_2 = y_1 d\omega_1 + y_2 d\omega_2$$
$$= -(T_x - T_\alpha) T_\alpha^{-2} dT_\alpha - (T_x - T_\beta) T_\beta^{-2} dT_\beta, \quad (31)$$

$$F_x = (T_x/T_\alpha) + \ln T_\alpha + (T_x/T_\beta) + \ln T_\beta + f(T_x). \quad (32)$$

If we set $F_x = 0$ for $T_\alpha = T_\beta = T_x$, we get

$$F_x = \ln(T_\alpha/T_x) + \ln(T_\beta/T_x) + (T_x/T_\alpha) + (T_x/T_\beta) - 2 \geq 0. \quad (33)$$

The potential F_x has the following properties:

$$(\partial F_x/\partial T_\alpha)_{T_\beta,T_x} = (T_\alpha - T_x)/T_\alpha^2$$
$$= -T_\alpha^{-2}(\partial F_x/\partial \Omega_1)_{\Omega_2,T_x}$$
$$= -T_\alpha^{-2}(\partial F_x/\partial \omega_1)_{\omega_2,T_x}, \quad (34)$$

i.e., F_x has a minimum at the point $T_\alpha = T_x$—just the point where $Y_1 = y_1 = 0$ [note that there is no loss of generality in computing $(\partial F_x/\partial T_\alpha)_{T_\beta,T_x}$ instead of, say, $(\partial F_x/\partial \Omega_1)_{\Omega_2,T_x}$—both derivatives vanish at the same point];

$$(\partial F_x/\partial T_\beta)_{T_\alpha,T_x} = (T_\beta - T_x)/T_\beta^2, \quad (35)$$

i.e., F_x also has a minimum at the point $T_\beta = T_x$—the point where $Y_2 = 0$;

$$(\partial F_x/\partial T_x)_{T_\alpha,T_\beta} = T_\alpha^{-1} + T_\beta^{-1} - 2T_x^{-1}, \quad (36)$$

i.e., F_x does *not* have a minimum at the point ($T_x = \langle T \rangle$) where $y_2 = 0$. In a neighborhood of T_x such that $T_\alpha \to T_x$ and $T_\beta \to T_x$, we observe that $F_x \to \frac{1}{2}\Theta$.

Consider now the case of constant T_β:

$$dF_\beta = Y_1 d\Omega_1 + Y_2 d\Omega_2 = y_1 d\omega_1 + y_2 d\omega_2$$
$$= -(T_x - T_\alpha) T_\alpha^{-2} dT_\alpha - (T_\alpha + T_\beta - 2T_x) T_x^{-2} dT_x. \quad (37)$$

The Pfaffian form dF_β is not a total differential; it can, however, be converted (at constant T_β) into a total differential by being multiplied by an integrating factor $\phi(T_\alpha, T_x)$. Instead of searching for an integrating factor, let us consider the more special case of constant T_α and T_β:

$$dF_{\alpha\beta} = Y_1 d\Omega_1 + Y_2 d\Omega_2 = y_1 d\omega_1 + y_2 d\omega_2$$
$$= (2T_x - T_\alpha - T_\beta) T_x^{-2} dT_x, \quad (38)$$

$$F_{\alpha\beta} = 2\{\ln T_x + (\langle T \rangle / T_x)\} + f(T_\alpha, T_\beta). \quad (39)$$

Let $F_{\alpha\beta} = F_{\alpha\beta}^*$ when $T_x = \langle T \rangle$; we have, then,

$$\Delta F_{\alpha\beta} \equiv F_{\alpha\beta} - F_{\alpha\beta}^* = 2\{\ln(T_x/\langle T \rangle) + (\langle T \rangle/T_x) - 1\} \geq 0, \quad (40)$$

$$(\partial \Delta F_{\alpha\beta}/\partial T_x)_{T_\alpha,T_\beta} = 2(T_x - \langle T \rangle)/T_x^2, \quad (41)$$

$$(\partial \Delta F_{\alpha\beta}/\partial T_\alpha)_{T_\beta,T_x} = (\partial \Delta F_{\alpha\beta}/\partial T_\beta)_{T_\alpha,T_x} = T_x^{-1} - \langle T \rangle^{-1}. \quad (42)$$

The quantity $\Delta F_{\alpha\beta}$ has a minimum at the point where $y_2 = 0$; it does not have minima at the points where $Y_1 = 0$ or $Y_2 = 0$. Li[8] has discussed the properties of Eq. (40) in the context of the *steady state versus transient states* language.

Consider finally the expression

$$F = 2\ln T_x - \ln T_\alpha - \ln T_\beta + (T_\alpha/T_x) + (T_\beta/T_x) - 2 \geq 0, \quad (43)$$

which has the following properties:

$$(\partial F/\partial T_x)_{T_\alpha,T_\beta} = 2(T_x - \langle T \rangle)/T_x^2, \quad (44)$$

$$(\partial F/\partial T_\alpha)_{T_\beta,T_x} = T_x^{-1} - T_\alpha^{-1}, \quad (45)$$

$$(\partial F/\partial T_\beta)_{T_\alpha,T_x} = T_x^{-1} - T_\beta^{-1}. \quad (46)$$

The quantity F thus has a minimum at each of the points $Y_1 = y_1 = 0$, $Y_2 = 0$, $y_2 = 0$; for our special case it is the perfect *thermokinetic potential*. Equation (43) may be generated from Eq. (33) by the transformation $T \to T^{-1}$; it may also be obtained from Eq. (39) by setting $f(T_\alpha, T_\beta) = -(\ln T_\alpha + \ln T_\beta + 2)$. (Note also the following points: Θ itself remains invariant to the transformation $T \to T^{-1}$; and, in a neighborhood of T_x such that $T_\alpha \to T_x$ and $T_\beta \to T_x$, $F \to \frac{1}{2}\Theta$.)

Let $dF)_\beta$ be the differential of F at constant T_β,

$$dF)_\beta = -(T_x - T_\alpha) T_\alpha^{-1} T_x^{-1} dT_\alpha - (T_\alpha + T_\beta - 2T_x) T_x^{-2} dT_x. \quad (47)$$

Observe that $dF)_\beta$ differs from dF_β by having at one place T_x^{-1} instead of T_α^{-1}; it thus seems unlikely that an integrating factor $\phi(T_\alpha, T_x)$ can be found such that $dF)_\beta = \phi(T_\alpha, T_x)\ dF_\beta$—in other words, we cannot find the perfect thermokinetic potential F by operating directly with dF_β.

POISEUILLE FLOW OF A GAS

The thermodynamic patterns that we have just observed for the flow of heat in wires are repeated in the case of the steady monothermal[3] Poiseuille flow of a gas. In Fig. 1 let α and β now represent constant pressure mass reservoirs (large bore cylinder-and-piston devices) connected by a capillary tube of length Λ and cross sectional area B, the entire setup being immersed in a thermostat of temperature T. If the apparatus is filled with a(n) (ideal) gas of viscosity η, the steady rate of flow of gas (mole sec^{-1}) into the α reservoir for given pressures P_α and P_β is[9]

$$\dot{n}_\alpha = B^2(P_\beta^2 - P_\alpha^2)/16\pi\Lambda\eta RT \tag{48}$$

for a substantial range of values for P_α, P_β, T. Now cut the capillary tube at the point x in Fig. 1 and insert another mass reservoir of pressure P_x—Fig. 2. Let $\dot{n}_\alpha$, $\dot{n}_\beta$, $\dot{n}_x$ represent the molar rate of influx of gas into the appropriate reservoir. By determining the steady state conditions that guarantee that $\dot{n}_x = 0$ for any arbitrary point x in Fig. 2, we in effect determine the steady state pressure distribution along the capillary tube in Fig. 1. Consider the following special case—Fig. 2: Let $x = \Lambda/2$ and choose the dimensions of the capillary such that $B^2/8\pi\Lambda\eta RT = 1$; we have, then, the relations

$$\dot{n}_\alpha + \dot{n}_\beta + \dot{n}_x = 0, \tag{49}$$

$$\dot{n}_\alpha = P_x^2 - P_\alpha^2, \qquad \dot{n}_\beta = P_x^2 - P_\beta^2, \tag{50}$$

$$\dot{n}_x = P_\alpha^2 + P_\beta^2 - 2P_x^2, \tag{51}$$

$$\Theta = -(\dot{n}_\alpha\mu_\alpha + \dot{n}_\beta\mu_\beta + \dot{n}_x\mu_x)/T \tag{52}$$

$$= \dot{n}_\alpha(\mu_x - \mu_\alpha)T^{-1} + \dot{n}_\beta(\mu_x - \mu_\beta)T^{-1} \equiv Y_1\Omega_1 + Y_2\Omega_2 \tag{53}$$

$$= \dot{n}_\alpha(\mu_\beta - \mu_\alpha)T^{-1} + \dot{n}_x(\mu_\beta - \mu_x)T^{-1} \equiv y_1\omega_1 + y_2\omega_2, \tag{54}$$

where $\mu = \mu(T, P)$ is the chemical potential of the gas in the appropriate mass reservoir. Under the assumed conditions of flow, Eqs. (53) and (54) take the forms

$$\Theta = (P_x^2 - P_\alpha^2)R\ln(P_x/P_\alpha) + (P_x^2 - P_\beta^2)R\ln(P_x/P_\beta), \tag{55}$$

$$= (P_x^2 - P_\alpha^2)R\ln(P_\beta/P_\alpha) + (P_\alpha^2 + P_\beta^2 - 2P_x^2) \times R\ln(P_\beta/P_x). \tag{56}$$

Current-Affinity Relations

Just as in the heat conduction case, we can, for fixed boundary conditions, express the currents $(\dot{n}_\alpha, \dot{n}_\beta, \dot{n}_x)$ as power series in the affinities (Ω_j or ω_j) with constant coefficients. For fixed T and P_x we get relations analogous to Eqs. (18) and (23). For fixed T and P_β we find that relations analogous to Eqs. (25) and (26) hold. The currents thus show second order independence at constant T, P_x, but only first order independence at constant T, P_β.

Thermokinetic Potentials

Proceeding as before, we see that

$$dF_x = Y_1 d\Omega_1 + Y_2 d\Omega_2 = y_1 d\omega_1 + y_2 d\omega_2$$
$$= -R(P_x^2 - P_\alpha^2)P_\alpha^{-1}dP_\alpha - R(P_x^2 - P_\beta^2)P_\beta^{-1}dP_\beta, \tag{57}$$

$$F_x/R = 2P_x^2\ln P_x - P_x^2\ln P_\alpha - P_x^2\ln P_\beta + \tfrac{1}{2}(P_\alpha^2 + P_\beta^2) - P_x^2 \geq 0, \tag{58}$$

$$dF_\beta/R = -(P_x^2 - P_\alpha^2)P_\alpha^{-1}dP_\alpha - (P_\alpha^2 + P_\beta^2 - 2P_x^2)P_x^{-1}dP_x, \tag{59}$$

$$dF_{\alpha\beta}/R = -(P_\alpha^2 + P_\beta^2 - 2P_x^2)P_x^{-1}dP_x, \tag{60}$$

$$F_{\alpha\beta}/R = P_x^2 - (P_\alpha^2 + P_\beta^2)\ln P_x + f(P_\alpha, P_\beta). \tag{61}$$

Once again, by trial and error, we discover a perfect potential;

$$F/R = P_x^2 - \tfrac{1}{2}(P_\alpha^2 + P_\beta^2) + P_\alpha^2\ln P_\alpha + P_\beta^2\ln P_\beta - (P_\alpha^2 + P_\beta^2)\ln P_x \geq 0, \tag{62}$$

$$R^{-1}(\partial F/\partial P_\alpha)_{T,P_\beta,P_x} = 2P_\alpha\ln(P_\alpha/P_x), \tag{63}$$

$$R^{-1}(\partial F/\partial P_\beta)_{T,P_\alpha,P_x} = 2P_\beta\ln(P_\beta/P_x), \tag{64}$$

$$R^{-1}(\partial F/\partial P_x)_{T,P_\alpha,P_\beta} = P_x^{-1}(2P_x^2 - P_\alpha^2 - P_\beta^2); \tag{65}$$

the potential F thus has a minimum at each of the points $\dot{n}_\alpha = 0$, $\dot{n}_\beta = 0$, $\dot{n}_x = 0$. In a neighborhood of $P_\alpha \to P_x$ and $P_\beta \to P_x$, we find that $F \to \tfrac{1}{2}\Theta$.

OBSERVATIONS

Consider an arbitrary steady state situation not involving magnetic or centrifugal fields, and resolve the rate of entropy production Θ into a given set of currents Y_i and affinities Ω_i:

$$\Theta = \sum_i Y_i\Omega_i \geq 0. \tag{66}$$

In the linear current–affinity region, the rate of entropy production is a perfect thermokinetic potential for any choice of boundary conditions[3]—Θ has a minimum at each point where one of the currents Y_i vanishes. Outside the linear current–affinity region, Θ is in general no longer a perfect potential: although Θ may have a minimum at some of the points $Y_i = 0 (i = 1, 2, \cdots)$, it will (in general) not have a minimum at *every* such point. For the heat conduction special case of Fig. 2, for example,

$$(\partial\Theta/\partial T_\alpha)_{T_\beta,T_x} = (T_\alpha - T_x)(T_\alpha + T_x)(T_\alpha^2 T_x)^{-1}, \tag{67}$$

$$(\partial\Theta/\partial T_x)_{T_\alpha,T_\beta} = (T_x^2 - T_\alpha T_\beta)(T_\alpha + T_x)(T_x^2 T_\alpha T_\beta)^{-1}; \tag{68}$$

Θ thus has a minimum at the point where $T_\alpha = T_z(y_1=0)$ but does not have a minimum at the point $(T_z = \langle T \rangle)$ where $y_2 = 0$.

The heat conduction and Poiseuille flow examples show that outside the linear current–affinity region it is possible to construct a family of thermokinetic potentials by operating with the Pfaffian forms $dF_\sigma \equiv \sum_i Y_i d\Omega_i$ for given boundary conditions σ. In general, for different boundary conditions we get different potentials F_σ. It may be possible to find a perfect thermokinetic potential F, but no well-defined algorithm exists for constructing such a function—it seems to be a matter of hit or miss.

Perfect thermokinetic potentials would be of great use if we could mechanically write them down (in the manner in which we can mechanically write down the equilibrium constant for a chemical reaction or the rate of entropy production for a steady process). From a perfect potential we could quickly determine the thermodynamic requirements for a given current to vanish under given boundary conditions: the condition $0 = (\partial F/\partial \Omega_i)_{\Omega_j,\sigma}$ is the thermodynamically necessary condition for the relation $Y_i = 0$ to be satisfied for fixed boundary conditions σ. Now if we have to know all about the currents Y_i (precisely how they depend on the affinities Ω_j and the boundary conditions σ) before we construct our potential F—as we have to when operating with the Pfaffian forms $dF_\sigma \equiv \sum_i Y_i d\Omega_i$—then we gain nothing for our efforts since we already know the conditions for which $Y_i = 0$. An analogous case would be one where in order to write down an equilibrium constant for a chemical reaction we had first to determine the precise kinetic mechanism of the forward and backward reactions—under such circumstances the equilibrium constant would be a trivial corollary following from our knowledge of the kinetic laws of behavior of the system and would not have any intrinsic interest of its own. Of course, it is precisely because we *can* write down equilibrium constants *without* solving the problem of exact kinetic mechanism that we find equilibrium constants so useful. Correspondingly, until we learn how to write down perfect thermokinetic potentials *without* first solving the kinetic problem of determining $Y_i = Y_i(\Omega_j, \sigma)$—until then, the practical utility of thermokinetic potentials will be severely limited. Here lies one of the great challenges of nonequilibrium thermodynamics.

Now to conclude with a couple of minor observations. In the linear current-affinity region, for given boundary conditions σ, the currents Y_i are exactly reversed when all the affinities Ω_j are reversed; we see that outside the linear region the exact reversal need no longer hold—Eqs. (19), (24), (27), (28). Note finally that there do exist some simple physical situations that have current–affinity relations devoid of second order independence, i.e., $L_{112} \neq L_{211}$, etc.[10]—Eqs. (25) and (26); note also that the choice of boundary conditions is very important in determining such matters. See also the comments in Ref. 11.

ACKNOWLEDGMENT

I am indebted to J.C.M. Li for an enlightening correspondence on matters pertaining to the thermokinetic potential.

[1] I. Prigogine, *Introduction to the Thermodynamics of Irreversible Processes* (Wiley, New York, 1961).
[2] S. de Groot and P. Mazur, *Non-Equilibrium Thermodynamics* (North-Holland, Amsterdam, 1962).
[3] R. J. Tykodi, *Thermodynamics of Steady States* (Macmillan, New York, 1967).
[4] F. C. Andrews, Ind. Eng. Chem. Fundamentals **6,** 48 (1967).
[5] C. F. Holmes and R. G. Mortimer, Ind. Eng. Chem. Fundamentals **6,** 321 (1967).
[6] R. J. Donnelly, R. Herman, and I. Prigogine, editors, *Non-Equilibrium Thermodynamics, Variational Techniques, and Stability* (University of Chicago Press, Chicago, 1966).
[7] J. C. M. Li, J. Appl. Phys. **33,** 616 (1962).
[8] J. C. M. Li, J. Phys. Chem. **66,** 1414 (1962).
[9] K. F. Herzfeld and H. M. Smallwood, *A Treatise on Physical Chemistry, Vol. 2, States of Matter,* edited by H. S. Taylor and S. Glasstone (Van Nostrand, Princeton, N.J., 1951), 3rd ed., p. 105.
[10] R. P. Rastogi, R. C. Srivastava, and K. Singh, Trans. Faraday Soc. **61,** 854 (1965).
[11] R. J. Tykodi, Bull. Chem. Soc. (Japan) **44,** 1001 (1971).

Physica **72** (1974) 341–354 © *North-Holland Publishing Co.*

THERMODYNAMICS OF STEADY STATES: A WEAK ENTROPY-PRODUCTION PRINCIPLE

R.J. TYKODI

Department of Chemistry, Southeastern Massachusetts University, North Dartmouth, Massachusetts 02747, USA

Received 28 November 1972
Revised 17 September 1973

Synopsis

Two classes of steady-state situations, a family of chemical reaction schemes and the simple gradient situations (heat conduction, diffusion, Poiseuille flow of a fluid), are exhibited such that each member of each class fails to satisfy the principle of minimum entropy production outside the linear current-affinity region but nevertheless satisfies unrestrictedly a weaker form of entropy-production principle. A necessary and sufficient condition for the validity of the weak entropy-production principle is established. The weak entropy-production principle is necessarily valid for all situations involving linear kinetic laws (whether involving magnetic and centrifugal fields or not) but is not necessarily valid for situations involving nonlinear kinetic laws (special case counterexamples can be constructed).

1. *Introduction.* In my studies of the thermodynamics of steady states[1–4]) I have always dealt with the structure of the space of steady states for systems with a given apparatus configuration and given boundary constraints; I have always compared steady states with one another. I continue to operate in the same fashion in this paper. This preamble is to alert the reader to the fact that I shall not be considering the relationship between a class of transient states and a final steady state toward which the transient states converge (after the relaxation or change of a boundary constraint), *i.e.*, I shall *not* be considering the problem of the temporal evolution of a system with fixed boundary constraints following upon a perturbation of those boundary constraints. The temporal evolution problem has been studied intensively by Prigogine and Glansdorff[5–8]). For a discussion of the relationship between my studies and those of Prigogine and Glansdorff see chapter 15 of ref. 1.

In my book-length study[1]) of the thermodynamics of steady states, I studied systems that were entirely resolvable into terminal parts and gradient parts[9]), systems that I shall hereafter refer to as archimedean[2]). For steady-state situations involving such systems the rate of entropy production $\Theta \equiv \dot{S}$ (system) + $\dot{S}$ (surroundings) can always be expressed in the form

$$\Theta = \sum_i Y_i \Omega_i \geq 0, \tag{1}$$

where the currents Y_i are (essentially) time derivatives of extensive thermodynamic quantities[10]) ($\dot{U}^{(r)}$ or $\dot{Q}^{(r)}$, $\dot{V}$, $\dot{n}^{(j)}$, *etc.*) and the affinities Ω_i are defined *via* eq. (1). I made much use of the following assumption, which I shall refer to as the strong entropy-production principle: if a system that is the site of $\nu + 1$ steady (independent) currents converges *via* a sequence of steady states to a state involving ν steady currents in such a way that the affinities conjugate to the ν nonvanishing currents are maintained constant, then the state with ν currents is a state of minimum rate of entropy production relative to the (steady) states with $\nu + 1$ currents.

The mathematical statement of the strong entropy-production principle takes the form

$$0 = (\partial\Theta/\partial Y_k)_{\Omega',\sigma}|_{Y_k=0}, \qquad 0 < (\partial^2\Theta/\partial Y_k^2)_{\Omega',\sigma}|_{Y_k=0}, \tag{2}$$

where the subscript σ refers to a given thermodynamic (equilibrium) reference state[1]) and the subscript Ω' means constant Ω_i for $i \neq k$. To shorten the notation, let

$$(\partial/\partial Y_k)\cdots|_{Y_k=0} \equiv (\delta/\tilde{\delta} Y_k)\cdots. \tag{3}$$

Statement (2) then takes the form

$$0 = (\delta\Theta/\tilde{\delta} Y_k)_{\Omega',\sigma}, \qquad 0 < (\delta^2\Theta/\tilde{\delta} Y_k^2)_{\Omega',\sigma}. \tag{4a, b}$$

Straightforward differentiation of eq. (1) leads to

$$(\delta\Theta/\tilde{\delta} Y_k)_{\Omega',\sigma} = \Omega_k(0) + \sum_i^{(k)} \Omega_i\,(\delta Y_i/\tilde{\delta} Y_k)_{\Omega',\sigma}, \tag{5}$$

$$(\delta^2\Theta/\tilde{\delta} Y_k^2)_{\Omega',\sigma} = 2\,(\delta\Omega_k/\tilde{\delta} Y_k)_{\Omega',\sigma} + \sum_i^{(k)} \Omega_i\,(\delta^2 Y_i/\delta Y_k^2)_{\Omega',\sigma}, \tag{6}$$

where the superior index on the summation sign means that the term with $i = k$ is to be omitted from the sum.

The basic problem of the thermodynamics of steady states is to determine the conditions of *migrational equilibrium* in a given spatial field[1]), *i.e.*, to determine the conditions for the vanishing of a current Y_k for a given reference state σ and a given set of affinities $\Omega_i = \Omega_i(*)$ $(i \neq k)$. Put more briefly: for a given reference state σ and a given set of affinities $\Omega_i = \Omega_i(*)$ $(i \neq k)$, what is the value of $\Omega_k = \Omega_k(0)$ required to ensure that $Y_k = 0$?

The strong entropy-production principle [eqs. (4) and (5)] supplies the following solution of the basic problem:

$$\Omega_k(0) = -\sum_i^{(k)} \Omega_i(*)\,(\delta Y_i/\tilde{\delta} Y_k)_{\Omega',\sigma}. \tag{7}$$

In a neighborhood of the reference state σ where the currents may be considered to be linear functions of the affinities,

$$Y_i = \sum_j L_{ij}(\sigma)\,\Omega_j, \tag{8}$$

$$\Theta = \sum_{i,j} L_{ij}\Omega_i\Omega_j \geq 0, \tag{9}$$

for systems not involving magnetic or centrifugal fields (such systems are called *ordinary*, whereas those involving magnetic or centrifugal fields are called *special*[10])) the strong entropy-production principle is valid[10]); consequently eq. (7) is, in this case, a complete solution to the problem of migrational equilibrium. I suggested in my book that the strong entropy-production principle (4) was probably valid well outside the linear current-affinity region. In two recent publications[3,4]) I analyzed a simple heat-conduction situation for which the linear current-affinity region was infinitesimal and which violated the strong entropy-production principle outside that infinitesimal region. I went on to show[3]) that steady-state situations fall into two classes, (i) those for which the strong entropy-production principle holds outside the linear current-affinity region and (ii) those for which it does not; and I offered a tentative criterion for distinguishing situations (i) from situations (ii).

At the present time, it appears that the most general statements we can make about all steady-state situations, with minimum fear of contradiction by Nature, are

$$0 < (\partial^2\Theta/\partial Y_k^2)_{\Omega',\sigma}, \tag{10a}$$

$$0 < (\partial\Omega_k/\partial Y_k)_{\Omega',\sigma}. \tag{10b}$$

The condition $0 < (\partial^2\Theta/\partial Y_k^2)_{\Omega',\sigma}$ means that, for a given reference state σ and a given set of affinities $\Omega_i = \Omega_i(*)$ $(i \neq k)$, the rate of entropy production Θ is a convex function of the current Y_k; since $\Theta \geq 0$, the convex property implies that, for the given conditions, a plot of Θ *vs.* Y_k will show a minimum (but that minimum may or may not occur at the point $Y_k = 0$). The convex character of the Θ *vs.* Y_k relation is in accord with our intuitive feelings about irreversibility and the rate of entropy production. The stability condition $0 < (\partial\Omega_k/\partial Y_k)_{\Omega',\sigma}$ is not independent of the convexity properties of Θ[10]), the two conditions $0 < (\partial^2\Theta/\partial Y_k^2)_{\Omega',\sigma}$ and $0 < (\partial\Omega_k/\partial Y_k)_{\Omega',\sigma}$ are indissolubly tied together for all steady-state situations.

Once we properly[10]) select our currents and affinities in a given steady-state situation we expect our affinities to function as "driving forces" for the flow of their conjugate currents, *i.e.*, we expect that $0 < (\partial Y_k/\partial \Omega_k)_{\Omega',\sigma}$; if the conditions of stability [eq. (10b)] hold, our expectations about the physical characteristics of our currents and affinities will be met.

Since I am satisfied[10]) that the two relations (10) stand or fall together for steady states, I shall refer to the pair of them as the *convexity relations* for steady-state situations. Now the convexity relations (10) hold in *all* cases; relation (4a) holds in *some* cases. Is there any additional thermodynamic constraint for steady states intermediate in strength between relations (4a) and (10)? There is, and it forms the basis for what I call a *weak principle of entropy production* for steady states. I shall show that the heat-conduction situation analyzed previously[3,4]) is but one member of a special class of steady-state situations, the simple gradient situations, all the members of which violate the strong entropy-production principle [eq. (4a)] outside the linear current-affinity region, but, nevertheless, satisfy at all times a weak entropy-production principle. I shall also show that the same sort of behavior is true of the members of a class of chemical-reaction schemes.

Although the weak entropy-production principle is more general than the strong entropy-production principle and holds in some domains where the strong principle fails, the weak principle is not universally valid (counterexamples are known). The only general class of steady-state situations necessarily satisfying the weak entropy-production principle is the class of situations (both ordinary and special) involving linear kinetic laws.

2. *The weak entropy-production principle.* Consider again the problem of migrational equilibrium in a (steady state) spatial field. The value of Ω_k guaranteeing that $Y_k = 0$ for the given conditions is, rigorously,

$$\Omega_k(0) = -\sum_i^{(k)} \Omega_i(*)\,(\delta Y_i/\tilde{\delta} Y_k)_{\Omega',\sigma} + (\delta\Theta/\tilde{\delta} Y_k)_{\Omega',\sigma}. \tag{11}$$

When eq. (4a) holds (strong entropy-production principle), eq. (11) reduces to eq. (7). Now suppose that, without knowing whether eq. (4a) is satisfied or not, we define a quantity $\Omega_k(*)$ by

$$\Omega_k(*) \equiv -\sum_i^{(k)} \Omega_i(*)\,(\delta Y_i/\tilde{\delta} Y_k)_{\Omega',\sigma}. \tag{12}$$

When eq. (4a) is satisfied, $\Omega_k(*) = \Omega_k(0)$; we wish to investigate now the relation between $\Omega_k(0)$ and $\Omega_k(*)$, and between the physical states defined by $\Omega_k(0)$ and $\Omega_k(*)$, for the case when eq. (4a) does not hold.

Let

$$\Theta = \sum_{i=1}^{\nu} Y_i \Omega_i \geq 0, \tag{13}$$

and set $\Omega_i = \Omega_i(*)$ for $i \neq k$. Since $Y_k = Y_k(\Omega_k, \Omega_i(*), \sigma)$, the relation $Y_k = 0$ determines a particular value $\Omega_k(0)$ of Ω_k. When $Y_k = 0$, we have

$$\Theta_k(0) \equiv \Theta(\Omega_k(0), \Omega_i(*), \sigma) = \sum_i^{(k)} Y_i(0)\,\Omega_i(*). \tag{14}$$

Now let $\Omega_k(*)$ be defined by eq. (12); sometimes, for convenience, I shall write eq. (12) as

$$0 = \sum_i \Omega_i(*)\,(\delta Y_i/\tilde{\delta} Y_k)_{\Omega',\sigma}, \tag{15}$$

[note that the summation index can take on the value k in eq. (15)]. Observe that whereas $Y_k(\Omega_k(0), \Omega_i(*), \sigma) \equiv 0$, $Y_k(\Omega_k(*), \Omega_i(*), \sigma) \neq 0$ unless $\Omega_k(*) = \Omega_k(0)$. Next set

$$\Theta_k(*) \equiv \Theta(\Omega_k(*), \Omega_i(*), \sigma) = \sum_i^{(k)} \Omega_i(*)\,[Y_i - Y_k(\delta Y_i/\tilde{\delta} Y_k)_{\Omega',\sigma}]; \tag{16}$$

it follows then that

$$\Theta_k(*) - \Theta_k(0) = \sum_i^{(k)} \Omega_i(*)\,\{Y_i - [Y_i(0) + Y_k(\delta Y_i/\tilde{\delta} Y_k)_{\Omega',\sigma}]\}. \tag{17}$$

If we consider that $Y_i = Y_i(Y_k, \Omega_j(*), \sigma)$ and expand each Y_i about the value $Y_i(0)$, we have, by Taylor's theorem with the remainder,

$$Y_i = Y_i(0) + Y_k(\delta Y_i/\tilde{\delta} Y_k)_{\Omega',\sigma} + \tfrac{1}{2} Y_k^2\,(\partial^2 Y_i/\partial Y_k^2)_{\Omega',\sigma}|_{Y_k=\lambda_i}, \tag{18}$$

where $0 < |\lambda_i| < |Y_k|$. Eq. (17) thus becomes

$$\Theta_k(*) - \Theta_k(0) = \tfrac{1}{2} Y_k^2 \sum_i^{(k)} \Omega_i(*)\,(\partial^2 Y_i/\partial Y_k^2)_{\Omega',\sigma}|_{Y_k=\lambda_i}. \tag{19}$$

In the event that $(\partial Y_i/\partial Y_k)_{\Omega',\sigma}$ does not vary when Y_k varies [at constant Ω_i $(i \neq k)$], *i.e.*, in the event that $0 \equiv (\partial^2 Y_i/\partial Y_k^2)_{\Omega',\sigma}$, we get $\Theta_k(*) - \Theta_k(0) = 0$.

From the relation

$$(\delta\Theta/\tilde{\delta} Y_k)_{\Omega',\sigma} = \Omega_k(0) + \sum_i^{(k)} \Omega_i(*)\,(\delta Y_i/\tilde{\delta} Y_k)_{\Omega',\sigma} = \Omega_k(0) - \Omega_k(*), \tag{20}$$

we see that when the strong entropy-production principle holds then $\Omega_k(*) = \Omega_k(0)$ and $\Theta_k(*) = \Theta_k(0) = \Theta_k\,(\text{min})$. When the strong entropy-production principle does not hold, it is still possible to have $\Theta_k(*) = \Theta_k(0) > \Theta_k\,(\mathrm{min})$. The relations $0 = Y_k$ and $0 = \sum_i \Omega_i(*)\,(\delta Y_i/\delta Y_k)_{\Omega',\sigma}$ each define a physical state for a given set of affinities $\Omega_i = \Omega_i(*)$ $(i \neq k)$. If these two states are identical, then the strong entropy-production principle holds and Θ for the common state is the minimum value compatible with the set $\Omega_i(*)$ $(i \neq k)$. If the two states are different, then the $0 = Y_k$ state will not yield the minimum value of Θ for the set $\Omega_i(*)$ $(i \neq k)$; it is possible, however, for the two different states to have exactly the same Θ values: $\Theta_k(*) = \Theta_k(0)$. This suggests that whereas the strong entropy-production principle [eq. (4a)] is of limited validity, the *weak* entropy-production principle

$$\Theta_k(*) = \Theta_k(0) \geq \Theta_k\,(\mathrm{min}), \qquad k = 1, 2, \ldots, \nu, \tag{21}$$

may have a wider range of validity. The condition (15), then, either locates the state of migrational equilibrium ($Y_k = 0$) or locates a state that has exactly the same rate of entropy production as the state of migrational equilibrium. It follows from eq. (19) that eq. (21) holds for any set of linear current-affinity relations, whether the situation be ordinary or special[10]).

Can we see the significance of eq. (21) in somewhat more physical terms? Our convexity relations (10) guarantee that, for the given conditions, the plot of Θ *vs.* Y_k will have a minimum and, hence, that the state $\Theta_k(0)$ will have a mirror-image state $\Theta_k(\dagger)$ on the other side of $\Theta_k\,(\mathrm{min})$ such that $\Theta_k(\dagger) = \Theta_k(0)$ [if the strong entropy-production principle holds, the state $\Theta_k(0)$ and its mirror-image state $\Theta_k(\dagger)$ fall together and are identical with one another and with $\Theta_k\,(\mathrm{min})$]. The physical content of eq. (21) is, then, the statement that the value of Ω_k in the mirror-image state, $\Omega_k = \Omega_k(\dagger)$, is equal to the value $\Omega_k(*)$ calculated *a priori* according to the recipe in eq. (12):

$$\Omega_k(\dagger) = \Omega_k(*); \tag{22}$$

i.e., eq. (12) locates the mirror-image state.

Whereas the strong entropy-production principle has interconnections[10]) with the convexity relations (10), the weak entropy-production principle does not. The convexity relations are more primitive than the weak entropy-production principle.

3. *A necessary and sufficient condition.* Let us consider the following theorem:

Theorem: for the case $Y_k(*) \neq 0$, the condition $0 \equiv (\partial^2 Y_i/\partial Y_k^2)_{\Omega',\sigma}$ $(i \neq k)$ is both a necessary and a sufficient condition for the validity of the relation $\Theta_k(*) = \Theta_k(0)$. The case $Y_k(*) = 0$ implies that $\Theta_k(*) = \Theta_k(0) = \Theta_k\,(\mathrm{min})$ and that the strong entropy-production principle [eq. (4a)] holds; it is thus not an interesting case in the present context.

The proof or disproof of the theorem is purely a mathematical matter. The sufficiency of the condition follows directly from eq. (19); the necessity of the condition, therefore, is the only part of the theorem that requires proof. Consider the case of just two currents [eq. (13) with $\nu = 2$]; this is by far the most important case in practice. For given Ω_1, σ consider the mirror-image states $\Theta_2(0)$, $\Theta_2(\dagger)$

$$\Theta_2(0) = \Theta_2(\dagger), \tag{23}$$

$$Y_1(0)\,\Omega_1 = Y_1(\dagger)\,\Omega_1 + Y_2(\dagger)\,\Omega_2(\dagger), \tag{24}$$

$$\Omega_2(\dagger) = -[Y_1(\dagger) - Y_1(0)]\,\Omega_1 Y_2(\dagger)^{-1} \qquad [Y_2(\dagger) \neq 0]. \tag{25}$$

Next, consider Y_1 to be a function of Ω_1 and Y_2 for a given reference state σ, *i.e.*,

$$Y_1 = F(\Omega_1, Y_2, \sigma). \tag{26}$$

Now let

$$Y_1(\dagger) = Y_1(0) + Y_2(\dagger)\left(\frac{\delta Y_1}{\delta Y_2}\right)_{\Omega_1,\sigma} + \tfrac{1}{2}Y_2(\dagger)^2\left(\frac{\partial^2 Y_1}{\partial Y_2^2}\right)_{\Omega_1,\sigma}\Bigg|_{Y_2=\lambda_1}, \tag{27}$$

with $0 < |\lambda_1| < |Y_2(\dagger)|$ (Taylor's theorem); then

$$\Omega_2(\dagger) = -\Omega_1\left[\left(\frac{\delta Y_1}{\delta Y_2}\right)_{\Omega_1,\sigma} + \tfrac{1}{2}Y_2(\dagger)\left(\frac{\partial^2 Y_1}{\partial Y_2^2}\right)_{\Omega_1,\sigma}\Bigg|_{Y_2=\lambda_1}\right] \tag{28}$$

and [see eq. (12)]

$$\Omega_2(\dagger) - \Omega_2(*) = -\tfrac{1}{2}\Omega_1 Y_2(\dagger)\,(\partial^2 Y_1/\partial Y_2^2)_{\Omega_1,\sigma}|_{Y_2=\lambda_1}. \tag{29}$$

For $\Omega_1 \neq 0$, $Y_2(\dagger) \neq 0$, the condition $\Theta_2(*) = \Theta_2(0)$ [*i.e.*, $\Omega_2(\dagger) = \Omega_2(*)$ and $Y_2(\dagger) = Y_2(*)$] requires that

$$0 = (\partial^2 Y_1/\partial Y_2^2)_{\Omega_1,\sigma}|_{Y_2=\lambda_1}. \tag{30}$$

Eq. (30), in turn, requires that eq. (27) reduce to

$$Y_1(\dagger) = Y_1(0) + Y_2(\dagger)\,(\delta Y_1/\delta Y_2)_{\Omega_1,\sigma}. \tag{31}$$

We see from eq. (26) that $Y_1(0) \equiv F(\Omega_1, 0, \sigma)$ and $(\delta Y_1/\delta Y_2)_{\Omega_1,\sigma} = (\partial F/\partial Y_2)_{\Omega_1,\sigma}|_{Y_2=0}$ are each functions of Ω_1, σ only, say $f_1(\Omega_1, \sigma)$ and $f_2(\Omega_1, \sigma)$, respectively. Eq. (31) is thus of the form

$$Y_1(\dagger) = f_1(\Omega_1, \sigma) + Y_2(\dagger) f_2(\Omega_1, \sigma). \tag{32}$$

Now for a fixed reference state σ, vary Ω_1, maintaining at all times, however, the constraint $\Theta_2(*) = \Theta_2(0)$. We thus find an infinite number of pairs $Y_1(\dagger)$, $Y_2(\dagger)$

for which eqs. (31) and (32) hold. Since [eq. (26)] we claim that there is only one generally valid functional form connecting Y_1, Y_2, Ω_1, σ, it must be the form of eq. (32). The condition $\Theta_2(*) = \Theta_2(0)$ forces us to conclude then that quite generally

$$Y_1 = f_1(\Omega_1, \sigma) + Y_2 f_2(\Omega_1, \sigma) \tag{33}$$

and, consequently, that

$$0 \equiv (\partial^2 Y_1/\partial Y_2^2)_{\Omega_1, \sigma}, \qquad Q.E.D. \tag{34}$$

I have been unsuccessful thus far in my attempts to construct a proof of the theorem for the case of three or more currents [eq. (13) with $\nu \geq 3$].

4. *Chemical-reaction schemes.* An example will help us to see the implications of section 3 more clearly. Consider a monothermal[1]) (isothermal) layout consisting of a constant-volume stirred-flow chemical reactor fed by individual mass reservoirs containing the chemical species A, B, X, respectively. Let the steady mass flow into or out of each reservoir be adjusted so that the chemical reaction inside the reactor just balances the loss or gain in the reservoir. The concentration of each species in the reactor, C_A, C_B, C_X (moles per liter) is then constant in time for a given setting of the conditions in the mass reservoirs. If $\dot{n}$ is the rate of inflow (moles per second) of mass into a reservoir where the molar chemical potential is μ, we have in the steady state[1])

$$\Theta = -T^{-1}(\dot{n}_A\mu_A + \dot{n}_B\mu_B + \dot{n}_X\mu_X). \tag{35}$$

Let us assume, because of the stirred-flow property of our reactor, that the chemical potentials in the reservoirs can be expressed (approximately) in terms of the concentrations in the reactor in the form

$$\mu_A = \mu_A\,(\mathrm{ref}) + RT \ln C_A, \tag{36}$$

etc., where R is the gas constant, T is the monothermal temperature of the layout, and μ_A (ref), *e.g.*, is a reference or standard value for the species in question.

Consider now the following family of chemical-reaction schemes:

$$\nu_A A \underset{k_2}{\overset{k_1}{\rightleftharpoons}} \nu_X X \underset{k_4}{\overset{k_3}{\rightleftharpoons}} \nu_B B, \tag{37}$$

where the ν's are stoichiometric coefficients and the rate constants k_i are defined according to the following convention:

$$\nu_Z Z \xrightarrow{k_i} \cdots \qquad \text{implies that} \qquad -\nu_Z^{-1}\dot{C}_Z = k_i C_Z^{\nu_Z} \tag{38}$$

for the indicated step in the reaction scheme. In the steady state we have the relation

$$\nu_A^{-1}\dot{n}_A + \nu_X^{-1}\dot{n}_X + \nu_B^{-1}\dot{n}_B = 0; \tag{39}$$

consequently,

$$\dot{n}_B = -(\nu_B/\nu_A)\,\dot{n}_A - (\nu_B/\nu_X)\,\dot{n}_X. \tag{40}$$

Let the volume of the chemical reactor be V. Divide eq. (35) through by RV and make use of eq. (40):

$$\begin{aligned}\Theta/RV &= (-\nu_A^{-1}\dot{n}_A/V)\,[(\nu_A\mu_A - \nu_B\mu_B)/RT]\\ &\quad + (-\nu_X^{-1}\dot{n}_X/V)\,[(\nu_X\mu_X - \nu_B\mu_B)/RT] \equiv Y_1\Omega_1 + Y_2\Omega_2,\end{aligned} \tag{41}$$

where $Y_1 \equiv (-\nu_A^{-1}\dot{n}_A/V)$ (a "normalized" or "reduced" current; Ω_1 is likewise a reduced affinity) and so forth. In order that the concentrations in the reactor vessel be time-invariant in the steady state we must have

$$-\nu_A^{-1}\dot{n}_A/V = k_1C_A^{\nu_A} - k_2C_X^{\nu_X}, \tag{42}$$

$$-\nu_X^{-1}\dot{n}_X/V = (k_2 + k_3)\,C_X^{\nu_X} - k_1C_A^{\nu_A} - k_4C_B^{\nu_B}. \tag{43}$$

Upon introducing the forms (36) and observing the standard chemical thermodynamic relations between differences in μ (ref) and ratios of the k_i's, we obtain finally

$$\begin{aligned}\Theta/RV &= Y_1\Omega_1 + Y_2\Omega_2\\ &= (k_1C_A^{\nu_A} - k_2C_X^{\nu_X})\ln(k_1k_3C_A^{\nu_A}/k_2k_4C_B^{\nu_B})\\ &\quad + [(k_2 + k_3)\,C_X^{\nu_X} - k_1C_A^{\nu_A} - k_4C_B^{\nu_B}]\ln(k_3C_X^{\nu_X}/k_4C_B^{\nu_B}).\end{aligned} \tag{44}$$

The rate constants k_i are functions of temperature alone, *i.e.*, $k_i = k_i(T)$; hence if we pick the state of species B as our reference state σ, then keeping T, C_A, C_B each constant is equivalent to constant Ω_1, σ in our general formalism. It follows then that

$$\begin{aligned}\left(\frac{\partial Y_1}{\partial Y_2}\right)_{\Omega_1,\sigma} &= \left(\frac{\partial Y_1}{\partial Y_2}\right)_{T,C_A,C_B} = \frac{(\partial Y_1/\partial C_X)_{T,C_A,C_B}}{(\partial Y_2/\partial C_X)_{T,C_A,C_B}}\\ &= -\frac{\nu_X k_2 C_X^{\nu_X-1}}{\nu_X(k_2 + k_3)\,C_X^{\nu_X-1}} = -\frac{k_2}{k_2 + k_3}\end{aligned} \tag{45}$$

and

$$0 \equiv (\partial^2 Y_1/\partial Y_2^2)_{T, C_A, C_B}. \tag{46}$$

So all the members of the family of chemical-reaction schemes (37) satisfy the weak entropy-production principle, regardless of the individual values of the stoichiometric coefficients ν.

The reaction scheme

$$2A \underset{k_2}{\overset{k_1}{\rightleftharpoons}} 2X \underset{k_4}{\overset{k_3}{\rightleftharpoons}} B \tag{47}$$

is of the form (37) so it satisfies eq. (46). Consider now the reaction scheme

$$A \underset{k_2'}{\overset{k_1'}{\rightleftharpoons}} X, \qquad 2X \underset{k_4}{\overset{k_3}{\rightleftharpoons}} B, \tag{48}$$

which has the same overall stoichiometry as (47) but has a different reaction mechanism. Thermodynamics requires that $(k_1'/k_2') = (k_1/k_2)^{\frac{1}{2}}$. If we resolve Θ/RV into currents and affinities for situations (47) and (48), we find that the affinities are the same, but the currents are different; for situation (48) we get

$$Y_1 \equiv -\dot{n}_A/2V = \tfrac{1}{2}(k_1'C_A - k_2'C_X), \tag{49}$$

$$Y_2 \equiv -\dot{n}_X/2V = k_3C_X^2 - k_4C_B + \tfrac{1}{2}(k_2'C_X - k_1'C_A), \tag{50}$$

$$\left(\frac{\partial Y_1}{\partial Y_2}\right)_{T, C_A, C_B} = \frac{(\partial Y_1/\partial C_X)_{T, C_A, C_B}}{(\partial Y_2/\partial C_X)_{T, C_A, C_B}} = \frac{-\frac{1}{2}k_2'}{2k_3C_X + \frac{1}{2}k_2'}, \tag{51}$$

$$\left(\frac{\partial^2 Y_1}{\partial Y_2^2}\right)_{T, C_A, C_B} = \frac{k_2'k_3}{(2k_3C_X + \frac{1}{2}k_2')^3} \neq 0. \tag{52}$$

We see from eq. (52) that the second derivative of Y_1 is not identically zero; furthermore, it cannot even vanish "accidentally" for the "right" value of λ_1 [see eqs. (19) and (29)]. There is thus no way for scheme (48) to satisfy the weak entropy-production principle. This example reinforces our belief that, for 2-current situations at least, the theorem stated in section 3 is an accurate representation of the facts.

Note that according to eq. (44) the concentrations C_A, C_X are exponentially related to the affinities Ω_1, Ω_2 at constant T, C_B; consequently the current-affinity equations appropriate to reaction scheme (37) are nonlinear. The reader may easily verify that outside the linear current-affinity neighborhood of the reference state T, C_B the reaction scheme (37) fails to satisfy the strong entropy-production principle [eq. (4a)].

5. *Triangular reaction scheme.* Form the reaction scheme of eq. (37) into a closed cycle:

$$\nu_A A \underset{k_2}{\overset{k_1}{\rightleftharpoons}} \nu_X X \underset{k_4}{\overset{k_3}{\rightleftharpoons}} \nu_B B \underset{k_6}{\overset{k_5}{\rightleftharpoons}} \nu_A A, \tag{53}$$

giving a closed triangular reaction scheme. For the steady-state model considered, the affinities for scheme (53) are the same as those in eq. (41); the current Y_2 is the same as that of eq. (43); and the current Y_1 [eq. (42)] changes to

$$-\nu_A^{-1}\dot{n}_A/V = (k_1 + k_6)\, C_A^{\nu_A} - k_2 C_X^{\nu_X} - k_5 C_B^{\nu_B}. \tag{54}$$

In addition, thermodynamics requires that $k_1k_3/k_2k_4 = k_6/k_5$. We find by direct calculation that eqs. (45) and (46) hold in exactly the same form for scheme (53) as for scheme (37). The triangular reaction scheme, therefore, satisfies the weak entropy-production principle.

The conventional triangular reaction scheme[11]) is the scheme of eq. (53) with $\nu_A = \nu_X = \nu_B = 1$. In dealing with the triangular reaction scheme, Rastogi *et al.*[11]) have used mole fractions as the variables in their mass-action and chemical potential expressions. The concentrations C_A, C_B, C_X are, in principle, independently variable whereas the mole fractions of A, B, X must add up to unity. This extra constraint on the mole fractions makes them rather inconvenient variables for carrying out the kinds of mathematical operations we have been using in our discussion of the weak entropy-production principle. For example, in mole fraction language the operation $(\partial/\partial Y_2)_{\Omega_1}$ for the triangular reaction scheme[11]) involves differentiation at constant temperature and at a constant ratio of the mole fractions of A and B.

The triangular reaction scheme expressed in mole-fraction language also satisfies the weak entropy-production principle (*via* the theorem of section 3). The summation constraint on the mole fractions, however, *can* (when thermodynamically necessary relations are sought among the phenomenological coefficients of the current-affinity equations) lead to spurious results; it is much easier, neater and safer to deal with reaction schemes in terms of concentrations instead.

6. *The simple gradient situations.* I recapitulate the essential features of the heat-conduction situation treated earlier[3,4]) and use it as a paradigm case for discussing the class of simple gradient situations. Stretch a wire of cross-sectional area B and of (exposed) length Λ between two thermostats α, β; insulate the wire in such a way that there are no lateral heat losses; lay off a coordinate x, in the direction $\alpha\beta$, along the wire ($0 < x < \Lambda$); call the situation just described "configuration I".

Now what about the temperature distribution along the wire in the steady state ($T_\alpha \neq T_\beta$)? We may approach this problem in the following way. Consider a point on the wire at a distance x from the α thermostat: configuration I; cut the wire at point x and let the two exposed faces of the wire communicate with a heat reservoir (thermostat) of temperature T_x; call this new situation "configuration II". When the inflow of heat into the reservoir at x across one face of the wire is just balanced by the outflow of heat across the other face, the temperature T_x of configuration II will correspond to the temperature at point x in configuration I and the two situations will be entirely equivalent. We can, therefore, analyze configuration II to determine the steady-state temperature distribution in configuration I. In the case of configuration II we have in the steady state

$$\dot{Q}_\alpha^{(r)} + \dot{Q}_\beta^{(r)} + \dot{Q}_x^{(r)} = 0, \tag{55}$$

$$\begin{aligned}\Theta &= \frac{\dot{Q}_\alpha^{(r)}}{T_\alpha} + \frac{\dot{Q}_\beta^{(r)}}{T_\beta} + \frac{\dot{Q}_x^{(r)}}{T_x} = \dot{Q}_\alpha^{(r)}\left(\frac{1}{T_\alpha} - \frac{1}{T_\beta}\right) + \dot{Q}_x^{(r)}\left(\frac{1}{T_x} - \frac{1}{T_\beta}\right) \\ &= Y_1\Omega_1 + Y_2\Omega_2,\end{aligned} \tag{56}$$

$$\left(\frac{\delta\Theta}{\delta Y_2}\right)_{\Omega_1,\beta} = \left(\frac{\delta\Theta}{\delta\dot{Q}_x^{(r)}}\right)_{T_\alpha, T_\beta} = \left(\frac{\delta\dot{Q}_\alpha^{(r)}}{\delta\dot{Q}_x^{(r)}}\right)_{T_\alpha, T_\beta}\left(\frac{1}{T_\alpha} - \frac{1}{T_\beta}\right) + \frac{1}{T_x} - \frac{1}{T_\beta}, \tag{57}$$

$$Y_1 \equiv \dot{Q}_\alpha^{(r)} = (B/x)\int_{T_\alpha}^{T_x} \varkappa(T)\,\mathrm{d}T = (B/x)\,[F(T_x) - F(T_\alpha)], \tag{58}$$

$$\begin{aligned}Y_2 \equiv \dot{Q}_x^{(r)} &= -\dot{Q}_\alpha^{(r)} - \dot{Q}_\beta^{(r)} \\ &= -\frac{B}{x}\int_{T_\alpha}^{T_x} \varkappa(T)\,\mathrm{d}T - \frac{B}{\Lambda - x}\int_{T_\beta}^{T_x} \varkappa(T)\,\mathrm{d}T \\ &= -(B/x)\,[F(T_x) - F(T_\alpha)] - [B/(\Lambda - x)]\,[F(T_x) - F(T_\beta)],\end{aligned} \tag{59}$$

where $\dot{Q}_\alpha^{(r)}$, *e.g.*, represents the rate of influx of heat to the reservoir (thermostat) at α, and $\varkappa$ is the thermal conductivity of the wire.

After some elementary algebraic manipulations, we find

$$Y_1 = -\frac{\Lambda - x}{\Lambda}\,Y_2 + \frac{B}{\Lambda}\,[F(T_\beta) - F(T_\alpha)] \tag{60}$$

and, consequently

$$\left(\frac{\partial Y_1}{\partial Y_2}\right)_{T_\alpha, T_\beta} = -\frac{\Lambda - x}{\Lambda}, \qquad 0 \equiv \left(\frac{\partial^2 Y_1}{\partial Y_2^2}\right)_{T_\alpha, T_\beta}; \tag{61}$$

the weak entropy-production principle therefore holds in this case. If we take state β as our reference state σ, then[4]) we can express T_α in terms of Ω_1 and T_β; hence eq. (60) is precisely of the form of eq. (33). For arbitrary temperature dependence of the thermal conductivity $\varkappa(T)$, the function of integration $F(T)$ can be a rather involved function of T and yet the currents Y_1, Y_2 satisfy the simple relation (60).

For the case of constant thermal conductivity, *i.e.*, $\varkappa \neq f(T)$, the relation $\dot{Q}_x^{(r)} = 0$ implies

$$T_x = \frac{\Lambda - x}{\Lambda} T_\alpha + \frac{x}{\Lambda} T_\beta, \tag{62}$$

and eq. (57) becomes

$$\begin{aligned} \left(\frac{\delta\Theta}{\tilde{\delta}\dot{Q}_x^{(r)}}\right)_{T_\alpha, T_\beta} &= -\frac{x(\Lambda - x)}{\Lambda^2} \frac{(T_\alpha - T_\beta)^2}{T_\alpha T_\beta T_x} \\ &= -\frac{x(\Lambda - x)}{\Lambda^2} \frac{r^2}{T_x(1 - \frac{1}{4}r^2)} \neq 0, \end{aligned} \tag{63}$$

where $r \equiv (T_\alpha - T_\beta)/\frac{1}{2}(T_\alpha + T_\beta)$. For this situation, then, the strong entropy-production principle is, strictly, never satisfied (as long as $T_\alpha \neq T_\beta$). [In the linear current-affinity region, however, $r \ll 1$ and, consequently, $0 \approx (\delta\Theta/\tilde{\delta}\dot{Q}_x^{(r)})_{T_\alpha, T_\beta}$.]

The situation that we analyzed in configuration II is but one of a class of simple gradient situations; similar results hold for the steady monothermal (isothermal) Poiseuille flow of a fluid through a capillary and for the steady monothermal diffusion of a given component through a stationary mixture.

In the 2-current case it is important to note that whereas linear current-affinity relations automatically satisfy eqs. (33) and (34) there are a surprising number of nonlinear current-affinity relations that also satisfy eqs. (33) and (34), the simple gradient situations and the chemical-reaction schemes of section 4, for example.

7. *Thermodynamics of steady states.* My present thoughts on the thermodynamic restrictions obeyed by steady-state archimedean systems are as follows: (i) it is always possible to express the rate of entropy production Θ in the form $\Theta = \sum_i Y_i \Omega_i$, where the currents Y_i are restricted to being certain kinds of time derivatives[10]); (ii) the relations $0 < (\partial^2\Theta/\partial Y_k^2)_{\Omega',\sigma}$, $0 < (\partial\Omega_k/\partial Y_k)_{\Omega',\sigma}$ are always true; (iii) for any equilibrium state σ, there is a neighborhood of σ (the linear current-affinity region) where the relation $0 = (\delta\Theta/\tilde{\delta} Y_k)_{\Omega',\sigma}$ holds with sufficient accuracy for any current Y_k; (iv) For situations meeting the criteria listed in ref. 3, the relation $0 = (\delta\Theta/\tilde{\delta} Y_k)_{\Omega',\sigma}$ holds unrestrictedly; (v) there exist classes of situa-

tions (the simple gradient situations, chemical reaction schemes, *etc.*) such that the members of each class fail to satisfy the relation $0 = (\delta\Theta/\tilde{\delta} Y_k)_{\Omega',\sigma}$ outside the linear current-affinity region, but nevertheless satisfy the relation $\Theta_k(*) = \Theta_k(0) \geq \Theta_k$ (min) unrestrictedly.

Conditions (iii) and (iv) apply to ordinary[10]) steady states only.

Outside the linear current-affinity region, steady-state situations have a rich thermodynamic structure. The wide diversity in behavior for such systems, as observed to date, has overtaxed the powers of any one entropy-production principle, although perhaps two or three progressively weaker principles might introduce a degree of taxonomic order into this fascinating field. All is not yet discovered, much useful work remains to be done.

Acknowledgement. I am indebted to the referee for calling to my attention the chemical-reaction schemes of section 4 and for suggesting several ways to improve the presentation in the paper.

REFERENCES

1) Tykodi, R.J., Thermodynamics of Steady States, Macmillan (New York, 1967).
2) Tykodi, R.J., Amer. J. Phys. **38** (1970) 586.
3) Tykodi, R.J., Bull. Chem. Soc. (Japan) **44** (1971) 1001.
4) Tykodi, R.J., J. chem. Phys. **57** (1972) 37.
5) Glansdorff, P. and Prigogine, I., Physica **30** (1964) 351.
6) Prigogine, I. and Glansdorff, P., Physica **31** (1965) 1242.
7) Non-Equilibrium Thermodynamics, Variational Techniques and Stability, R.J. Donnelly, R. Herman and I. Prigogine, eds., University of Chicago Press (Chicago, 1966).
8) Glansdorff, P. and Prigogine, I., Thermodynamic Theory of Structure, Stability and Fluctuations, Wiley-Interscience (London, 1971).
9) Ref. 1, ch. 1.
10) Ref. 1, ch. 3.
11) Rastogi, R.P., Srivastava, R.C. and Singh, K., Trans. Fraraday Soc. **61** (1965) 854.

BULLETIN OF TEH CHEMICAL SOCIETY OF JAPAN, VOL. 52 (2), 564—570 (1979) [Vol. 52, No. 2

Thermodynamics of Steady States: "Resistance Change" Transitions in Steady-state Systems

R. J. TYKODI

Department of Chemistry, Southeastern Massachusetts University, North Dartmouth, Massachusetts 02747, U.S.A.

(Received August 1, 1977)

Some one-current steady-state systems showing a discontinuous change in flow resistance at a point of instability—forced vaporization of carbon tetrachloride, Bénard instability in a horizontal layer of fluid heated from below, Taylor instability in the Couette flow of a liquid between coaxial rotating cylinders—are analyzed from a thermodynamic point of view: the behavior of the rate of entropy production and of the local potential (thermokinetic potential) for the system at the point of instability is explored. A multi-current situation involving a flip-flop current *vs.* voltage relation for the flow of electric current across a porous charged membrane is also commented on briefly.

Continuing my studies[1–6] of the thermodynamics of steady-state systems, I examine here the thermodynamic properties of several one-current steady-state systems, each of which develops an instability in its flow behavior and undergoes a discontinuous change in flow resistance. The "resistance change" transition at a point of instability for a steady-state system is in some respects analogous[7–9] to a phase transition in a thermo-static system. I shall first briefly discuss the "resistance change" transition in general terms; and I shall then discuss specifically the forced vaporization of carbon tetrachloride, Bénard instability in a horizontal layer of fluid heated from below, and Taylor instability in the Couette flow of a liquid between coaxial rotating cylinders. Finally I shall comment briefly on a multi-current example showing a flip-flop current-voltage relation for the flow of electric current across a porous charged membrane.

1. General Considerations (One-current Situations). Consider a fluid in a state of thermo-static equilibrium characterized by the variables T_σ, P_σ, $\mu_\sigma^{(1)}, \cdots$—a fluid system in the equilibrium state σ, for short. Relabel the set of equilibrium state variables in the following fashion: $\{T_\sigma, P_\sigma, \mu_\sigma^{(1)}\cdots\} \equiv \{\sigma_1, \sigma_2, \sigma_3, \cdots\} \equiv \{\sigma_i\}$. By changing some of the boundary conditions, induce the steady flow of a current Y in the fluid system. Use the equilibrium state σ as a reference state; the current Y will then depend on the departure of variables of the type σ_i from their reference state values. For each state of steady current flow, the rate of entropy production $\theta \equiv \dot{S}(\text{system}) + \dot{S}(\text{surroundings})$ takes the form

$$\theta = Y\Omega \geq 0, \tag{1}$$

where Ω is the affinity[1] conjugate to the current Y; both Y and Ω measure, in a certain sense, the "distance" of the steady-flow state from the reference equilibrium state σ. In addition to the rate of entropy production, another thermodynamic concept, the local potential (also called the generalized entropy production or the thermokinetic potential[5,10,11])— first introduced by Glansdorff and Prigogine[7,12,13]— is of use in discussing "resistance change" transitions. When the Pfaffian differential form (for a one-current situation)

$$\mathrm{d}F_\sigma = Y\mathrm{d}\Omega \tag{2}$$

is integrable under the given experimental conditions, the resulting quantity F_σ is the local potential:[5,7,10–13]

$$F_\sigma = \int Y\mathrm{d}\Omega = \int_0^Y Y(\partial\Omega/\partial Y)_\sigma \mathrm{d}Y. \tag{3}$$

The local potential F_σ (when it exists) is never negative, has a minimum at the point where $Y=0$, and satisfies the condition

$$0 < (\partial^2 F_\sigma/\partial\Omega^2)_\sigma|_{Y=0} = (\partial Y/\partial\Omega)_\sigma|_{Y=0}. \tag{4}$$

Glansdorff and Prigogine[7,12,13] found the local potential concept to be very useful for discussing the temporal evolution of a system with fixed boundary conditions as it passed through a sequence of transient states toward some stable steady state. I discussed in an earlier paper[5] the role played by the local potential (thermokinetic potential) in thermodynamic analyses of *all steady state* situations. In this paper I am somewhat more interested in the behavior of the rate of entropy production at "resistance change" transitions than in the behavior of the local potential; I shall, however, keep both quantities in view.

Consider again a fluid system in the equilibrium state σ. By successive manipulations of the boundary conditions establish a series of steady-flow states in the system with monotonically increasing values of the current Y. Observe θ as Y increases. If the plot of θ *vs.* Y shows a jump or a kink— a point of discontinuity in θ or in $\partial\theta/\partial Y$— we have a point of instability, a point at which the fluid system undergoes a "resistance change" transition. It is the thermodynamic properties of the fluid system at the point of instability that I intend to discuss— in macroscopic phenomenological terms.

Let $\partial\Omega/\partial Y$ be the (differential) flow resistance, let the subscript c refer to the point of instability, and let ΔZ be the jump in an arbitrary property Z at the point of instability, *i.e.* let $\Delta Z \equiv Z(Y_c+0) - Z(Y_c-0)$. We have then

$$\Delta\theta = Y_c\Delta\Omega, \tag{5}$$

$$\Delta\partial\theta/\partial Y = \Delta\Omega + Y_c\Delta\partial\Omega/\partial Y, \tag{6}$$

$$\Delta\partial^2\theta/\partial Y^2 = 2\Delta\partial\Omega/\partial Y + Y_c\Delta\partial^2\Omega/\partial Y^2, \tag{7}$$

$$\Delta\partial^m\theta/\partial Y^m = m\Delta\partial^{m-1}\Omega/\partial Y^{m-1} + Y_c\Delta\partial^m\Omega/\partial Y^m \quad (m = 2, 3, 4, \cdots). \tag{8}$$

In the canonical procedure, some parts of the boundary are maintained at the reference state values σ_i, and other parts are varied in condition so as to control the current Y. The partial derivatives in Eqs. 6—8

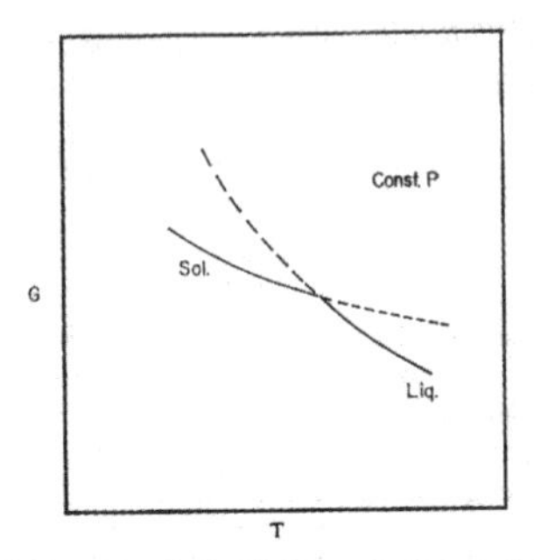

Fig. 1. Relative stability of the solid and liquid phases, at constant pressure, in a one-component system.

are thus at constant σ_t where appropriate.

"Resistance change" transitions in steady state systems are in some respects analogous[7-9] to thermostatic phase transitions where the phase with the lower Gibbs function value is the more stable one (see Fig. 1). Prigogine[7,8,12,13] considers the local potential F_σ to be the analog of the Gibbs function for "resistance change" transitions; for one-current steady-state situations there is no advantage in considering F_σ rather than θ, and I shall devote primary consideration to θ.

For stable steady states it is always true that[6,14]

$$\partial^2\theta/\partial Y^2 > 0, \quad \partial\Omega/\partial Y > 0; \tag{9}$$

and for "resistance change" transitions where both the upper $(Y>Y_c)$ and lower $(Y<Y_c)$ branches have metastable extensions beyond Y_c we anticipate that

$$\Delta\theta = 0, \quad \Delta\Omega = 0, \quad \Delta\partial\theta/\partial Y < 0, \tag{10}$$

i.e. we expect a kink in the θ *vs.* Y plot reflecting [Eq. 6] a discontinuous change in the flow resistance with $\Delta\partial\Omega/\partial Y<0$: pipes spring leaks, vigorously rubbed paper tears, screwdrivers often shear the slotted heads of screws— in short, *processes usually take* the *path* of *least resistance.* If, however, the transition point is a point of intrinsic instability for one or both of the upper and lower branches, *i.e.* if one or both of the branches do not have metastable extensions beyond Y_c, then Eq. 10 must be modified— more about this later.

Rather than continuing in general terms, it will be better if we look at the special cases that we are interested in.

2. *Forced Vaporization of Carbon Tetrachloride.* Consider a pure liquid in equilibrium with its vapor in a glass cell in a thermostat of temperature T. Let P_0 be the equilibrium vapor pressure of the liquid at temperature T. Reduce the pressure of the vapor to a pressure $P<P_0$, and establish a steady rate of vaporization of the liquid:

$$\text{liquid}(T, P_0) \rightarrow \text{vapor}\ (T, P). \tag{11}$$

The rate of entropy production for the forced vaporization process[15] is

$$\theta = -\dot{G}/T = \dot{\xi}[(\mu_{\text{liq}}-\mu_{\text{vap}})T^{-1}] = Y\Omega, \tag{12}$$

where $\dot{\xi}$ is the reaction velocity (amount vaporized per unit time) for Reaction 11 and I have neglected

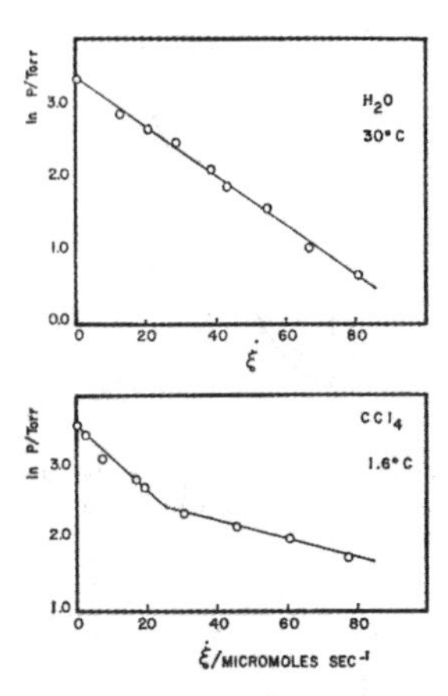

Fig. 2. Plots of $\ln P$ *versus* $\dot{\xi}$ for the steady forced vaporization of water at 30 °C and of carbon tetrachloride at 1.6 °C; P in Torr and $\dot{\xi}$ in μmol s^{-1}. Data of Alty and Nicoll.[17]

kinetic energy terms. Treating the vapor as an ideal gas $\{\mu_{\text{liq}}-\mu_{\text{vap}}=RT\ln(P_0/P)\}$ we have

$$\theta/R = \dot{\xi}\ln(P_0/P), \tag{13}$$

where R is the gas constant. Steady vaporization has been studied by Alty[15-17] and by Erikson;[15,18,19] nonsteady vaporization has been studied by Spangenberg and Rowland.[20]

Figure 2 shows the data of Alty and Nicoll[17] for the steady forced vaporization of water and of carbon tetrachloride; the current is given in micromoles per second and the pressure is measured in Torr. There is clear evidence in the carbon tetrachloride data for a discontinuous change in flow resistance at $\dot{\xi}_c=$ 25 μmol/s; the two flow regimes show the following behavior:

$$\ln P = \ln P_0 - A\dot{\xi}, \qquad (0\leq\dot{\xi}\leq\dot{\xi}_c), \tag{14}$$

$$\ln P = \ln P_* - B\dot{\xi}, \qquad (\dot{\xi}_c\leq\dot{\xi}), \tag{15}$$

$$\theta/R = A\dot{\xi}^2, \quad F_\sigma/R = \frac{1}{2}A\dot{\xi}^2, \qquad (0\leq\dot{\xi}\leq\dot{\xi}_c), \tag{16}$$

$$\theta/R = B\dot{\xi}^2 + (A-B)\dot{\xi}_c\dot{\xi},$$

$$F_\sigma/R = \frac{1}{2}B\dot{\xi}^2 + \frac{1}{2}(A-B)\dot{\xi}_c^2, \qquad (\dot{\xi}_c\leq\dot{\xi}), \tag{17}$$

where $-A$ and $-B$ are the slopes of the two line segments and P_* is the intercept on the line $\dot{\xi}=0$ of the (extrapolated) line segment with slope $-B$. Values for A and B, in seconds per micromole, are $A=0.045$, $B=0.013$. Figure 3 is a graph of $10^6\theta/R$ *vs.* $\dot{\xi}$ for the carbon tetrachloride data, calculated from Eqs. 16 and 17 after insertion of the appropriate numerical values of A, B, and $\dot{\xi}_c$. The dashed and dotted curves are potential hysteresis curves extending the observed behavior into unstable regions. We see that at the

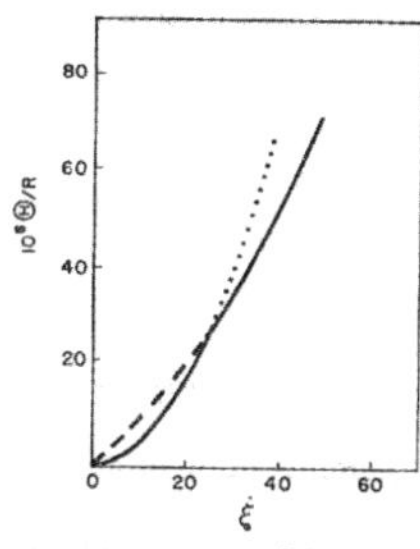

Fig. 3. Plot of $10^3\theta/R$ *versus* $\dot{\xi}$ for the steady forced vaporization of carbon tetrachloride at 1.6 °C. Data af Alty and Nicoll.[17]

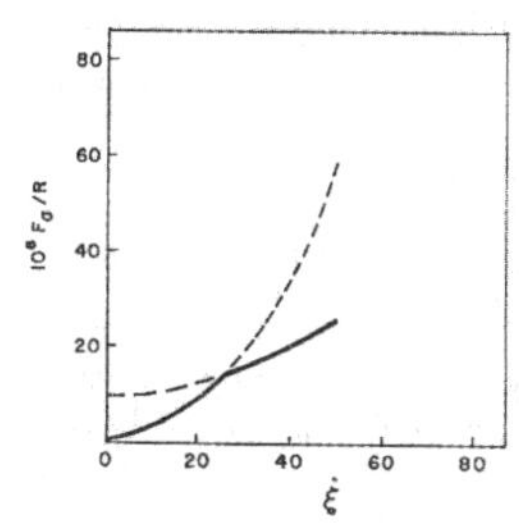

Fig. 4. Plot of $10^6F_\sigma/R$ *versus* $\dot{\xi}$ for the steady forced vaporiization of carbon tetrachloride at 1.6 °C. Data of Alty and Nicoll.[17]

transition point

$$\Delta\{\partial(\theta/R)/\partial\dot{\xi}\}_T = (B-A)\dot{\xi}_c < 0, \tag{18}$$

i.e. the discontinuous change is to a configuration of lower rate of entropy production.

Figure 4 shows that the behavior of the local potential F_σ at the transition point is somewhat similar to the behavior of θ: the branch with the lower value of F_σ is the more stable. In the case of F_σ, however, the metastable extension of the upper branch does not go through the origin of coordinates.[7]

In the carbon tetrachloride case, for fixed T, we have the transition point values θ_c, $\dot{\xi}_c$, P_c; if we repeat the experiment at another value of T we produce a new set of transition point values. In general then the thermodynamic data are the values of T, θ_c, $\dot{\xi}_c$, and P_c at the transition point (for a given apparatus configuration); and, from a thermodynamic point of view, we are interested in derivatives of the type $d\theta_c/d\dot{\xi}_c$, $d\dot{\xi}_c/dT$, $dP_c/d\dot{\xi}_c$, *etc.* Whereas in the somewhat analogous case of a phase transition in a thermo-static system we have the Clapeyron equation ($dP/dT = \Delta S/\Delta V$) to help us out in determining the slope of the phase coexistence line, in the present case we have no such useful general result.

2.1 Expectations and Queries: For the steady forced vaporization of carbon tetrachloride we have

$$\Delta\partial^2(\theta/R)/\partial\dot{\xi}^2 = 2(B-A) < 0, \tag{19}$$

$$\Delta\partial^n(\theta/R)/\partial\dot{\xi}^n = 0 \quad (n = 3, 4, 5, \cdots). \tag{20}$$

Equations 20 are consequences of the linear form of the given current-affinity relation; they, therefore, cannot be expected to have any *general* validity. Relation 19 will merit further investigation; is it generally true that

$$\Delta\partial^2\theta/\partial Y^2 < 0\,? \tag{21}$$

The physics of the steady forced vaporization process supplies us with some reasonable expectations concerning the behavior of the transition point variables. If the dominant feature of the transition point is the temperature gradient in the liquid in the vicinity of the liquid-vapor interface, then, since the heat of vaporization is a decreasing function of the temperature, we expect to find that

$$d\dot{\xi}_c/dT > 0, \tag{22}$$

to establish the same sort of temperature gradient at a higher temperature we have to pump off the vapor faster because the heat of vaporization now has a lower value. Note that

$$\frac{d(\theta_c/R)}{dT} = \frac{d\dot{\xi}_c}{dT}(\ln P_0 - \ln P_c) + \dot{\xi}_c\left(\frac{d\ln P_0}{dT} - \frac{d\ln P_c}{dT}\right). \tag{23}$$

Now we expect the coefficient A in Eq. 14 to be a decreasing function of temperature,[15,18,19,21] so, given Eq. 22, $d\ln P_c/dT$ should be positive. It is conceivable, then, that $d\ln P_c/dT$ might be large enough to make the right hand side of Eq. 23 vanish, leading to the result that, for a given liquid and a given apparatus configuration,

$$\theta_c = \text{const.} \tag{24}$$

Relation 24 would imply, for this type of experiment, an intrinsic limit to the rate of entropy production that the initial liquid configuration could support; in order to pass beyond the critical rate of entropy production, the liquid would have to change its configuration. An observation analogous to this is that small droplets of liquid have an intrinsic limit (at low to moderate pressures) to the degree of superheat that they can sustain.[22]

If Relation 24 were found to hold, then it would follow that for forced vaporization

$$d\theta_c/dY_c = 0 \tag{25}$$

and, consequently, that

$$d\Omega_c/dY_c = -\Omega_c/Y_c < 0. \tag{26}$$

It is hard to see how Relation 26 could possibly be true, and in section 3 I show that it is very unlikely the Relation 24 can hold true.

The carbon tetrachloride data in Fig. 2 clearly show a "resistance change" transition; the studies of Spangenberg and Rowland[20] indicate that such behavior should regularly occur in forced vaporization experiments; yet the data for water in Fig. 2 and other published results of forced vaporization experiments[15,16,18,19] do not clearly show evidence of "resistance change" transitions. What controls the presence or absence of such transitions? I shall discuss this matter at the end of the section on Bénard instability, but, to anticipate that discussion, what we need in forced vaporization experiments are a closer spacing of points in the $\ln P$ *vs.* $\dot{\xi}$ plots and more attention given to the depth of the liquid layer as an experimental parameter.

3. Bénard Instability. Sandwich a thin layer of fluid between two heat reservoirs (a, b) of temperatures T_a and T_b, with reservoir a above and b below the layer of fluid. By making $T_b \neq T_a$ we can generate a steady flow of heat through the fluid layer from one reservoir to the other. Characterize a reference thermo-static equilibrium state of the fluid by the variables T_a, P_a. Make T_b progressively larger than T_a. A sequence of states of steady heat flow results which ultimately shows an instability: thermal expansion causes the fluid to be less dense at the bottom of the layer than at the top — the layer becomes "top heavy;" eventually conductive heat flow gives way to convective heat flow. The sudden onset of a pattern of convective heat flow is referred to as the Bénard instability;[23,24] it is another example of a "resistance change" transition.

Let $T_b > T_a$ and let $\dot{Q}$ be the rate of influx of heat to the upper reservoir (reservoir a); take state a (T_a, P_a) as the reference state, then

$$\theta = Y\Omega = \dot{Q}\left(\frac{1}{T_a} - \frac{1}{T_b}\right) = \dot{Q}(T_b - T_a)/T_a T_b, \tag{27}$$

$$(\partial\Omega/\partial Y)_{T_a} = (1/T_b)^2[\partial(T_b - T_a)/\partial\dot{Q}]_{T_a}. \tag{28}$$

Schmidt and Milverton[25] immersed two circular brass plates, a fixed distance apart, into a tank of water. They passed a steady electric current I through a resistor of resistance ω affixed to the underside of the lower plate. If f is the fraction of the electrical energy dissipated per unit time by the resistor that passes directly from the lower plate to the upper one through the intervening layer of water, then

$$\dot{Q} = f\omega I^2, \tag{29}$$

$$\theta = \dot{Q}\left(\frac{1}{T_a} - \frac{1}{T_b}\right) = \frac{f\omega I^2(T_b - T_a)}{T_a T_b}, \tag{30}$$

$$T_0^2\theta/f\omega = I^2(T_b - T_a)(T_0^2/T_a T_b), \tag{31}$$

where T_0 is an additional reference temperature introduced for computational convenience. Figure 5 is a plot of $(T_b - T_a)(T_0^2/T_a T_b)$ *vs.* I^2, with $T_0 = 291$ K, of the data listed for Experiment 4 in the paper of Schmidt and Milverton;[25] the distance of separation between the plates was 5.5 mm. There is again clear evidence for a "resistance change" transition, and a plot of $T_0^2\theta/f\omega$ *vs.* I^2 would be qualitatively of the same shape as that of Fig. 3: at the point of transition the configuration of lower rate of entropy production is the more stable, and Relations 9, 10, and 21 are valid. Similarly a plot of $T_0^2 F_\theta/f\omega$ *vs.* I^2 would be qualitatively of the same shape as that of Fig. 4.

In the case of Bénard instability the thermodynamic variables characterizing the transition point are θ_c, $\dot{Q}_c$, $(T_a^{-1} - T_b^{-1})_c$ or $(T_b - T_a)_c$, T_a, and P_a (depending on the experimental setup, P_a may or may not be an independent variable). We are again interested in quantitites such as $\partial\theta_c/\partial\dot{Q}_c$, $\partial\Omega_c/\partial Y_c$, $\partial\dot{Q}_c/\partial T_a$, *etc.* We

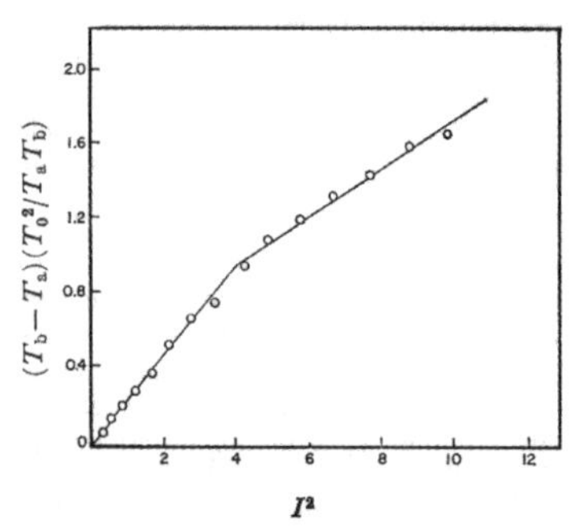

Fig. 5. Plot of $(T_b - T_a)(T_0^2/T_a T_b)$ *versus* I^2, with $T_0 = 291$ K. Data of Schmidt and Milverton.[25]

get some help from hydrodynamic stability theory[23] in this case. The (dimensionless) Rayleigh number $R_\#$ is defined to be

$$R_\# = g\alpha(T_b - T_a)\lambda^3/\kappa\nu, \tag{32}$$

where g is the gravitational acceleration, λ is the thickness of the fluid layer, and α, κ, and ν are the coefficients of volume expansion, thermometric conductivity, and kinematic viscosity, respectively. Stability theory[23] shows that at the point of Bénard instability

$$R_{\#c} = 1708. \tag{33}$$

In the conductive heat flow regime $\dot{Q} = D(T_b - T_a) \times \kappa\rho C_V/\lambda$, where D is the area through which the heat current flows, ρ is the density of the fluid, and C_V is the specific heat of the fluid; consequently

$$\theta_c \approx D(1708\nu/g\alpha T_a)^2(\kappa^3\rho C_V/\lambda^7), \tag{34}$$

where I have set $T_a T_b \approx T_a^2$. Since ν, α, κ, ρ, and C_V are all functions of the reference temperature T_a, the rate of entropy production θ_c for a given fluid with fixed values of D and λ, is apt to be a complicated function of T_a; and a relation such as Eq. 24 is not apt to be satisfied.

3.1 Forced Vaporization Reconsidered: With respect to the hydrodynamic stability of a layer of liquid, heating from below and cooling from above are much the same thing. The instability in the forced vaporization of carbon tetrachloride should thus be of the same nature as the Bénard instability, and the sensitivity of the Bénard instability to the thickness of the fluid layer [Eq. 32] indicates the advisability of treating the depth of the vaporizing liquid as an experimental variable in the forced vaporization case. (The forced vaporization problem is a more complicated one than the Bénard problem in that the temperature gradient in the vaporizing liquid has a complicated 3-dimensional structure[20] whereas the temperature gradient in the Bénard case is a simple 1-dimensional one.)

The data of Alty and Nicoll[17] displayed in Fig. 2 were all gathered in the same apparatus at the same fixed depth for the liquid layer. At the point of instability for the carbon tetrachloride the difference between the thermostat temperature T and the temperature T_s at the surface of the vaporizing liquid was $T - T_s = 7.5$ K. If we assume the geometric factors of the two experiments displayed in Fig. 2 to be the same (same apparatus, same depth of liquid, same 3-dimensional structure for the thermal gradients) and if we scale the water data according to Eq. 33, relative to the carbon tetrachloride data, *i.e.* if we say that

$$[\alpha(T-T_s)/\kappa\nu]_{c,\,H_2O} = [\alpha(T-T_s)/\kappa\nu]_{c,\,CCl_4}, \qquad (35)$$

we find a predicted value of $T-T_s \approx 59$ K at the point of instability for water under the given experimental conditions at 30 °C— such a value is far outside the range of experimental conditions displayed in Fig. 2.

In Alty's other experiments[16] and in Erikson's experiments[18,19] the depth of the liquid layer was an uncontrolled variable, so the resulting data were not gathered in such a way as to highlight the onset of Bénard-type instability. Note that Erikson did see evidence of convection currents in some of his experiments.[18] As I mentioned previously, what is needed in forced vaporization experiments is strict control of the depth-of-liquid variable — and a closer spacing of experimental points in plots of $\ln P$ *vs.* $\dot{\xi}$ (for a fixed depth of liquid). Also, as I mentioned in my discussion of Eq. 34, Relation 24 is unlikely to have any validity whatsoever.

4. Taylor Instability. Place a sample of liquid between two coaxial cylinders; put the device in thermal communication with a thermostat of temperature T, and rotate one of the cylinders at a constant angular velocity γ by exerting on it a torque N. (The case of simultaneous rotation of both cylinders is also of interest, but I do not consider it in this paper.) The simple Couette flow between the cylinders ultimately becomes unstable at $\gamma=\gamma_c$, and toroidal Taylor vortices form at the point of instability[24,26–28]— another example of a "resistance change" transition. As in the previous case, the instability is correlated with the critical value of a dimensionless combination of fluid properties — the Taylor number.[26]

The rate of entropy production for a given angular velocity γ is

$$\theta = \gamma(N/T) = Y\Omega. \qquad (36)$$

To analyze the Taylor instability we need experimental data in the form of N *vs.* γ plots (plots of driving torque *versus* angular velocity) so as to be able to see discontinuities in N and/or $\partial N/\partial\gamma$ at $\gamma=\gamma_c$. The usual experimental procedure,[26–28] however, is to measure the torque on the *stationary cylinder* as a function of the angular velocity γ. The measurements of Donnelly[28] show a discontinuity in the torque on the stationary cylinder at γ_c, *i.e.* the fluid shows a discontinuous increase in apparent viscosity at the point of instability. What do these results imply concerning the variation of N with γ? Until we have some direct measurements or dependable calculations of the N, γ relationship, we cannot be sure of the implication; we can, however, explore some of the possibilities.

Suppose that the N, γ relation turns out to be similar in nature to the torque-on-the-stationary-cylinder, γ relation: suppose that $\Delta N \neq 0$, $\Delta\partial N/\partial\gamma > 0$, and the configuration of *higher* rate of entropy production (and of higher local potential) is the more stable one beyond γ_c. What would be the thermodynamic implications of such a result? The idea here is to pursue the analogy to thermo-static phase transitions. If both phases have metastable regimes extending out beyond the transition point (Fig. 1, *e.g.*), then we are dealing with a problem of *relative stability* and the phase with the smaller Gibbs function value is the more stable. But it is also possible to have a phase transition point such that one of the phases reaches an *absolute limit of stability* at the transition point and has no metastable existence out beyond the transition point — the order-disorder transition in β-brass, for example.[29] For "resistance change" transitions, then, where each configuration has a metastable extension beyond the point of instability (in one direction or the other) the question of relative stability is decided by having the configuration of lower rate of entropy production (or lower local potential) more stable (Fig. 3, *e.g.*). But if a given configuration approaches an inflection point in the θ, Y plot, *i.e.* $\partial^2\theta/\partial Y^2$ approaches zero, it approaches an intrinsic limit of stability [see Eq. 9]: at the point where $\partial^2\theta/\partial Y^2=0$ the configuration must change to some other stable configuration; it is no longer a question of relative stability but an absolute requirement for a change of configuration. Under such circumstances $\Delta\partial\Omega/\partial Y$ could just as well be positive as negative, and we could even have a discontinuity in Ω (and F_σ): $\Delta\Omega\neq 0$ ($\Delta F_\sigma \neq 0$). The requirement for such a point of intrinsic instability would be

$$0 = \partial^2\theta/\partial Y^2 = 2(\partial\Omega/\partial Y) + Y_c(\partial^2\Omega/\partial Y^2), \qquad (37)$$

and we should have

$$Y_c = -2(\partial\Omega/\partial Y)/(\partial^2\Omega/\partial Y^2) \qquad (38)$$

as $Y \to Y_c - 0$.

If the transition point is a point of intrinsic instability for both the upper ($Y > Y_c$) and lower ($Y < Y_c$) branches, we cannot say anything *a priori* about $\Delta\theta$ or $\Delta\partial\Omega/\partial Y$. If the transition point is a point of intrinsic instability for the lower branch and if the upper branch has a metastable extension below Y_c, then we can make the following observations relative to the transition point: i) it is not possible, under these circumstances, to have $\Delta\theta < 0$ since the extension of the upper branch into the region $0 \le Y < Y_c$ would be more stable (would have a lower θ value) than the experimental lower branch; ii) it is not possible to have $\Delta\theta=0$ and $\Delta\partial\theta/\partial Y > 0$ for the same reason as in i)— the extension of the upper branch into the region $0 \le Y < Y_c$ would be more stable than the experimental lower branch; iii) it is possible, but not likely, to have $\Delta\theta=0$ and $\Delta\partial\Omega/\partial Y < 0$ — it is improbable to expect the lower branch to just "happen" to intersect the upper branch at the point of intrinsic instability; iv) it is possible, and likely, to have $\Delta\theta > 0$ and either $\Delta\partial\Omega/\partial Y > 0$ or $\Delta\partial\Omega/\partial Y < 0$ — at the point of intrinsic instability it is highly probable that the lower branch will be some distance away from the upper branch, hence a jump discontinuity in θ (and in Ω) — and in F_σ — will occur, with no necessary restriction on the sign of the change in slope $\Delta\partial\Omega/\partial Y$. Note that if $\Delta\partial\Omega/\partial Y > 0$ then there *must* be a discontinuity in θ (and in Ω) — see Fig. 6.

If the N, γ relation for the Taylor instability proves to have $\Delta N \neq 0$ at $\gamma=\gamma_c$ the most likely explanation

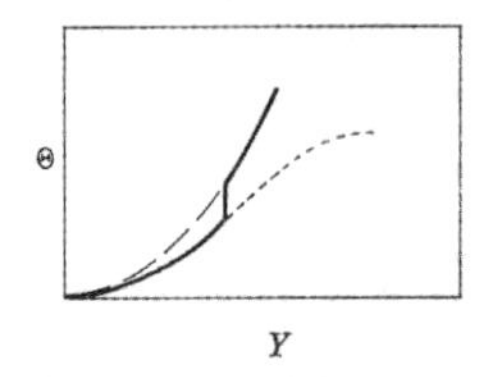

Fig. 6. Example: point of intrinsic instability with $\Delta\theta>0$ and $\Delta\partial\Omega/\Omega Y>0$.

will be that the transition point is a point of intrinsic instability for the lower branch. Points of intrinsic instability are special cases of the "bifurcations" and "catastrophes" discussed by Nicolis and Prigogine.[30]

5. A Multi-current Situation — Flip-Flop Current vs. Voltage Relation. Interpose a sintered glass membrane between aqueous sodium chloride solutions of concentrations 0.01 and 0.1 M; place inert electrodes on opposite sides of the membrane and impose a voltage $\Delta\psi$ between the electrodes; exert an excess pressure ΔP on the more concentrated solution. Keep the concentration difference and the pressure difference across the membrane fixed and measure the electric current I induced by the impressed voltage $\Delta\psi$. (Maintain the entire system at a constant temperature T.)

Let subscript 1 indicate water and subscript 2 indicate sodium chloride, and let a single prime designate the more concentrated solution and a double prime designate the less concentrated solution. Then, if the experiment is conducted in a steady-state fashion, we have[31]

$$T\theta = -(\dot{n}_1'\mu_1' + \dot{n}_2'\mu_2' + \dot{n}_1''\mu_1'' + \dot{n}_2''\mu_2'') + I\Delta\psi, \tag{39}$$

and

$$\begin{aligned} T\theta &= \dot{n}_1'(\mu_1''-\mu_1') + \dot{n}_2'(\mu_2''-\mu_1') + I\Delta\psi \\ &= Y_1\Omega_1 T + Y_2\Omega_2 T + Y_3\Omega_3 T, \end{aligned} \tag{40}$$

since $\dot{n}_1'+\dot{n}_1''=0$ and $\dot{n}_2'+\dot{n}_2''=0$. If we treat the sodium chloride solutions as "ideal" we can say that

$$\begin{aligned} \mu_1' - \mu_1'' &= \int_{P''}^{P'} \bar{V}_1\,\mathrm{d}P + RT\ln(X_1'/X_1'') \\ &= \langle\bar{V}_1\rangle\Delta P + RT\ln(X_1'/X_1''), \end{aligned} \tag{41}$$

$$\begin{aligned} \mu_2' - \mu_2'' &= \int^{P'} \bar{V}_2\,\mathrm{d}P + RT\ln(C_2'/C_2'') \\ &= \langle\bar{V}_2\rangle\Delta P + RT\ln(C_2'/C_2''), \end{aligned} \tag{42}$$

where X_i is the mole fraction, $\bar{V}_i$ the partial molar volume, $\langle\bar{V}_i\rangle$ the average partial molar volume over the pressure interval ΔP, and C_i the concentration (moles per liter) of component i. Let $\dot{n}_1'\langle\bar{V}_1\rangle + \dot{n}_2'\langle\bar{V}_2\rangle \approx \dot{V}'$ and evaluate X_1'/X_1'' by setting molarities approximately equal to molalities. Impose the restriction on $\dot{n}_1'$ and $\dot{n}_2'$ that they keep the ratio n_1'/n_2' constant (*i.e.* constant C_2'). Upon substitution of the appropriate numerical values, we get

$$T\theta \approx -\dot{V}'\Delta P - 1.4\,\dot{n}_2'RT + I\Delta\psi. \tag{43}$$

Experiments measuring the current-voltage relationship for this type of experiment were carried out by H. Jahnke and reported on by Franck.[32] Figure 7 shows the I *vs.* $\Delta\psi$ relation for various fixed pressure differences for one series of experiments carried out by Jahnke. The rather spectacular form of the I, $\Delta\psi$ relation is explained by Kobatake[8] in the following way: "... under the external conditions studied here, the Poiseuille pressure flow transports fluid from the more concentrated to the less concentrated solution, while the electro-osmotic flow caused by the potential gradient tends to carry fluid in the opposite direction. With increasing $\Delta\psi$, the electro-osmotic flow becomes appreciable, outweighs the pressure flow, and eventually changes the direction of mass flow from negative to positive. Calculations show that this change of the direction of mass flow occurs discontinuously when $\Delta\psi$ reaches a certain value, when the pressure difference ΔP is larger than a critical value ΔP_c. Correspondingly, the average salt concentration in the membrane is lowered, *i.e.* the membrane is occupied with the less concentrated solution... The effect of this change in concentration in the membrane is reflected in the I *vs.* $\Delta\psi$ relationships depicted in [Fig. 7]."

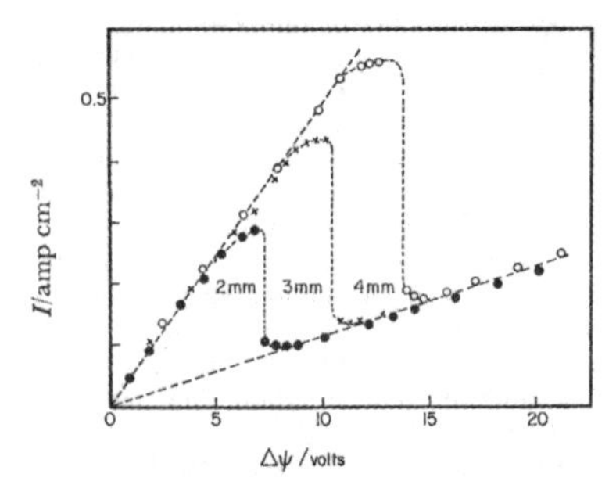

Fig. 7. Plots of I *versus* $\Delta\psi$ for the system of a sintered glass membrane and NaCl solutions of 0.01 and 0.1 M with various fixed pressure differences. Data of Jahnke as reported by Franck.[32]

The results reported by Franck[32] do not include data for $\dot{V}'$ and $\dot{n}_2'$ so we cannot actually calculate $T\theta$ for the experiment. It seems reasonable to expect, however, that at the critical value $\Delta\psi_c$ of the voltage $T\theta$ will actually show a discontinuous *decrease*. If such is indeed the case, it must be because the point $\Delta\psi = \Delta\psi_c$ is a point of intrinsic instability ($\partial^2\theta/\partial Y^2 \to 0$) for *both* the upper ($\Delta\psi = \Delta\psi_c + 0$) and lower ($\Delta\psi = \Delta\psi_c - 0$) branches of the I *vs.* $\Delta\psi$ relation, *i.e. neither* branch has a metastable extension beyond the point $\Delta\psi = \Delta\psi_c$.

Kobatake[8] has analyzed Jahnke's experiments in terms of the local potential:

$$TF_\sigma = T\int \Sigma_i Y_i \mathrm{d}\Omega_i = \int_0^{\Delta\psi} I\mathrm{d}\Delta\psi + f(\Delta P, \Delta\mu_2). \tag{44}$$

By an elementary theorem of the calculus[33] the integral in Eq. 44 is continuous at the point $\Delta\psi_c$, so TF_σ does not have a discontinuity at the flip-flop point.

6. Summary. I have considered the behavior

of the rate of entropy production θ and the local potential F_σ for several one-current steady-state situations showing "resistance change" transitions — forced vaporization of carbon tetrachloride, Bénard instability, and Taylor instability. For these cases the behavior of the local potential is qualitatively similar to the behavior of the rate of entropy production. The complexity of the thermodynamic relations pertaining to the transition point depends upon whether the branches intersecting at the transition point have metastable extensions beyond that point. I also considered (briefly) a multi-current situation showing a flip-flop current-voltage relation at the transition point; here θ and F_σ seemingly show different kinds of behavior— F_σ is continuous at the transition point whereas θ appears to undergo a discontinuous change (it seems that *neither* of the branches intersecting at the transition point has a metastable extension beyond that point).

References

1) R. J. Tykodi, "Thermodynamics of Steady States," Macmillan, New York (1967).
2) R. J. Tykodi, *J. Chem. Phys.*, **47**, 1879 (1967).
3) R. J. Tykodi, *Am. J. Phys.*, **38**, 586 (1967).
4) R. J. Tykodi, *Bull. Chem. Soc. Jpn.*, **44**, 1001 (1971).
5) R. J. Tykodi, *J. Chem. Phys.*, **57**, 37 (1972).
6) R. J. Tykodi, *Physica*, **72**, 341 (1974).
7) "Non-Equilibrium Thermodynamics, Variational Techniques and Stability," ed by R. J. Donnelly, R. Herman, and I. Prigogine, University of Chicago Press, Chicago (1966), pp. 3—16.
8) Y. Kobatake, *Physica*, **48**, 301 (1970).
9) "Fluctuations, Instabilities, and Phase Transitions," ed by T. Riste, Plenum, New York (1975).
10) J. C. M. Li, *J. Appl. Phys.*, **33**, 616 (1962).
11) J. C. M. Li, *J. Phys. Chem.*, **66**, 1414 (1962).
12) P. Glansdorff and I. Prigogine, *Physica*, **30**, 351 (1964).
13) P. Glansdorff and I. Prigogine, "Thermodynamic Theory of Structure, Stability and Fluctuations," Wiley-Interscience, London (1971).
14) Ref. 1, Chap. 3.
15) Ref. 1, Chap. 2.
16) T. Alty, *Proc. R. Soc. London, Ser. A*, **131**, 554 (1931).
17) T. Alty and F. Nicoll, *Can. J. Res.*, **4**, 547 (1931).
18) T. A. Erikson and R. J. Tykodi, *J. Chem. Phys.*, **33**, 46 (1960).
19) T. A. Erikson and R. J. Tykodi, J. of Heat Transfer, Trans. ASME, Vol. 91, Series C. Number 2 (May 1969), p. 221.
20) W. G. Spangenberg and W. R. Rowland, *Phys. Fluids*, **4**, 743 (1961).
21) R. J. Tykodi and T. A. Erikson, *J. Chem. Phys.*, **31**, 1521 (1959).
22) R. C. Reid, *Am. Scientist*, **64**, 146 (1976).
23) S. Chandrasekhar, "Hydrodynamic and Hydromagnetic Stability," Oxford U. P., Oxford (1961), Chap. II.
24) Ref. 7, pp. 165—197.
25) R. J. Schmidt and S. W. Milverton, *Proc. R. Soc. London, Ser. A*, **152**, 586 (1935).
26) Ref. 23, Chap. VII.
27) G. I. Taylor, *Proc. R. Soc. London, Ser. A*, **157**, 546 (1936).
28) R. J. Donnelly, *Proc. R. Soc. London, Ser. A*, **246**, 312 (1958).
29) E. A. Guggenheim, "Thermodynamics," 2nd ed, North-Holland Pub. Co., Amsterdam (1950), pp. 276—285.
30) G. Nicolis and I. Prigogine, "Self-Organization in Nonequilibrium Systems," Wiley, New York (1977).
31) Ref. 1, Chaps. 4, 13.
32) U. F. Franck, Z. Elektrochem., *Ber. Bunsenges. Phys. Chem.*, **67**, 657 (1963).
33) D. V. Widder, "Advanced Calculus," 2nd ed, Prentic-Hall, Englewood Cliffs (1961), p. 178.

Thermodynamics of steady states: Is the entropy-production surface convex in the thermodynamic space of steady currents?

R. J. Tykodi

Department of Chemistry, Southeastern Massachusetts University, North Dartmouth, Massachusetts 02747

(Received 16 May 1983; accepted 26 July 1983)

If the rate of entropy production Θ is a convex function of the steady currents Y_i in a steady-state situation, then the quantities $\partial^2\Theta/\partial Y_k \partial Y_i$ and $\partial\Omega_i/\partial Y_k$ (Ω_i is the affinity conjugate to the steady current Y_i) must meet certain requirements, and Θ itself must be an "inflationary" function of the steady currents Y_i, i.e., $\Theta(\lambda Y_i) > \Theta(Y_i)$ for $\lambda > 1$ and $\{Y_i\} \neq \{0\}$. A heat-conduction counterexample that does not satisfy the convexity conditions is displayed.

INTRODUCTION

Continuing my studies[1-8] of the thermodynamics of steady-state systems, I examine here some properties of the entropy-production surface in the thermodynamic space of steady currents.

In my book-length study[1] of the thermodynamics of steady states, I studied systems that were entirely resolvable into terminal parts and gradient parts.[9] For steady-state situations involving such systems, the rate of entropy production $\Theta \equiv \dot{S}(\text{system}) + \dot{S}(\text{surroundings})$ can always be expressed in the form

$$\Theta = \sum_i Y_i \Omega_i \geq 0 , \tag{1}$$

where the currents Y_i are (essentially) time derivatives of extensive thermodynamic quantities[10] ($\dot{U}^{(r)}$ or $\dot{Q}^{(r)}$, $\dot{V}$, $\dot{n}^{(j)}$, etc.) and the affinities Ω_i are defined via Eq. (1). If we produce a given steady state by starting from a reference thermodynamic (equilibrium) state σ and then adjusting some of the boundary conditions so as to induce the steady currents Y_i, we find that the resulting steady state is fully determined by the set of steady currents Y_i and the properties of the equilibrium state σ.[10]

I use the following notational conventions in this paper.

(i) Dummy indices on summations $(i, j, k, \ldots)$ always run from 1 to n;

(ii) $\partial/\partial Y_k)_{Y'} \equiv \partial/\partial Y_k)_{Y',\sigma}$,

where the subscript Y' means constant Y_i for $i \neq k$, i.e., I do not explicitly remind the reader of the dependence on the properties of the reference equilibrium state σ;

(iii) $K_{ik} \equiv (\partial\Omega_i/\partial Y_k)_{Y'}$, $\quad K_{ikj} = \partial^2\Omega_i/\partial Y_j \partial Y_k$;

(iv) $\Theta_{ik} \equiv \partial^2\Theta/\partial Y_k \partial Y_i$ (evidently $\Theta_{ik} = \Theta_{ki}$);

(v) $F(X_i) \equiv F(X_1, X_2, \ldots, X_n)$;

(vi) I indicate the determinant of the matrix whose elements are the quantities a_{ik} by $\det(a_{ik})$.

CONVEXITY RELATIONS

From the observation that $\Theta = \Theta(Y_i, \sigma)$ we note that for a given reference state σ the rate of entropy production Θ is a well-defined function of the steady currents Y_i. For a fixed reference state σ, then, set up a rectangular coordinate system in the n-dimensional space of steady currents Y_i; indicate a point in this space of steady currents by $\{Y_i\}$. In a similar fashion, set up a rectangular coordinate system in the $(n+1)$-dimensional space spanned by the Y_i axes and an entropy-production axis Θ; indicate a point in this *thermodynamic space of steady currents* by $\{Y_i, \Theta\}$; and construct the entropy-production surface $\Theta(Y_i)$ in the $[(n+1)$-dimensional$]$ thermodynamic space of steady currents. Now it would be intuitively pleasing if $\Theta(Y_i)$ were always a convex function of the steady currents Y_i, in particular if $\Theta(Y_i)$ were twice differentiable and were always of such a nature that

$$0 < \Theta_{ii} \equiv \partial^2\Theta/\partial Y_i^2 \quad (i = 1, 2, \ldots, n) . \tag{2}$$

Let us look at some of the conditions that $\Theta(Y_i)$ would have to satisfy if it were a convex function of its arguments.[11,12] If $\Theta(Y_i)$ *is* a convex (twice differentiable) function of its arguments, then[11]

$$\sum_{i,k} \Theta_{ik}(Y_i)\xi_i\xi_k \geq 0 \quad (\xi_j \equiv Y_j + y_j, \text{ with } -\infty < y_j < +\infty) \tag{3}$$

for all ξ_i, ξ_k and every allowable steady state $\{Y_i\}$; the positive definite nature of the quadratic form (3) requires[12] that

$$\left.\begin{array}{l} 0 < \det(\Theta_{ik}) \quad (i, k = 1, 2, \ldots, n) , \\ 0 < \text{each principal minor of } \det(\Theta_{ik}) . \end{array}\right\} \tag{4}$$

Result (4) thus includes requirement (2). Since $\Theta = \sum_i Y_i \Omega_i$, then

$$\Theta_{jk} = K_{kj} + K_{jk} + \sum_i Y_i K_{ikj} \quad (i, j, k = 1, 2, \ldots, n) , \tag{5}$$

and relation (5) imposes [via Eq. (4)] some restrictions on the quantities K_{ij} and K_{ijk}.

Further consequences of the hypothetical convex nature of $\Theta(Y_i)$ are the relations [for the case $\{Y_i\} \neq \{0\}$]

$$\Theta(\mu Y_i) < \mu\Theta(Y_i) < \Theta(Y_i) , \quad 0 < \mu < 1 , \tag{6}$$

$$\Theta(\lambda Y_i) > \lambda\Theta(Y_i) > \Theta(Y_i) , \quad \lambda > 1 , \tag{7}$$

where μ and λ are mere scaling parameters. Relations (6) and (7) reflect the supposed geometry of the entropy-production surface in the thermodynamic space of steady

 0021-9606/84/041652-04$02.10

currents: Construct the straight line (ray) starting at the origin $\{0, 0\}$ $[\Theta(0)=0]$ and passing through the point $\Theta(Y_i^\dagger)$ of the surface; if the entropy-production surface is convex, then all those points $[\{Y_i, \Theta'\}$, with $\Theta'=\mu\Theta(Y_i^\dagger)]$ on the line that are nearer the origin than the surface point $\Theta(Y_i^\dagger)$ will lie entirely "inside" and "above" the surface [Eq. (6)], whereas all those points $[\{Y_i, \Theta''\}$, with $\Theta''=\lambda\Theta(Y_i^\dagger)]$ on the line that are farther from the origin than the surface point $\Theta(Y_i^\dagger)$ will lie entirely "outside" and "below" the surface [Eq. (7)].[13]

Equations (6) and (7) show us that [for $\{Y_i\}\neq\{0\}$]

$$\Theta(\mu Y_i)<\Theta(Y_i)<\Theta(\lambda Y_i)\,, \quad 0<\mu<1\,, \quad 1<\lambda\,; \tag{8}$$

if $\Theta(Y_i)$ is a convex function of its arguments, then if we "inflate" the currents Y_i by suitable scaling $\mu Y_i \to Y_i$ or $Y_i\to\lambda Y_i$, we increase the value of the rate of entropy production. Thus if $\Theta(Y_i)$ is a convex function, then it is also an "inflationary" function of its arguments.

THE QUADRATIC FORM $D(Y_i)$

Let us compare two neighboring points $\Theta(Y_i)$ and $\Theta[(1+\beta)Y_i]$ $(0<\beta\ll 1)$ on the Θ surface (in the thermodynamic space of steady currents). As we pass from one point to the other, $Y_i\to(1+\beta)Y_i$ and $\Omega_i\to\Omega_i+d\Omega_i$. Now we have

$$d\Omega_i=\sum_k(\partial\Omega_i/\partial Y_k)_{Y^*}dY_k=\sum_k K_{ik}dY_k \tag{9}$$

and

$$dY_k=(1+\beta)Y_k-Y_k=\beta Y_k\,, \tag{10}$$

so

$$d\Omega_i=\beta\sum_k K_{ik}Y_k\,. \tag{11}$$

We see, therefore, that

$$\begin{aligned}\Theta[(1+\beta)Y_i]&=\sum_i(1+\beta)Y_i(\Omega_i+d\Omega_i)\\&=(1+\beta)\sum_i Y_i\Omega_i+(1+\beta)\sum_i Y_i d\Omega_i\\&=(1+\beta)\Theta(Y_i)+\beta(1+\beta)\sum_{i,k}Y_iK_{ik}Y_k\,.\end{aligned} \tag{12}$$

Now if $\Theta(Y_i)$ is a convex function of its arguments, then [Eqs. (7) and (12)] it must follow that:

$$D(Y_i)\equiv\sum_{i,k}Y_iK_{ik}Y_k>0 \quad \text{for } \{Y_i\}\neq\{0\}\,. \tag{13}$$

Under the convexity hypothesis, Eq. (13) holds for each (physically accessible) point $\{Y_i\}$ of the space of steady currents, so relation (13), therefore, must *always* be true. If Eq. (13) holds, then in a neighborhood of an arbitrary point $\{Y_i\}$, the quadratic form

$$\sum_{i,k}\xi_iK_{ik}(Y_i)\xi_k>0\,, \quad \xi_j\equiv Y_j+y_j\,, \quad \text{with } |y_j/Y_j|\ll 1 \tag{14}$$

will be positive definite for all values of ξ_i, ξ_k in the neighborhood. By an argument[12] analogous to that used for form (3), we conclude that the convexity hypothesis for $\Theta(Y_i)$ implies that

$$\left.\begin{aligned}&0<\det(K'_{ik})\,, \quad (i,k=1,2,\ldots,n)\,,\\&0<\text{each principal minor of }\det(K'_{ik})\,,\end{aligned}\right\} \tag{15}$$

where $K'_{ik}\equiv\frac{1}{2}(K_{ik}+K_{ki})$. The quantity $D(Y_i)$ is closely related to the local potential (thermokinetic potential) F_σ introduced by Prigogine and Glansdorff[5,8,14-16]:

$$\begin{aligned}F_\sigma&\equiv\sum_i\int Y_i d\Omega_i=\sum_i\int Y_i\sum_k K_{ik}dY_k\\&=\int\sum_{i,k}Y_iK_{ik}dY_k=\sum_{i,k}\langle Y_iK_{ik}\rangle Y_k\,,\end{aligned} \tag{16}$$

where $\langle Y_iK_{ik}\rangle$ is the average value of the product Y_iK_{ik} over the interval from the reference state σ to the state $\{Y_i\}$. In a suitable neighborhood of the origin $\{0\}$, the affinities Ω_i are linear functions of the steady currents:

$$\Omega_i=\sum_k K_{ik}(0)Y_k\,, \tag{17}$$

where $K_{ik}(0)=(\partial\Omega_i/\partial Y_k)_{Y^*}|\{Y_i\}=\{0\}$. In the linear current-affinity neighborhood of the origin,

$$\Theta=D=\sum_{i,k}Y_iK_{ik}(0)Y_k \tag{18}$$

and

$$\langle Y_iK_{ik}\rangle=\tfrac{1}{2}Y_iK_{ik}(0)\,, \tag{19}$$

so

$$F_\sigma=\tfrac{1}{2}\sum_{i,k}Y_iK_{ik}(0)Y_k=\tfrac{1}{2}D=\tfrac{1}{2}\Theta\,. \tag{20}$$

HOMOGENEITY CONSIDERATIONS

From $\Theta=\sum_i Y_i\Omega_i$, we see that

$$\sum_i Y_i(\partial\Theta/\partial Y_i)_{Y^*}=\Theta+\sum_{i,k}Y_iK_{ik}Y_k=\Theta+D\,. \tag{21}$$

We can also rearrange Eq. (12) somewhat:

$$\begin{aligned}\Theta[(1+\beta)Y_i]&=(1+\beta)\Theta(Y_i)+\beta(1+\beta)D(Y_i)\\&=(1+\beta)\Theta(Y_i)[1+\beta(D/\Theta)]\\&=(1+\beta)^2\left(\frac{1+q\beta}{1+\beta}\right)\Theta(Y_i)\,,\end{aligned} \tag{22}$$

where $q(Y_i)\equiv D/\Theta$. As $\{Y_i\}\to\{0\}$, $D\to\Theta$, and $q\to1$, so we see that Θ is asymptotically homogeneous of degree 2 in the Y_i in a neighborhood of the origin $\{0\}$. Furthermore, we note that for β sufficiently small $(0<\beta\ll1)$ $1+q\beta\approx(1+\beta)^q$ and [Eq. (22)]

$$\Theta[(1+\beta)Y_i]\approx(1+\beta)^{1+q}\Theta(Y_i)\,; \tag{23}$$

we can, therefore, say that Θ is quasihomogeneous of degree $1+q$ in a differential neighborhood of a point $\{Y_i\}$ [remembering of course that $q=q(Y_i)$]. Equation (23) *does* yield the correct form of Eq. (21):

$$\sum_i Y_i(\partial\Theta/\partial Y_i)_{Y^*}=(1+q)\Theta=\Theta+D\,. \tag{21'}$$

SUMMARY OF CONVEXITY RELATIONS

The hypothesis that $\Theta(Y_i)$ is a convex function of its arguments thus leads to the relations

$$\sum_{i,k} Y_i \Theta_{ik} Y_k > 0 \,, \tag{3}$$

$$\left.\begin{array}{l} 0<\det(\Theta_{ik}) \,, \\ 0<\text{each principal minor of } \det(\Theta_{ik}) \,, \end{array}\right\} \tag{4}$$

$$\Theta(\mu Y_i) \le \mu\Theta(Y_i) \le \Theta(Y_i) \,, \qquad 0 \le \mu < 1 \,, \tag{6}$$

$$\Theta(\lambda Y_i) \ge \lambda\Theta(Y_i) \ge \Theta(Y_i) \,, \qquad \lambda \ge 1 \,, \tag{7}$$

$$D \equiv \sum_{i,k} Y_i K_{ik} Y_k \ge 0 \,, \tag{13}$$

$$\left.\begin{array}{l} 0<\det(K'_{ik}) \,, \\ 0<\text{each principal minor of } \det(K'_{ik}) \,, \end{array}\right\} \tag{15}$$

$$\Theta[(1+\beta)Y_i] = (1+\beta)^2 \left(\frac{1+q\beta}{1+\beta}\right)\Theta(Y_i) \quad q = D/\Theta \,,$$

$$0<\beta \ll 1 \,, \tag{22}$$

$$\approx (1+\beta)^{1+q}\Theta(Y_i) \,. \tag{23}$$

We also have the identity

$$\sum_i Y_i(\partial\Theta/\partial Y_i)_{Y'} = \Theta + D \,. \tag{21}$$

HEAT-CONDUCTION COUNTEREXAMPLE

We can think of many cases where $\Theta(Y_i)$ is clearly a convex function of the steady currents (in the linear current-affinity region, for example), but is $\Theta(Y_i)$ necessarily always a convex function of its arguments? I shall show, by means of a counterexample, that $\Theta(Y_i)$ need not *always* be convex.

Take a length of wire of uniform cross section and let the ends communicate with heat reservoirs of temperatures T_α and T_β. Cut the wire at its midpoint and let both of the cut ends communicate with a heat reservoir of temperature T_x. Consider states of steady heat flow along the wire segments (no lateral heat losses) for given T_α, T_β, T_x. Let the (steady) rates of influx of heat into the reservoirs be $\dot{Q}_\alpha$, $\dot{Q}_\beta$, $\dot{Q}_x$. Choose the material of the wires and the dimensions of the wire segments so that the thermal conductivity times the length over the area for each segment equals 1 (for a suitable choice of units), and assume that the thermal conductivity is independent of temperature. For these choices and assumptions it follows that[4-6]:

$$\dot{Q}_\alpha = T_x - T_\alpha \,, \quad \dot{Q}_\beta = T_x - T_\beta \,, \quad \dot{Q}_x = T_\alpha + T_\alpha - 2T_x \,. \tag{24}$$

Now we see that in a steady state

$$\Theta = \frac{\dot{Q}_\alpha}{T_\alpha} + \frac{\dot{Q}_\beta}{T_\beta} + \frac{\dot{Q}_x}{T_x} \,, \quad \dot{Q}_\alpha + \dot{Q}_\beta + \dot{Q}_x = 0 \,. \tag{25}$$

We can think of any particular steady state as having arisen from a reference equilibrium state with $T_\alpha = T_\beta$ and $T_x = T_\beta$ via the changes T_α $(= T_\beta) \to T_\alpha$ (arbitrary) and T_x $(= T_\beta) \to T_x$(arbitrary). Treating T_β as our reference temperature, then, we get [by eliminating $\dot{Q}_\beta$ between Eqs. (25)]

$$\Theta = \dot{Q}_\alpha\left(\frac{1}{T_\alpha} - \frac{1}{T_\beta}\right) + \dot{Q}_x\left(\frac{1}{T_x} - \frac{1}{T_\beta}\right)$$

$$= Y_1\Omega_1 + Y_2\Omega_2 \,. \tag{26}$$

Let us keep T_β constant and explore the structure of $\Theta(\dot{Q}_\alpha, \dot{Q}_x)$—is $\Theta(\dot{Q}_\alpha, \dot{Q}_x)$ necessarily always convex?

First let us look at requirement (15). We note that

$$\begin{aligned} K_{11} &= (\partial\Omega_1/\partial Y_1)_{Y_2,a} = (\partial\Omega_1/\partial\dot{Q}_\alpha)_{\dot{Q}_x, T_\beta} = 2/T_\alpha^2 \,, \\ K_{12} &= (\partial\Omega_1/\partial\dot{Q}_x)_{\dot{Q}_\alpha, T_\beta} = 1/T_\alpha^2 \,, \\ K_{21} &= (\partial\Omega_2/\partial\dot{Q}_\alpha)_{\dot{Q}_x, T_\beta} = 1/T_x^2 \,, \\ K_{22} &= (\partial\Omega_2/\partial\dot{Q}_x)_{\dot{Q}_\alpha, T_\beta} = 1/T_x^2 \,; \end{aligned} \tag{27}$$

and we find that $K_{11} > 0$, $K_{22} > 0$. Next we check the quantity

$$\det(K'_{ik}) = K_{11}K_{22} - \left(\frac{K_{12} + K_{21}}{2}\right)^2 \,,$$

is it always nonnegative? From Eq. (27) we see that

$$K_{11}K_{22} - \tfrac{1}{4}(K_{12} + K_{21})^2 = \frac{2}{T_\alpha^2 T_x^2} - \frac{1}{4}\left(\frac{1}{T_\alpha^2} + \frac{1}{T_x^2}\right)^2$$

$$= \frac{1}{4T_\alpha^4 T_x^4}[4T_\alpha^2 T_x^2 - (T_x^2 - T_\alpha^2)^2] \,. \tag{28}$$

When the ratio of the larger to the smaller of T_x and T_α is greater than about 2.5, $\det(K'_{ik})$ will be negative; therefore, for the heat-conduction case that we are considering, we can find temperatures T_α, T_β, T_x such that $\Theta(\dot{Q}_\alpha, \dot{Q}_x)$ is *not* a convex function of its arguments.

To reinforce the conclusion we have just arrived at, let us look at the quantities Θ_{22} and Θ_{11}:

$$\begin{aligned} \Theta_{22} &= 2K_{22} + Y_1K_{122} + Y_2K_{222} \\ &= \frac{2}{T_x^2} + (T_x - T_\alpha)\frac{2}{T_x^3} + (T_\alpha + T_\beta - 2T_x)\frac{2}{T_x^3} \\ &= \frac{2}{T_\alpha^3 T_x^3}(T_x - T_\alpha)^2(T_x^2 + T_xT_\alpha + T_\alpha^2) + T_\beta T_\alpha^3 \ge 0 \,; \end{aligned} \tag{29}$$

Θ_{22} thus meets requirements (2) and (4).

$$\begin{aligned} \Theta_{11} &= 2K_{11} + Y_1K_{111} + Y_2K_{211} \\ &= \frac{4}{T_\alpha^2} + (T_x - T_\alpha)\frac{8}{T_\alpha^3} + (T_\alpha + T_\beta - 2T_x)\frac{2}{T_x^3} \\ &= \frac{2}{T_\alpha^3 T_x^3}(4T_x^4 - 2T_\alpha T_x^3 - 2T_xT_\alpha^3 + T_\alpha^4 + T_\beta T_\alpha^3) \\ &= \frac{2}{T_\alpha^3 T_x^3}(2T_x - T_\alpha)(2^{1/3}T_x - T_\alpha) \\ &\quad \times (2^{2/3}T_x^2 + 2^{1/3}T_xT_\alpha + T_\alpha^2) + T_\beta T_\alpha^3 \,. \end{aligned} \tag{30}$$

Let $\Theta_{11}^* \equiv T_\alpha^3 T_x^3 \Theta_{11}/2$; obviously $\operatorname{sign}\Theta_{11}^* = \operatorname{sign}\Theta_{11}$. Clearly $\Theta_{11}^* > 0$ when $T_x > T_\alpha/2^{1/3}$ and when $T_x < T_\alpha/2$; if Θ_{11}^* is negative anywhere in the interval $T_\alpha/2 \le T_x \le T_\alpha/2^{1/3}$, it must reach a minimum in that interval. Now

$$\begin{aligned} (\partial\Theta_{11}^*/T_x)_{T_\alpha, T_\beta} &= 16T_x^3 - 6T_\alpha T_x^2 - 2T_\alpha^3 \\ &= 2(8T_x^3 - 3T_\alpha T_x^2 - T_\alpha^3) \\ &= 2T_\alpha^3(8z^3 - 3z^2 - 1) \,, \end{aligned} \tag{31}$$

where $z \equiv T_x/T_\alpha$. By trial and error, we find that a root of the equation $8z^3 - 3z^2 - 1 = 0$ is $z \approx 0.66$. Thus (for given T_α, T_β) Θ_{11}^* reaches its minimum value when $T_x \approx 0.66T_\alpha$:

$$\Theta_{11}^*(\min) = T_\alpha^3(T_\beta - 0.136T_\alpha) \,, \tag{32}$$

so Θ_{11}^* (and consequently Θ_{11}) can be negative for $T_\beta < 0.136 T_\alpha$. For the case in hand, then, Θ_{11} can violate conditions (2) and (4), and $\Theta(\dot{Q}_\alpha, \dot{Q}_x)$ need not always be convex.

Finally, let me display the explicit form of $\Theta(\dot{Q}_\alpha, \dot{Q}_x)$:

$$T_\beta \Theta = \frac{\dot{Q}_\alpha(2\dot{Q}_\alpha + \dot{Q}_x)}{T_\beta - 2\dot{Q}_\alpha - \dot{Q}_x} + \frac{\dot{Q}_x(\dot{Q}_\alpha + \dot{Q}_x)}{T_\beta - \dot{Q}_\alpha - \dot{Q}_x} \; ; \tag{33}$$

we see that for small values of the currents ($|\dot{Q}_\alpha|$, $|\dot{Q}_x| \ll T_\beta$) the rate of entropy production Θ is asymptotically homogeneous of degree 2 in the steady currents $\dot{Q}_\alpha$, $\dot{Q}_x$.

CONCLUSION

We explored the requirements that the rate of entropy production $\Theta(Y_i)$ had to satisfy to be a convex function of its arguments. We then saw that a heat-conduction example exhibited behavior at some points $\{Y_i\}$ that failed to meet those convexity requirements. We must conclude, I think, that it is *not* necessary that $\Theta(Y_i)$ *always* be a convex function of its arguments.

Once again[4-6,8] the rate of entropy production has failed to give any hint of a universal pattern of behavior persisting outside the linear current-affinity region.

APPENDIX

In exploring the properties of the heat-conduction example of the text, I noticed a curious thing: When I randomly selected a set of number triples (T_α, T_β, T_x), for each triple of the set the value of Θ stood between the values of the "dual"[10] quantities $D \equiv \sum_{i,k} Y_i K_{ik} Y_k$ and $B \equiv \sum_{i,k} \Omega_i L_{ik} \Omega_k$, where $L_{ik} \equiv (\partial Y_i / \partial \Omega_k)_{\Omega', \sigma}$; i.e., for no member of the set of randomly selected number triples was it true that the relations $B \le \Theta \le D$ and $D \le \Theta \le B$ were both false.

The preceding observation suggests the following conjecture, which I shall call *the sandwich conjecture*:

For any steady-state situation, at each allowable (physically realizable) point $\{Y_i\}$ of the corresponding space of steady currents the value of Θ stands between the values of B and D.

I started this paper with one question; I now end it with another: Is *the sandwich conjecture* true?

[1]R. J. Tykodi, *Thermodynamics of Steady States* (Macmillan, New York, 1967).
[2]R. J. Tykodi, J. Chem. Phys. **47**, 1879 (1967).
[3]R. J. Tykodi, Am. J. Phys. **38**, 586 (1970).
[4]R. J. Tykodi, Bull. Chem. Soc. Jpn. **44**, 1001 (1971).
[5]R. J. Tykodi, J. Chem. Phys. **57**, 37 (1972).
[6]R. J. Tykodi, Physica **72**, 341 (1974).
[7]R. J. Tykodi, J. Non-Equilib. Thermodyn. **2**, 193 (1977).
[8]R. J. Tykodi, Bull. Chem. Soc. Jpn. **52**, 564 (1979).
[9]Reference 1, Chap. 1.
[10]Reference 1, Chap. 3.
[11]G. H. Hardy, J. E. Littlewood, and G. Polya, *Inequalities*, 2nd ed. (Cambridge University, Cambridge, 1952), Chap. 3.
[12]H. Cramer, *Mathematical Methods of Statistics* (Princeton University, Princeton, 1946), Chap. 11.

[13]For some general mathematical properties of convex surfaces, see, for example, H. Busemann, *Convex Surfaces* (Interscience, New York, 1958). Note that the point $\{0\}$, i.e., $Y_i = 0$ for $i = 1, 2, \ldots, n$ represents a state of ordinary thermodynamic equilibrium, the thermodynamic reference state σ in fact. In the reference state σ, not only are all the currents zero ($Y_i = 0$) but all the affinities are zero as well ($\Omega_i = 0$ for $i = 1, 2, \ldots, n$): The values of the affinities reflect the differences in intensive thermodynamic quantities between terminal-part (Ref. 1) intensive variables and the corresponding reference state (σ) intensive variables; when all the currents vanish, all the differences in intensive variables vanish also. In a suitable neighborhood of the origin $\{0\}$, the affinities are expressible as linear functions of the currents $\Omega_i = \sum_k K_{ik}(0) Y_k$, and Θ has the form $\Theta = \sum_{i,k} Y_i K_{ik}(0) Y_k$—see the section entitled *The Quadratic Form* $D(Y_i)$.

[14]P. Glansdorff and I. Prigogine, Physica **30**, 351 (1964).
[15]I. Prigogine and P. Glansdorff, Physica **31**, 1242 (1965).
[16]P. Glansdorff and I. Prigogine, *Thermodynamic Theory of Structure, Stability, and Fluctuations* (Wiley–Interscience, London, 1971).

J. Non-Equilib. Thermodyn.
Vol. 16 (1991), pages 267–279

Offprint

Thermodynamics of Steady States: Definitions, Relations and Conjectures

R. J. Tykodi
Dept. of Chemistry, University of Massachusetts Dartmouth,
North Dartmouth, MA, U.S.A.

Registration Number 534

Abstract

A steady-state situation has associated with it a set of steady currents Y_i, a set of affinities Ω_i, a rate of entropy production $\Theta = \Sigma_i Y_i \Omega_i$, sets of phenomenological coefficients L_{ik}, K_{ik} for which $L_{ik} \equiv (\partial Y_i / \partial \Omega_k)_{\Omega',\sigma}$, $K_{ik} \equiv (\partial \Omega_i / Y_k)_{Y',\sigma}$, and auxiliary dissipation functions B, D, and θ for which $B \equiv \Sigma_{i,j} \Omega_i L_{ij} \Omega_j$, $D \equiv \Sigma_{i,j} Y_i K_{ij} Y_j$, and $\theta \equiv \Sigma_{i,j,k} \Omega_i L_{ik} Y_j K_{jk}$. Some identities connecting the quantities Y_i, Ω_i, L_{ik}, K_{ik}, Θ, B, D, θ are displayed; and some additional, non-identical, relations among the same quantities are conjectured. The conjectured relations are tested for a set of heat-conduction examples with non-linear current-affinity relations. A prior conjecture, the sandwich conjecture, is shown to fail for one of the heat-conduction examples.

1. Introduction

Continuing my studies [1–9] of the thermodynamics of steady-state systems, *I* present some general relations (identities) for steady-state systems, and *I* propose as conjectures some additional relations that seem to hold at least in the case of some simple heat-conduction situations having non-linear current-affinity relations.

In my book-length study [1] of the thermodynamics of steady states, *I* studied systems that were entirely resolvable into terminal parts and gradient parts [10]. For steady-state situations involving such systems, the rate of entropy production $\Theta \equiv \dot{S}(\text{system}) + \dot{S}(\text{surroundings})$ can always be expressed in the form

$$\Theta = \Sigma_i Y_i \Omega_i \geq 0\,, \tag{1}$$

where the currents Y_i are (essentially) time derivatives of extensive thermodynamic quantities [11] ($\dot{U}^{(r)}$ or $\dot{Q}^{(r)}$, $\dot{V}$, $\dot{n}^{(j)}$, etc.) and the affinities Ω_i are defined via equation (1). If we produce a given steady state by starting from a reference thermodynamic (equilibrium) state σ and then adjusting some of the boundary conditions so as to induce the steady currents Y_i, we find that the resulting steady state is fully determined by the properties of the equilibrium state σ and either the set of steady currents Y_i or the set of affinities Ω_i.

2. Definitions

I here define a number of quantities that are of use in treating steady-state situations.

$$L_{ik} \equiv (\partial Y_i / \partial \Omega_k)_{\Omega', \sigma}, \quad K_{ik} \equiv (\partial \Omega_i / \partial Y_k)_{Y', \sigma}, \tag{2}$$

where the subscript $\Omega'(Y')$ means constant $\Omega_j(Y_j)$ for $j \neq k$;

$$y_k \equiv \Sigma_i \Omega_i L_{ik}, \quad \omega_k \equiv \Sigma_i Y_i K_{ik}; \tag{3}$$

$$l_{ik} \equiv (\partial y_i / \partial \Omega_k)_{\Omega', \sigma}, \quad k_{ik} \equiv (\partial \omega_i / \partial Y_k)_{Y', \sigma}; \tag{4}$$

$$B \equiv \Sigma_i y_i \Omega_i = \Sigma_{i,k} \Omega_i L_{ik} \Omega_k, \quad D \equiv \Sigma_i Y_i \omega_i = \Sigma_{i,k} Y_i K_{ik} Y_k,$$

$$\theta \equiv \Sigma_i y_i \omega_i = \Sigma_{i,j,k} \Omega_i L_{ik} Y_j K_{jk}. \tag{5}$$

The quantities L_{ik} and K_{ik} are the phenomenological coefficients much talked about in discussions of the thermodynamics of nonequilibrium situations [1]. I call the $y_k(\omega_k)$ quasi-currents (quasi-affinities). I call the l_{ik} and k_{ik} quasi-coefficients. I call B, D, and θ auxiliary dissipation functions, and I sometimes refer to Θ as the prime dissipation function.

3. General relations

There are a number of useful and interesting relations that are direct consequences of the above definitions.

$$(\partial \Theta / \partial \Omega_k)_{\Omega', \sigma} = Y_k + y_k, \quad (\partial \Theta / \partial Y_k)_{Y', \sigma} = \Omega_k + \omega_k; \tag{6}$$

$$\Sigma_i \Omega_i (\partial \Theta / \partial \Omega_i)_{\Omega', \sigma} = \Theta + B, \quad \Sigma_i Y_i (\partial \Theta / \partial Y_i)_{Y', \sigma} = \Theta + D; \tag{7}$$

$$\Sigma_i [(\partial \Theta / \partial Y_i)_{Y', \sigma} - \Omega_i][(\partial \Theta / \partial \Omega_i)_{\Omega', \sigma} - Y_i] = \Sigma_i \omega_i y_i = \theta; \tag{8}$$

$$\Sigma_i (\partial \Theta / \partial Y_i)_{Y', \sigma} (\partial \Theta / \partial \Omega_i)_{\Omega', \sigma} = \Theta + B + D + \theta; \tag{9}$$

$$L_{ik} + l_{ik} = L_{ki} + l_{ki}, \quad K_{ik} + k_{ik} = K_{ki} + k_{ki}; \tag{10}$$

$$\Sigma_i y_i d\Omega_i = \Sigma_i \Omega_i dY_i, \quad \Sigma_i \omega_i dY_i = \Sigma_i Y_i d\Omega_i; \tag{11}$$

$$Y_k = \Sigma_i \omega_i L_{ik}, \quad \Omega_k = \Sigma_i y_i K_{ik}; \tag{12}$$

$$\Sigma_i L_{ki} K_{ij} = \delta_{kj}, \quad \Sigma_i L_{ik} K_{ji} = \delta_{kj}, \tag{13}$$

where δ_{kj} is the Kronecker delta, i.e. $\delta_{kj}=0$ for $k \neq j$ and $\delta_{kj}=1$ for $k=j$.

Relations (10) are consequences of the observation that $\partial^2\Theta/\partial Y_k \partial Y_i = \partial^2\Theta/\partial Y_i \partial Y_k$ and $\partial^2\Theta/\partial\Omega_k\partial\Omega_i = \partial^2\Theta/\partial\Omega_i\partial\Omega_k$. Relations (11) follow from the observation that $d\Theta = \Sigma_i(\partial\Theta/\partial\Omega_i)_{\Omega',\sigma}d\Omega_i = \Sigma_i(Y_i+y_i)d\Omega_i$, $d\Theta = \Sigma_i Y_i d\Omega_i + \Sigma_i \Omega_i dY_i$, $d\Theta = \Sigma_i(\partial\Theta/\partial Y_i)_{Y',\sigma}dY_i = \Sigma_i(\Omega_i+\omega_i)dY_i$. Relations (12) follow from relations (11), and relations (13) follow from equations (12) and (3).

Relations (13) can be expressed in matrix notation. Let L be the matrix of the coefficients L_{ik}, i.e. $L=[L_{ik}]$; let K be the matrix of the coefficents K_{ik}, i.e. $K=[K_{ik}]$; and let I be the identity matrix, i.e. $I=[\delta_{ik}]$. Then

$$LK = KL = I. \tag{14}$$

Each matrix is the inverse of the other: $K = L^{-1}$ and $L = K^{-1}$.

Relations (13) *have some interesting consequences. Note that*

$$\begin{aligned} 1 &= \Sigma_i L_{ki}K_{ik} = L_{kk}K_{kk} + \sum_{i \neq k} L_{ki}K_{ik}, \\ 1 &= \Sigma_i L_{ik}K_{ki} = L_{kk}K_{kk} + \sum_{i \neq k} L_{ik}K_{ki} \end{aligned} \tag{15}$$

and, consequently, that

$$\sum_{i \neq k} L_{ki}K_{ik} = \sum_{i \neq k} L_{ik}K_{ki}. \tag{16}$$

For a 2-current steady-state situation, equation (16) gives

$$L_{21}K_{12} = L_{12}K_{21}. \tag{17}$$

For a 3-current situation, (16) gives several relations of the type

$$L_{12}K_{21} - L_{21}K_{12} + L_{13}K_{31} - L_{31}K_{13} = 0 \quad \text{(etc.)}. \tag{18}$$

For a 4-current situation, (16) gives several relations of the type

$$L_{12}K_{21} - L_{21}K_{12} + L_{13}K_{31} - L_{31}K_{13} + L_{14}K_{41} - L_{41}K_{14} = 0 \quad \text{(etc.)}. \tag{19}$$

4. Linear current-affinity neighborhood of σ

Let $\{Y_i\}$ represent the set of currents Y_i, let $\{\Omega_i\}$ represent the set of affinities Ω_i, and let $\{0\}$ be the set with zero elements.

In a neighborhood of the equilibrium state σ, the currents can be expressed as linear functions of the affinities and vice versa:

$$Y_k = \Sigma_i \Omega_i L_{ik}(0), \quad \Omega_k = \Sigma_i Y_i K_{ik}(0), \tag{20}$$

where

$$L_{ik}(0) \equiv (\partial Y_i / \partial \Omega_k)_{\Omega', \sigma} |_{\{\Omega_j\} = \{0\}} \quad \text{and}$$
$$K_{ik}(0) \equiv (\partial \Omega_i / \partial Y_k)_{Y', \sigma} |_{\{Y_j\} = \{0\}} . \tag{21}$$

In a linear current-affinity neighborhood of the equilibrium state σ, the following relations hold for ordinary (as opposed to special) steady-state situations [1]:

$$L_{ik}(0) = L_{ki}(0), \quad K_{ik}(0) = K_{ki}(0); \tag{22}$$

$$y_k = Y_k, \ \omega_k = \Omega_k, \ l_{ik} = L_{ik}, \ k_{ik} = K_{ik}; \tag{23}$$

$$B = D = \theta = \Theta; \tag{24}$$

$$(\delta \Theta / \tilde{\delta} Y_k)_{\Omega', \sigma} = 0 \qquad k = 1, 2, \ldots, \tag{25}$$

where $(\delta / \tilde{\delta} Y_k)_{\Omega', \sigma}$ means $(\partial / \partial Y_k)_{\Omega', \sigma} |_{Y_k = 0}$. Equations (22) are the well-known Onsager reciprocal relations [1].

5. Conjectures

There are a few plausible relations that might turn out to be of general validity. In the next section *I* shall test these conjectures for a set of simple heat-conduction examples characterized by non-linear current-affinity relations.

$$\frac{1}{2}(B + D) \geq \Theta; \tag{26}$$

$$L_{ki} K_{ik} = L_{ik} K_{ki} \qquad i, k = 1, 2, \ldots \text{ (any number of currents)}. \tag{27}$$

Relation (27) is necessarily true for 2-current situations; *I* conjecture that it continues to be true when the number of currents is expanded to 3, 4, 5, ...

As a generalization of equation (25) to the non-linear current-affinity region, *I* conjecture that there always exists a (possibly several) master dissipation function χ, depending on the prime dissipation function Θ and the auxiliary dissipation functions B, D, and θ, such that

$$\chi = \chi(\Theta, B, D, \theta, \sigma), \tag{28}$$

Θ, B, D and θ entering in suitably symmetric fashion into the function χ, and

$$(\delta \chi / \tilde{\delta} Y_k)_{\Omega', \sigma} = 0 \qquad k = 1, 2, \ldots \tag{29}$$

The master dissipation function χ is to be an extremum at each of the places where one of the currents Y_k vanishes. It should also be the case that $\chi \to \Theta$ as B, D, and θ each, simultaneously, $\to \Theta$.

6. A set of heat-conduction examples

Consider next a series of heat-conduction examples with non-linear current-affinity relations. Place heat reservoirs at the corners of an equilateral triangle and connect the reservoirs to each other by straight-line wire segments, the wires being insulated against lateral heat losses. Let λ_{ij} = thermal conductivity × area / length for the wire segment connecting reservoirs i and j, and let the thermal conductivity of each wire segment be independent of temperature. Then, for the triangular arrangement of heat reservoirs, we have:

$$\begin{aligned}
\dot{Q}_1^{(r)} &= \lambda_{12}(T_2 - T_1) + \lambda_{13}(T_3 - T_1)\,, \\
\dot{Q}_2^{(r)} &= \lambda_{12}(T_1 - T_2) + \lambda_{23}(T_3 - T_2)\,, \\
\dot{Q}_3^{(r)} &= \lambda_{13}(T_1 - T_3) + \lambda_{23}(T_2 - T_3)\,;
\end{aligned} \tag{30}$$

$$\dot{Q}_1^{(r)} + \dot{Q}_2^{(r)} + \dot{Q}_3^{(r)} = 0\,; \tag{31}$$

$$\begin{aligned}
\Theta = \frac{\dot{Q}_1^{(r)}}{T_1} + \frac{\dot{Q}_2^{(r)}}{T_2} + \frac{\dot{Q}_3^{(r)}}{T_3} &= \dot{Q}_1^{(r)}\left(\frac{1}{T_1} - \frac{1}{T_3}\right) + \\
\dot{Q}_2^{(r)}\left(\frac{1}{T_2} - \frac{1}{T_3}\right) &= Y_1\Omega_1 + Y_2\Omega_2\,,
\end{aligned} \tag{32}$$

where $\dot{Q}_i^{(r)}$ is the rate of influx of heat into the ith reservoir and [in equation (32)] T_3 has been chosen as the reference equilibrium state σ.

Rewrite equations (30) as

$$\begin{aligned}
\dot{Q}_1^{(r)} &= -(\lambda_{12} + \lambda_{13})T_1 + \lambda_{12}T_2 + \lambda_{13}T_3\,, \\
\dot{Q}_2^{(r)} &= \lambda_{12}T_1 - (\lambda_{12} + \lambda_{23})T_2 + \lambda_{23}T_3\,, \\
\dot{Q}_3^{(r)} &= \lambda_{13}T_1 + \lambda_{23}T_2 - (\lambda_{13} + \lambda_{23})T_3\,,
\end{aligned} \tag{33}$$

and rename the coefficient of T_j in the expression for $\dot{Q}_i^{(r)}$ to be a_{ij}; then

$$\dot{Q}_i^{(r)} = \Sigma_j a_{ij} T_j \tag{34}$$

and the matrix of the coefficients a_{ij} is symmetric:

$$a_{ij} = a_{ji}\,. \tag{35}$$

Let A be the determinant of the matrix $[a_{ij}]$, let A_{33} be the minor associated with element a_{33}, and let the cofactor of element a_{ij} in A_{33} be $|A_{33}|_{ij}$; then, because of equation (35),

$$|A_{33}|_{ij} = |A_{33}|_{ji} \tag{36}$$

– the two cofactors are but transposes of each other.

From the definitions of (2), we get

$$L_{ij} = \left(\frac{\partial Y_i}{\partial \Omega_j}\right)_{\Omega',\sigma} = \left(\frac{\partial \dot{Q}_i^{(r)}}{\partial \Omega_j}\right)_{T',T_3} = \frac{(\partial \dot{Q}_i^{(r)}/\partial T_j)_{T',T_3}}{(\partial \Omega_j/\partial T_j)_{T',T_3}} = -a_{ij}T_j^2 \,. \tag{37}$$

Equation (37) follows from (34) and the observation that $\Omega_j = T_j^{-1} - T_3^{-1}$ [equation (32)].

Let the determinant of the matrix L be $|L|$, and let the cofactor of element L_{ik} in $|L|$ be $|L|_{ik}$; similarly, let the determinant of the matrix K be $|K|$, and let the cofactor of element K_{ik} in $|K|$ be $|K|_{ik}$. For the case being considered, it follows that

$$|L| = A_{33}\prod_{k=1}^{2}(-T_k^2) \quad \text{and} \quad |L|_{ij} = |A_{33}|_{ij}\prod_{\substack{k=1\\k\neq j}}^{2}(-T_k^2)\,. \tag{38}$$

Since the matrices L and K are inverses of each other, it is true that

$$K_{ij} = |L|_{ji}/|L| = -|A_{33}|_{ji}/A_{33}T_i^2\,. \tag{39}$$

Writing out relations (37) and (39) explicitly gives

$$L_{12} = -a_{12}T_2^2, \quad L_{21} = -a_{21}T_1^2\,; \tag{40}$$

$$K_{12} = -|A_{33}|_{21}/A_{33}T_1^2, \quad K_{21} = -|A_{33}|_{12}/A_{33}T_2^2\,. \tag{41}$$

Note that relations (40) and (41) do satisfy the necessary relation of 2-current situations, equation (17):

$$L_{12}K_{21} = L_{21}K_{12}\,. \tag{42}$$

Relation (42) is trivially true for 2-current situations because, in that case,

$$K_{ij} = |L|_{ji}/|L| = L_{ij}/|L| \qquad (\text{for } i \neq j)\,; \tag{43}$$

and relation (42) becomes

$$L_{12}L_{21}/|L| = L_{21}L_{12}/|L|. \tag{44}$$

Note also that, except for an infinitesimal neighborhood of the reference state T_3, $(T_1 \approx T_2 \approx T_3)$,

$$L_{12} \neq L_{21} \quad \text{and} \quad K_{12} \neq K_{21}. \tag{45}$$

To produce a 3-current situation, put heat reservoirs at the corners of a regular tetrahedron and connect the reservoirs to each other by straight-line wire segments of the kind used for the 2-current situation.

To produce a 4-current situation, put heat reservoirs at the corners of a regular tetrahedron and put a fifth heat reservoir at the center of the tetrahedron; connect the reservoirs to each other by straight-line wire segments of the kind used for the 2-current situation.

For the 3-current situation we have

$$\dot{Q}_1^{(r)} = \lambda_{12}(T_2 - T_1) + \lambda_{13}(T_3 - T_1) + \lambda_{14}(T_4 - T_1), \tag{46}$$

etc.;

and for the 4-current situation we have

$$\dot{Q}_1^{(r)} = \lambda_{12}(T_2 - T_1) + \lambda_{13}(T_3 - T_1) + \lambda_{14}(T_4 - T_1) + \lambda_{15}(T_5 - T_1), \tag{47}$$

etc.

Putting the heat currents into the form of (33), we get

$$\dot{Q}_1^{(r)} = -(\lambda_{12} + \lambda_{13} + \lambda_{14})T_1 + \lambda_{12}T_2 + \lambda_{13}T_3 + \lambda_{14}T_4, \tag{48}$$

etc.;

for the 3-current situation and

$$\dot{Q}_1^{(r)} = -(\lambda_{12} + \lambda_{13} + \lambda_{14} + \lambda_{15})T_1 + \lambda_{12}T_2 + \lambda_{13}T_3 + \lambda_{14}T_4 + \lambda_{15}T_5, \tag{49}$$

etc.;

for the 4-current situation.

Let there be n heat reservoirs for an $(n-1)$-current situation, and let T_n be the reference equilibrium state. Then

$$\sum_{i=1}^{n} \dot{Q}_i^{(r)} = 0, \tag{50}$$

$$\Theta = \sum_{i=1}^{n} \dot{Q}_i^{(r)}/T_i = \sum_{i=1}^{n-1} \dot{Q}_i^{(r)}(T_i^{-1} - T_n^{-1}) = \sum_{i=1}^{n-1} Y_i \Omega_i, \tag{51}$$

$$\dot{Q}_i^{(r)} = \sum_{j=1}^{n} a_{ij} T_j, \tag{52}$$

and the matrix of the coefficients a_{ij} is symmetric in each case:

$$a_{ij} = a_{ji}. \tag{53}$$

Let A be the determinant of the matrix $[a_{ij}]$, let A_{nn} be the minor associated with element a_{nn}, and let the cofactor of element a_{ij} in A_{nn} be $|A_{nn}|_{ij}$; then

$$|A_{nn}|_{ij} = |A_{nn}|_{ji} \tag{54}$$

because of (53) – the two cofactors are but transposes of each other.

From the definitions of (2), we get [see equation (37)]

$$L_{ij} = -a_{ij} T_j^2. \tag{55}$$

We also observe that [see equations (38) and (39)]

$$|L| = A_{nn} \prod_{k=1}^{n} (-T_k^2), \quad |L|_{ij} = |A_{nn}|_{ij} \prod_{\substack{k=1 \\ k \neq j}}^{n} (-T_k^2). \tag{56}$$

and

$$K_{ij} = |L|_{ji}/|L| = -|A_{nn}|_{ji}/A_{nn} T_i^2. \tag{57}$$

Therefore, for 3-current and 4-current situations (as well as for 2-current situations) we have

$$L_{ij} K_{ji} = L_{ji} K_{ij}, \tag{58}$$

although, except for an infinitesimal neighborhood of the reference state T_n ($T_i \approx T_n, \quad i = 1, 2, \ldots, n-1$),

$$L_{ij} \neq L_{ji} \quad \text{and} \quad K_{ij} \neq K_{ji}. \tag{59}$$

It is also true that, for the heat-current examples we are considering,

$$L_{12}L_{23}L_{31} = L_{13}L_{32}L_{21} \qquad \text{(3 currents)} \tag{60}$$

and

$$L_{12}L_{23}L_{34}L_{41} = L_{14}L_{43}L_{32}L_{21} \qquad \text{(4 currents)}\,. \tag{61}$$

Special-case 2-current situation

For the 2-current situation, choose the thermal conductivity and the dimensions of the straight-line wire segments so that, in appropriate units, each $\lambda_{ij} = 1$; then

$$Y_1 = \dot{Q}_1^{(r)} = -2T_1 + T_2 + T_3, \quad \Omega_1 = T_1^{-1} - T_3^{-1}; \tag{62}$$

$$Y_2 = \dot{Q}_2^{(r)} = T_1 - 2T_2 + T_3, \quad \Omega_2 = T_2^{-1} - T_3^{-1}; \tag{63}$$

$$L_{11} = 2T_1^2, \quad L_{12} = -T_2^2, \quad K_{11} = (2/3)T_1^{-2}, \quad K_{12} = (1/3)T_1^{-2}; \tag{64}$$

$$L_{21} = -T_1^2, \quad L_{22} = 2T_2^2, \quad K_{21} = (1/3)T_2^{-2}, \quad K_{22} = (2/3)T_2^{-2}. \tag{65}$$

Let $T_1 = \alpha T_3$ and $T_2 = \beta T_3$; then

$$\Theta = \alpha + \frac{1}{\alpha} + \beta + \frac{1}{\beta} + \frac{\beta}{\alpha} + \frac{\alpha}{\beta} - 6\,, \tag{66}$$

$$B = 4 + \frac{\alpha^2}{\beta} + \frac{\beta^2}{\alpha} + \alpha^2 + \beta^2 - 3\alpha - 3\beta - \frac{\alpha}{\beta} - \frac{\beta}{\alpha}, \tag{67}$$

$$D = 4 + \frac{1}{\alpha^2} + \frac{1}{\beta^2} + \frac{\beta}{\alpha^2} + \frac{\alpha}{\beta^2} - \frac{3}{\alpha} - \frac{3}{\beta} - \frac{\beta}{\alpha} - \frac{\alpha}{\beta}, \tag{68}$$

$$\begin{aligned} \theta = {} & \alpha^2\left(\frac{2}{\alpha} - \frac{1}{\beta} - 1\right)\left[\frac{2}{3\alpha^2}(1 + \beta - 2\alpha) + \frac{1}{3\beta^2}(1 + \alpha - 2\beta)\right] \\ & + \beta^2\left(\frac{2}{\beta} - \frac{1}{\alpha} - 1\right)\left[\frac{1}{3\alpha^2}(1 + \beta - 2\alpha) + \frac{2}{3\beta^2}(1 + \alpha - 2\beta)\right], \end{aligned} \tag{69}$$

$$\begin{aligned} \Delta & \equiv \frac{1}{2}(B + D) - \Theta \\ & = 10 + \frac{1}{2}\left(\alpha^2 + \frac{1}{\alpha^2} + \beta^2 + \frac{1}{\beta^2} + \frac{\alpha^2}{\beta} + \frac{\beta}{\alpha^2} + \frac{\beta^2}{\alpha} + \frac{\alpha}{\beta^2}\right) - \\ & \qquad 2\left(\frac{\beta}{\alpha} + \frac{\alpha}{\beta}\right) - \frac{5}{2}\left(\alpha + \frac{1}{\alpha}\right) - \frac{5}{2}\left(\beta + \frac{1}{\beta}\right). \end{aligned} \tag{70}$$

Each of the relations (66)–(70) is symmetric in the variables α, β; furthermore, the following transformations hold:

$$\text{if } \alpha \to 1/\alpha \text{ and } \beta \to 1/\beta, \quad \text{then } \Theta \to \Theta \text{ and } \Delta \to \Delta\,; \tag{71}$$

$$\text{if } \alpha \to 1/\alpha \text{ and } \beta \to 1/\beta, \quad \text{then } B \to D \text{ and } D \to B\,. \tag{72}$$

For the special case $\alpha = 3$, $\beta = \frac{1}{2}$, we have $\Theta = \frac{18}{3}$, $B = \frac{44}{3}$, $D = \frac{21}{3}$; and in this case the sandwich conjecture [9] – that Θ always lies between B and D – is false.

With respect to the conjecture that $\Delta \geq 0$ [equation (27)], we note that, for constant β, $\Delta \to \infty$ as $\alpha \to 0$ or $\alpha \to \infty$. For the special case $\beta = 1$, Δ has a minimum at $\alpha = 1$ $(T_1 = T_2 = T_3)$, i.e. $\Delta(\alpha = 1, \ \beta = 1) = 0$ and $\Delta(\alpha \neq 1, \beta = 1) > 0$. For the special case $\beta = 2$, Δ has a minimum in the near neighborhood of $\alpha = 1.5225$, at which point $\Delta = +0.167$.

It seems quite likely, therefore, that conjecture (27) holds for the example being considered.

With respect to the master dissipation function χ, when looking at the structure of $(\delta\chi/\tilde{\delta} Y_1)_{\Omega_2, T_3}$, observe that

$$Y_1 = 0 \text{ when } \alpha = \frac{1}{2}(\beta + 1) \quad \left[\text{i.e. when } T_1 = \frac{1}{2}(T_2 + T_3)\right]. \tag{73}$$

Note also that

$$\left.\frac{\partial}{\partial Y_1}\right|_{Y_1 = 0} = \left(\frac{\partial Y_1}{\partial T_1}\right)^{-1} \left.\frac{\partial}{\partial T_1}\right|_{Y_1 = 0} = \frac{1}{2}\frac{1}{T_3}\left.\frac{\partial}{\partial \alpha}\right|_{\alpha = \frac{1}{2}(\beta + 1)}. \tag{74}$$

So, for $(\delta/\tilde{\delta} Y_1)_{\Omega_2, T_3}$ to be zero, it is sufficient to have $\partial/\partial\alpha|_{\alpha = \frac{1}{2}(\beta+1)}$ equal to zero:

$$0 = (\partial\chi/\partial\alpha)_\beta|_{\alpha = \frac{1}{2}(\beta + 1)} \quad \text{implies} \quad (\delta\chi/\tilde{\delta} Y_1)_{\Omega_2, T_3} = 0\,. \tag{75}$$

For the purpose of discussing possible functional forms for $\chi(\Theta, B, D, \theta, \sigma)$, let $f_z(\Theta, B, D, \theta, \sigma)$ mean $f(\ldots) \to z$ as B, D, θ each $\to \Theta$, and let $\xi = \xi(\sigma)$. Then there are two large classes of possible functional forms for χ:

$$\chi_1 = f_\Theta + \xi f_0, \quad \chi_2 = f_\Theta f_1^\xi\,. \tag{76}$$

Let $[\ldots, \ldots, \ldots]$ represent a symmetric combination of the displayed quantities. Then my own inclinations for symmetry constraints on the functions f_Θ, f_0, and f_1 are:

$$f_\Theta = \Theta \quad \text{or} \quad f_\Theta([\Theta, B, D, \theta]) \quad \text{or} \quad f_\Theta([B, D], [\Theta, \theta])\,, \tag{77}$$

$$f_0 = f_0([B, D], [\Theta, \theta]), \tag{78}$$

$$f_1 = f_1([B, D], [\Theta, \theta]). \tag{79}$$

Some possible candidate functions $f(\ldots)$ that I have considered are:

$$f_\Theta\colon\ \Theta, \frac{1}{4}(\Theta + B + D + \theta), (\Theta BD\theta)^{1/4}; \tag{80}$$

$$f_0\colon\ BD - \Theta\theta,\ B + D - (\Theta + \theta), \frac{B}{D} + \frac{D}{B} - \frac{\Theta}{\theta} - \frac{\theta}{\Theta}, \ln f_1; \tag{81}$$

$$f_1\colon\ \frac{BD}{\Theta\theta}, \frac{B+D}{\Theta+\theta}, \frac{BD^{-1}+DB^{-1}}{\Theta\theta^{-1}+\theta\Theta^{-1}}, \frac{1}{4}\left(\frac{B}{D} + \frac{D}{B} + \frac{\Theta}{\theta} + \frac{\theta}{\Theta}\right), e^{f_0}. \tag{82}$$

One way of testing the forms (76) is to select appropriate candidate functions $f(\ldots)$ and to evaluate ξ, using the relation $0 = (\partial\chi/\partial\alpha)_\beta|_{\alpha=\frac{1}{2}(\beta+1)}$, for several values of β: if ξ remains the same for all values of β, the tested form is a true master dissipation function for the system analyzed. I tested many combinations of the candidate functions listed in (80)–(82) for the values $\beta = 2$ and $\beta = 3$; none of the forms passed the test – the one that came closest to giving satisfaction was

$$\chi_1 = \Theta + \xi(B + D - \Theta - \theta),\ \xi(\beta = 2) = 0.6267,\ \xi(\beta = 3) = 0.6467. \tag{83}$$

I have, for now, suspended my search for a master dissipation function.

7. Conclusion

The thermodynamics of steady states deals with interrelations among currents Y_i, affinities Ω_i, phenomenological coefficients L_{ik}, K_{ik}, the rate of entropy production Θ, and auxiliary dissipation functions B, D, θ. I displayed some of the many identities that these quantities satisfy and listed some conjectured relations. I showed that for a class of heat-conduction examples having non-linear current-affinity relations the conjectured relations are valid, and I explored the possibility of finding a master dissipation function for one of the examples – the results thus far obtained neither prove nor disprove the existence of a master dissipation function for the case in question. I also showed, by means of a counter-example, that the sandwich conjecture is false.

8. Appendix

I cite here some additional observations pertaining to the special-case 2-current situation.

Note that for relations (66)–(69), at constant β, as $\alpha \to \infty$ we have

$$\Theta \to \alpha\left(1+\frac{1}{\beta}\right) \to \infty\,, \tag{84}$$

$$B \to \alpha\left(\frac{\alpha}{\beta}+\alpha-3-\frac{1}{\beta}\right) \to \infty\,, \tag{85}$$

$$D \to \frac{\alpha}{\beta}\left(\frac{1}{\beta}-1\right) \to \pm\infty \quad (\beta \neq 1)\,, \tag{86}$$

$$\theta \to \frac{\alpha}{3\beta}\left[4-2\beta-\frac{(1+\beta)\alpha^2}{\beta^2}\right] \to -\infty\,; \tag{87}$$

and as $\alpha \to 0$ the limiting relations hold

$$\Theta \to \frac{1}{\alpha}(1+\beta) \to \infty\,, \tag{88}$$

$$B \to \frac{\beta}{\alpha}(\beta-1) \to \pm\infty \quad (\beta \neq 1)\,, \tag{89}$$

$$D \to \frac{1}{\alpha}\left(\frac{1}{\alpha}+\frac{\beta}{\alpha}-3-\beta\right) \to \infty\,, \tag{90}$$

$$\theta \to \frac{1}{3\alpha}\left\{4(1+\beta)-\beta^2\left[\frac{1+\beta}{\alpha^2}+\frac{2}{\beta^2}(1-2\beta)\right]\right\} \to -\infty\,. \tag{91}$$

For each of the auxiliary functions B, D, and θ, therefore, there are values of α and β for which the function is negative.

Whereas the old sandwich conjecture,

$$\max(B, D) \geq \Theta \geq \min(B, D)\,, \tag{92}$$

turned out not to be generally valid, *I* now propose a new sandwich conjecture:

$$\frac{1}{2}(B+D) \geq \Theta \geq \theta\,. \tag{93}$$

Although the functions B, D, and θ individually can change sign, it appears that, for the model under investigation, the combinations $B+D$ and $\Theta-\theta$ are of one sign only:

$$B+D \geq 0 \quad \text{and} \quad \Theta-\theta \geq 0\,. \tag{94}$$

Relations (94) suggest additional forms for the master dissipation function χ:

$$\chi_2 = \frac{1}{2}(B+D) + \xi(\Theta - \theta), \quad \xi(2) = 0.6415, \quad \xi(3) = 0.6207; \tag{95}$$

$$\chi_3 = \frac{1}{2}\left[\Theta + \frac{1}{2}(B+D)\right] + \xi\left[\frac{1}{2}(B+D) - \theta\right], \quad \xi(2) = 0.6336, \quad \xi(3) = 0.6324, \tag{96}$$

where $\chi_3 = \frac{1}{2}(\chi_1 + \chi_2)$.

Results (95) and (96) suggest that perhaps the symmetry constraints (77)–(79) on the functions f_Θ, f_0, and f_1 should be relaxed.

References

[1] Tykodi, R. J., Thermodynamics of Steady States, Macmillan, New York, 1967.
[2] Tykodi, R. J., On the Thermocouple in a Magnetic Field, J. Chem. Phys., 47 (1967), 1879.
[3] Tykodi, R. J., The Porous Plug Experiment – A Lot of Mileage on a Little Gas, Am. J. Phys., 38 (1970), 586.
[4] Tykodi, R. J., Non-equilibrium Thermodynamics: Steady State Thermodynamic Relations in the Non-linear Current-Affinity Region, Bull. Chem. Soc. Jpn., 44 (1971), 1001.
[5] Tykodi, R. J., Nonequilibrium Thermodynamics: Current-Affinity Relations and Thermokinetic Potentials, J. Chem. Phys., 57 (1972), 37.
[6] Tykodi, R. J., Thermodynamics of Steady States: A Weak Entropy-Production Principle, Physica, 72 (1974), 341.
[7] Tykodi, R. J., Thermodynamics of Steady States: Additional Non-Equilibrium Effects, J. Non-Equilib. Thermodyn., 2 (1977), 193.
[8] Tykodi, R. J., Thermodynamics of Steady States: "Resistance Change" Transitions in Steady-state Systems, Bull. Chem. Soc. Jpn., 52 (1979), 564.
[9] Tykodi, R. J., Thermodynamics of Steady States: Is the Entropy-Production Surface Convex in the Thermodynamic Space of Steady Currents? J. Chem. Phys., 80 (1984), 1652.
[10] Reference [1], Chap. 1.
[11] Reference [1], Chap. 3.

Paper received: 1991-6-14
Paper accepted: 1991-6-21

Professor Ralph J. Tykodi
Department of Chemistry
University of Massachusetts Dartmouth
North Dartmouth, Massachusetts 02747, USA

SCIENTIFIC PUBLICATIONS OF RALPH J. TYKODI

1. S.V.R. Mastrangelo, R.J. Tykodi, and J.G. Aston, Preparation of Highly Compressed Samples for Adsorption Studies
   J. Am. Chem. Soc. **75**, 5430 (1953).
2. R.J. Tykodi, Thermodynamics of Adsorption
   J. Chem. Phys. **22**, 1647 (1954).
3. R.J. Tykodi, On First Order Transitions in Adsorption Systems
   J. Phys. Chem. **59**, 383 (1955).
4. R.J. Tykodi, J.G. Aston, and G.D.L. Schreiner, Thermodynamics of Neon Adsorbed on Titanium Dioxide
   J. Am. Chem. Soc. **77**, 2168 (1955).
5. J.G. Aston, R.J. Tykodi, and W.A. Steele, Tests of a Simplified General Model for the Adsorption of Rare Gases on Solids
   J. Phys. Chem. **59**, 1053 (1955).
6. J.G. Aston, S.V.R. Mastrangelo, and R.J. Tykodi, Adsorption of Helium on Titanium Dioxide
   J. Chem. Phys. **23**, 1633 (1955).
7. R.J. Tykodi, Adsorption at Low Coverage
   Trans. Faraday Soc. **54**, 918 (1958).
8. R.J. Tykodi, Thermodynamics, Stationary States, and Steady-Rate Processes. I. Introduction and Migrational Equilibrium in Isothermal and Monothermal Fields
   J. Chem. Phys. **31**, 1506 (1959).
9. R.J. Tykodi and T.A. Erikson, Thermodynamics, Stationary States, and Steady-Rate Processes. II. Migrational Equilibrium in Thermal Fields
   J. Chem. Phys. **31**, 1510 (1959).
10. R.J. Tykodi, Thermodynamics, Stationary States, and Steady-Rate Processes. III. The Thermocouple Revisited
    J. Chem. Phys. **31**, 1517 (1959).
11. R.J. Tykodi and T.A. Erikson, IV. Steady Monothermal Rate Processes
    J. Chem. Phys. **31**, 1521 (1959).
12. R.J. Tykodi, Thermodynamics, Stationary States, and Steady-Rate Processes. V. A Collection of Corollaies
    J. Chem. Phys. **33**, 40 (1960).
13. R.J. Tykodi and T.A. Erikson, Thermodynamics, Stationary States, and Steady-Rate Processes. VI. Forced Vaporization of Selected Liquids
    J. Chem. Phys. **33**, 46 (1960).
14. R.J. Tykodi, Irreversible Thermodynamics
    Chem. Eng. **68**, 233 (Nov. 1961).

15. R.J. Tykodi, A Toy Model for Multilayer Adsorption
Z. fur phys. Chem., N.F., **43**, 260 (1964).
16. R.J. Tykodi, Monolayer Adsorption with Nearest Neighbor Interactions
Bull. Chem. Soc. Japan **39**, 1627 (1966).
17. R.J. Tykodi, *Thermodynamics of Steady States* (Macmillan, New York, 1967).
18. R.J. Tykodi, Thermodynamics and Classical Relativity
Am. J. Phys. **35**, 250 (1967).
19. R.J. Tykodi, On the Thermocouple in a Magnetic Field
J. Chem. Phys. **47**, 1879 (1967).
20. R.J. Tykodi, Thermodynamics - Thermotics
Ind. Eng. Chem. **60**(11), 22 (1968).
21. T.A. Erikson and R.J. Tykodi, Migrational Properties for the Steady Forced Vaporization of Water
J. of Heat Transfer, Trans, ASME, Vol. **91**, Series C, Number 2, page 221.
22. R.J. Tykodi, The Porous Plug Experiment - A Lot of Mileage on a Little Gas
Am. J. Phys. **38**, 586 (1970).
23. R.J. Tykodi, Non-Equilibrium Thermodynamics: Steady State Thermodynamic Relations in the Non-Linear Current - Affinity Region
Bull. Chem. Soc. Japan **44**, 1001 (1971).
24. R.J. Tykodi, A Global Thermodynamic Inequality
J. Stat. Physics **4**, 417 (1971).
25. R.J. Tykodi, Non-Equilibrium Thermodynamics: Current-Affinity Relations and Thermokinetic Potentials
J. Chem. Phys. **57**, 37 (1972).
26. R.J. Tykodi, Statistical Thermodynamics and Adsorption
Bull. Chem. Soc. Japan **45**, 2993 (1972).
27. R.J. Tykodi and E.P. Hummel, On the Equation of State for Gases
Am. J. Phys. **41**, 340 (1973).
28. R.J. Tykodi, Thermodynamics of Steady States: A Weak Entropy-Production Principle
Physica **72**, 341 (1974).
29. R.J. Tykodi, The Equation of State for Gases
Am. J. Phys. **42**, 794 (1974).
30. R.J. Tykodi, editor, *A Taste of Science* (Technomic Pub. Co., Westport, Conn., 1975).
31. R.J. Tykodi, Negative Kelvin Temperatures: Some Anomalies and a Speculation
Am. J. Phys. **43**, 271 (1975).
32. R.J. Tykodi, A Thermodynamic Derivation of Sackur's Equation for Osmotic Pressure
Z. fur phys. Chem. N.F. **96**, 9 (1975).
33. R.J. Tykodi, Refereeing Technical Articles
Chem. and Eng. News, May 10, 1976, page 3.
34. R.J. Tykodi, Negative Kelvin Temperatures
Am. J. Phys. **44**, 997 (1976).
35. R.J. Tykodi, Thermodynamics of Steady States: Additional Non-Equilibrium Effects
J. Non-Equilib. Thermodyn. **2**, 193 (1977).
36. R.J. Tykodi, Quasi-Carnot Cycles, Negative Kelvin Temperatures, and the Laws of Thermodynamics
Am. J. Phys. **46**, 354 (1978).

37. R.J. Tykodi, Thermodynamics of Steady States: "Resistance Change" Transitions in Steady-State Systems
Bull. Chem. Soc. Japan **52**, 564 (1979).
38. P.T. Landsberg, R.J. Tykodi, and A.-M. Tremblay, Systematics of Carnot Cycles at positive and negative Kelvin temperatures
J. Phys. A. Math. Gen. **13**, 1063 (1980).
39. R.J. Tykodi, On Euler's Theorem for Homogeneous Functions and Proofs Thereof
J. Chem. Educ. **59**, 557 (1982).
40. R.J. Tykodi, What Do We Measure in Moles?
J. Chem. Educ. **60**, 782 (1983).
41. R.J. Tykodi, Thermodynamics of steady states: Is the entropy-production surface convex in the thermodynamic space of steady currents?
J. Chem. Phys. **80**, 1652 (1984).
42. R.J. Tykodi, Toward a more rational terminology
J. Chem. Educ. **62**, 241 (1985).
43. R.J. Tykodi, Thermodynamics and Reactions in the Dry Way
J. Chem. Educ. **63**, 107 (1986).
44. R.J. Tykodi, Periodic table
Chem. and Eng. News, Vol. **64**, No. 15, 14 April 1986, page 3.
45. R.J. Tykodi, A Better Way of Dealing with Chemical Equilibria
J. Chem. Educ. **63**, 582 (1986).
46. R.J. Tykodi, Annotating Reaction Equations
J. Chem. Educ. **64**, 243 (1987).
47. R.J. Tykodi, Throw the Crutches Away! Let the Students walk!
J. Chem. Educ. **64**, 884 (1987).
48. R.J. Tykodi, Reaction Stoichiometry
J. Chem. Educ. **64**, 958 (1987).
49. R.J. Tykodi, The Ground State Electronic Structure for Atoms and Monatomic Ions
J. Chem. Educ. **64**, 943 (1987). [Erratum: J. Chem. Educ. **65**, 237 (1988)]
50. R.J. Tykodi, Suggestions for IUPAC
Chem. and Eng. News, Vol. **66**, No. 23 , 6 June 1988, page 3.
51. R.J. Tykodi, Estimated Thermochemical Properties of Some Noble-Gas Monoxides and Difluorides
J. Chem. Educ. **65**, 981 (1988).
52. R.J. Tykodi, A Thermochemical Note on the Bonding in Metallic Crystals
J. Chem. Educ. **66**, 306 (1989).
53. R.J. Tykodi, Identifying Polar and Nonpolar Molecules
J. Chem. Educ. **66**, 1007 (1989).
54. R.J. Tykodi, Amending the IUPAC Green Book
J. Chem. Educ. **66**, 1064 (1989).
55. R.J. Tykodi, In Praise of Thiosulfate
J. Chem. Educ. **67**, 146 (1990).
56. R.J. Tykodi, Equilibrium constant symbol
Chemistry International **12**, 48 (1990).
57. R.J. Tykodi, "An Instructive Gibbs-Function Problem" Revisited
J. Chem. Educ. **67**, 383 (1990).

58. R.J. Tykodi, A Catalog of Reactions for General Chemistry
J. Chem. Educ. **67**, 665 (1990).
59. R.J. Tykodi, In Praise of Copper
J. Chem. Educ. **68**, 106 (1991)
60. R.J. Tykodi, Thermodynamics of an Incompressible Solid and a Thermodynamic Functional Determinant
J. Chem. Educ. **68**, 830 (1991).
61. R.J. Tykodi, Thermodynamics of Steady States: Definitions, Relations, and Conjectures
J. Non-Equilib. Thermodyn. **16**, 267 (1991).
62. R.J. Tykodi, On Using Incomplete Theories as Cataloging Schemes: Aufbau, Abbau, and VSEPR
J. Chem. Educ. **71**, 273 (1994).
63. R.J. Tykodi, Student Grades: Factoring In Optional Supplementary work
J. Chem. Educ. **71**, 518 (1994).
64. R.J. Tykodi, Spontaneity, Accessibility, Irreversibility, "Useful Work"
J. Chem. Educ. **72**, 103 (1995).
65. R.J. Tykodi, The Gibbs Function, Spontaneity, and Walls
J. Chem. Educ. **73**, 398 (1996). [Erratum: J. Chem. Educ. **75**, 258 (1998)]
66. R.J. Tykodi, Using Model Systems to Demonsttrate Instances of Mathematical Inequalities
Am. J. Phys. **64**, 644 (1996).
67. R.J. Tykodi, An Exceptional Theoretical Process
J. Chem. Educ. **74**, 286 (1997).
68. R.J. Tykodi, *Thermodynamics of Systems in Nonequilibrium States* (Thinkers' Press, Davenport, Iowa, 2002).

NAME INDEX

SUBJECT INDEX

IN-PRESS ADDENDUM TO BOOK I, SECTION 14

While the book was in the process of publication, I developed some additional material for systems subject to sonic pulses; and I decided to include that material as an Addendum at the end of the book. *N.B* The material in this Addendum is not covered by the general index.

Transverse Pulses on a Stretched Filament

In a manner analogous to that of the last article in Section 14 of Book I, I wish to consider a stretched filament (rubber band, metal wire, etc.) of length L which is being traversed by a train of transverse pulses, the pulses traveling along the filament at the speed c_{tr}. To make the analysis interesting, let's put the stretched filament into an ideal gas medium of temperature T, pressure p, and let the filament be capable of sorbing (adsorbing and/or absorbing) the gas. Let's use the same sort of rod-and-bob pulse generators as we used in the article in Section 14 to keep the stretched filament in a (time-average) steady state in which there are always $\dot{N}Lc_{tr}^{-1}$ pulses traversing the filament, where $\dot{N}$ is the frequency with which the pulse generators introduce pulses onto the filament. Let the pulses propagate in a horizontal plane, parallel to the bench top on which the apparatus (a cylinder containing the ideal gas, the filament stretched between the two end faces, and the pulse generators) rests. Let the rod-and-bob pulse generators swing in a plane to which the line of the stretched (but otherwise undisturbed) filament is perpendicular. Let the last bit of the arc traced out by the swing of the bob just as it strikes the filament be tangent to the plane of propagation of the pulses along the filament.

Let the pulse generators on the left-hand side of the filament all start at a height h_L above the plane of propagation of the pulses along the filament, and let each bob (of mass m) rebound to a height h'_L above that plane of propagation. The energy imparted to the filament via a pulse from a generator is ε:

$$\varepsilon \equiv -mg\left(h'_L - h_L\right) \equiv -mg\Delta_L h, \tag{1}$$

just as in the previous article. Let the left-hand pulse generators feed energy to the stretched filament, and let the right-hand generators remove the same amount of energy from the filament, so that the filament is in a steady (time average) state. The bob of each right-hand side generator rests against the stretched filament, so $h_R = 0$. A traveling pulse kicks a right-side bob (of mass m) to a height h'_R above the plane of propagation of the pulses. The energy added to the filament (the energy is actually removed from the filament) by a right-hand side generator is

$$-mg\Delta_R h = -mg\left(h'_R - h_R\right) = -mg\left(h'_R - 0\right) = -mgh'_R < 0. \tag{2}$$

For the situation as described,

$$\Delta_L h + \Delta_R h = 0; \tag{3}$$

and the number of pulses traveling along the filament in the steady state is

$$\begin{gathered}\text{\# of traveling pulses on the} \\ \text{filament in the steady state} = \dot{N}Lc_{tr}^{-1}.\end{gathered} \tag{4}$$

Let there be a piston in a small side chamber to the main chamber by means of which we can vary the pressure in the gas. Let the entire apparatus (diathermic walls) reside in an air thermostat of temperature T. Let the amount n_g of free gas be in migrational equilibrium with the amount n_s of the sorbed gas, and let us examine the effect of the pulsed field on the gas state–sorbed state equilibrium.

Start in a steady state with $\dot{N}Lc_{tr}^{-1}$ pulses traversing the stretched filament, gas pressure p, temperature T. Increase the gas pressure to $p+dp$, and in differentially tailored fashion change the amounts of gas and sorbed material by dn_g, dn_s, with

$$dn_g + dn_s = 0. \tag{5}$$

The first law for the changes just described says that

$$\overline{U}_g dn_g + \overline{U}_s dn_s = dQ + p\overline{V}_g dn_g + d(\varepsilon \dot{N}Lc_{tr}^{-1}) \tag{6}$$

where $\overline{U}_s$ is the partial molar internal energy of the sorbed material, and dQ is the heat supplied to the sytem by the air thermostat. In this case, we have to do our best to evaluate dQ directly. Upon rearranging the terms of Eq.(6) and making use of Eq.(5), we have

$$\left(\overline{H}_s - \overline{H}_g\right) - d\left(\varepsilon \dot{N}Lc_{tr}^{-1}\right) / dn_s = dQ / dn_s, \tag{7}$$

where $\overline{H}_s \equiv \overline{U}_s + p\overline{V}_s \approx \overline{U}_s$.

For the entropy change of the system, we have

$$dS / dn_s = \overline{S}_s - \overline{S}_g; \tag{8}$$

for the entropy change of the surroundings attributable to the process, we have

$$\frac{d_{\uparrow}S_{sur}}{dn_s} = -\frac{1}{T}\frac{dQ}{dn_s} = -\frac{1}{T}\left(\overline{H}_s - \overline{H}_g\right) + \frac{1}{T}\frac{d\left(\varepsilon \dot{N}Lc_{tr}^{-1}\right)}{dn_s}. \tag{9}$$

Then,

$$\frac{d\Gamma_{\uparrow}}{dn_s} = \overline{S}_s - \overline{S}_g - \frac{1}{T}\left(\overline{H}_s - \overline{H}_g - \frac{d\left(\varepsilon \dot{N}Lc_{tr}^{-1}\right)}{dn_s}\right); \tag{10}$$

and the Thomson-Gibbs Corollary gives, as the condition for migrational equilibrium between the gas state and the sorbed state in the steady-state pulsed field,

$$0 = \mu_s - \mu_g - \frac{d\left(\varepsilon \dot{N}Lc_{tr}^{-1}\right)}{dn_s}. \tag{11}$$

If the third term on the right-hand side of Eq.(11) is not zero, there will be coupling between the gas state–sorbed state equilibrium and the pulses traveling along the filament.

Enhanced Filtering Due to Sonic Pulses

At the 73rd annual Society of Rheology meeting in Bethesda, Maryland, in October 2001, D. L. Feke, Dept. of Chemical Engineering, Case Western Reserve University, reported on the use of an acoustic field of standing sound waves to enhance the filtering effect of a porous medium. A porous medium through which there flowed a fluid containing small particles of foreign matter was exposed to an acoustic field of standing sound waves; the acoustic field interacted with the porous medium and the small particles in such a way that the filtering effect of the medium was enhanced to the point where particles one hundredth the size of the nominal pore size of the medium were held back (i.e. "filtered") by the acoustic-field-porous-medium system. "Such an acoustically aided filter offers little resistance to the fluid that flows through it, yet collects particles as efficiently as a much finer filter does. And once the filter has done its job, the trapped particles can be released with the flip of a switch that cuts off the signal."—Physics Today, **55**(1), 9 (2002).

The ideas of the last subsection to Section 14 of Book I seem to have some relevance to the case studied by Feke: the particles are nominally free to pass through the porous medium (the particles are much smaller in size than the pore size of the medium), yet they do not do so, i.e. the particles behave as though they were in a condition of migrational equilibrium in the acoustic field.

Analysis

Let's carry through a naive thermodynamic analysis of the situation investigated by Feke. Model the situation as in Fig. 1: let the fluid with suspended particles flow between terminal parts at α, β, γ, the porous medium being close to the γ terminal part. Let the walls of the apparatus be of diathermic material, and let the entire apparatus be placed in a heat bath (thermostat) of temperature T. Let the rate of influx of heat to the heat bath be $\dot{Q}^{(r)}$. Let pulsed sonic energy flow, from the IN port to the OUT port, through the straight length of tubing containing the porous medium at the rate $\varepsilon\left(\dot{N}\lambda c_s^{-1}\right)$, where ε is the energy per pulse, $\dot{N}$ is the number of pulses per unit time, λ is the distance between the pulse generators as indicated in Fig. 1, and c_s is the (mean) speed of sound along the path of length λ—just as in the ACOUSTIC PULSE subsection of SECTION 14 of Book I.

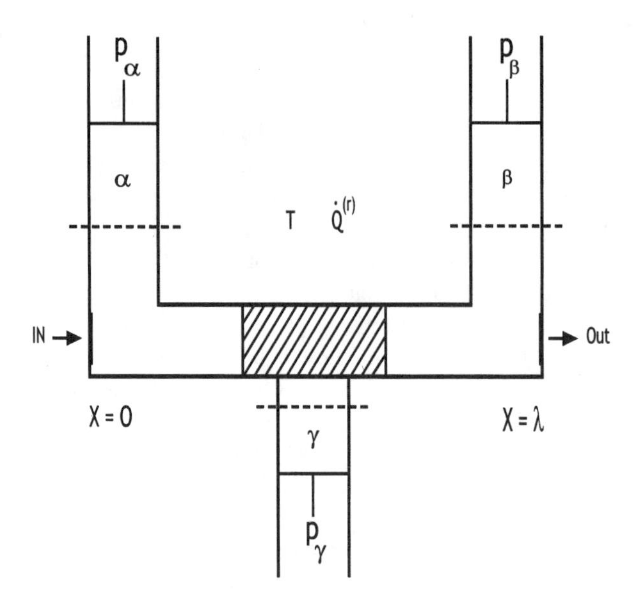

Fig 1. A fluid with suspended particles flows between terminal parts at α, β, γ. There is a porous medium in the straight part of the tubing between the IN-PORT and the OUT-PORT, close to the γ terminal part. Sonic pulses introduced into the sytem influence the filtering effect of the porous medium.

Treat the particles as macroscopic objects. Then the intrinsic properties (specific properties or property densities) of the fluid and particle components of the system [call the components component 1 (fluid) and component 2 (particles)] will each be constant throughout the system (neglect the macroscopic kinetic energy of the fluid and of the particles).

Let C_ω $(\omega = \alpha, \beta, \gamma)$ be the number concentration of particles in the fluid in terminal part ω. Establish a steady state for the flow of components 1 and 2 from terminal part α into terminal parts β and γ with the steady flow of pulsed energy as indicated. The number concentrations will be such that

$$C_\gamma \geq C_\alpha \geq C_\beta . \tag{1}$$

Let a star superscript $\left(^*\right)$ refer to properties of a particle. Let the volume occupied by a particle of the suspension be v^*, then

$$V_\omega = V_\omega^{(1)} + V_\omega^{(2)}, \ V_\omega^{(2)} = C_\omega V_\omega v^*, \ V_\omega^{(1)} = V_\omega - C_\omega V_\omega v^* \ ; \tag{2}$$

the total volume of the particles in the terminal parts of the system stays constant as the particles are distributed over the terminal parts:

$$\sum C_\omega V_\omega v^* = \text{constant or } v^* \sum C_\omega dV_\omega = 0 \tag{3}$$

where the sum is over the terminal parts $(\omega = \alpha, \beta, \gamma)$.

Let the system be in a steady state such that

$$\sum \dot{V}_\omega^{(1)} = 0 \text{ and } \sum \dot{V}_\omega^{(2)} = 0 \quad , \tag{4}$$

and let the pulsed acoustic energy flowing through the straight length of tubing containing the porous medium be $\varepsilon\left(\dot{N}\lambda c_s^{-1}\right)$. Now instantaneously increase the energy per pulse by an amount $d\varepsilon$. As the system relaxes to a new steady state, the changes attributable to the change in energy per pulse will be connected via the first law (assume that the specific properties of the fluid, e.g. $u^{(1)}$ and $s^{(1)}$, and the properties of the particles, e.g. $U^{(*)}$ and $S^{(*)}$, are everywhere the same in the suspension and are unaffected by the change in the energy per pulse):

$$dU^{(1)} + dU^{(2)} = dQ - \sum_\omega p_\omega dV_\omega + \left(\dot{N}\lambda c_s^{-1}\right)d\varepsilon \tag{5}$$

or

$$dH^{(1)} + dH^{(2)} = dQ + \left(\dot{N}\lambda c_s^{-1}\right)d\varepsilon \tag{6}$$

Under our assumptions, $dH^{(1)} + dH^{(2)} = 0$, so

$$-dQ = \left(\dot{N}\lambda c_s^{-1}\right)d\varepsilon \ . \tag{7}$$

There is no change in the intrinsic (bound) entropy of the system, i.e.

$$dS^{(1)} + dS^{(2)} = 0; \tag{8}$$

however there may be some configurational entropy associated with the suspension, say $S(config)$, and $dS(config)$ may not be equal to zero.

The entropy change in the surroundings attributable to the process is

$$d_\uparrow S_{sur} = -\frac{1}{T}dQ = \frac{\dot{N}\lambda}{Tc_s}d\varepsilon \ . \tag{9}$$

Then

$$d\Gamma_\uparrow = dS(config) + d_\uparrow S_{sur} = dS(config) + \frac{\dot{N}\lambda}{Tc_s}d\varepsilon \tag{10}$$

and

$$\frac{d\Gamma_\uparrow}{d\varepsilon} = \frac{dS(config)}{d\varepsilon} + \frac{\dot{N}\lambda}{Tc_s} \tag{11}$$

Then, the Thomson-Gibbs Corollary gives

$$\frac{dS(config)}{d\varepsilon} = -\frac{\dot{N}\lambda}{Tc_s} \le 0 \; . \tag{12}$$

Let

$$S(config) = S\left(C_\alpha, C_\gamma\right) \tag{13}$$

and

$$dS(config) = \left.\frac{\partial S}{\partial C_\alpha}\right)_{C_\gamma} dC_\alpha + \left.\frac{\partial S}{\partial C_\gamma}\right)_{C_\alpha} dC_\gamma \; , \tag{14}$$

since [Eq.(3)] only two of the concentrations are independent. Then,

$$\frac{dS(config)}{d\varepsilon} = \left.\frac{\partial S}{\partial C_\alpha}\right)_{C_\gamma} \frac{dC_\alpha}{d\varepsilon} + \left.\frac{\partial S}{\partial C_\gamma}\right)_{C_\alpha} \frac{dC_\gamma}{d\varepsilon} = -\frac{\dot{N}\lambda}{Tc_s} \tag{15}$$

and

$$\frac{dC_\gamma}{d\varepsilon} = \frac{\frac{\dot{N}\lambda}{Tc_s} + \left[\left.\frac{\partial S}{\partial C_\alpha}\right)_{C_\gamma} \frac{dC_\alpha}{d\varepsilon}\right]}{-\left.\frac{\partial S}{\partial C_\gamma}\right)_{C_\alpha}} \tag{16}$$

Let's assume that $S\left(C_\alpha, C_\gamma\right)$ is symmetric in its arguments and that

$$\left.\frac{\partial S}{\partial C}\right) < 0 \; ; \tag{17}$$

then we conclude that

$$\frac{dC_\gamma}{d\varepsilon} > 0 \;\text{ if }\; \frac{dC_\alpha}{d\varepsilon} < 0 \; . \tag{18}$$

Our conclusion, then, is that the suspension must have a configurational entropy that behaves in the required way for there to be enhanced filtering due to the sonic pulses.